Nonlinear Dynamics and Applications—In Honor of Professor Osvaldo Anibal Rosso's 70th Birthday

Nonlinear Dynamics and Applications—In Honor of Professor Osvaldo Anibal Rosso's 70th Birthday

Guest Editors

José María Amigó
Fernando Montani

MDPI

Basel • Beijing • Wuhan • Barcelona • Belgrade • Novi Sad • Cluj • Manchester

Guest Editors

José María Amigó
Centro de Investigación
Operativa
Universidad Miguel
Hernández
Elche
Spain

Fernando Montani
Physics Department
Universidad Nacional de La
Plata (UNLP)
La Plata
Argentina

Editorial Office
MDPI AG
Grosspeteranlage 5
4052 Basel, Switzerland

This is a reprint of the Special Issue, published open access by the journal *Entropy* (ISSN 1099-4300), freely accessible at: https://www.mdpi.com/journal/entropy/special_issues/9K9BA658DA.

For citation purposes, cite each article independently as indicated on the article page online and as indicated below:

Lastname, A.A.; Lastname, B.B. Article Title. *Journal Name* **Year**, *Volume Number*, Page Range.

ISBN 978-3-7258-4735-8 (Hbk)
ISBN 978-3-7258-4736-5 (PDF)
https://doi.org/10.3390/books978-3-7258-4736-5

Contents

About the Editors

José María Amigó

José María Amigó is a Professor of Applied Mathematics at the Department of Statistics, Mathematics and Informatics of the Miguel Hernández University in Elche (Spain) and is also affiliated with its Operations Research Centre. He holds a PhD in Theoretical Physics from the University of Göttingen (Germany). His scientific interests include dynamical systems, time series analysis, and information theory, having published around 150 research articles on these topics and a monograph on "Permutation Complexity in Dynamical Systems", as well as presented more than 100 communications at international conferences.

Fernando Montani

Fernando Montani is a professor of physics at the Universidad Nacional de La Plata (UNLP) and an independent researcher at CONICET in Argentina. He holds dual PhDs in Computational Neuroscience from Imperial College London and in Theoretical Physics from UNLP. His research uses nonlinear dynamics, information theory, neurocomputational modeling, and neural networks to investigate neural coding and brain information processing dynamics, with a core focus on applications for neurodegenerative diseases. This work is supported by around 50 publications on neuronal modeling, nonlinear physics, and time series analysis.

Preface

This Special Issue of Nonlinear Dynamics and Applications honors the 70th birthday of Professor Osvaldo A. Rosso, a distinguished, proactive pioneer whose groundbreaking work has significantly shaped the field of nonlinear dynamics and its interdisciplinary applications. This collection pays tribute to his remarkable career, defined by relentless innovation, intellectual generosity, a collegial spirit, and transformative contributions. It features cutting-edge research that captures the breadth and depth of his scientific legacy. This Special Issue highlights recent advances in nonlinear dynamics, seamlessly bridging theoretical rigor with practical applications. It encompasses foundational methodologies—from chaos theory and symbolic reconstruction to entropy-based complexity measures—and their implementation across diverse domains, including physics, neuroscience, signal processing, and network science.

José María Amigó and Fernando Montani
Guest Editors

entropy

MDPI

Editorial

Nonlinear Dynamics and Applications

José M. Amigó [1,*] and Fernando Montani [2]

1 Centro de Investigación Operativa, Universidad Miguel Hernández, 03202 Elche, Spain
2 Instituto de Física de La Plata (IFLP), CONICET-UNLP, La Plata B1900, Buenos Aires, Argentina; f.montani@fisica.unlp.edu.ar
* Correspondence: jm.amigo@umh.es

Nonlinear dynamics is the study of dynamical systems in finite dimensions, whether in discrete or continuous time, in which the evolution equation (a difference or differential equation, respectively) is not linear in the state variables [1,2]. A potential result of nonlinear dynamics is sensitivity to initial conditions (i.e., small perturbations of the initial condition can lead to large divergences in the corresponding trajectories in the state space over time), making such systems unpredictable on a given time scale [3]. Sensitivity to initial conditions, popularized as the butterfly effect, is a hallmark of deterministic chaos [4–6]. As the well-known example of the logistic parabolas shows, it suffices a quadratic term in the evolution equation for a (in this case, one-dimensional) system to exhibit, for suitably chosen parameters, a mind-wobbling complex behavior [7,8]. And low-dimensional chaos is just the entry point to an increasingly weird and fascinating world as the dimension increases [9,10].

Although nonlinear dynamics is as old as Newtonian dynamics (as its name reveals), it is generally acknowledged that the formulation of theoretical properties and qualitative methods started with the work of Poincaré on the three-body problem [11,12], and continued with Lyapunov [13], Birkhoff [14], Kolmogorov [15] and many other towering figures of pure and applied mathematics. The list of new, seminal concepts introduced in mathematics in the development of nonlinear dynamics is impressive: recurrence, homoclinic and heteroclinic orbits and tangles, invariant measures, ergodicity, Lyapunov exponents, metric and topological entropies, attractors, chaos, bifurcations, fractals, Julia sets, horseshoes, time-delay embedding, and many more [16,17]. In fact, the conceptual framework of nonlinear dynamics belongs to the core of modern mathematics. Furthermore, nonlinear dynamics has important intersections with other fundamental areas of mathematics such as ordinary differential equations, ergodic theory, statistics and probability theory.

Traditionally, real world applications of nonlinear dynamics have focused on the understanding of complex phenomena in nature. To mention a few representative examples, nonlinear dynamics is used in physics to study turbulence, weather evolution, nonlinear oscillations and chaotic synchronization [18–20]. In biology, we can find applications to population dynamics, epidemiology and heart dynamics [21–23]. And in engineering, it help in understanding mechanical vibrations, design control systems and electrical circuits [24,25]. More recently, the applications of nonlinear dynamics have also been extended to social sciences, e.g., economy and sociology, especially as data-driven analyses [26]. At the same time, new areas of research have emerged, such as chaotic cryptography [27], computational neuroscience [28] and nonlinear time series analysis [29]. In sum, in many complex phenomena amenable to the dynamical approach, nonlinear dynamics has enriched our understanding, bringing light to old and new problems alike.

Precisely, the research of Professor Osvaldo Rosso falls into nonlinear time series analysis, nonlinear dynamics, information theory, complex networks and their applications to physics, biology and medical sciences. In this regard, let us highlight his work on the differentiation between deterministic chaos and randomness [30], as well as the characterization of signals by means of several generally called entropy–complexity planes [31]. In doing so, he has also proven to be a driving force in the application of ordinal methods to the analysis of time series [32]. Professor Rosso was nominated among the top 5% of researchers with high impact in biological applications in the years 2022, 2023 and 2024.

In addition to the vast and influential work of Professor Rosso on the nonlinear and complexity sciences, we would also like to underline his leadership and mentoring, especially in South America, where he has nurtured a large and very active community of students and collaborators in the field of nonlinear time series analysis with a strong emphasis on applications. In recognition of his distinguished and productive scientific career, this Special Issue on "Nonlinear Dynamics and Applications" is dedicated to him on the occasion of his 70th birthday (15 October 2024).

This Special Issue comprises eighteen contributions, one of them being a review article and the rest being research articles. For the sake of this Preface, the research articles have been grouped into five topics: nonlinear time series analysis, dynamical systems and models, physics, chaos synchronization, and networks. In the following, we briefly describe all these articles.

Knowing the scientific trajectory of Professor Rosso, it is not surprising that a significant part of the contributions deal with time series analysis. Thus, Guisande et al. examine in [33] awake human intracranial electroencephalography (iEEG) data from normal brain regions to explore how biological sex influences these dynamics. The iEEG data were analyzed using permutation entropy and statistical complexity in the time domain and power spectrum calculations in the frequency domain. This study emphasizes the importance of considering sex as a biological variable in brain dynamics research, which is essential for improving the diagnosis and treatment of neurological and psychiatric disorders. Zunino et al. [34] quantify the similarity between two time series with the permutation Jensen–Shannon distance (JSD), i.e., the square root of the Jensen–Shannon divergence of the time series in the ordinal representation, with the probability distributions being estimated by the frequencies of the corresponding ordinal patterns. The authors show that the JSD is a powerful and robust metric for identifying the precise temporal scales at which two distinct time series exhibit ordinal similarity. Valverde et al. [35] study the differential behavior of the ECG during occlusion of both the left anterior descending (LAD) and right anterior coronary artery (RCA), respectively, in subjects with acute myocardial ischemia. To this end, the authors use ordinal patterns and locate the resulting signals in the causal complexity plane. The authors find that the locations for subjects with LAD ischemia, RCA ischemia and controls are different, making it possible to discriminate the three groups. A similar study is performed by Duarte et al. [36], this time to study sleep stages with generalized weighted permutation entropy (GWPE) and standard permutation entropy. Comparison of the results shows that GWPE significantly enhances sleep stage differentiation, which makes GWPE a valuable tool for the diagnosis of sleep disorders. The list of contributions to the topic of time series analysis continues with [37]. In this article, Olivares et al. show how ordinal patterns and metrics derived from them can be used to assess the effectiveness of detrending methods. As an interesting application, the authors apply their method to data representing the evolution of delays in major European and US airports, obtaining operational conclusions about how delays propagate in the two systems. Lange and Hauhs study the complexity of environmental time series in [38]. Specifically, the authors analyze selected long-term data from three headwater catchments in Lower Saxony (Germany). The metrics used, based on ordinal pattern statistics, comprise permutation entropy, permutation complexity and Fisher information, as well as their generalized versions (q-entropy and α-entropy). The analysis tools include Tarnopolski diagrams, ordinal pattern position slopes, horizontal visibility graphs and the exponent of the decay of the degree distribution. As a result, the authors characterize the dynamics of those systems. Finally, Suriano et al. [39] investigate the temporal evolution of cryptocurrency time series using information measures such as complexity, entropy, and Fisher information, with the objective of differentiating between various levels of randomness and chaos. Complexity–entropy causality plane (CECP) analysis reveals that daily cryptocurrency series with lengths of two years or less exhibit chaotic behavior, while those longer than two years display stochastic behavior.

The following contributions belong to the topic of dynamical systems and models. In the context of evolution forecasting of epidemic outbreaks from deterministic compartmental models, Rojas-Venegas et al. [40] extend the classical deterministic compartmental models to a quantum-like formalism to explore whether the uncertainty of epidemic forecasts is also shaped by the stochastic nature of epidemic processes. The authors conclude that the stochasticity of contagion and recovery processes poses a natural constraint for the uncertainty of epidemic forecasts. Quintanilha Valente et al. [41] explore the escape dynamics of bistable systems influenced by multiplicative noise. To this end, the authors use a generalized stochastic calculus framework to derive an analytical expression for the escape rate that is corroborated with numerical simulations. The results have applications in physics, chemistry and biology. The study of dynamical models continues with the article in [42], where Al-Saffar and Kim use a tumor-immune growth model to investigate how immunotherapy affects its dynamical characteristics. Their model is an extension of the prey–predator model of tumor cells and immune cells that includes periodic immunotherapy, nonlinear damping of cancer cells and the dynamics of a healthy cell population. The authors also study the effects of the model parameters.

As for the articles on physics, Monteoliva et al. [43] use information theory to investigate notable features of the quantum degree of mixedness (QDM) in a finite model of interacting fermions. The authors find that QDM exhibits a strong dependence on the total fermion number, with varying trends across different temperature regimes. Their results have potential implications for diverse fields ranging from condensed matter physics to quantum information science. In the context of non-collisional space plasmas, Consolini and De Michelis [44] explore irreversibility at ion and sub-ion scales during a magnetospheric current disruption event. Their results clearly demonstrate the irreversible nature of fluctuations at those scales. Hubay et al. [45] study turbulent flows over street canyons. The authors apply the quadrant method to transform time series of fluctuating velocities into symbolic sequences, enabling the investigation of their information content via word frequency and normalized entropy levels. Their results indicate that noisy periodic sequences exhibit entropy distributions qualitatively similar to those of the measured and simulated data.

Synchronization in nonlinear systems is a phenomenon of great interest. Lin and Pattanayak introduce in [46] the Difference Time Series Peak Complexity (DTSPC) algorithm, a technique that uses entropy as a tool to quantitatively measure synchronization. The authors show with coupled Lorenz systems that this technique reveals and quantifies the complexity of chaos synchronization, also capturing transitional behaviors between different regimes. Precisely, the article [47] contributed by Leyva et al. deals with transitions to synchronization, too, but of a particular kind: explosive transitions. Specifically, the authors propose using the ordinal pattern transition (OPT) entropy measured at sentinel central nodes as a potential predictor of explosive transitions to synchronization in networks of various dynamical systems.

Their results show that the OPT entropic measure surpasses traditional early warning signal measures and could be a valuable tool for predicting critical transitions.

Among the many tools to study symbolic sequences, Hamming distance is tailored to the binary ones. Schneider and Zanette [48] study the structural properties of networks formed by random sets of bit strings; two bit strings are connected by a network link when they are sufficiently similar to each other, i.e., when their Hamming distance is below a certain threshold. Their results have to do with the degree distribution and the conditions for the existence of a giant component, as well as with clustering, assortativity and mean geodesic distance in this kind of network. Dec et al. [49] introduce a novel method of exploring linguistic networks by mapping word-adjacency networks to time series and applying multifractal analysis techniques. Their results indicate that the time series derived from clustering coefficients, when following the natural reading order, exhibits multifractal characteristics, revealing inherent complexity in textual organization. An analogous analysis based on the node degrees does not show such rich behaviors.

Finally, Sepúlveda and Amigó [50] provide a broad overview of a selection of generalized entropies and other entropic quantities, as well as their applications to data analysis and machine learning. This review also provides a brief historical account of the concept of entropy across several disciplines, its axiomatic formulation and a bibliometric survey of publications on this topic and their impact.

In conclusion, the above articles fittingly showcase the diversity of topics embraced by nonlinear dynamics, while also suggesting timely lines of research.

Acknowledgments: The Guest Editors are very grateful to all authors for their excellent contributions to this Special Issue, to the anonymous peer-reviewers for their constructive feedback, and to the Editorial Assistants of the journal *Entropy* for their support.

Conflicts of Interest: The authors declare no conflicts of interests.

References

1. Robinson, R.C. *An Introduction to Dynamical Systems*; Pearson Prentice Hall: Upper Saddle River, NJ, USA, 2004.
2. Datseris, G.; Parlitz, U. *Nonlinear Dynamics*; Springer Nature: Cham, Switzerland, 2022.
3. Lorenz, E.N. Deterministic nonperiodic flow. *J. Atmos. Sci.* **1963**, *20*, 130–141. [CrossRef]
4. Devaney, R.L. *An Introduction to Chaotic Dynamical Systems*, 2nd ed.; Westview Press: Boulder, CO, USA, 2003.
5. Ott, E. *Chaos in Dynamical Systems*; Cambridge University Press: Cambridge UK, 2012.
6. Strogatz, S.H. *Nonlinear Dynamics and Chaos*, 2nd ed.; Westview Press: Boulder, CO, USA, 2015.
7. May, R.M. Simple mathematical models with very complicacted dynamics. *Nature* **1976**, *261*, 459–467. [CrossRef] [PubMed]
8. Feigenbaum, M.J. Quantitative universality for a class of nonlinear transformations. *J. Stat. Phys.* **1978**, *19*, 25–52. [CrossRef]
9. Gleick, J. *Chaos: Making a New Science*; Viking Books: New York, NY, USA, 1987.
10. Peitgen, H.O.; Jürgen, H.; Saupe, D. *Chaos and Fractals*; Springer: New York, NY, USA, 1992.
11. Poincaré, H. Sur le problème des trois corps et les équations de la dynamique. *Acta Math.* **1890**, *13*, A3–A270.
12. Barrow-Green, J. *Poincare and the Three Body Problem*; History of Mathematics; American Mathematical Society: Providence, RI, USA, 1996; Volume 11.
13. Lyapunov, A.M. The general problem of the stability of motion. *Int. J. Control* **1992**, *55*, 531–773. [CrossRef]
14. Birkhoff, G.D. Proof of the Ergodic Theorem. *Proc. Natl. Acad. Sci. USA* **1931**, *17*, 656–660. [CrossRef]
15. Kolmogorov, A.N. Entropy per unit time as a metric invariant of automorphisms. *Dokl. Russ. Acad. Sci. USA* **1959**, *124*, 754–755.
16. Shilnikov, L.P; Shilnikov, A.L.; Turaev, D.V.; Chua, L.O. *Methods of Qualitative Theory in Nonlinear Dynamics, Part I*; World Scientific: Singapore, 1998.
17. Shilnikov, L.P; Shilnikov, A.L.; Turaev, D.V.; Chua, L.O. *Methods of Qualitative Theory in Nonlinear Dynamics, Part 2*; World Scientific: Singapore, 1998.
18. Takens, F. Detecting strange attractors in turbulence, In *Dynamical Systems and Turbulence*; Rand, D., Young, L.S., Eds.; Lecture Notes in Mathematics; Springer: Berlin/Heidelberg, Germany, 1981; Volume 898, pp. 366–381.

19. Hayashi, C. *Nonlinear Oscillations in Physical Systems*; Princeton University Press: Princeton, NJ, USA, 1964.
20. Pecora, L.M.; Carroll, T.L. Synchronization in chaotic systems. *Rev. Phys. Lett.* **1990**, *64*, 821–824. [CrossRef]
21. Murray, J.D. *Mathematical Biology I: An Introduction*; Springer: New York, NY, USA, 2002.
22. Murray, J.D. *Mathematical Biology II: Spatial Models and Biomedical Applications: An Introduction*; Springer: New York, NY, USA, 2002.
23. Amigó, J.M.; Small, M. Mathematical methods in medicine: Neuroscience, cardiology and pathology. *Philos. Trans. R. Soc. A* **2017**, *375*, 20170016. [CrossRef]
24. Van der Pohl, B.; van der Mark, J. Frequency demultiplication. *Nature* **1927**, *120*, 363–364. [CrossRef]
25. Chua, L. O.; Matsumoto, T.; Komuro, M. The Double Scroll. *IEEE Trans. Circuits Syst.* **1985**, *32*, 798–818.
26. De Gooijer, J.G. *Elements of Nonlinear Time Series Analysis and Forecasting*; Springer Nature: Cham, Switzerland, 2017.
27. Amigó, J.M.; Kocarev, L.; Szczepanski, J. Theory and Practice of Chaotic Cryptography. *Phys. Lett. A* **2007**, *366*, 211–216. [CrossRef]
28. Izhikevich, E.M. *Dynamical Systems in Neuroscience*; The MIT Press: Cambridge, MA, USA, 2007.
29. Kantz, H.; Schreiber, T. *Nonlinear Time Series Analysis*, 2nd ed.; Cambridge University Press: Cambridge, MA, USA, 2003.
30. Rosso, O.A.; Larrondo, H.A.; Martín, M.T.; Plastino, A.; Fuentes, M.A. Distinguishing noise from chaos. *Phys. Rev. Lett.* **2007**, *99*, 154102. [CrossRef] [PubMed]
31. Rosso, O.A.; Olivares, F.; Plastino, A. Noise versus Chaos in a causal Fisher-Shannon plane. *Pap. Phys.* **2015**, *7*, 070006. [CrossRef]
32. Amigó, J.M.; Rosso, O.A. Ordinal methods: Concepts, applications, new developments, and challenges. *Chaos* **2023**, *33*, 080401. [CrossRef]
33. Guisande, N.; Rosso, O.A.; Montani, F. Neural Dynamics Associated with Biological Variation in Normal Human Brain Regions. *Entropy* **2024**, *26*, 828. [CrossRef] [PubMed]
34. Zunino, L.; Porte, X.; Soriano, M.C. Identifying Ordinal Similarities at Different Temporal Scales. *Entropy* **2024**, *26*, 1016. [CrossRef]
35. Valverde, E.R.; Vampa, V.; Rosso, O.A.; Arini, P.D. Spatio-Temporal Analysis of Acute Myocardial Ischaemia Based on Entropy–Complexity Plane. *Entropy* **2025**, *27*, 8. [CrossRef]
36. Duarte, C.D.; Pacheco, M.; Iaconis, F.R.; Rosso, O.A.; Gasaneo, G.; Delrieux, C.A. Statistical Complexity Analysis of Sleep Stages. *Entropy* **2025**, *27*, 76. [CrossRef]
37. Olivares, F.; Marín-Rodríguez, F.J.; Acharya, K.; Zanin, M. Evaluating Methods for Detrending Time Series Using Ordinal Patterns, with an Application to Air Transport Delays. *Entropy* **2025**, *27*, 230. [CrossRef]
38. Lange, H.; Hauhs, M. Complexity Analysis of Environmental Time Series. *Entropy* **2025**, *27*, 381. [CrossRef] [PubMed]
39. Suriano, M.; Caram, L.F.; Caiafa, C.; Merlino, H.D.; Rosso, O.A. Information Theory Quantifiers in Cryptocurrency Time Series Analysis. *Entropy* **2025**, *27*, 450. [CrossRef] [PubMed]
40. Rojas-Venegas, J.A.; Gallarta-Sáenz, P.; Hurtado, R.G.; Gómez-Gardeñes, J.; Soriano-Paños, D. Quantum-Like Approaches Unveil the Intrinsic Limits of Predictability in Compartmental Models. *Entropy* **2024**, *26*, 888. [CrossRef]
41. Quintanilha Valente, S.C.; da Costa Lima Bruni, R.; González Arenas, Z.; Barci, D.G. Effects of Multiplicative Noise in Bistable Dynamical Systems. *Entropy* **2025**, *27*, 155. [CrossRef] [PubMed]
42. Al-Saffar, A.; Kim, E.-J. Sustainability of a Three-Species Predator–Prey Model in Tumor-Immune Dynamics with Periodic Treatment. *Entropy* **2025**, *27*, 264. [CrossRef]
43. Monteoliva, D.; Plastino, A.; Plastino, A.R. Information Theoretical Analysis of Quantum Mixedness in a Finite Model of Interacting Fermions. *Entropy* **2025**, *27*, 37. [CrossRef]
44. Consolini, G.; De Michelis, P. Irreversibility and Energy Transfer at Non-MHD Scales in a Magnetospheric Current Disruption Event. *Entropy* **2025**, *27*, 260. [CrossRef]
45. Hubay, C.Á.; Papp, B.; Kalmár-Nagy, T. Turbulent Flow in Street Canyons: A Complexity Approach. *Entropy* **2025**, *27*, 488. [CrossRef]
46. Lin, Z.; Pattanayak, A.K. A Chaos Synchronization Diagnostic: Difference Time Series Peak Complexity (DTSPC). *Entropy* **2024**, *26*, 1085. [CrossRef]
47. Leyva, I.; Almendral, J.A.; Letellier, C.; Sendiña-Nadal, I. Local Predictors of Explosive Synchronization with Ordinal Methods. *Entropy* **2025**, *27*, 113. [CrossRef]
48. Schneider, D.M.; Zanette, D.H. The Structure of Bit-String Similarity Networks. *Entropy* **2025**, *27*, 57. [CrossRef] [PubMed]
49. Dec, J.; Dolina, M.; Drożdż, S.; Kluszczyński, R.; Kwapień, J.; Stanisz, T. Exploring Word-Adjacency Networks with Multifractal Time Series Analysis Techniques. *Entropy* **2025**, *27*, 356. [CrossRef] [PubMed]
50. Sepúlveda-Fontaine, S.A.; Amigó, J.M. Applications of Entropy in Data Analysis and Machine Learning: A Review. *Entropy* **2024**, *26*, 1126. [CrossRef] [PubMed]

entropy

MDPI

Article

Neural Dynamics Associated with Biological Variation in Normal Human Brain Regions

Natalí Guisande [1], Osvaldo A. Rosso [1,2] and Fernando Montani [1,*]

[1] Instituto de Física de La Plata (IFLP), CONICET-UNLP, La Plata B1900, Buenos Aires, Argentina; guisande.natali@fisica.unlp.edu.ar (N.G.); oarosso@gmail.com (O.A.R.)
[2] Instituto de Física, Universidade Federal de Alagoas (UFAL), BR 104 Norte km 97, Maceió 57072-970, Brazil
* Correspondence: f.montani@fisica.unlp.edu.ar

Abstract: The processes involved in encoding and decoding signals in the human brain are a continually studied topic, as neuronal information flow involves complex nonlinear dynamics. This study examines awake human intracranial electroencephalography (iEEG) data from normal brain regions to explore how biological sex influences these dynamics. The iEEG data were analyzed using permutation entropy and statistical complexity in the time domain and power spectrum calculations in the frequency domain. The Bandt and Pompe method was used to assess time series causality by associating probability distributions based on ordinal patterns with the signals. Due to the invasive nature of data acquisition, the study encountered limitations such as small sample sizes and potential sources of error. Nevertheless, the high spatial resolution of iEEG allows detailed analysis and comparison of specific brain regions. The results reveal differences between sexes in brain regions, observed through power spectrum, entropy, and complexity analyses. Significant differences were found in the left supramarginal gyrus, posterior cingulate, supplementary motor cortex, middle temporal gyrus, and right superior temporal gyrus. This study emphasizes the importance of considering sex as a biological variable in brain dynamics research, which is essential for improving the diagnosis and treatment of neurological and psychiatric disorders.

Keywords: human iEEG; biological sex differences; shannon entropy; statistical complexity; information theory; power spectral density (PSD); brain dynamics; MNI open iEEG atlas; normal brain regions

1. Introduction

A dynamic system is characterized by its evolving state over time, which can be fully described by a set of state variables at any given moment. These models allow for the analysis and prediction of system behavior under different inputs or conditions. Studying and understanding dynamic systems provides valuable insights into their behavior, allowing for predictions and even the control or optimization of their performance in various applications.

Dynamical systems can exhibit different dynamic properties, such as stability, periodicity, or chaos. Chaotic systems exhibit sensitivity to initial conditions, known as the butterfly effect, where small changes in initial conditions can lead to significant differences in the system's behavior over time [1]. This is a fundamental characteristic shared by various physical systems, including neural networks, with the factors contributing to this sensitivity yet to be fully understood.

The dynamics of the brain show complex nonlinear characteristics, and their underlying mechanisms are still not well understood. While concrete evidence of chaos in cerebral dynamics, from a mathematical standpoint, has primarily been observed at the level of axons, individual cells, and paired cells, findings suggest that brain signals may exhibit chaotic patterns across all levels of their hierarchy [2–4].

Neural information processing is a challenging topic that requires an understanding of the intricate mechanisms underlying neuronal activity. One key aspect of this process is the nonlinear nature of neuronal dynamics, leading to complex phenomena such as chaos, synchronization, and bifurcations. These phenomena have significant implications for neural coding and computation. Analyzing the dynamics of nonlinear systems is not a straightforward task, especially when there are no available differential equations to model the system under study, as is the case with the brain.

One method of capturing the brain's activity over time is through temporal recordings of its electrical activity via intercellular local field potentials (LFPs). These recordings reflect the extracellular electrical activity of nearby neurons and are obtained using deep electrodes inserted into the brain, resulting in intracranial electroencephalography (iEEG) signals, which are temporal voltage series.

The application of information theory to analyze iEEG signals enhances the understanding of brain activity by providing quantitative and objective measures. It enables the extraction of temporal characteristics and quantification of the information contained in the brain signals. Furthermore, information theory allows investigation of the complexity and organization of brain signals, revealing underlying patterns and organizational structures.

Utilizing ordinal patterns to compute measures based on permutation entropy enables the extraction of causal characteristics from signals and the assessment of nonlinear dynamics within these systems. The approach to computing the probability distribution function (PDF) from ordinal patterns, as proposed by Bandt and Pompe (BP), is widely embraced for electroencephalography (EEG) analysis, including iEEG, and has demonstrated superior results compared to conventional analyses [5].

Extensive studies have been conducted on non-intracranial EEG signals, such as the analysis of time series in epilepsy EEG [6,7], the distinction between brain death and coma [8], and the study of states of consciousness with the use of anesthetics [9], sleep stages [5,10–12], and signal discrimination [13–15]. In iEEG, this method has been applied to epilepsy for seizure prediction [16], detecting the epileptogenic focus [17], identifying preictal markers [18], extracting different characteristics for identifying the epileptogenic focus [17], and differentiating epileptic signals in an unsupervised manner [19–21]. These findings underscore the potential of permutation analysis in delving into diverse facets of EEG signal processing and analysis, empowered researchers to investigate the causal characteristics of signals, and evaluate the nonlinear dynamics within neural systems.

A growing body of research indicates sex-related differences in neural activity recorded in human EEG data [22–44]. Quantifiers derived from information theory, including markers based on the entropy of EEG background activity [45], have shown potential for identifying these sex differences. Several approaches have been proposed to exploit this potential, such as estimating differences in brain status through entropy measurements [46] and using permutation entropy to extract features for gender identification in emotional-based EEG datasets [47] and non-stationary EEG signals [48].

Applying these analyses to iEEG facilitates a deeper understanding of how men and women encode information during cognitive processes. A recent development involves creating a database to compile iEEG data from normal brain regions [49–51]. Given the impracticality of screening asymptomatic patients using invasive and costly methods, this database, comprising epileptic patients with identified epileptogenic foci and distinct healthy (non-affected) regions, currently provides the closest approximation for studying signals from normal areas.

iEEG signals, which record the brain's electrical activity, are obtained through electrodes implanted in the cranial area or brain tissue. These signals offer superior spatial resolution and less tissue attenuation compared to EEGs, facilitating the precise localization of specific brain regions. They reflect the activity of ensembles or populations of neurons responsible for encoding sensory, motor, or cognitive information. Adopting the perspective of neural population [52] allows for investigating how the brain integrates and combines the activity of multiple neurons or brain regions to generate a coherent and efficient representation of internal or external stimuli.

This study aims to investigate how biological sex differences influence the way the brain encodes information through iEEG signals and to explore their impact on neuronal encoding dynamics. To achieve this, the study employs information theory quantifiers based on ordinal patterns that preserve causal coding features, using the Bandt and Pompe method. Complete signals are analyzed using the complexity-entropy causality plane and power spectral density. In each region, the mean, median, and Mann–Whitney U test were applied to entropy and complexity to compare between sexes. The analysis is based on iEEG signals sourced from healthy sectors of the human brain sourced from the Open iEEG Atlas [49–51].

The objective is to enhance the understanding of sex-based differences in brain activity through a comprehensive exploration of local field potentials. This involves the segregation of signals by sex, region, and hemisphere, followed by their analysis and the overlaying of results for graphical comparisons and statistical tests. Ultimately, the purpose of this study is to broaden the foundation for future research in neuroscience, thereby advancing comprehension of the observed diversity in the human brain and the influence of biological sex on cerebral dynamics.

2. Methods

2.1. Data Source and Preprocessing Methods

The data for this study were obtained from the MNI Open iEEG Atlas [49–51] database, which features recordings of intracranial activity in normal brain regions during various states, including quiet wakefulness with eyes closed, non-REM sleep (N2 and N3 stages), and REM sleep.

The signals were recorded from 1772 channels across 106 patients with focal epilepsy, using only those located in gray matter and considered "normal" (i.e., distant from epileptic regions). Various types of intracerebral electrodes were used, including Dixi, homemade MNI, AdTech electrodes, and AdTech subdural strips and grids.

The dataset includes patient information, such as sex, channel type, hemisphere, channel name, channel position, and channel region. Artifacts were detected and removed from the signals, and all signals were resampled at 200 samples per second to ensure consistency. Power-line interference was minimized using an adaptive filter. All channels were zero-padded to a length of 68 s (13,600 samples) to maintain uniform length regardless of the number of segments.

The electrodes were placed in a common stereotactic space to facilitate patient activity comparison and accumulation of results from multiple subjects. To obtain further information about the database and acquisition methods, the reader can refer to the following references [49–51].

2.2. Clustering of iEEG Data

The signals were separated by hemisphere, region, and sex, so that each region was analyzed in both hemispheres, with female and male analyzed separately. Only eyes-closed wakefulness data were analyzed in this study because there were no skipped channels or interruptions to consider when analyzing causality.

To compare male and female behavior within the same region, a condition was imposed requiring at least five patients of each sex. This was achieved by analyzing the signals without distinguishing between the electrodes used, as otherwise, fewer than two regions per hemisphere had more than five patients of each sex. All cases where this minimum statistic was not met were excluded from the analysis.

Figure 1 provides a schematic representation of the brain, highlighting the localized regions that surpass the statistical restriction, as well as the corresponding electrodes used for channel analysis. The graphics were created based on the nodes of the surfaces and the electrode positions provided by the atlas [49–51].

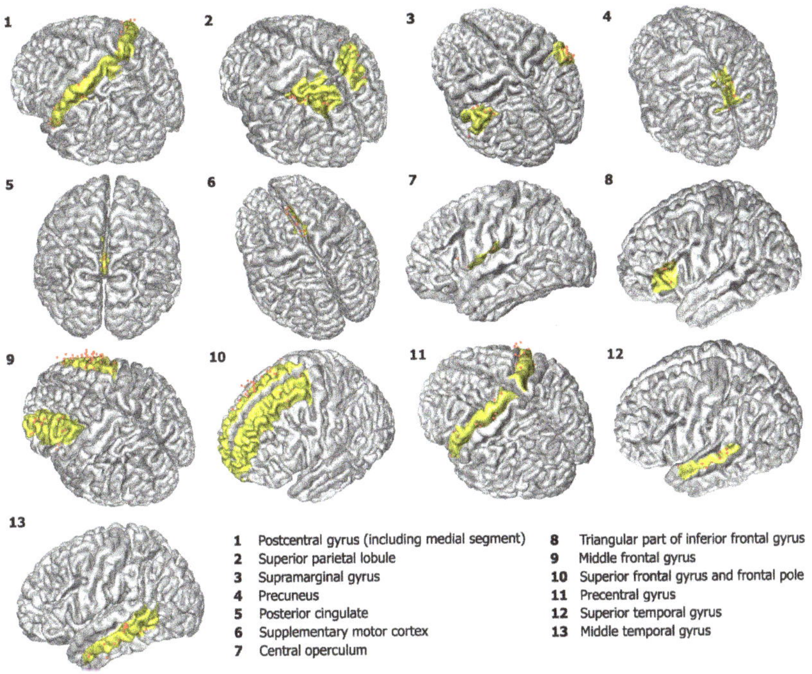

Figure 1. Schema of the analyzed region locations along with the corresponding electrodes used. The yellow area represents the region of interest, and the red dots indicate the positions of the electrodes.

2.3. *Exploring iEEG Dynamics with Information Theory Tools*

The study of neural systems and their dynamics is crucial for research in neuroscience. Understanding patterns and structures of neural activity can provide insights into brain function and potential mechanisms underlying neurological disorders. Information theory provides tools for analyzing neuronal activity by associating a probability distribution function (PDF) with the time series (TS) of brain signals. The Bandt and Pompe (BP) method offers a robust approach for processing nonlinear and noisy data while accounting for causality. This article uses the BP method to extract causal features from the signal and explore brain dynamics in both males and females. The iEEG TS is analyzed to quantify the information contained in observed neural activity by calculating Shannon entropy and statistical complexity. The following section discusses the information theory tools employed to analyze iEEG data, with a focus on the BP methodology used to obtain a probability distribution function for each signal while considering causality. Entropy and statistical complexity are then calculated from these distributions.

2.3.1. Examining Ordinal Structure and Quantifying Information Using the Bandt-Pompe Approach

Quantifying the information contained in observed neural activity is essential for analyzing neural systems. This can be achieved by computing the Shannon entropy and statistical complexity of a TS associated with a PDF [53]. However, directly calculating a histogram from the signal may lead to information loss concerning the temporal causality of the underlying dynamical system. Fortunately, the BP approach offers a robust alternative for analyzing nonlinear and noisy data, including neural activity [54]. This method utilizes the ordinal structure of the TS to derive a PDF of ordinal patterns, which captures the underlying structure of the TS without assuming any specific model. By comparing the relative values of the points in the TS rather than the amplitudes, this methodology constructs the PDF associated with the TS using two parameters: embedding dimension and embedding delay.

In this study, the BP technique is used to estimate the PDF of an iEEG TS, $\chi(t) = \{x_t;\ t = 1, \cdots, M\}$, comprising M observations, as described in [54–57]. The BP approach involves transforming the TS into symbolic sequences and identifying ordinal structures by considering all possible permutations of the series values within fixed-size windows. The resulting PDF of ordinal patterns is obtained by calculating histograms of patterns in the signal that reflect causality.

The TS is partitioned into $n = M - (D-1)\tau$ overlapping segments, each with a predetermined embedding dimension (D) and embedding delay (τ). Each segment is represented by a D-dimensional vector of values, from which the permutation of index numbers is determined. Every possible permutation of order D (with $D!$ permutations) is considered, and its relative frequency is computed to generate a histogram of ordinal patterns. This distribution reflects the structure of the TS and can be used to compute the Shannon entropy and statistical complexity of the observed neural activity.

Finally, to calculate the PDF associated with the TS, the following expression can be used:

$$p_j(\Pi_j) = \frac{\text{number of partitions of type } \Pi_j \text{ in } \pi_i}{n},$$

where $p_j(\Pi_j)$ represents the relative frequency of the j-th ordinal pattern Π_j, and π_i denotes the sequence of all ordinal patterns for all partitions. Note that the estimated PDF P is discrete since it is derived from a histogram. Additionally, the TS analyzed, $\chi(t)$, should be much longer than the number of possible ordinal patterns ($M >> D!$ where M is the number of samples in the TS) to ensure good statistical results when applying this technique. Readers interested in further information and examples of the BP methodology can refer to the following references: [53–58].

2.3.2. Quantifying Information-Theoretic Measures: Entropy and Complexity Calculations

Quantifying information-theoretic measures is crucial for understanding the dynamics of a system represented by its TS. Although permutation Shannon entropy effectively measures randomness in a system, it may not fully capture ordinal structures. Therefore, measures of statistical or structural complexity are needed for a more comprehensive characterization of the system's dynamics [59].

In this study, the BP methodology is used to estimate the underlying PDF of a TS, $\chi(t)$, and two information-theoretic quantifiers are calculated using the ordinal patterns method to detect and quantify ordinal structures in the TS. These quantifiers enable a comparison of the dynamics of physiological signals in the human brain between biological sexes.

Shannon entropy measures randomness in a TS and is calculated using its PDF. In the discrete case, the Shannon entropy of a TS with a corresponding PDF $P \equiv p_{j\{j=1,\cdots,N\}}$ is given by the equation $S[P] = -\sum_{j=1}^{N} p_j \log_2(p_j)$, where N is the number of possible states and p_j is the probability of each state. However, large changes in P over a small range have minimal impact on the Shannon entropy, as it is a "global" measure.

To more accurately characterize the system's dynamics, measures of statistical or structural complexity are needed. Statistical complexity, developed by López-Ruiz et al. [60], captures subtle differences in the dynamics of the systems under study. This study uses the version of the metric developed by Martin-Rosso-Plastino, known as MPR complexity [61,62]. From this point on, the term "complexity" refers specifically to this metric.

Statistical complexity is computed using $C = \mathcal{Q}_J[P, P_e] \cdot H[P]$. The normalized Shannon entropy H, is defined as $H[P] = \frac{S[P]}{S_{max}}$, where $S_{max} = S[P_e] = \log_2 N$. The disequilibrium ($\mathcal{Q}_J = \mathcal{Q}_0\mathcal{J}$) is a function of the Jensen–Shannon divergence, given as $\mathcal{J}[P, P_e] = H[\frac{P+P_e}{2}] - \frac{H[P]}{2} - \frac{H[P_e]}{2}$, and a normalization constant $\mathcal{Q}_0$. Here, P_e refers to the uniform distribution. To obtain the statistical complexity, $\mathcal{Q}_J$ and H are multiplied together.

Statistical complexity and entropy were calculated for channels within the same region, hemisphere, and biological sex. To traverse the TS, non-overlapping time windows smaller than the original signals were utilized. A window size of 15 s was chosen to balance the need for maximum causality ($M >> D!$) with minimal loss of information in neural encoding [63]. This choice corresponded to 3000 samples, given the signals' sampling frequency of 200 Hz.

The BP methodology was applied to each window to obtain a PDF that accounts for the causality of the signal. For this analysis, $D = 6$ and $\tau = 1$ were used. Complexity and entropy were then calculated from these PDFs. The values were accumulated for patients meeting the specified conditions, and the results were plotted for both males and females. In each region, the mean, median, and Mann–Whitney U test were utilized to quantify any differences found between the results for males and females.

To further validate the condition $M >> D!$, the same analysis was conducted with an embedding dimension of $D = 5$, using the same time window and $\tau = 1$.

2.4. Investigating Frequency Bands in iEEG Signals: Power Spectral Density Analysis Using the Welch Method

In the study of neuroscientific phenomena, power spectral density (PSD) is frequently utilized as a measure of the distribution of power in a signal across different frequencies. This method is commonly applied to electroencephalography (EEG) and intracranial EEG (iEEG) signals to gain insight into underlying neural processes. PSD analysis can be employed in iEEG to examine the frequency content of LFPs and identify neural processes associated with specific frequency bands.

The Welch method [64] is a widely used approach for estimating PSD. It involves dividing the signal into overlapping segments, applying a window function to each segment to reduce spectral leakage, computing a periodogram for each segment, and then averaging the periodograms to estimate the PSD. A periodogram is a mathematical tool that breaks down a signal into its component frequencies, helping to identify which frequencies are present and their strengths. The Fourier transform is commonly employed to compute periodograms, decomposing signals into sine and cosine waves of varying amplitude and frequency.

Welch's approach is particularly useful for analyzing non-stationary signals as it allows for the estimation of the PSD at multiple time points. To estimate the PSD of iEEG signals using the Welch method, the iEEG signal is first divided into overlapping segments of equal length. A window function, such as the Hamming window, is then applied to each segment to reduce spectral leakage, and the periodogram of each segment is computed using the fast Fourier transform (FFT) method. The periodograms are then averaged across the segments to obtain an estimate of the PSD. For further information on these methods and their implementation, refer to the literature cited in Refs. [65,66].

In this work, Welch's method was used to estimate the PSD of the iEEG. Specifically, the magnitude of the discrete-time Fourier transform of 59 overlapping blocks of 2 s duration and 1 s step, weighted by a Hamming window, was averaged to obtain the spectral density in each channel. The resulting spectral density in each channel was then normalized to a total power, ensuring independence from the amplitude of the iEEG signal. The selection of these parameters was based on the work by the creators of the MNI Open iEEG Atlas (see Ref. [49–51]).

3. Results

This section presents the results of Shannon entropy and statistical complexity across different brain regions in the left and right hemispheres of both male and female individuals, using embedding dimensions of $D = 6$ and $D = 5$, with a time window of 15 s and a delay of $\tau = 1$. Additionally, power spectra and the complexity-entropy plane are shown for those regions where significant differences between sexes were identified using the Mann–Whitney U test.

The first step involved examining the age distribution of the patients to account for any potential age-related differences in the results. Table 1 displays the number of patients (n), mean age, and standard deviation for each region. Within each region, the age ranges for males and females were similar and fell within the standard deviation, which supported the assumption that no significant disturbances in the calculated quantifiers arose from age differences.

Table 1. Mean ages (in years) of patients in each analyzed region, along with their respective standard deviation. 'n' represents the number of patients.

Left Hemisphere						
		Female			**Male**	
Region	n	**Mean**	**std**	n	**Mean**	**std**
Superior parietal lobule	5	30	10	5	40	16
Supramarginal gyrus	7	27	9	7	40	12
Precuneus	5	26	9	7	30	12
Posterior cingulate	5	30	11	5	40	15
Supplementary motor cortex	5	29	7	8	37	9
Central operculum	5	24	6	8	35	9
Triangular part of inferior frontal gyrus	6	30	10	7	35	7
Middle frontal gyrus	10	29	8	13	36	6
Superior frontal gyrus and frontal pole	5	26	7	8	38	7
Precentral gyrus	6	25	6	12	30	11
Superior temporal gyrus	12	30	11	5	36	4
Middle temporal gyrus	13	30	10	9	35	7
Right Hemisphere						
		Female			**Male**	
Region	n	**mean**	**std**	n	**mean**	**std**
Postcentral gyrus (including medial segment)	7	20	10	7	30	11
Middle frontal gyrus	6	40	15	17	34	8
Superior temporal gyrus	6	30	12	5	30	10
Middle temporal gyrus	8	30	11	8	35	11

Differences in Shannon entropy and statistical complexity across different brain regions in the left and right hemispheres of male and female individuals are presented in Table 2. The analysis was conducted using the embedding dimension $D = 6$. Only regions with a minimum of five female and five male patients were included in the analysis. For each region, the mean and median values of Shannon entropy and statistical complexity were calculated separately for males and females. The statistical significance of the differences between male and female values was assessed using the Mann–Whitney U test, with the results presented as p-values and h-index. Significant differences in both Shannon entropy and statistical complexity between males and females were observed in certain brain regions, as highlighted in the table. Conversely, the remaining regions showed no significant differences between sexes for either measure.

Table 2. Comparison of Shannon entropy and statistical complexity between males and females across different brain regions using $D = 6$.

Left Hemisphere												
	Shannon Entropy						Statistical Complexity					
Region	Female		Male		Test		Female		Male		Test	
	Mean	Median	Mean	Median	*p*-Value	h	Mean	Median	Mean	Median	*p*-Value	h
Superior parietal lobule	0.35	0.35	0.37	0.37	1×10^{-5}	1	0.64	0.59	0.69	0.70	0.05	0
Supramarginal gyrus	0.37	0.38	0.36	0.36	0.02	1	0.68	0.67	0.74	0.75	2×10^{-6}	1
Precuneus	0.36	0.35	0.34	0.35	0.04	1	0.73	0.74	0.73	0.78	0.05	0
Posterior cingulate	0.33	0.34	0.31	0.31	0.01	1	0.77	0.78	0.81	0.81	0.01	1
Supplementary motor cortex	0.39	0.39	0.37	0.36	0.01	1	0.71	0.70	0.72	0.75	0.05	0
Central operculum	0.36	0.38	0.37	0.37	0.81	0	0.69	0.71	0.69	0.70	0.70	0
Triangular part of inferior frontal gyrus	0.34	0.35	0.36	0.36	0.03	1	0.75	0.75	0.73	0.73	0.19	0
Middle frontal gyrus	0.35	0.35	0.35	0.36	0.31	0	0.75	0.76	0.74	0.75	0.89	0
Superior frontal gyrus and frontal pole	0.36	0.36	0.36	0.37	0.92	0	0.72	0.71	0.70	0.72	0.50	0
Precentral gyrus	0.38	0.39	0.36	0.38	0.15	0	0.70	0.71	0.71	0.70	0.96	0
Superior temporal gyrus	0.36	0.36	0.36	0.36	0.64	0	0.66	0.68	0.65	0.66	0.43	0
Middle temporal gyrus	0.36	0.37	0.37	0.36	0.64	0	0.66	0.66	0.69	0.71	3×10^{-3}	1

Right Hemisphere												
	Shannon Entropy						Statistical Complexity					
Region	Female		Male		Test		Female		Male		Test	
	mean	median	mean	median	*p*-value	h	mean	median	mean	median	*p*-value	h
Postcentral gyrus (including medial segment)	0.37	0.37	0.38	0.38	0.07	0	0.66	0.65	0.68	0.68	0.29	0
Middle frontal gyrus	0.35	0.35	0.36	0.37	0.08	0	0.76	0.78	0.74	0.76	0.10	0
Superior temporal gyrus	0.36	0.35	0.37	0.37	0.06	0	0.74	0.74	0.65	0.66	3×10^{-7}	1
Middle temporal gyrus	0.34	0.35	0.35	0.35	0.12	0	0.73	0.76	0.71	0.73	0.06	0

The following figures display the PSD and complexity-entropy causality plane for various brain regions in the left and right hemispheres. Each figure shows results from both male and female patients, normalized to the total power in each channel. The format is consistent across all figures: (A) and (B) present the median PSD for female and male patients respectively (where the shaded area represents the interquartile range , IQR), (C) shows the median PSD for both sexes, and (D) illustrates the complexity-entropy causality plane for each sex using $D = 6$. Each scatter plot includes a boxplot for each sex, showing the median, IQR, outliers, and the notch, which indicates the 95% confidence interval of the median.

The figures correspond to the brain regions highlighted in Table 2, which showed significant differences between males and females based on the statistical test with $D = 6$. These regions include the superior parietal lobule (Figure 2), supramarginal gyrus (Figure 3), precuneus (Figure 4), posterior cingulate (Figure 5), supplementary motor cortex (Figure 6), the triangular part of the inferior frontal gyrus (Figure 7), and the middle temporal gyrus in the left hemisphere (Figure 8), as well as the superior temporal gyrus in the right hemisphere (Figure 9).

In Table 2, using $D = 6$, it can be seen that differences between the entropy and statistical complexity results for both biological sexes were observed in half of the analyzed regions, as quantified by a Mann–Whitney test. The normalized spectral analysis of the signals also revealed differences between the median results for both sexes. The figures provide a qualitative analysis, showing not only the medians but also the dispersion of the data through interquartile ranges (IQRs).

Superior Parietal Lobule Region in the Left Hemisphere

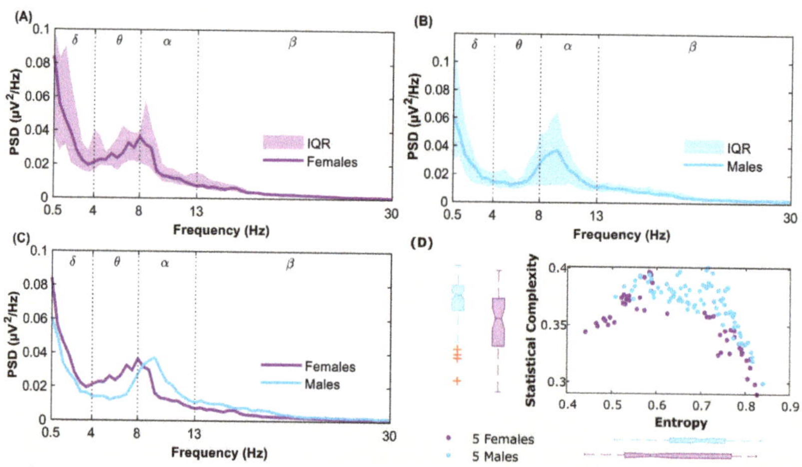

Figure 2. Power spectral density (PSD) and complexity-entropy causality plane of the superior parietal lobule region in the left hemisphere. (**A**,**B**) PSD of female and male patients, respectively, with the solid line representing the median and the shaded area representing the interquartile range (IQR). (**C**) Median PSD of females and males. (**D**) Complexity-entropy causality plane using an embedding dimension $D = 6$, a time delay $\tau = 1$, and a temporal window of 15 s for females and males. The boxplot shows the median, IQR, outliers, and the notch, which corresponds to the 95% confidence interval of the median.

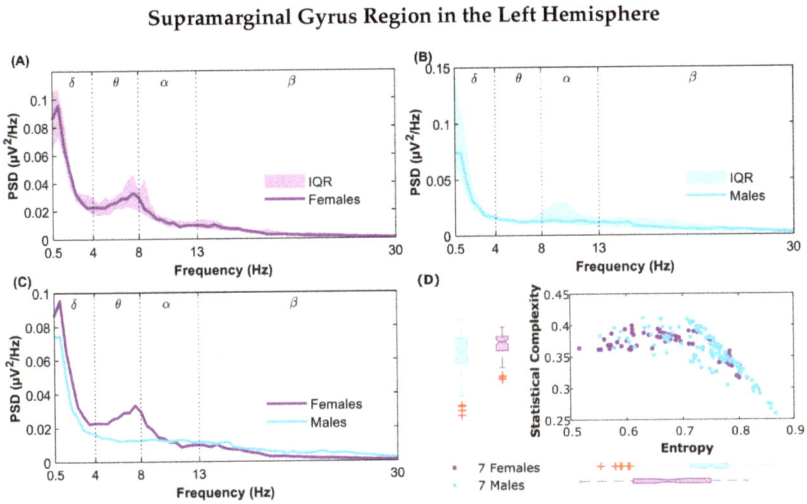

Figure 3. Power spectral density (PSD) and complexity-entropy causality plane of the supramarginal gyrus region in the left hemisphere. (**A**,**B**) PSD of female and male patients, respectively, with the solid line representing the median and the shaded area representing the interquartile range (IQR). (**C**) Median PSD of females and males. (**D**) Complexity-entropy causality plane using an embedding dimension $D = 6$, a time delay $\tau = 1$, and a temporal window of 15 s for females and males. The boxplot shows the median, IQR, outliers, and the notch, which corresponds to the 95% confidence interval of the median.

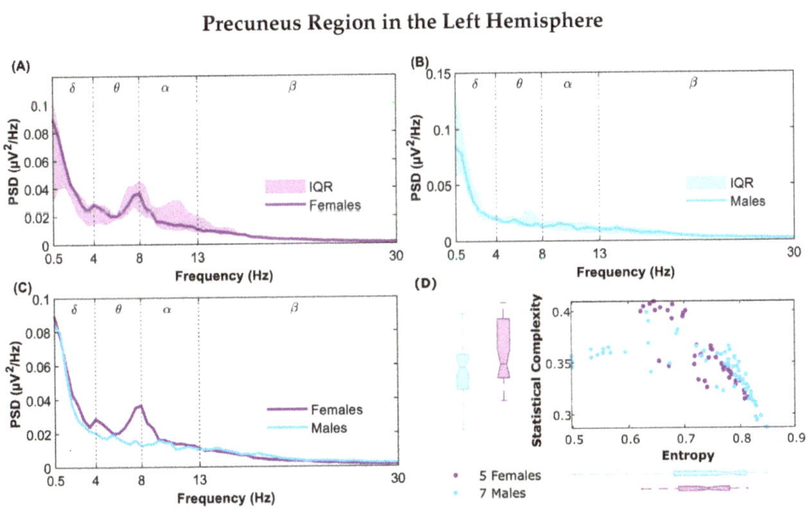

Figure 4. Power spectral density (PSD) and complexity-entropy causality plane of the precuneus region in the left hemisphere. (**A**,**B**) PSD of female and male patients, respectively, with the solid line representing the median and the shaded area representing the interquartile range (IQR). (**C**) Median PSD of females and males. (**D**) Complexity-entropy causality plane using an embedding dimension $D = 6$, a time delay $\tau = 1$, and a temporal window of 15 s for females and males. The boxplot shows the median, IQR, outliers, and the notch, which corresponds to the 95% confidence interval of the median.

Posterior Cingulate Region in the Left Hemisphere

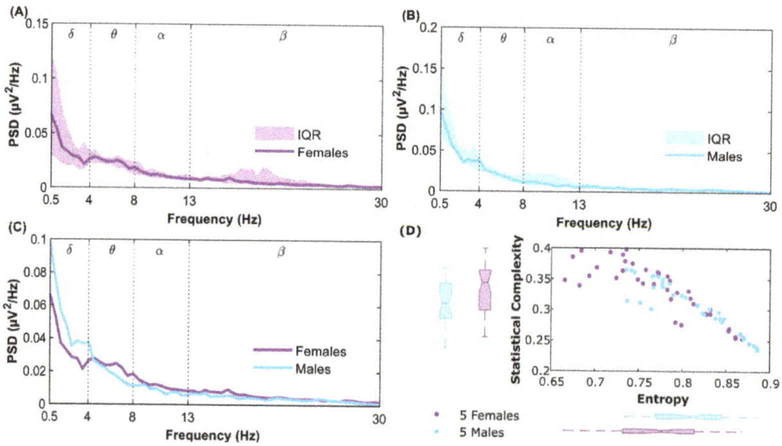

Figure 5. Power spectral density (PSD) and complexity-entropy causality plane of the posterior cingulate region in the left hemisphere. (**A**,**B**) PSD of female and male patients, respectively, with the solid line representing the median and the shaded area representing the interquartile range (IQR). (**C**) Median PSD of females and males. (**D**) Complexity-entropy causality plane using an embedding dimension $D = 6$, a time delay $\tau = 1$, and a temporal window of 15 s for females and males. The boxplot shows the median, IQR, outliers, and the notch, which corresponds to the 95% confidence interval of the median.

Supplementary Motor Cortex in the Left Hemisphere

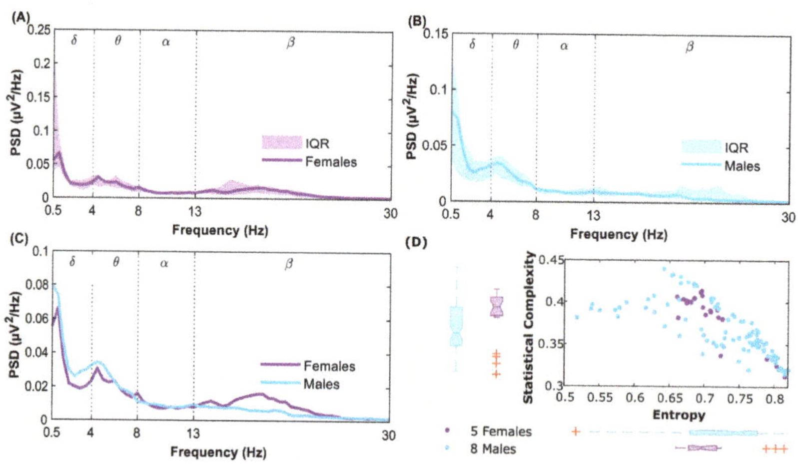

Figure 6. Power spectral density (PSD) and complexity-entropy causality plane of the supplementary motor cortex in the left hemisphere. (**A**,**B**) PSD of female and male patients, respectively, with the solid line representing the median and the shaded area representing the interquartile range (IQR). (**C**) Median PSD of females and males. (**D**) Complexity-entropy causality plane using an embedding dimension $D = 6$, a time delay $\tau = 1$, and a temporal window of 15 s for females and males. The boxplot shows the median, IQR, outliers, and the notch, which corresponds to the 95% confidence interval of the median.

Triangular Part of Inferior Frontal Gyrus in the Left Hemisphere

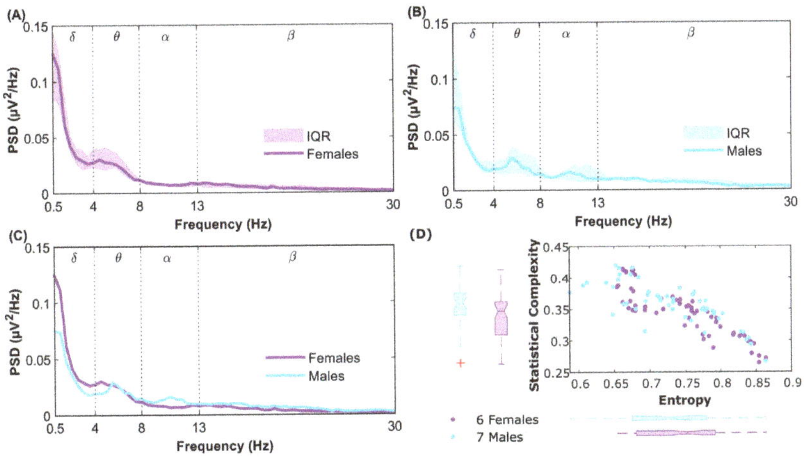

Figure 7. Power spectral density (PSD) and complexity-entropy causality plane of the triangular part of the inferior frontal gyrus in the left hemisphere. (**A**,**B**) PSD of female and male patients, respectively, with the solid line representing the median and the shaded area representing the interquartile range (IQR). (**C**) Median PSD of females and males. (**D**) Complexity-entropy causality plane using an embedding dimension $D = 6$, a time delay $\tau = 1$, and a temporal window of 15 s for females and males. The boxplot shows the median, IQR, outliers, and the notch, which corresponds to the 95% confidence interval of the median.

Middle Temporal Gyrus in the Left Hemisphere

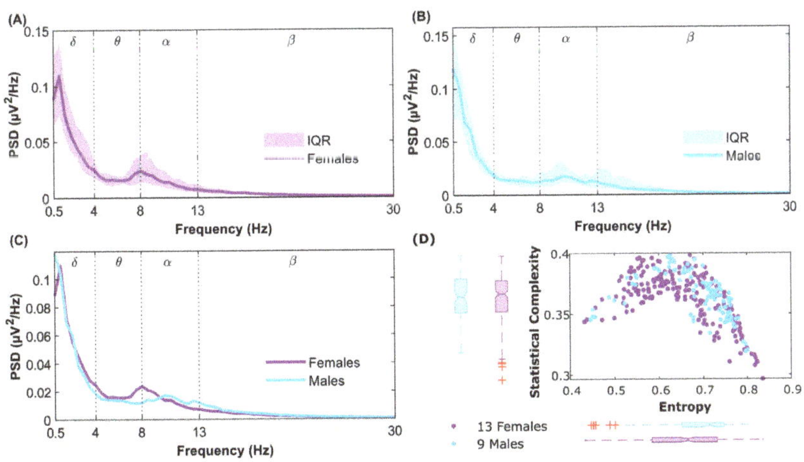

Figure 8. Power spectral density (PSD) and complexity-entropy causality plane of the middle temporal gyrus in the left hemisphere. (**A**,**B**) PSD of female and male patients, respectively, with the solid line representing the median and the shaded area representing the interquartile range (IQR). (**C**) Median PSD of females and males. (**D**) Complexity-entropy causality plane using an embedding dimension $D = 6$, a time delay $\tau = 1$, and a temporal window of 15 s for females and males. The boxplot shows the median, IQR, outliers, and the notch, which corresponds to the 95% confidence interval of the median.

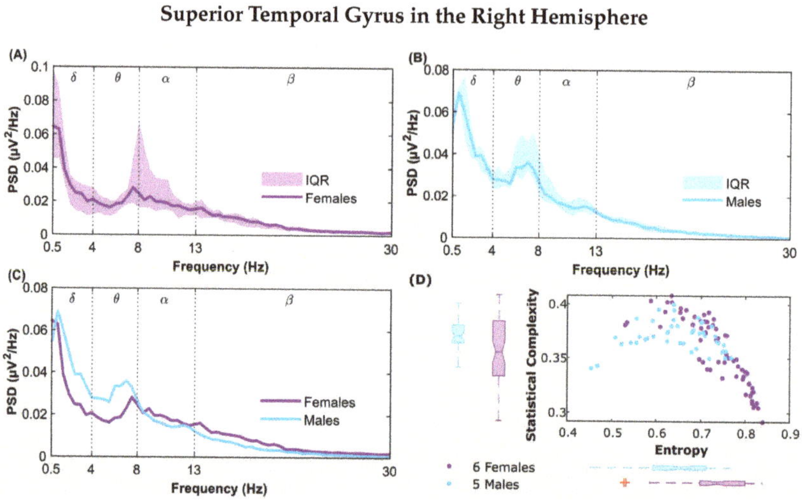

Figure 9. Power spectral density (PSD) and complexity-entropy causality plane of the superior temporal gyrus in the right hemisphere. (**A**,**B**) PSD of female and male patients, respectively, with the solid line representing the median and the shaded area representing the interquartile range (IQR). (**C**) Median PSD of females and males. (**D**) Complexity-entropy causality plane using an embedding dimension $D = 6$, a time delay $\tau = 1$, and a temporal window of 15 s for females and males. The boxplot shows the median, IQR, outliers, and the notch, which corresponds to the 95% confidence interval of the median.

In the left hemisphere, seven regions with significant differences were identified for $D = 6$. In the superior parietal lobule, the median complexity and entropy values showed disjoint notches in the boxplots, and the statistical test indicated that the entropy values come from different distributions (Figure 2D). Figure 2C shows that the PSD had a peak for both men and women. For women, this peak was located at the boundary of the $\theta - \alpha$ bands, while for men, it was shifted towards the α band.

The supramarginal gyrus region exhibited differences in the PSD spectra. A peak was observed in the spectrum for women at the end of the θ band, while the male spectrum lacked peaks in both the θ and γ bands (Figure 3C). The complexity-entropy plane also showed differences, with notched medians indicating distinct distributions for men and women (Figure 3D), as confirmed by statistical testing.

Similar patterns were observed in the precuneus region. The female PSD displayed a peak between the $\theta - \alpha$ bands, whereas no peaks were seen in the male PSD within these regions (Figure 4C). Although the notches in the medians had some overlap (Figure 4D), the statistical test revealed differences in entropy values between men and women.

In the posterior cingulate region (Figure 5), both spectra showed a small peak in the boundary between the δ and θ bands, with the male peak shifted to slightly lower frequencies (Figure 5C). Regarding complexity and entropy (Figure 5D), while some overlap was observed between the notches in entropy, the values for complexity were well separated. Statistical tests indicated that the distributions for both quantifiers were different.

The supplementary motor cortex presented a peak at the beginning of the θ band in both spectra. However, as frequencies increased, differences emerged: the female spectrum showed a peak in the β band, which was absent in the male spectrum (Figure 6C). Although the medians for complexity and entropy were well differentiated with non-overlapping notches, statistical tests revealed significant differences only for entropy (Figure 6D).

In the triangular part of the inferior frontal gyrus region, both spectra showed a peak in the θ band, but the female peak was shifted toward lower frequencies. Additionally, the male PSD featured a second peak centered in the α band which was not present in the female spectrum (Figure 7). Despite overlapping notches in the medians for complexity and entropy, statistical tests indicated differences between sexes for entropy.

The middle temporal gyrus region showed differences between the θ and α bands in the PSD. A peak in the female spectrum occurred at the boundary between these bands, whereas the male spectrum displayed two peaks in the α band (Figure 8C). While entropy medians were well differentiated, complexity medians presented overlapping notches (Figure 8D). Statistical tests revealed significant differences in complexities, but not in entropies.

In the right hemisphere, among the four regions with acceptable statistics, only the superior temporal gyrus region displayed differences for $D = 6$. Both spectra showed a peak, with the female peak located between the θ and α bands, and the male peak shifted to the left in the θ band with higher amplitude (Figure 9C). Non-overlapping notches for both entropy and complexity were observed (Figure 9D), but statistical tests revealed significant differences only for complexity.

To improve the statistics of BP and ensure that the length of the time series (M) was much greater than $D!$, an embedding dimension of $D = 5$ was used with the same time window. Results for mean, median, and statistical tests of Shannon entropy and statistical complexity using $D = 5$ are presented in Table 3.

With $D = 5$, the regions of the supramarginal gyrus, posterior cingulate, supplementary motor cortex, and middle temporal gyrus in the left hemisphere showed differences between males and females. Shannon entropy differences were detected in the first three regions, while all four regions displayed differences in statistical complexity. In the right hemisphere, differences were observed in two regions: the superior temporal gyrus, which showed differences in both quantifiers, and the middle temporal gyrus, which showed differences in entropy.

In summary, significant differences were observed in both $D = 6$ and $D = 5$ analyses in the following regions: the supramarginal gyrus, the posterior cingulate, the supplementary motor cortex, and the middle temporal gyrus in the left hemisphere, and the superior temporal gyrus in the right hemisphere.

Table 3. Comparison of Shannon entropy and statistical complexity between males and females across different brain regions using $D = 5$.

Left Hemisphere												
	Shannon Entropy						Statistical Complexity					
Region	Female		Male		Test		Female		Male		Test	
	Mean	Median	Mean	Median	p-Value	h	Mean	Median	Mean	Median	p-Value	h
Superior parietal lobule	0.27	0.30	0.28	0.28	0.98	0	0.68	0.63	0.73	0.74	0.09	0
Supramarginal gyrus	0.28	0.29	0.26	0.26	0.001	1	0.71	0.71	0.77	0.78	2×10^{-6}	1
Precuneus	0.26	0.25	0.25	0.24	0.10	0	0.77	0.78	0.77	0.81	0.07	0
Posterior cingulate	0.23	0.24	0.20	0.21	0.01	1	0.81	0.81	0.84	0.84	0.01	1
Supplementary motor cortex	0.29	0.30	0.27	0.26	0.05	1	0.74	0.73	0.76	0.78	0.05	1
Central operculum	0.27	0.29	0.28	0.28	0.90	0	0.73	0.75	0.73	0.73	0.74	0
Triangular part of inferior frontal gyrus	0.25	0.25	0.26	0.26	0.05	0	0.78	0.79	0.76	0.77	0.15	0
Middle frontal gyrus	0.25	0.25	0.25	0.26	0.30	0	0.78	0.79	0.78	0.79	0.96	0
Superior frontal gyrus and frontal pole	0.27	0.26	0.27	0.27	0.58	0	0.76	0.75	0.74	0.76	0.52	0
Precentral gyrus	0.29	0.29	0.27	0.29	0.49	0	0.74	0.74	0.74	0.74	0.99	0
Superior temporal gyrus	0.28	0.29	0.28	0.28	0.79	0	0.69	0.72	0.69	0.70	0.47	0
Middle temporal gyrus	0.28	0.29	0.27	0.27	0.10	0	0.69	0.70	0.72	0.75	0.003	1
Right Hemisphere												
	Shannon Entropy						Statistical Complexity					
Region	Female		Male		Test		Female		Male		Test	
	mean	median	mean	median	p-value	h	mean	median	mean	median	p-value	h
Postcentral gyrus (including medial segment)	0.29	0.29	0.29	0.30	0.64	0	0.69	0.69	0.72	0.72	0.30	0
Middle frontal gyrus	0.25	0.25	0.26	0.26	0.07	0	0.80	0.81	0.77	0.79	0.10	0
Superior temporal gyrus	0.26	0.26	0.29	0.29	2×10^{-4}	1	0.77	0.77	0.69	0.70	4×10^{-7}	1
Middle temporal gyrus	0.24	0.24	0.26	0.26	0.05	1	0.77	0.79	0.75	0.76	0.05	0

4. Discussions

In summary, the findings of this study indicate that biological sex-based differences in brain function can be detected through iEEG signal analysis. These variations are evident in multiple brain regions, affecting both spectral characteristics and measures of complexity and entropy.

Further research is needed to explore the underlying causes of these differences. The invasive nature of iEEG signal collection prevents its application to healthy subjects, resulting in a restricted amount of LFP data from normal brain regions. However, despite constraints on patient numbers per region, iEEG analysis offers unparalleled spatial resolution compared to EEG, enabling the capture of region-specific LFPs and allowing for comparisons across sexes. Additionally, signals collected from electrodes implanted directly in the brain minimize attenuation caused by the skull and other tissues, a common issue with EEG, leading to less interference. These advantages come with challenges, as the "healthy" regions analyzed originate from the brains of epilepsy patients. This not only limits the dataset size but may also introduce bias, as the signals are taken from regions deemed "normal", rather than from truly "healthy" individuals.

To address potential factors influencing the results, a strategy was employed to adjust for age within the analyzed sample, despite its limited size. The dataset was subsampled to include five patients for each region, aiming to minimize age differences between males and females while reducing standard deviations. The results of this analysis are detailed in this Section for the interested reader. The tables for $D = 6$ show that the number of regions exhibiting differences increased under this restriction (see Tables 4 and 5) in both hemispheres. Additionally, Table 6 provides the results of a significance analysis using the Benjamini–Hochberg FDR correction for $D = 5$ and $D = 6$, applied to both the entire patient cohort and the subsets created to minimize differences and deviations. Reducing age differences in the dataset helps to minimize the impact of age-related variables or confounding factors that could introduce noise or skew the results. By reducing age variability between groups (i.e., males and females), the analysis can more directly focus on the effects of the variables of interest, reducing the influence of age as a potential source of noise or bias. This approach also increases sensitivity to variations attributable to biological sex, reinforcing the current findings.

Neuroscience research must increasingly acknowledge the vast diversity present in human brains. A key step in this direction is adopting a binary, proportional, and inclusive segmentation based on common biological classifications of male and female. To better understand brain dynamics, it is essential to reduce biases in sample selection and analyze the data both collectively and separately. Future research should also investigate whether these differences persist when considering perceived gender identity and hormonal backgrounds, as these factors can significantly influence brain structure and function. Incorporating these variables could provide a more comprehensive understanding of brain dynamics and help mitigate biases in neuroscience research.

This study found that both Shannon entropy and MPR statistical complexity can serve as sensitive biomarkers for biological sex in iEEG from awake patients. Recently, we proposed a framework constructed with Renyi entropy and its generalized statistical complexity, where we detected sex-based differences related to scale-free phenomena in non-REM sleep stages N1 and N2, as well as REM sleep [67].

These findings underscore the necessity of addressing biases in experimental design. Including both males and females in the study population is crucial for ensuring that results are representative and not skewed by gender imbalances. Differences in average brain dynamics between sexes highlight the importance of inclusivity and equality in research. Despite advancements in gender equality, preclinical neuroscience often inadvertently excludes women, leading to biased data and conclusions. Thus, considering biological sex and gender differences is essential for obtaining a comprehensive understanding of brain function. Incorporating these variables into future studies could enhance our understanding of brain dynamics and contribute to reducing biases in neuroscience research.

Table 4. Mean ages (in years) of patients selected to minimize differences in mean values between males and females for each analyzed region, along with their respective standard deviations.

Left Hemisphere				
	Female		Male	
Region	**Mean**	**sd**	**Mean**	**sd**
Superior parietal lobule	29	10	41	16
Supramarginal gyrus	30	9	32	10
Precuneus	26	9	29	8
Posterior cingulate	28	11	38	15
Supplementary motor cortex	29	7	33	10
Central operculum	24	6	31	8
Triangular part of inferior frontal gyrus	33	10	33	7
Middle frontal gyrus	31	9	31	6
Superior frontal gyrus and frontal pole	26	7	34	6
Precentral gyrus	26	6	26	6
Superior temporal gyrus	36	13	36	4
Middle temporal gyrus	34	8	34	5
Right Hemisphere				
	Female		**Male**	
Region	**mean**	**sd**	**mean**	**sd**
Postcentral gyrus (including medial segment)	27	10	27	8
Middle frontal gyrus	37	16	37	8
Superior temporal gyrus	32	13	33	10
Middle temporal gyrus	36	9	36	13

Table 5. Comparison of Shannon entropy and statistical complexity between males and females across different brain regions, using D = 6 for the five patients of each sex selected to minimize differences in mean age values and standard deviations. h: null hypothesis; sd: standard deviation. Light gray shading indicates regions with sex differences in one quantifier, while bold gray shading represents regions where both quantifiers showed sex differences.

Left Hemisphere																	
	Shannon Entropy								Statistical Complexity								
Region	Female			Male			Test		Female			Male			Test		
	Mean	Median	sd	Mean	Median	sd	p-Value	h	Mean	Median	sd	Mean	Median	sd	p-Value	h
Superior parietal lobule	0.64	0.59	0.13	0.69	0.705	0.031	0.05	0	0.352	0.355	0.026	0.370	0.375	0.020	1×10^{-5}	1
Supramarginal gyrus	0.675	0.657	0.090	0.76	0.758	0.051	3×10^{-6}	1	0.366	0.368	0.022	0.355	0.354	0.035	0.09	0
Precuneus	0.732	0.742	0.055	0.72	0.779	0.115	0.27	0	0.359	0.350	0.029	0.343	0.348	0.020	0.08	0
Posterior cingulate	0.774	0.777	0.056	0.81	0.810	0.047	0.01	1	0.333	0.343	0.042	0.306	0.307	0.039	0.01	1
Supplementary motor cortex	0.708	0.697	0.041	0.63	0.686	0.095	0.38	0	0.385	0.393	0.028	0.374	0.378	0.022	0.02	1
Central operculum	0.69	0.71	0.11	0.681	0.690	0.074	0.45	0	0.361	0.376	0.041	0.373	0.376	0.021	0.71	0
Triangular part of inferior frontal gyrus	0.749	0.751	0.071	0.73	0.732	0.068	0.37	0	0.343	0.346	0.045	0.353	0.354	0.033	0.18	0
Middle frontal gyrus	0.734	0.738	0.052	0.71	0.736	0.069	0.27	0	0.367	0.365	0.029	0.377	0.378	0.022	0.04	1
Superior frontal gyrus and frontal pole	0.722	0.714	0.073	0.67	0.689	0.107	0.02	1	0.362	0.362	0.043	0.367	0.375	0.027	0.86	0
Precentral gyrus	0.690	0.702	0.072	0.74	0.772	0.131	4×10^{-3}	1	0.389	0.392	0.027	0.333	0.354	0.098	3×10^{-5}	1
Superior temporal gyrus	0.689	0.673	0.059	0.65	0.663	0.100	0.16	0	0.377	0.383	0.022	0.359	0.363	0.019	2×10^{-4}	1
Middle temporal gyrus	0.578	0.556	0.088	0.67	0.682	0.088	3×10^{-5}	1	0.376	0.378	0.016	0.372	0.372	0.015	0.23	0
Right Hemisphere																	
	Shannon Entropy								Statistical Complexity								
Region	Female			Male			Test		Female			Male			Test		
	mean	median	sd	mean	median	sd	p-value	h	mean	median	sd	mean	median	sd	p-value	h
Postcentral gyrus (including medial segment)	0.625	0.640	0.091	0.697	0.701	0.084	1×10^{-3}	1	0.378	0.381	0.026	0.363	0.369	0.025	0.04	1
Middle frontal gyrus	0.765	0.778	0.071	0.735	0.802	0.056	0.01	1	0.349	0.355	0.044	0.326	0.320	0.039	5×10^{-3}	1
Superior temporal gyrus	0.731	0.734	0.071	0.650	0.657	0.082	1×10^{-6}	1	0.359	0.366	0.030	0.370	0.369	0.014	0.2	0
Middle temporal gyrus	0.71	0.79	0.12	0.73	0.77	0.10	0.9	0	0.331	0.313	0.036	0.341	0.339	0.023	0.1	0

Table 6. Results of the Benjamini–Hochberg FDR correction and significance analysis. Light gray shading indicates cases where sex differences were identified.

Region	D = 6				D = 6 (5 Patients)				D = 5			
	H		C		H		C		H		C	
	pv_c	Ho	pv_c	Ho	pv_c	Ho	pv_c	Ho	pv_c	Ho	pv_c	Ho
Left Hemisphere												
Superior parietal lobule	0.11	F	2.10×10^{-4}	T	0.09	F	2.10×10^{-4}	T	0.17	F	1.18×10^{-4}	T
Supramarginal gyrus	2.27×10^{-5}	T	0.13	F	2.40×10^{-5}	T	0.14	F	1.62×10^{-5}	T	0.14	F
Precuneus	0.36	F	0.13	F	0.36	F	0.14	F	0.16	F	0.14	F
Posterior cingulate	0.03	T	0.04	T	0.02	T	0.03	T	0.04	T	0.03	T
Supplementary motor cortex	0.41	F	0.06	F	0.43	F	0.05	F	0.13	F	0.05	F
Central operculum	0.45	F	0.76	F	0.48	F	0.76	F	0.85	F	0.76	F
Triangular part of inferior frontal gyrus	0.41	F	0.22	F	0.43	F	0.24	F	0.24	F	0.24	F
Middle frontal gyrus	0.36	F	0.11	F	0.36	F	0.08	F	0.99	F	0.09	F
Superior frontal gyrus and frontal pole	0.05	F	0.86	F	0.04	T	0.86	F	0.64	F	0.86	F
Precentral gyrus	0.01	T	2.10×10^{-4}	T	0.01	T	2.10×10^{-4}	T	0.99	F	1.40×10^{-4}	T
Superior temporal gyrus	0.26	F	9.33×10^{-4}	T	0.26	F	9.33×10^{-4}	T	0.63	F	7.00×10^{-4}	T
Middle temporal gyrus	1.58×10^{-4}	T	0.26	F	1.60×10^{-4}	T	0.26	F	0.02	T	0.28	F

Region	D = 6				D = 6 (5 patients)				D = 5			
	H		C		H		C		H		C	
	pv_c	Ho	pv_c	Ho	pv_c	Ho	pv_c	Ho	pv_c	Ho	pv_c	Ho
Right Hemisphere												
Postcentral gyrus (including medial segment)	0.36	F	0.13	F	4.00×10^{-3}	T	0.08	F	0.44	F	0.34	F
Middle frontal gyrus	0.18	F	0.13	F	0.02	T	0.02	T	0.17	F	0.15	F
Superior temporal gyrus	5.18×10^{-6}	T	0.13	F	0.00	T	0.25	F	6.88×10^{-6}	T	6.88×10^{-6}	T
Middle temporal gyrus	0.12	F	0.16	F	0.90	F	0.15	F	0.13	F	0.10	F

5. Conclusions

This study examined biological sex differences in statistical complexity MPR, Shannon entropy, and power spectral density (PSD) across various regions of the brain. Significant variations in Shannon entropy and statistical complexity were detected between males and females in certain brain areas, with these differences persisting across multiple analyses. The regions where these distinctions were observed, both in power spectra analyses and when computing entropy and complexity with embedding dimensions $D = 5$ and $D = 6$, include the supramarginal gyrus, posterior cingulate, supplementary motor cortex, and middle temporal gyrus in the left hemisphere, and in the right hemisphere, the superior temporal gyrus.

Overall, this study supports the notion that biological sex-based distinctions exist in brain function. Furthermore, it suggests that statistical complexity, Shannon entropy, and PSD can serve as sensitive biomarkers for identifying these variations. Differences related to biological sex and scale-free dynamics have also been found in this atlas when analyzing non-REM stages N2 and N3, as well as REM sleep, in another study [67]. Moreover, due to limited data, this study did not consider the impact of gender identity on the outcomes, and the number of patients by area was limited. Future investigations will aim for a more comprehensive analysis, including filtering by frequency bands and exploring high-frequency oscillation behavior to compare neural activity between sexes.

This research underscores the importance of recognizing sex as a biological variable when designing and conducting studies in the field of neuroscience.

Author Contributions: The authors contributed equally to the design of this research as well as to the writing of this paper. All authors have read and agreed to the published version of the manuscript.

Funding: Funding was provided by CONICET, Argentina, through grant PIP 2023-2025 11220220100001CO, and by Universidad Nacional de La Plata, Argentina, through project 80120230100001LP.

Institutional Review Board Statement: Not applicable.

Data Availability Statement: The dataset used in this study was sourced from the Montreal Neurological Institute (MNI) Open iEEG Atlas, which is publicly accessible at https://mni-open-ieegatlas.research.mcgill.ca/, accessed on June 2024. This resource is a collaborative effort involving the Montreal Neurological Institute and Hospital, the Grenoble-Alpes University Hospital, and the Centre hospitalier de l'Université de Montréal. The Atlas provides a comprehensive collection of normal intracranial EEG data. For more information about this dataset, refer to the references listed in the bibliography [49–51].

Acknowledgments: Acknowledgment is given to OpenAI's artificial intelligence, specifically ChatGPT, for its invaluable assistance in translating and refining the English wording.

Conflicts of Interest: The authors declare that the research was conducted in the absence of any commercial or financial relationships that could be construed as a potential conflict of interest.

References

1. Azar, A.T.; Serranot, F.E.; Vaidyanathan, S. Sliding Mode Stabilization and Synchronization of Fractional Order Complex Chaotic and Hyperchaotic Systems. In *Mathematical Techniques of Fractional Order Systems;* Elsevier: Amsterdam, The Netherlands, 2018; p. 283. [CrossRef]
2. Pereda, E.; Quiroga, R.Q.; Bhattacharya, J. Nonlinear multivariate analysis of neurophysiological signals. *Prog. Neurobiol.* **2005**, *77*, 1–37. [CrossRef] [PubMed]
3. Falahian, R.; Dastjerdi, M.M.; Molaie, M.; Jafari, S.; Gharibzadeh, S. Artificial neural network-based modeling of brain response to flicker light. *Nonlinear Dyn.* **2015**, *81*, 1951–1967. [CrossRef]
4. Korn, H.; Faure, P. Is there chaos in the brain? II. Experimental evidence and related models. *Comptes Rendus Biol.* **2003**, *326*, 787–840. [CrossRef]
5. Mateos, D.M.; Gómez-Ramírez, J.; Rosso, O.A. Using time causal quantifiers to characterize sleep stages. *Chaos Solitons Fractals* **2021**, *146*, 110798. [CrossRef]
6. Mateos, D.; Diaz, J.; Lamberti, P. Permutation Entropy Applied to the Characterization of the Clinical Evolution of Epileptic Patients under PharmacologicalTreatment. *Entropy* **2014**, *16*, 5668–5676. [CrossRef]

7. Li, J.; Yan, J.; Liu, X.; Ouyang, G. Using Permutation Entropy to Measure the Changes in EEG Signals During Absence Seizures. *Entropy* **2014**, *16*, 3049–3061. [CrossRef]
8. Zhu, L.; Cui, G.; Cao, J.; Cichocki, A.; Zhang, J.; Zhou, C. A Hybrid System for Distinguishing between Brain Death and Coma Using Diverse EEG Features. *Sensors* **2019**, *19*, 1342. [CrossRef]
9. Jordan, D.; Stockmanns, G.; Kochs, E.F.; Pilge, S.; Schneider, G. Electroencephalographic Order Pattern Analysis for the Separation of Consciousness and Unconsciousness. *Anesthesiology* **2008**, *109*, 1014–1022. [CrossRef]
10. Nicolaou, N.; Georgiou, J. The Use of Permutation Entropy to Characterize Sleep Electroencephalograms. *Clin. EEG Neurosci.* **2011**, *42*, 24–28. [CrossRef] [PubMed]
11. Bandt, C. A New Kind of Permutation Entropy Used to Classify Sleep Stages from Invisible EEG Microstructure. *Entropy* **2017**, *19*, 197. [CrossRef]
12. Kuo, C.E.; Liang, S.F. Automatic stage scoring of single-channel sleep EEG based on multiscale permutation entropy. In Proceedings of the 2011 IEEE Biomedical Circuits and Systems Conference (BioCAS), San Diego, CA, USA, 10–12 November 2011; IEEE: Piscataway, NJ, USA, 2011. [CrossRef]
13. Liu, X.; Wang, G.; Gao, J.; Gao, Q. A Quantitative Analysis for EEG Signals Based on Modified Permutation-Entropy. *IRBM* **2017**, *38*, 71–77. [CrossRef]
14. Baravalle, R.; Guisande, N.; Granado, M.; Rosso, O.A.; Montani, F. Characterization of Visuomotor/Imaginary Movements in EEG: An Information Theory and Complex Network Approach. *Front. Phys.* **2019**, *7*, 115. [CrossRef]
15. Yang, Y.; Zhang, H.; Lee, S. EEG Signal Discrimination with Permutation Entropy. In *Brain Informatics*; Springer: Berlin/Heidelberg, Germany, 2021. [CrossRef]
16. Yang, Y.; Zhou, M.; Niu, Y.; Li, C.; Cao, R.; Wang, B.; Yan, P.; Ma, Y.; Xiang, J. Epileptic Seizure Prediction Based on Permutation Entropy. *Front. Comput. Neurosci.* **2018**, *12*, 55. [CrossRef]
17. Zhu, G.; Li, Y.; Wen, P.P.; Wang, S.; Xi, M. Epileptogenic focus detection in intracranial EEG based on delay permutation entropy. In *AIP Conference Proceedings*; AIP: Haldenwang, Germany, 2013. [CrossRef]
18. Granado, M.; Collavini, S.; Baravalle, R.; Martinez, N.; Montemurro, M.A.; Rosso, O.A.; Montani, F. High-frequency oscillations in the ripple bands and amplitude information coding: Toward a biomarker of maximum entropy in the preictal signals. *Chaos Interdiscip. J. Nonlinear Sci.* **2022**, *32*, 093151. [CrossRef]
19. Akter, M.S.; Islam, M.R.; Iimura, Y.; Sugano, H.; Fukumori, K.; Wang, D.; Tanaka, T.; Cichocki, A. Multiband entropy-based feature-extraction method for automatic identification of epileptic focus based on high-frequency components in interictal iEEG. *Sci. Rep.* **2020**, *10*, 7044. [CrossRef]
20. Zhu, G.; Li, Y.; Wen, P.; Wang, S. Classifying Epileptic EEG Signals with Delay Permutation Entropy and Multi-scale K-Means. In *Signal and Image Analysis for Biomedical and Life Sciences*; Springer International Publishing: Cham, Switzerland, 2014; pp. 143–157. [CrossRef]
21. Nicolaou, N.; Georgiou, J. Detection of epileptic electroencephalogram based on Permutation Entropy and Support Vector Machines. *Expert Syst. Appl.* **2012**, *39*, 202–209. [CrossRef]
22. Smith, S.M. Discrimination between Electro-Encephalograph Recordings of Normal Females and Normal Males. *Ann. Eugen.* **1952**, *17*, 344–350. [CrossRef]
23. Davidson, R.J.; Schwartz, G.E.; Pugash, E.; Bromfield, E. Sex differences in patterns of EEG asymmetry. *Biol. Psychol.* **1976**, *4*, 119–137. [CrossRef]
24. Ray, W.J.; Newcombe, N.; Semon, J.; Cole, P.M. Spatial abilities, sex differences and EEG functioning. *Neuropsychologia* **1981**, *19*, 719–722. [CrossRef] [PubMed]
25. Dijk, D.J.; Beersma, D.G.M.; Bloem, G.M. Sex Differences in the Sleep EEG of Young Adults: Visual Scoring and Spectral Analysis. *Sleep* **1989**, *12*, 500–507. [CrossRef]
26. Reynolds, C.F.; Kupfer, D.J.; Thase, M.E.; Frank, E.; Jarrett, D.B.; Coble, P.A.; Hoch, C.C.; Buysse, D.J.; Simons, A.D.; Houck, P.R. Sleep, gender, and depression: An analysis of gender effects on the electroencephalographic sleep of 302 depressed outpatients. *Biol. Psychiatry* **1990**, *28*, 673–684. [CrossRef]
27. Marosi, E.; Harmony, T.; Becker, J.; Bernal, J.; Reyes, A.; Rodriguez, M.; Fernandez, T. Sex Differences in EEG Coherence in Normal Children. *Int. J. Neurosci.* **1993**, *72*, 115–121. [CrossRef] [PubMed]
28. Ehlers, C.; Kupfer, D. Slow-wave sleep: Do young adult men and women age differently? *J. Sleep Res.* **1997**, *6*, 211–215. [CrossRef] [PubMed]
29. Corsi-Cabrera, M.; Ramos, J.; Guevara, M.A.; Arce, C.; Gutierrez, S. Gender Differencesm in the Eeg During Cognitive Activity. *Int. J. Neurosci.* **1993**, *72*, 257–264. [CrossRef]
30. Wada, Y.; Takizawa, Y.; Zheng-Yan, J.; Yamaguchi, N. Gender Differences in Quantitative EEG at Rest and during Photic Stimulation in Normal Young Adults. *Clin. Electroencephalogr.* **1994**, *25*, 81–85. [CrossRef] [PubMed]
31. Corsi-Cabrera, M. Effect of 38 h of total sleep deprivation on the waking EEG in women: Sex differences. *Int. J. Psychophysiol.* **2003**, *50*, 213–224. [CrossRef]
32. Papageorgiou, C.C.; Nanou, E.D.; Tsiafakis, V.G.; Capsalis, C.N.; Rabavilas, A.D. Gender related differences on the EEG during a simulated mobile phone signal. *NeuroReport* **2004**, *15*, 2557–2560. [CrossRef]
33. Razumnikova, O.M. Gender differences in hemispheric organization during divergent thinking: An EEG investigation in human subjects. *Neurosci. Lett.* **2004**, *362*, 193–195. [CrossRef] [PubMed]

34. Silas, J.; Levy, J.P.; Nielsen, M.K.; Slade, L.; Holmes, A. Sex and individual differences in induced and evoked EEG measures of action observation. *Neuropsychologia* **2010**, *48*, 2417–2426. [CrossRef]
35. Tomescu, M.; Rihs, T.; Rochas, V.; Hardmeier, M.; Britz, J.; Allali, G.; Fuhr, P.; Eliez, S.; Michel, C. From swing to cane: Sex differences of EEG resting-state temporal patterns during maturation and aging. *Dev. Cogn. Neurosci.* **2018**, *31*, 58–66. [CrossRef]
36. van Putten, M.J.A.M.; Olbrich, S.; Arns, M. Predicting sex from brain rhythms with deep learning. *Sci. Rep.* **2018**, *8*, 3069. [CrossRef] [PubMed]
37. Tosun, P.D.; Dijk, D.J.; Winsky-Sommerer, R.; Abasolo, D. Effects of Ageing and Sex on Complexity in the Human Sleep EEG: A Comparison of Three Symbolic Dynamic Analysis Methods. *Complexity* **2019**, *2019*, 1–12. [CrossRef]
38. Ujma, P.P.; Konrad, B.N.; Simor, P.; Gombos, F.; Körmendi, J.; Steiger, A.; Dresler, M.; Bódizs, R. Sleep EEG functional connectivity varies with age and sex, but not general intelligence. *Neurobiol. Aging* **2019**, *78*, 87–97. [CrossRef]
39. Moctezuma, L.A.; Molinas, M. Sex differences observed in a study of EEG of linguistic activity and resting-state: Exploring optimal EEG channel configurations. In Proceedings of the 2019 7th International Winter Conference on Brain-Computer Interface (BCI), Gangwon, Republic of Korea, 18–20 February 2019; IEEE: Piscataway, NJ, USA, 2019. [CrossRef]
40. Bao, L.Q.; Qiu, J.L.; Tang, H.; Zheng, W.L.; Lu, B.L. Investigating Sex Differences in Classification of Five Emotions from EEG and Eye Movement Signals. In Proceedings of the 2019 41st Annual International Conference of the IEEE Engineering in Medicine and Biology Society (EMBC), Berlin, Germany, 23–27 July 2019; IEEE: Piscataway, NJ, USA, 2019. [CrossRef]
41. Arnegard, M.E.; Whitten, L.A.; Hunter, C.; Clayton, J.A. Sex as a Biological Variable: A 5-Year Progress Report and Call to Action. *J. Women's Health* **2020**, *29*, 858–864. [CrossRef] [PubMed]
42. Markovic, A.; Kaess, M.; Tarokh, L. Gender differences in adolescent sleep neurophysiology: A high-density sleep EEG study. *Sci. Rep.* **2020**, *10*, 15935. [CrossRef] [PubMed]
43. Cave, A.E.; Barry, R.J. Sex differences in resting EEG in healthy young adults. *Int. J. Psychophysiol.* **2021**, *161*, 35–43. [CrossRef] [PubMed]
44. Davidov, A.; Razumnikova, O.; Bakaev, M. Nature in the Heart and Mind of the Beholder: Psycho-Emotional and EEG Differences in Perception of Virtual Nature Due to Gender. *Vision* **2023**, *7*, 30. [CrossRef]
45. Al-Qazzaz, N.K.; Ali, S.H.M.; Ahmad, S.A. Entropy-Based EEG Markers for Gender Identification of Vascular Dementia Patients. In Proceedings of the 3rd International Conference for Innovation in Biomedical Engineering and Life Sciences, Kuala Lumpur, Malaysia, 6–7 December 2019; Springer: Cham, Switzerland, 2021; pp. 121–128. [CrossRef]
46. Hu, J. An approach to EEG-based gender recognition using entropy measurement methods. *Knowl.-Based Syst.* **2018**, *140*, 134–141. [CrossRef]
47. Al-Qazzaz, N.K.; Sabir, M.K.; Ali, S.H.B.M.; Ahmad, S.A.; Grammer, K. Complexity and Entropy Analysis to Improve Gender Identification from Emotional-Based EEGs. *J. Healthc. Eng.* **2021**, *2021*, 1–17. [CrossRef]
48. Wang, P.; Hu, J. A hybrid model for EEG-based gender recognition. *Cogn. Neurodynamics* **2019**, *13*, 541–554. [CrossRef]
49. Frauscher, B.; von Ellenrieder, N.; Zelmann, R.; Dolezalová, I.; Minotti, L.; Olivier, A.; Hall, J.; Hoffmann, D.; Nguyen, D.K.; Kahane, P.; et al. Atlas of the normal intracranial electroencephalogram: Neurophysiological awake activity in different cortical areas. *Brain* **2018**, *141*, 1130–1144. [CrossRef] [PubMed]
50. Frauscher, B.; von Ellenrieder, N.; Zelmann, R.; Rogers, C.; Nguyen, D.K.; Kahane, P.; Dubeau, F.; Gotman, J. High-Frequency Oscillations in the Normal Human Brain. *Ann. Neurol.* **2018**, *84*, 374–385. [CrossRef] [PubMed]
51. von Ellenrieder, N.; Gotman, J.; Zelmann, R.; Rogers, C.; Nguyen, D.K.; Kahane, P.; Dubeau, F.; Frauscher, B. How the Human Brain Sleeps: Direct Cortical Recordings of Normal Brain Activity. *Ann. Neurol.* **2020**, *87*, 289–301. [CrossRef] [PubMed]
52. Panzeri, S.; Montani, F.; Notaro, G.; Magri, C.; Peterson, R.S. Population Coding. In *Analysis of Parallel Spike Trains*; Springer: Midtown Manhattan, NY, USA, 2010; pp. 303–319. [CrossRef]
53. Olivares, F.; Souza, L.; Legnani, W.; Rosso, O.A. Informational Time Causal Planes: A Tool for Chaotic Map Dynamic Visualization. *Nonlinear Systems–Theoretical Aspects and Recent 621 Applications*; IntechOpen: London, UK, 2020; pp. 1–19. [CrossRef]
54. Bandt, C.; Pompe, B. Permutation Entropy: A Natural Complexity Measure for Time Series. *Phys. Rev. Lett.* **2002**, *88*, 174102. [CrossRef]
55. Rosso, O.A.; Masoller, C. Detecting and quantifying temporal correlations in stochastic resonance via information theory measures. *Eur. Phys. J. B* **2009**, *69*, 37–43. [CrossRef]
56. Montani, F.; Rosso, O.A. Entropy-Complexity Characterization of Brain Development in Chickens. *Entropy* **2014**, *16*, 4677–4692. [CrossRef]
57. Rosso, O.A.; Masoller, C. Detecting and quantifying stochastic and coherence resonances via information-theory complexity measurements. *Phys. Rev. E* **2009**, *79*, 040106. [CrossRef]
58. Pessa, A.A.B.; Ribeiro, H.V. ordpy: A Python package for data analysis with permutation entropy and ordinal network methods. *Chaos* **2021**, *31*, 063110. [CrossRef]
59. Feldman, D.P.; Crutchfield, J.P. Measures of statistical complexity: Why? *Phys. Lett. A* **1998**, *238*, 244–252. [CrossRef]
60. López-Ruiz, R.; Mancini, H.; Calbet, X. A statistical measure of complexity. *Phys. Lett. A* **1995**, *209*, 321–326. [CrossRef]
61. Martin, M.; Plastino, A.; Rosso, O. Generalized statistical complexity measures: Geometrical and analytical properties. *Phys. A Stat. Mech. Its Appl.* **2006**, *369*, 439–462. [CrossRef]
62. Lamberti, P.; Martin, M.; Plastino, A.; Rosso, O. Intensive entropic non-triviality measure. *Phys. A Stat. Mech. Its Appl.* **2004**, *334*, 119–131. [CrossRef]

63. Panzeri, S.; Petersen, R.S.; Schultz, S.R.; Lebedev, M.; Diamond, M.E. The Role of Spike Timing in the Coding of Stimulus Location in Rat Somatosensory Cortex. *Neuron* **2001**, *29*, 769–777. [CrossRef] [PubMed]
64. Welch, P. The use of fast Fourier transform for the estimation of power spectra: A method based on time averaging over short, modified periodograms. *IEEE Trans. Audio Electroacoust.* **1967**, *15*, 70–73. [CrossRef]
65. Stoica, P.; Moses, R.L. Spectral Analysis of Signals. In *Spectral Analysis of Signals*; Prentice Hall: Hoboken, NJ, USA, 2005; Chapter 1–2, pp. 1–2, 22, 49–50.
66. Hayes, M.H. Statistical Digital Signal Processing and Modeling. In *Statistical Digital Signal Processing and Modeling*; John Wiley and Sons: Hoboken, NJ, USA, 1996; Chapter 8, pp. 415–420.
67. Guisande, N.; Montani, F. Rényi entropy-complexity causality space: A novel neurocomputational tool for detecting scale-free features in EEG/iEEG data. *Front. Comput. Neurosci.* **2024**, *18*, 1342985. [CrossRef]

entropy

MDPI

Article

Identifying Ordinal Similarities at Different Temporal Scales

Luciano Zunino [1,2], Xavier Porte [3] and Miguel C. Soriano [4,*]

1 Centro de Investigaciones Ópticas (CONICET La Plata-CIC-UNLP), Gonnet 1897, La Plata, Argentina; lucianoz@ciop.unlp.edu.ar
2 Departamento de Ciencias Básicas, Facultad de Ingeniería, Universidad Nacional de La Plata (UNLP), La Plata 1900, Argentina
3 Institute of Photonics, Department of Physics, University of Strathclyde, 99 George Street, Glasgow G1 1RD, UK; javier.porte-parera@strath.ac.uk
4 Instituto de Física Interdisciplinar y Sistemas Complejos (IFISC, UIB-CSIC), Campus Universitat de les Illes Balears, E-07122 Palma de Mallorca, Spain
* Correspondence: miguel@ifisc.uib-csic.es

Abstract: This study implements the permutation Jensen–Shannon distance as a metric for discerning ordinal patterns and similarities across multiple temporal scales in time series data. Initially, we present a numerically controlled analysis to validate the multiscale capabilities of this method. Subsequently, we apply our methodology to a complex photonic system, showcasing its practical utility in a real-world scenario. Our findings suggest that this approach is a powerful tool for identifying the precise temporal scales at which two distinct time series exhibit ordinal similarity. Given its robustness, we anticipate that this method could be widely applicable across various scientific disciplines, offering a new lens through which to analyze time series data.

Keywords: time series; symbolic analysis; ordinal patterns; permutation entropy; Jensen–Shannon divergence; permutation Jensen–Shannon distance; multiscale analysis; ordinal similarity; chaotic semiconductor laser; delayed optical feedback

1. Introduction

It is widely known that quantifying the similarity (or dissimilarity) between two time series is an essential task for clustering and classification purposes [1]. Because of this main reason, a lot of methodologies have been developed to determine how much an arbitrary time series resembles another one. Actually, this issue can be addressed from very different perspectives: there are shape-, edit-, feature-, and structure-based measures [2]. Some of them are more computationally inspired, while others are based on physical notions. To the best of our knowledge, up to now, there is no optimal algorithm for estimating this concept in practice. It has also been demonstrated that the performance of similarity quantifiers can be highly reduced when time series with different sampling frequencies are contrasted [3,4], when there exist nonlinear dependencies between them [5] and/or in the presence of outliers [6]. Hence, a measure able to deal robustly with different types of data is sought.

The permutation Jensen-Shannon distance (PJSD) has recently been proposed within this realm [7]. It is a versatile and conceptually simple ordinal metric tool that, thanks to its noise robustness and invariance under scaling of the data, is particularly suited for the analysis of real-world signals [8]. The PJSD takes advantage of the Jensen-Shannon divergence (JSD) [9], a widely accepted method for assessing the dissimilarity between two probability distributions, and of the ordinal coarse-grained representation introduced more than 20 years ago by Bandt and Pompe (BP) [10]. The flexibility of the JSD to different distributional data types, together with the proven efficiency of the ordinal patterns for identifying equivalent dynamics [11] and for time series clustering [12], allow us to conjecture that the proposed fusion represents a useful addition to the repertoire of approaches intended to estimate the degree of similarity between two arbitrary time series.

In this work, we put special focus on the PJSD's ability to test similarity between two time series at different time scales. Numerical and experimental analyses are included to illustrate this fact. The results obtained confirm that the PJSD robustly identifies the time scales that maximize the similarity between two arbitrary signals. Consequently, this ordinal metric, implemented through a multiscale scheme, offers an efficient alternative to characterize how similar two signals recorded at different sampling rates are. Taking into account that widely implemented similarity measures, like dynamic time warping, are strongly affected when facing this challenge [3,4], we consider our finding relevant enough and of potential interest for the time series analysis community.

The rest of this paper is structured as follows. A brief presentation of the PJSD is first given in Section 2. Then, in Section 3, a numerically controlled analysis is developed to illustrate how the multiscale PJSD approach works. After that, a more complex practical application comparing time series of semiconductor laser experiments is performed in Section 4. The main conclusions obtained from this study are finally summarized in the last Section 5.

2. Permutation Jensen–Shannon Distance

The PJSD can be estimated by calculating the square root of the normalized JSD between the ordinal probability distributions associated with the two time series under comparison [7]. Defined in such a way, it is a metric able to quantify the degree of discernability between two arbitrary time series from an ordinal perspective.

The JSD is a measure of the distance between two arbitrary probability distributions, $P = \{p_1, \ldots, p_n\}$ and $Q = \{q_1, \ldots, q_n\}$, given by

$$\mathrm{JSD}(P,Q) = S\left(\frac{P+Q}{2}\right) - \frac{1}{2}S(P) - \frac{1}{2}S(Q), \tag{1}$$

where S is the Shannon entropy function $S(P) = -\sum_{i=1}^{n} p_i \ln p_i$, and, as usual, the convention $0 \ln 0 = 0$ is assumed in accordance with its mathematical limit. The JSD is always a well-defined and bounded quantity [9] that achieves its minimum possible value, i.e., 0, when identical probability distributions are compared, while its maximum potential value, i.e., $\ln 2$, is obtained whenever the supports of P and Q are disjoints (that is, $p_i q_i = 0$ for $i = 1, \ldots, n$). It has also been shown that $[\mathrm{JSD}(P,Q)]^{1/2}$ satisfies all the formal properties needed to be a metric [13]. Further statistical properties and theoretical interpretations of the JSD can be found in Ref. [14].

The estimation of the Jensen–Shannon distance, $[\mathrm{JSD}(P,Q)]^{1/2}$, between two time series requires us first to know the corresponding probability distributions, P and Q, associated with the two time series under analysis. This task is not straightforward nor simple, and it deserves careful attention [15]. We address it by implementing the BP mapping method. BP propose mapping a continuous-valued time series into a discrete series of ordinal symbols or ordinal patterns. Perhaps the most relevant property related to this symbolization scheme is the fact that, as stated by Amigó et al. [16], "ordinal patterns are not symbols ad hoc but they actually encapsulate qualitative information about the temporal structure of the underlying data". That is, the presence of underlying temporal correlations in the dynamics of the process that generates the time series is naturally considered in the BP recipe. Next, we will briefly summarize the discretization of time series via ordinal patterns. For further technical details interested readers are referred to the reviews [16–20]. Given a real-valued time series $X = \{x_t \in \mathbb{R}, t = 1, \ldots, L\}$, vectors of equally spaced D values of the form $(x_s, x_{s+\tau}, \ldots, x_{s+(D-1)\tau})$ with $s = 1, \ldots, L^* = L - (D-1)\tau$ are mapped to one of the $D!$ possible ordinal permutations of the same size that describe the order relation between these elements. For example, the vector $(2.5, 4.7, 0.6)$ is mapped to the ordinal or permutation pattern $(2,3,1)$, replacing each element in the original vector with its respective ranking in the subset. Assigning a symbol π_i to each ordinal pattern, the original time series is mapped to the coarse-graining sequence $Y = \{y_s \in \Pi_D, s = 1, \ldots, L^*\}$, with $\Pi_D = \{\pi_1, \pi_2, \ldots, \pi_{D!}\}$ representing the set of permutations of length D. Just for illustrative purposes, $\Pi_2 = \{\pi_1 = (1,2), \pi_2 = (2,1)\}$, and $\Pi_3 = \{\pi_1 = (1,2,3), \pi_2 = (1,3,2), \pi_3 = (2,1,3), \pi_4 = (2,3,1), \pi_5 = (3,1,2), \pi_6 = (3,2,1)\}$. Estimating the probability of each ordinal pattern $p(\pi_i)$ based on its relative frequency of occurrence in the symbolized sequence, an associated ordinal probability distribution (OPD) can be then obtained as follows:

$$P_\pi^{D,\tau} = \{p(\pi_i), i = 1, \ldots, D!\}. \tag{2}$$

Two parameters have to be fixed: the number of elements in the permutation patterns D (called the order or embedding dimension, $D \geq 2$ with $D \in \mathbb{N}$) and the time separation τ between the elements in the subsequence (called lag or embedding delay, $\tau \in \mathbb{N}$). Consecutive data are considered if $\tau = 1$, while τ-spaced data samples are analyzed if $\tau \geq 2$. A toy example is included below to illustrate the role played by the lag τ. Given the short time series $X = \{1.1, 2.1, 4.3, 3.2, 6.7, 0.5, 10.4, 8.9\}$ and fixing $D = 3$ and $\tau = 2$, the first vector $(x_1, x_3, x_5) = (1.1, 4.3, 6.7)$ is mapped to the ordinal pattern $\pi_1 = (1,2,3)$. The second three-dimensional vector is $(x_2, x_4, x_6) = (2.1, 3.2, 0.5)$, and $\pi_4 = (2,3,1)$ is its associated ordinal pattern. Finally, the following two remaining vectors, $(x_3, x_5, x_7) = (4.3, 6.7, 10.4)$ and $(x_4, x_6, x_8) = (3.2, 0.5, 8.9)$, are mapped to the permutations $\pi_1 = (1,2,3)$ and $\pi_3 = (2,1,3)$, respectively. Consequently, the symbolic sequence $Y = \{\pi_1, \pi_4, \pi_1, \pi_3\}$ is obtained when applying the BP coarse-graining procedure with parameters $D = 3$ and $\tau = 2$, and the OPD turns out to be $P_\pi^{3,2} = \{p(\pi_1) = 0.5, p(\pi_2) = 0, p(\pi_3) = 0.25, p(\pi_4) = 0.25, p(\pi_5) = 0, p(\pi_6) = 0\}$ for this simple numerical example.

On the one hand, the condition $L \gg D!$, with L representing the number of data in the original time series, is required for a robust estimation of $P_\pi^{D,\tau}$. It is also clear that larger values of D offer improved characterization of the system dynamics. Actually, the order D has to exceed a lower bound D_{min} to successfully resolve the underlying temporal structures for data from high-dimensional systems [21]. On the other hand, a value of the lag $\tau = 1$ is often used in discrete systems, and also when the chosen sampling frequency is the optimal one to characterize the underlying dynamics of continuous systems [18]. However, this arbitrary choice can lead to erroneous conclusions, especially for systems with scale-dependent dynamics [22]. A multiscale analysis, by analyzing how descriptors of the OPD change with τ, gives a more complete picture in these instances [23–25], also providing a practical and efficient alternative for identifying time delays from stochastic and chaotic models [26,27].

Among the different statistics that can be computed from the resulting OPD given by Equation (2), the permutation entropy [10] is undoubtedly the most representative and widely implemented quantifier. Defined as the Shannon entropy of this ordinal distribution, $S\left(P_\pi^{D,\tau}\right) = -\sum_{i=1}^{D!} p(\pi_i) \log p(\pi_i)$, it quantifies the variety of permutation patterns in the ordinal sequence obtained from a time series. However, many other much more complex descriptors of the OPD, which try to characterize some particular aspect of it, have been proposed. Without being exhaustive, we can mention permutation statistical complexity [28], Rényi permutation entropy [29] and permutation Fisher's information measure [30].

The PJSD is defined as the normalized Jensen–Shannon distance between the OPDs, P_π^{D,τ_1} and Q_π^{D,τ_2}, associated with two arbitrary time series, i.e.,

$$\left[\mathrm{JSD}\left(P_\pi^{D,\tau_1}, Q_\pi^{D,\tau_2}\right) / \ln(2)\right]^{1/2}. \tag{3}$$

The ordinal mapping of different time series is compared using the PJSD. It is worthy to highlight here that, when estimating the PJSD, the order D chosen to implement the BP symbolization recipe should be the same for both time series in order to have the same number of possible permutation patterns in the OPDs to be compared. However, different lags, τ_1 and τ_2, can be selected, opening the possibility of contrasting the ordinal similarity of only one or two time series at two different temporal scales. The potentialities of this multiscale approach are explored in this article.

3. An Illustrative Numerical Example

We focus our analysis on nonlinear systems in the chaotic regime, where the dynamics involve multiple time scales. In particular, we use the chaotic Mackey–Glass (MG) system as an example, which is a paradigmatic case of a high-dimensional (strange) chaotic attractor [31,32]. The evolution of the MG system, denoted by $x(t)$, is described by the following delay differential equation:

$$\dot{x}(t) = -ax(t) + \frac{bx(t-\tau_S)}{1+x(t-\tau_S)^c}, \tag{4}$$

where $a = 0.1$, $b = 0.2$ and $c = 10$ are standard parameters [32], and the delay $\tau_S = 30$ sets the system into the chaotic regime [32,33].

We numerically integrate Equation (4) with an integration step of $\Delta t = 0.01$, generating long time series, which are subsequently sampled for our analysis. Figure 1a presents a time series of the MG system in the chaotic regime, where we indicate different sampling intervals with either crosses or empty circles. The corresponding subsampled time series are shown in Figures 1b,c, where two different sampling intervals, $t_b = 18 \cdot 10^2 \Delta t$ (crosses, x_b) and $t_c = 24 \cdot 10^2 \Delta t$ (circles, x_c), are chosen for illustrative purposes. As shown in Figure 1, the time series generated with different sampling intervals maintain the oscillatory behavior of the original MG dynamics, but the similarity between Figure 1b,c cannot be readily observed by the naked eye.

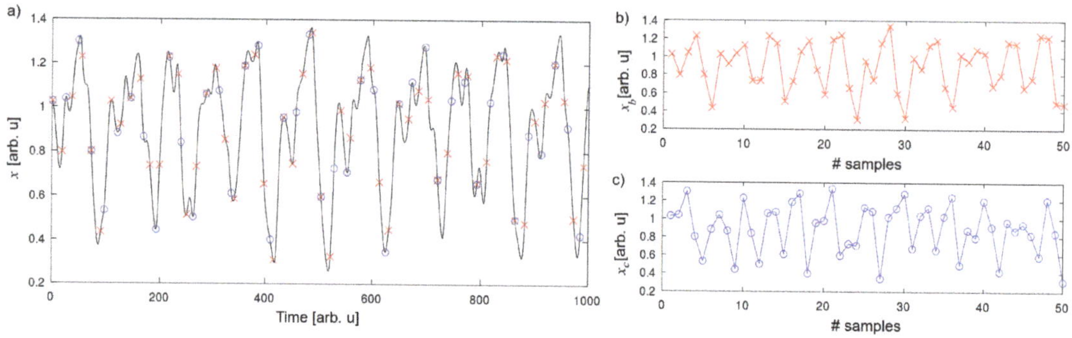

Figure 1. (**a**) Time series of MG system. Red crosses indicate sampling of $t_b = 18 \cdot 10^2 \Delta t$ and blue empty circles sampling of $t_c = 24 \cdot 10^2 \Delta t$. (**b**) Example of sequence extracted with sampling t_b. (**c**) Example of sequence extracted with sampling t_c.

To visualize the properties of the PJSD, we apply this measure to the MG time series presented in Figure 1. For this purpose, the PJSD is computed for the sampled time series x_b and x_c using varying symbolization lags, τ_1 and τ_2. The PJSD is expected to capture the similarity between x_b and x_c for certain ratios of the symbolization lags and sampling intervals. We present the evaluation of the PJSD for the sampled MG time series in Figure 2, computed for several embedding dimensions, D, ranging from 3 to 6. In this representation, larger similarities between the analyzed time series correspond to blue colors (PJSD values close to zero, shown on a logarithmic scale). As demonstrated in Figure 2, the PJSD measure successfully recovers the similarity between the MG time series x_b and x_c for several values of the symbolization lags. The similarity of the time series becomes evident when $\tau_1/\tau_2 = 4/3$, a ratio that precisely compensates for the difference in sampling times, $t_b/t_c = 3/4$, from the original MG time series. In other words, the minima of the computed PJSD occur when $\tau_1 \cdot t_b = \tau_2 \cdot t_c$. Consequently, smaller estimated PJSD values are observed not only for $\tau_1 = 4$ and $\tau_2 = 3$, but also for multiples of them: $\tau_1 = 8$ and $\tau_2 = 6$, $\tau_1 = 12$ and $\tau_2 = 9$, and so on. This behavior acts as a double check for identifying the temporal scales under which the time series under analysis are truly similar.

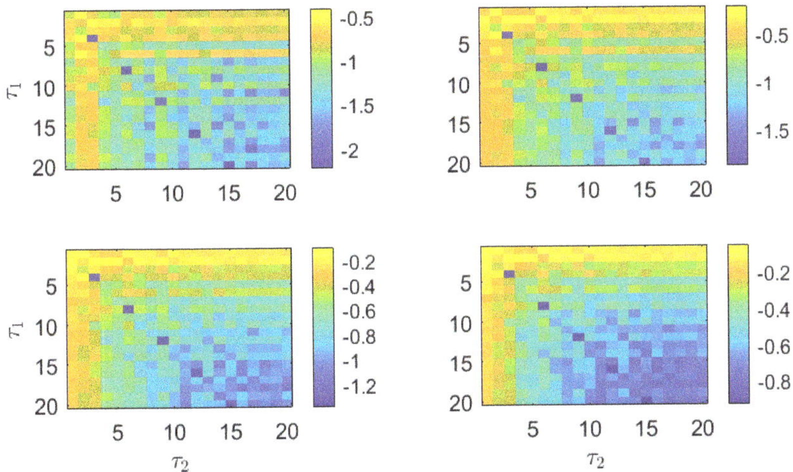

Figure 2. PJSD estimations in logarithmic base 10 scale from two numerical realizations of Mackey–Glass oscillator operating in chaotic regime ($\tau_S = 30$) at different sampling intervals $t_b = 18 \cdot 10^2\Delta t$ and $t_c = 24 \cdot 10^2\Delta t$. Order D increases from 3 to 6 (from upper left to lower right plots), and lags τ_1 and τ_2 vary from 1 to 20. Time series of length $L = 12{,}500$ data points are considered in analysis.

In addition to recovering the similarity between the MG time series, the PJSD values shown in Figure 2 exhibit other trends worth discussing. The PJSD values systematically decrease as τ_1 and τ_2 increase, corresponding to the yellow colors in the top left corners and blue colors in the bottom right corners of the different panels in Figure 2. This trend can be explained by examining the underlying ordinal pattern probabilities upon which the PJSD is computed. For instance, Figure 3 shows the ordinal pattern probabilities for the analyzed MG time series at selected values of τ_1 and τ_2. When $\tau_1/\tau_2 = 4/3$, the measured ordinal pattern probabilities for x_b and x_c are equal, resulting in PJSD values close to zero. Otherwise, two scenarios arise: either different probabilities are measured when τ_1 or τ_2 is small, or distributions approaching equiprobability are observed when τ_1 and τ_2 are both large. This analysis is further supported by examining the ordinal pattern probabilities for higher embedding dimensions D, as shown, for example, in Figure 4 for $D = 4$, where the probabilities for x_b and x_c are equal when $\tau_1/\tau_2 = 4/3$. Finally, it can also be observed that the ordinal pattern probabilities for both signals are clearly different if the original sampling intervals are considered, i.e., when $\tau_1 = \tau_2 = 1$.

As additional and more robust confirmation of the similarity between the OPDs for $\tau_1 = 4$ and $\tau_2 = 3$, a surrogate analysis is performed. More precisely, the PJSD of the original time series is compared against the distributions of the estimated PJSD values for 1000 independent shuffled realizations of the original records when the same lags are considered. Just for the sake of comparison, the same analysis is repeated for large values of the lag ($\tau_1 = 20$ and $\tau_2 = 20$). The results obtained are briefly summarized in Figures 5 and 6. On the one hand, when $\tau_1 = 4$ and $\tau_2 = 3$ (Figure 5), it can be concluded that the estimated PJSD value for the original MG time series is significantly lower than those obtained for their shuffled counterparts if larger values of the order D are considered ($D = 5$ and $D = 6$). On the other hand, when $\tau_1 = 20$ and $\tau_2 = 20$ (Figure 6), the estimated PJSD value is significantly higher than those associated with shuffled realizations, independently of the order D. Based on these findings, the similarity and dissimilarity between the original MG signals can be robustly concluded in the former and latter cases, respectively.

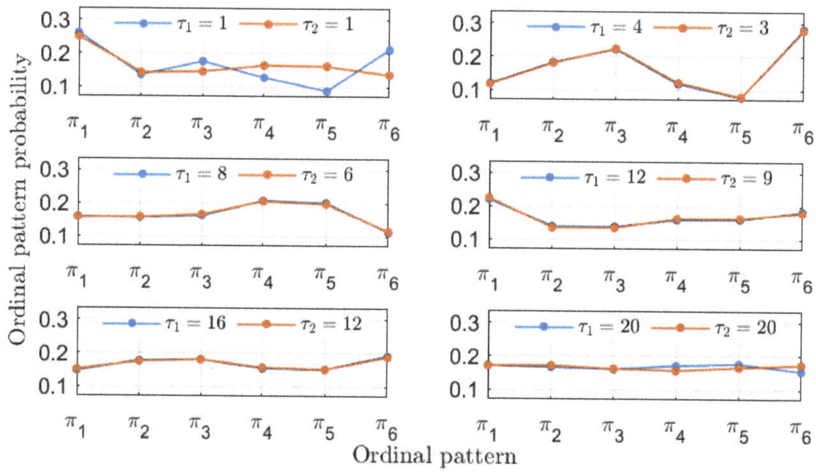

Figure 3. Ordinal pattern probabilities with $D = 3$ for the two numerical realizations of the Mackey–Glass oscillator operating in a chaotic regime ($\tau_S = 30$) at different sampling intervals $t_b = 18 \cdot 10^2 \Delta t$ and $t_c = 24 \cdot 10^2 \Delta t$. Particular choices of the symbolization lags τ_1 and τ_2 associated with these two signals are considered in each subplot.

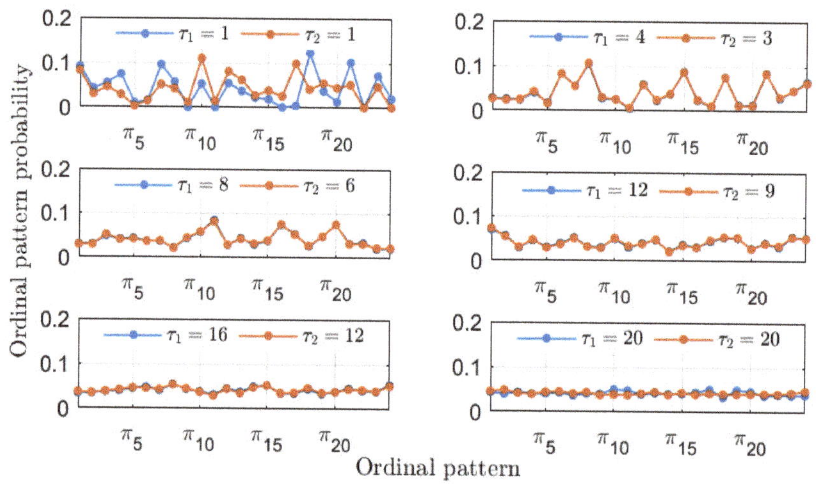

Figure 4. The same as in Figure 3 but with $D = 4$.

It is worth remarking here that the running time to estimate the PJSD between the MG signals ($L =$ 12,500) for the 400 considered combinations of the lags τ_1 and τ_2 is around a second. Thus, the proposed multiscale scheme is fast enough, paving the way for real-time analysis. Obviously, the time complexity increases for larger values of the order D and for longer time series. Please see Appendix A for further details.

In summary, we have validated the PJSD methodology for identifying similarities between different time series in a controlled numerical example generated from a single parameter set with varying sampling intervals. In the next section, we extend the application of the PJSD methodology to a numerical setting where time series are generated from different parameter sets, as well as to experimental data from a laser system.

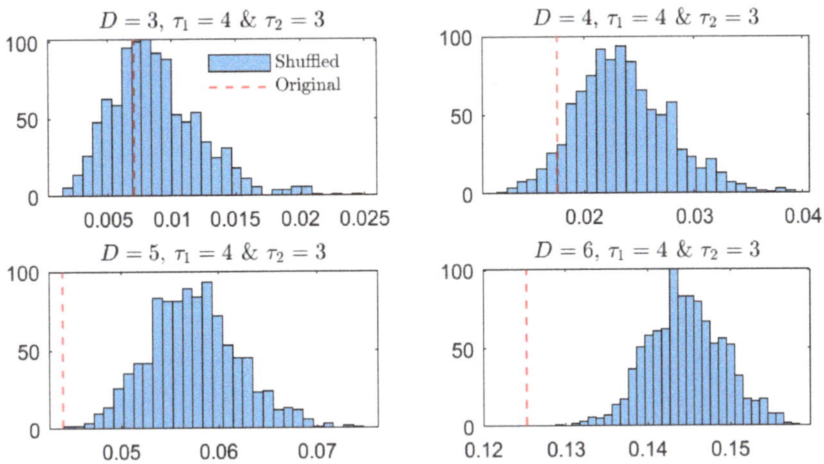

Figure 5. Distribution of 1000 estimated PJSD values from shuffled realizations of original MG time series for different orders D when lags τ_1 and τ_2 are equal to 4 and 3, respectively. Red vertical dashed line indicates estimated PJSD value for the original MG signals.

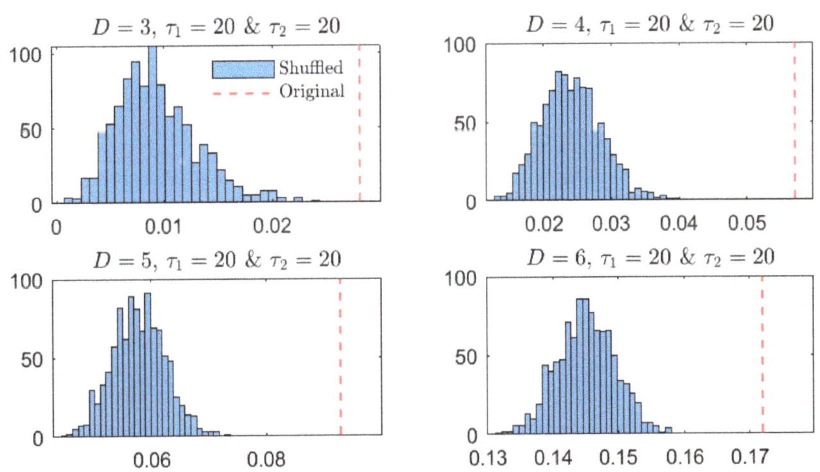

Figure 6. The same as in Figure 5 but for lags $\tau_1 = 20$ and $\tau_2 = 20$.

4. Practical Application: Semiconductor Lasers Subject to Optical Feedback

Semiconductor lasers can exhibit rich dynamical behavior when they are subjected to external perturbations such as optical feedback or optical injection [34]. In particular, semiconductor lasers with optical feedback are typically considered a paradigmatic physical system to observe complex dynamical behavior experimentally [35]. Several time scales are involved in the resulting complex dynamics, and often, the precise characterization is not straightforward due to the nonlinear interactions that occur in the semiconductor laser [26].

The complexity of the intensity emitted by the semiconductor laser subject to optical feedback makes it an ideal case to analyze the validity of the PJSD metric. In the following, we first proceed to describe the system under study, both theoretically and experimentally. We then present the results of the PJSD metric applied to numerical and experimental time series of the laser subject to feedback.

4.1. Theoretical Model

A single-mode semiconductor laser subject to moderate optical feedback can typically be described by the following Lang–Kobayashi (LK) equations [36,37],

$$\dot{E}(t) = \frac{1 + \mathrm{i}\,\alpha}{2}\, G_{\mathrm{N}}\, n(t)\, E(t) + \kappa E(t - \Lambda) e^{-\mathrm{i}\omega_0 \Lambda}, \tag{5}$$

$$\dot{n}(t) = p_{exc}\, J_{\mathrm{th}} - \gamma\, n(t) - [\Gamma + G_{\mathrm{N}}\, n(t)] |E(t)|^2, \tag{6}$$

where E and n are the complex electric field amplitude and the carrier number above the threshold, respectively. The equations are normalized such that $P(t) = |E(t)|^2$ is the number of photons. The optical feedback term in the field equation includes the delay time Λ and the feedback rate κ. The other parameters in Equation (5) are the linewidth enhancement factor (α), the differential optical gain ($G_{\mathrm{N}} = 2.142 \cdot 10^4\ \mathrm{s}^{-1}$) and the laser solitary frequency (ω_0). In Equation (6), p_{exc} is the excess pump current over the threshold ($\frac{J}{J_{\mathrm{th}}} - 1$), $J_{\mathrm{th}} = 1.552 \cdot 10^{17}\ \mathrm{s}^{-1}$ denotes the pump threshold current in units of the electron charge, $\gamma = 0.909 \cdot 10^9\ \mathrm{s}^{-1}$ is the carrier decay rate, and $\Gamma = 0.357 \cdot 10^{12}\ \mathrm{s}^{-1}$ is the cavity decay rate. The parameter values were chosen according to Refs. [34,37], except for $\alpha = 3$, $\Lambda = 10\ \mathrm{ns}^{-1}$, in order to obtain dynamical behavior similar to the experimental behavior [38].

The relaxation oscillation (RO) frequency is the natural resonance of the semiconductor laser, which results from light–matter interactions. In this model, the RO is given by $f_{RO} = \frac{1}{2\pi}\sqrt{G_{\mathrm{N}} p_{exc} J_{th}}$. In turn, the optical feedback induces a frequency shift in the emitted optical frequency, where the maximum feedback-induced frequency shift is $\Delta f_{fb} \sim \alpha\kappa/2\pi$. As shown in Ref. [38], similar dynamics can be observed for different laser and feedback conditions as long as the ratio between f_{RO} and Δf_{fb} is kept constant. Here, similarity refers to the phenomenon of observing laser time series with equivalent statistical properties but with different time scales. We control the ratio between f_{RO} and Δf_{fb} by changing the values of p_{exc} and κ, which are readily accessible to the experimentalist. To be precise, the time scale of the delay time Λ will also need to be adjusted by a factor of $\sqrt{p_{exc}}$ to achieve similar laser dynamics.

4.2. Description of Experimental Setup

Figure 7 depicts the scheme of the experimental setup, achieved by employing standard fiber-based telecommunication components. The semiconductor laser diode in the experimental setup has an emission wavelength around 1550 nm. This laser emits in a single longitudinal mode, with side-mode suppression ratios larger than 40 dB, and has a threshold current of I_{th} = 12.08 mA at a working temperature of 22 °C. The dashed red line in Figure 7 encloses the external cavity of round-trip time Λ, which depends on the length of the fiber components of the external cavity feedback loop. Such a fiber-optic external cavity has characteristic round-trip times of 10 to 100 ns. In the external cavity loop, a maximum feedback rate of $\kappa_{max} \simeq 70\ \mathrm{ns}^{-1}$ was estimated [38]. Time series were acquired using a photodiode with a 12.5 GHz bandwidth and a 16 GHz analog bandwidth oscilloscope with a sampling rate of 40 Gigasamples/s.

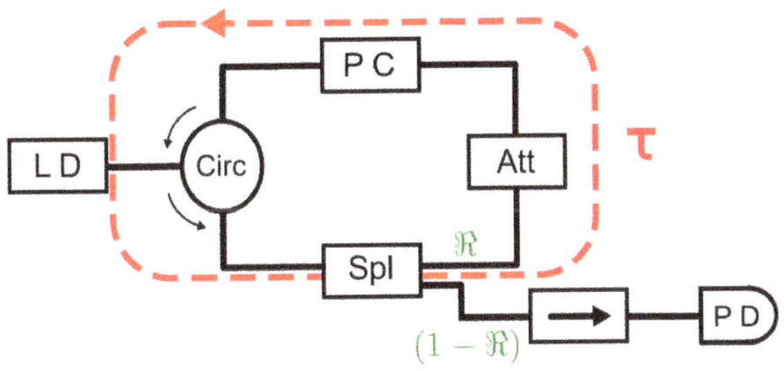

Figure 7. Scheme of experimental setup to study feedback dynamics. LD: laser diode; Circ: optical circulator; PC: polarization controller; Att: optical attenuator; Spl: one-by-two intensity splitter with $\Re = 0.95$ and $(1 - \Re) = 0.05$ splitting ratios; $\rightarrow$: optical isolator; and PD: photodiode.

4.3. Numerical Results

Motivated by the experimental settings, we consider two numerically generated time series that have a constant sampling rate but are generated using different laser and feedback parameters. Similarity of two laser time series, A and B, is expected to be found when $\kappa_A/\sqrt{p_{excA}} = \kappa_B/\sqrt{p_{excB}}$ and $\sqrt{p_{excA}} \cdot \Lambda_A = \sqrt{p_{excB}} \cdot \Lambda_B$ are fulfilled [38]. In particular, we choose, for illustrative purposes, the following two parameter sets: $p_{excA} = 0.25$, $\kappa_A = 50$ ns^{-1}, $\Lambda_A = 11.2$ ns, and p_{excB}=1, $\kappa_B = 100$ ns^{-1}, $\Lambda_B = 5.6$ ns, respectively. As a result, the two numerically generated time series should be equivalent for sampling intervals that correspond to a ratio 2 to 1, i.e., $\sqrt{1/p_{excA}}$ to $\sqrt{1/p_{excB}}$.

The estimations of the PJSD for the two numerically generated laser time series are presented in Figure 8. Independently of the embedding dimension D, low values of the PJSD are obtained when the symbolization lags τ_1 and τ_2 follow the 2-to-1 ratio. Similarity of the compared laser time series is recovered for several combinations of the symbolization lags, including $\tau_1 = 4$ and $\tau_2 = 2$; $\tau_1 = 6$ and $\tau_2 = 3$; and $\tau_1 = 8$ and $\tau_2 = 4$.

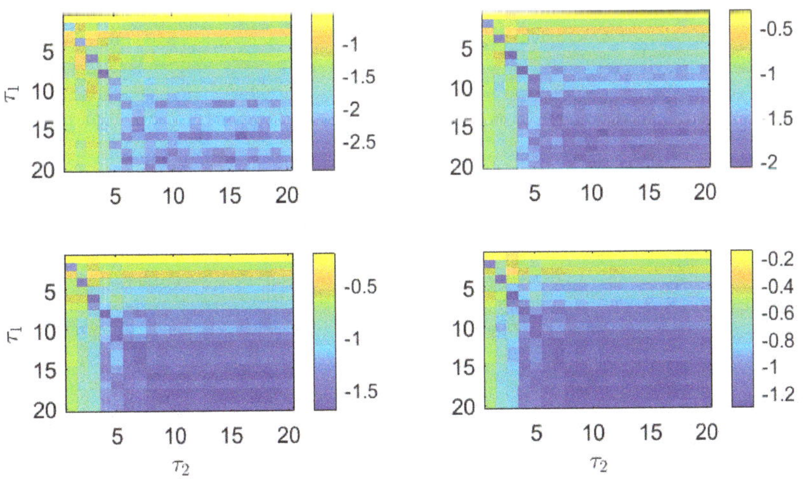

Figure 8. PJSD estimations in logarithmic base 10 scale from two numerical realizations of LK equations ($L = 10^5$ data points) at different time scales but with equivalent statistical properties. Order D increases from 3 to 6 (from upper left to lower right plots), and lags τ_1 and τ_2 vary from 1 to 20. Numerical laser time series are subsampled to $10/\Gamma \simeq 28$ ps prior to analysis.

In agreement with the results presented for the MG time series in Figure 2, large PJSD values (yellow colors) tend to appear for small values of the symbolization lags, and low

PJSD values (blue colors) appear for large values of the symbolization lags. As shown in Figure 9, low values of the PSJD metric can be obtained either when the ordinal pattern probabilities are computed following a 2-to-1 ratio or when the symbolization lags are large, obtaining quasi-equiprobable distributions.

The ordinal pattern probabilities computed using $D = 3$ are relatively simple. It is also interesting to observe the probabilities obtained using a larger embedding dimension. Figure 10 presents the ordinal pattern probabilities computed using $D = 4$. In this case, the similarity of the numerical laser time series is recovered for the proper 2-to-1 ratio of the symbolization lags, although the observed probabilities are significantly more involved.

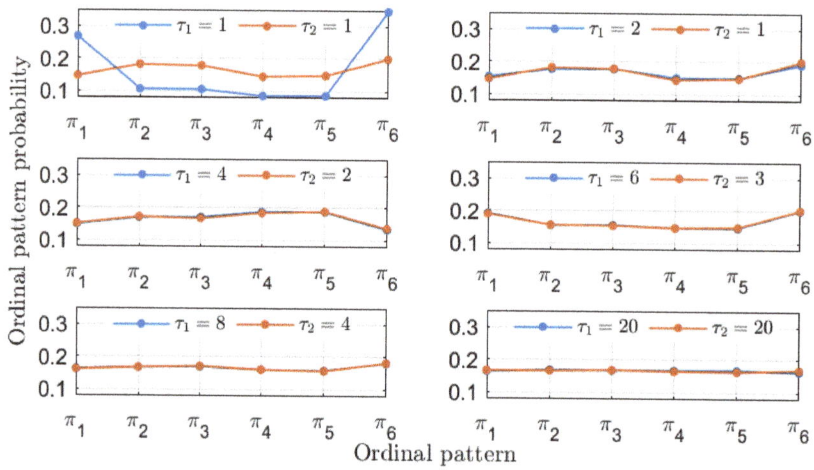

Figure 9. Ordinal pattern probabilities with $D = 3$ for the two numerical simulations of the LK equations. Particular choices of the symbolization lags τ_1 and τ_2 associated with these two signals are considered in each subplot.

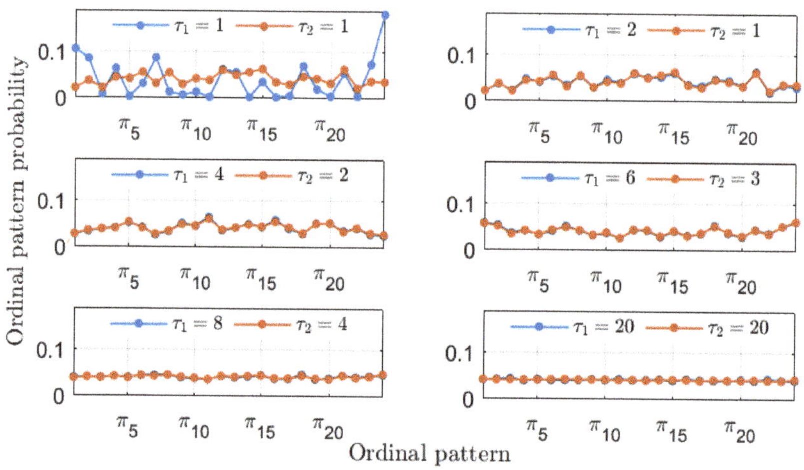

Figure 10. The same as in Figure 9 but with $D = 4$.

4.4. Experimental Results

We now analyze two experimental time series of a semiconductor laser subject to optical feedback. Identifying the similarity of experimental time series is a challenging task, as the data may contain noise and other non-idealities. Considering the dynamical study of this laser system, we can identifying similarity when the two laser time series A and B fulfill the conditions $\kappa_A/\sqrt{p_{excA}} = \kappa_B/\sqrt{p_{excB}}$ and $\sqrt{p_{excA}} \cdot \Lambda_A = \sqrt{p_{excB}} \cdot \Lambda_B$. In this case, we analyze time series with the parameter sets p_{excA}=0.57 ($\sqrt{p_{excA}} \simeq 3/4$), $\Lambda_A = 99.95$ ns and $p_{excB} = 1$, $\Lambda_B = 75.18$ ns ($\Lambda_B \simeq 3/4\Lambda_A$), respectively. Accordingly, the feedback strengths are set to $\kappa_B = 4/3\kappa_A$ such that the same ratio between f_{RO} and Δf_{fb} is maintained for the two experimental laser time series.

We present the PJSD estimations of the two experimental laser time series in Figure 11. A slanted blue line, corresponding to a 4-to-3 ratio between the symbolization lags τ_1 and τ_2, can be observed for all embedding dimensions. This result implies that the same dynamical features can be reproduced at different time scales when the laser and feedback parameters are properly adjusted. Since semiconductor lasers can be used in optical communications for secure key distribution [39–41], similarity could be exploited, e.g., to transmit secure keys at different rates.

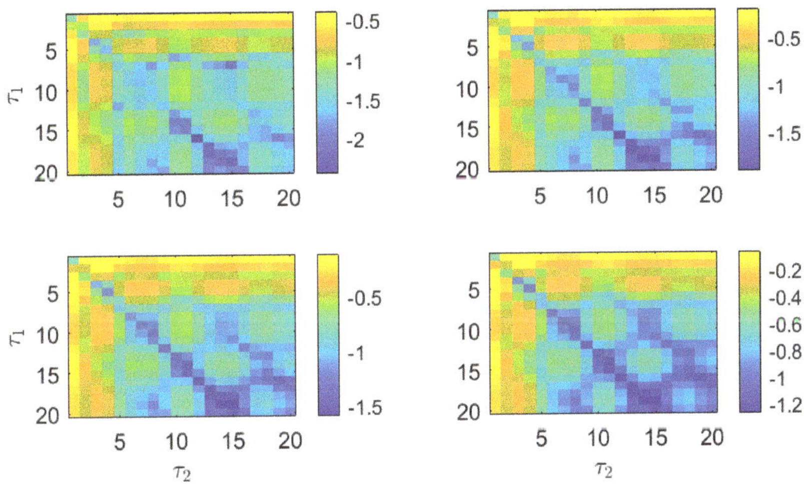

Figure 11. PJSD estimations in logarithmic base 10 scale from two experimentally obtained signals ($L = 10^5$ data points) with dynamical behaviors similar to numerical counterparts. Order D increases from 3 to 6 (from upper left to lower right plots), and lags τ_1 and τ_2 vary from 1 to 20. Experimental laser time series are acquired with sampling interval of 25 ps for analysis.

The similarity between the two experimental laser time series is further evidenced by the ordinal pattern probabilities displayed in Figures 12 and 13. Here, given the limited experimental precision, the 4-to-3 ratio for the similarity is not exact, and the 5-to-4 ratio also yields similar ordinal pattern probabilities. The temporal precision of identifying similarity is bounded by the sampling interval of 25 ps corresponding to the acquisition oscilloscope. Nevertheless, the existence of similarity between the two experimental laser time series can be readily identified given the results in Figure 11.

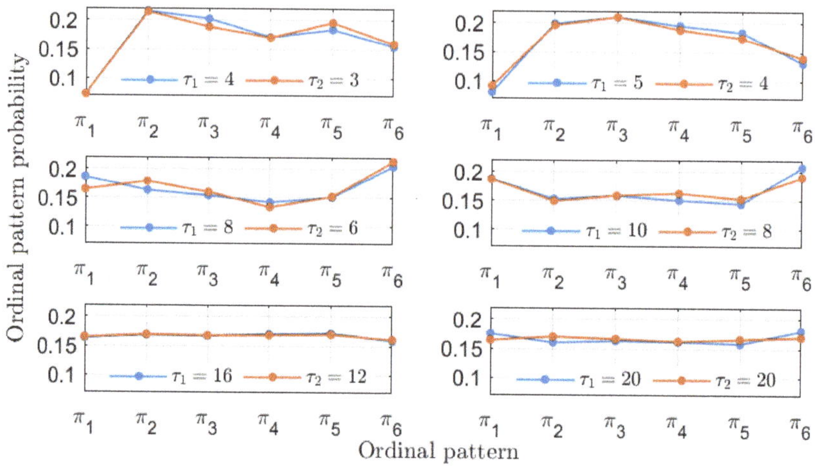

Figure 12. Ordinal patterns probabilities with $D = 3$ for the two experimental signals. Particular choices of the symbolization lags τ_1 and τ_2 associated with these two signals are considered in each subplot.

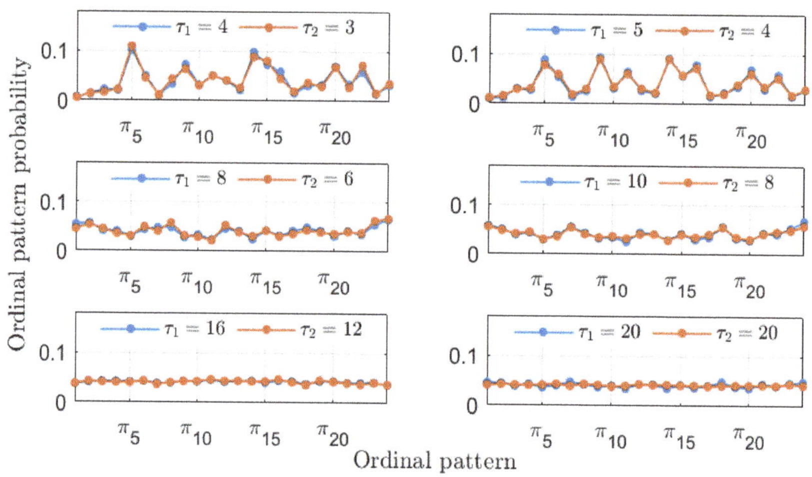

Figure 13. The same as in Figure 12 but with $D = 4$.

5. Conclusions

The ability of the PJSD to characterize the ordinal similarity of two time series at different temporal scales has been carefully analyzed in this work. Through numerical and experimental analyses, it has been shown that the PJSD offers a simple and flexible approach for identifying the sampling rates that minimize the distance between two arbitrary sequences from an ordinal perspective. This finding stands in stark contrast to what happens with other popular similarity measures, such as dynamic time warping, which show reduced performance when dealing with this issue.

Given its versatility, robustness to noise and outliers, and invariance under data scaling, it is reasonable to predict that the proposed multiscale PJSD approach could be of utility in analyses of real-world data from heterogeneous scientific fields. We encourage interested researchers and practitioners to implement this tool in order to confirm this hypothesis.

Author Contributions: Methodology, L.Z. and M.C.S.; Software, L.Z.; Data curation, X.P. and M.C.S.; Writing—original draft, L.Z. and M.C.S.; Writing—review & editing, L.Z., X.P. and M.C.S. All authors have read and agreed to the published version of the manuscript.

Funding: This research was funded by Consejo Nacional de Investigaciones Científicas y Técnicas (CONICET), Argentina, and by the Spanish State Research Agency (MICIU/AEI/10.13039/501100011033), through the Severo Ochoa and María de Maeztu Program for Centers and Units of Excellence in R&D (CEX2021-001164-M) and the INFOLANET project (PID2022-139409NB-I00).

Data Availability Statement: The raw data supporting the conclusions of this article will be made available by the authors on request.

Acknowledgments: The authors acknowledge fruitful discussions with Ingo Fischer, Daniel Brunner and Claudio R. Mirasso.

Conflicts of Interest: The authors declare no conflicts of interest. The funders had no role in the design of the study; in the collection, analyses, or interpretation of the data; in the writing of the manuscript; or in the decision to publish the results.

Abbreviations

The following abbreviations are used in this manuscript:

PJSD	Permutation Jensen–Shannon distance
JSD	Jensen–Shannon divergence
BP	Bandt and Pompe
OPD	Ordinal probability distribution
MG	Mackey–Glass
LK	Lang–Kobayashi
RO	Relaxation oscillation

Appendix A. Computational Time

With the aim of obtaining a better handle on the computational cost of the multiscale scheme proposed in this work, the time necessary to perform this multiscale analysis by running a self-made MATLAB (Version 9.5, R2018b) script was calculated. Being more precise, a pair of arbitrary time series (Gaussian white noise in the following analysis) of length L were generated, and then, the running time to estimate the PJSD between these two signals for the 400 combinations of lags τ_1 and τ_2 with $1 \leq \tau_1, \tau_2 \leq 20$ were obtained. This analysis was repeated for 100 independent pairs of simulations, varying the time series length L ($L \in \{10^3, 5 \cdot 10^3, 10^4, 2 \cdot 10^4, 4 \cdot 10^4, 6 \cdot 10^4, 8 \cdot 10^4, 10^5\}$) and the order D ($D \in \{3, 4, 5, 6\}$). The results obtained for the running times (in seconds) as a function of length L for the different orders D are illustrated in Figure A1. The mean and standard deviation (presented as error bars), from the 100 independent pairs of numerical simulations, were plotted. The time complexity increases linearly with the time series length L, with a slope that depends on the order D. Obviously, higher computational costs are observed as D increases, since the number of possible ordinal patterns is larger. It is important to highlight that even in the worst case, i.e., $D = 6$ and $L = 10^5$, the average time required to estimate the PJSD for the 400 combinations of lags τ_1 and τ_2 is around 8 seconds. Consequently, the proposed methodology is fast enough and can be implemented for the analysis of large databases.

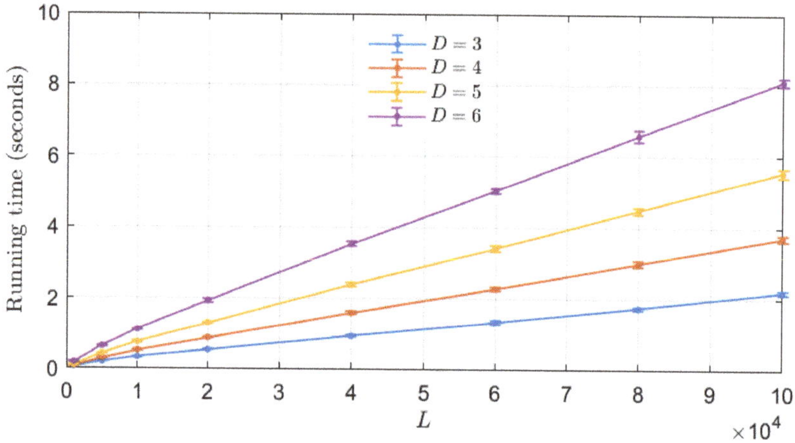

Figure A1. Running time to estimate the multiscale PJSD as a function of the time series length L for different orders D. The mean and standard deviation (presented as error bars) of the running time, from 100 independent estimations of the PJSD between two Gaussian white noise numerical realizations with order D and lags τ_1 and τ_2 between 1 and 20 (400 combinations), are plotted.

References

1. Serrà, J.; Arcos, J.L. An empirical evaluation of similarity measures for time series classification. *Knowl.-Based Syst.* **2014**, *67*, 305–314. [CrossRef]
2. Górecki, T.; Łuczak, M.; Piasecki, P. An exhaustive comparison of distance measures in the classification of time series with 1NN method. *J. Comput. Sci.* **2024**, *76*, 102235. [CrossRef]
3. Jiang, Y.; Qi, Y.; Wang, W.K.; Bent, B.; Avram, R.; Olgin, J.; Dunn, J. EventDTW: An improved dynamic time warping algorithm for aligning biomedical signals of nonuniform sampling frequencies. *Sensors* **2020**, *20*, 2700. [CrossRef]
4. Li, W.; He, R.; Liang, B.; Yang, F.; Han, S. Similarity measure of time series with different sampling frequencies based on context density consistency and dynamic time warping. *IEEE Signal Process. Lett.* **2023**, *30*, 1417–1421. [CrossRef]
5. Zhang, B.; An, B. Clustering time series based on dependence structure. *PLoS ONE* **2018**, *13*, e0206753. [CrossRef]
6. Doan, T.; Phan, T.; Nguyen, P.; Than, K.; Visani, M.; Takasu, A. Partial ordered Wasserstein distance for sequential data. *Neurocomputing* **2024**, *595*, 127908. [CrossRef]
7. Zunino, L.; Olivares, F.; Ribeiro, H.V.; Rosso, O.A. Permutation Jensen-Shannon distance: A versatile and fast symbolic tool for complex time-series analysis. *Phys. Rev. E* **2022**, *105*, 045310. [CrossRef]
8. Zunino, L. Revisiting the characterization of resting brain dynamics with the permutation Jensen–Shannon distance. *Entropy* **2024**, *26*, 432. [CrossRef]
9. Lin, J. Divergence measures based on the Shannon entropy. *IEEE Trans. Inf. Theory* **1991**, *37*, 145–151. [CrossRef]
10. Bandt, C.; Pompe, B. Permutation entropy: A natural complexity measure for time series. *Phys. Rev. Lett.* **2002**, *88*, 174102. [CrossRef]
11. Parlitz, U.; Suetani, H.; Luther, S. Identification of equivalent dynamics using ordinal pattern distributions. *Eur. Phys. J. Spec. Top.* **2013**, *222*, 553–568. [CrossRef]
12. Cánovas, J.S.; Guillamón, A.; Ruiz-Abellón, M.C. Using permutations for hierarchical clustering of time series. *Entropy* **2019**, *21*, 306. [CrossRef]
13. Endres, D.; Schindelin, J. A new metric for probability distributions. *IEEE Trans. Inf. Theory* **2003**, *49*, 1858–1860. [CrossRef]
14. Grosse, I.; Bernaola-Galván, P.; Carpena, P.; Román-Roldán, R.; Oliver, J.; Stanley, H.E. Analysis of symbolic sequences using the Jensen-Shannon divergence. *Phys. Rev. E* **2002**, *65*, 041905. [CrossRef] [PubMed]
15. Daw, C.S.; Finney, C.E.A.; Tracy, E.R. A review of symbolic analysis of experimental data. *Rev. Sci. Instrum.* **2003**, *74*, 915–930. [CrossRef]
16. Amigó, J.M.; Keller, K.; Unakafova, V.A. Ordinal symbolic analysis and its application to biomedical recordings. *Phil. Trans. R. Soc. A* **2015**, *373*, 20140091. [CrossRef]
17. Amigó, J.M.; Keller, K.; Kurths, J. Recent progress in symbolic dynamics and permutation complexity—Ten years of permutation entropy. *Eur. Phys. J. Spec. Top.* **2013**, *222*, 241–247. [CrossRef]
18. Zanin, M.; Olivares, F. Ordinal patterns-based methodologies for distinguishing chaos from noise in discrete time series. *Commun. Phys.* **2021**, *4*, 190. [CrossRef]
19. Leyva, I.; Martínez, J.H.; Masoller, C.; Rosso, O.A.; Zanin, M. 20 years of ordinal patterns: Perspectives and challenges. *Europhys. Lett.* **2022**, *138*, 31001. [CrossRef]

20. Amigó, J.M.; Rosso, O.A. Ordinal methods: Concepts, applications, new developments, and challenges—In memory of Karsten Keller (1961–2022). *Chaos* **2023**, *33*, 080401. [CrossRef]
21. Kottlarz, I.; Parlitz, U. Ordinal pattern-based complexity analysis of high-dimensional chaotic time series. *Chaos* **2023**, *33*, 053105. [CrossRef]
22. Olivares, F.; Zunino, L. Multiscale dynamics under the lens of permutation entropy. *Phys. A Stat. Mech. Its Appl.* **2020**, *559*, 125081. [CrossRef]
23. Zunino, L.; Soriano, M.C.; Rosso, O.A. Distinguishing chaotic and stochastic dynamics from time series by using a multiscale symbolic approach. *Phys. Rev. E* **2012**, *86*, 046210. [CrossRef]
24. Parlitz, U.; Berg, S.; Luther, S.; Schirdewan, A.; Kurths, J.; Wessel, N. Classifying cardiac biosignals using ordinal pattern statistics and symbolic dynamics. *Comput. Biol. Med.* **2012**, *42*, 319–327. [CrossRef]
25. Bandt, C. Small order patterns in big time series: A practical guide. *Entropy* **2019**, *21*, 613. [CrossRef]
26. Soriano, M.C.; Zunino, L.; Rosso, O.A.; Fischer, I.; Mirasso, C.R. Time scales of a chaotic semiconductor laser with optical feedback under the lens of a permutation information analysis. *IEEE J. Quantum Electron.* **2011**, *47*, 252–261. [CrossRef]
27. Soriano, M.C.; Zunino, L. Time-delay identification using multiscale ordinal quantifiers. *Entropy* **2021**, *23*, 969. [CrossRef]
28. Rosso, O.A.; Larrondo, H.A.; Martin, M.T.; Plastino, A.; Fuentes, M.A. Distinguishing noise from chaos. *Phys. Rev. Lett.* **2007**, *99*, 154102. [CrossRef]
29. Zhao, X.; Shang, P.; Huang, J. Permutation complexity and dependence measures of time series. *Europhys. Lett.* **2013**, *102*, 40005. [CrossRef]
30. Olivares, F.; Plastino, A.; Rosso, O.A. Contrasting chaos with noise via local versus global information quantifiers. *Phys. Lett. A* **2012**, *376*, 1577–1583. [CrossRef]
31. Mackey, M.C.; Glass, L. Oscillation and chaos in physiological control systems. *Science* **1977**, *197*, 287–289. [CrossRef] [PubMed]
32. Farmer, J.D. Chaotic attractors of an infinite-dimensional dynamical system. *Phys. D Nonlinear Phenom.* **1982**, *4*, 366–393. [CrossRef]
33. Grassberger, P.; Procaccia, I. Measuring the strangeness of strange attractors. *Phys. D Nonlinear Phenom.* **1983**, *9*, 189–208. [CrossRef]
34. van Tartwijk, G.H.M.; Lenstra, D. Semiconductor lasers with optical injection and feedback. *Quantum Semiclassical Opt. J. Eur. Opt. Soc. Part B* **1995**, *7*, 87. [CrossRef]
35. Ohtsubo, J. *Semiconductor Lasers: Stability, Instability and Chaos*; Springer: Berlin/Heidelberg, Germany, 2017; Volume 111.
36. Lang, R.; Kobayashi, K. External optical feedback effects on semiconductor injection laser properties. *IEEE J. Quantum Electron.* **1980**, *16*, 347–355. [CrossRef]
37. Ahlers, V.; Parlitz, U.; Lauterborn, W. Hyperchaotic dynamics and synchronization of external-cavity semiconductor lasers. *Phys. Rev. E* **1998**, *58*, 7208–7213. [CrossRef]
38. Porte, X.; Soriano, M.C.; Fischer, I. Similarity properties in the dynamics of delayed-feedback semiconductor lasers. *Phys. Rev. A* **2014**, *89*, 023822. [CrossRef]
39. Vicente, R.; Mirasso, C.R.; Fischer, I. Simultaneous bidirectional message transmission in a chaos-based communication scheme. *Opt. Lett.* **2007**, *32*, 403–405. [CrossRef]
40. Uchida, A. *Optical Communication with Chaotic Lasers: Applications of Nonlinear Dynamics and Synchronization*; Wiley-VCH: Weinheim, Germany, 2012.
41. Porte, X.; Soriano, M.C.; Brunner, D.; Fischer, I. Bidirectional private key exchange using delay-coupled semiconductor lasers. *Opt. Lett.* **2016**, *41*, 2871–2874. [CrossRef]

entropy

MDPI

Article

Spatio-Temporal Analysis of Acute Myocardial Ischaemia Based on Entropy–Complexity Plane

Esteban R. Valverde [1,2], Victoria Vampa [3,*], Osvaldo A. Rosso [4,5] and Pedro D. Arini [1,6,*]

1 Grupo de Sistema Cardiovascular, Instituto de Ingeniería Biomédica (IIBM), Facultad de Ingeniería, Universidad de Buenos Aires, Buenos Aires C1063, Argentina; evalverde@fi.uba.ar
2 Grupo de Neurociencia de Sistemas, Instituto de Fisiología y Biofísica "Bernardo Houssay" (IFIBIO-Houssay), Departamento de Ciencias Fisiológicas, Facultad de Medicina, CONICET-Universidad de Buenos Aires, Buenos Aires C1121, Argentina
3 Departamento de Ciencias Básicas, Facultad de Ingeniería, Universidad Nacional de La Plata, La Plata B1900, Argentina
4 Instituto de Física, Universidade Federal de Alagoas (UFAL), Maceió 57072-970, Brazil; oarosso@gmail.com
5 Instituto de Física La Plata (IFLP), CONICET-Universidad Nacional de La Plata, La Plata B1900, Argentina
6 Grupo de Investigación en Cardioseñales, Instituto Argentino de Matemática "Alberto P. Calderón" (IAM), CONICET, Buenos Aires C1425, Argentina
* Correspondence: victoria.vampa@ing.unlp.edu.ar (V.V.); pedro.arini@conicet.gov.ar (P.D.A.)

Abstract: Myocardial ischaemia is a decompensation of the oxygen supply and demand ratio, often caused by coronary atherosclerosis. During the initial stage of ischaemia, the electrical activity of the heart is disrupted, increasing the risk of malignant arrhythmias. The aim of this study is to understand the differential behaviour of the ECG during occlusion of both the left anterior descending (LAD) and right anterior coronary artery (RCA), respectively, using spatio-temporal quantifiers from information theory. A standard 12-lead ECG was recorded for each patient in the database. The control condition was obtained initially. Then, a percutaneous transluminal coronary angioplasty procedure (PTCA), which encompassed the occlusion/reperfusion period, was performed. To evaluate information quantifiers, the Bandt and Pompe permutation method was used to estimate the probability distribution associated with the electrocardiographic vector modulus. Subsequently, we analysed the positioning in the $\mathcal{H} \times \mathcal{C}$ causal plane for the control and ischaemia. In LAD occlusion, decreased entropy and increased complexity can be seen , i.e., the behaviour is more predictable with an increase in the degree of complexity of the system. RCA occlusion had the opposite effects, i.e., the phenomenon is less predictable and exhibits a lower degree of organisation. Finally, both entropy and complexity decrease during the reperfusion phase in LAD and RCA cases.

Keywords: ECG permutation entropy; dynamic electrocardiography; PTCA

1. Introduction

Cardiac ischaemia occurs when the blood flow that nourishes and oxygenates the heart muscle is insufficient to satisfy the demands of cellular metabolism. This results in a lack of oxygen (anoxia) in heart cells. Consequently, both contractile function and the removal of metabolic waste products are impaired. For a short time after the onset of anoxia, certain reversible ischaemic changes occur in the internal structure of affected cells, altering depolarisation and repolarisation, i.e., slowing conduction velocity, delaying activation times, and prematurely ending ventricular repolarisation. Regarding the electrocardiographic signal, cardiac ischaemia results in notable distortions of the ST-T and

QRS complexes compared to normal traces using a 12-lead standard ECG. In this sense, myocardial ischaemia has been identified as a risk factor, as it increases the likelihood of malignant ventricular arrhythmia and sudden cardiac death.

One promising approach to assess ischaemia is the use of information theory tools. The following investigations examine acute ischaemia specifically using measures of entropy. These comprise models of coronary artery occlusion in animals [1] and patients with positive exercise tests [2,3].

Lemire et al. [1] applied the wavelet transform to compute the entropy of the electrocardiographic signal across several frequency levels. Using the Frank lead system, they analysed data obtained from the orthogonal ECG of five pigs. They used two drugs to accentuate ischaemia and a comparison between both was realised in the context of a single and long duration coronary occlusion. They proposed this new marker, which is independent of the ST segment changes and can effectively detect ischaemic states.

With a clinical approach, Farahabadi et al. [2] proposed entropy as a diagnostic indicator for the detection of ischaemia. Healthy subjects (n = 10) and patients with a positive stress test (n = 10) were recorded on a 12-lead ECG. Four techniques were employed to compare entropy in electrocardiographic signals between healthy and patient subjects, utilising both the spatial and wavelet domains. They showed that the presented method based on wavelet sub-bands outperforms the others. Subsequently, Rabbani et al. [3] later confirmed, during the exercise test, this measure as an estimator of the significant ST-segment deviation. These patients (n = 40) had exhibited ischaemic signs based on their initial diagnosis by a medical practitioner.

In terms of other measures, a recent study by Calderon-Juarez et al. [4] investigated the immediate effect of acute ischaemia on the non-linear dynamical characteristics of heart rate variability (HRV) during acutely induced ischaemia through the PTCA. They utilised the STAFF III database, the same as the one we use in our study, where a single prolonged balloon inflation was performed in one of the major coronary arteries depending on the specific case and assessed HRV using both traditional measures and recurrence quantification analysis, a method for non-linear data analysis aimed at investigating dynamic systems. The findings confirmed the presence of non-linear behaviour in a significant proportion of the HRV time series post-PTCA. The authors concluded that this non-linear phenomenon does not necessarily align with the observed changes in those measures.

Quantifiers such as entropy and complexity are important in analysing time series. In biomedical signals, these metrics facilitate the identification of irregularities, revealing hidden structures and a more comprehensive understanding of complex biological processes. Their application contributes to advancements in areas such as diagnostics and disease monitoring.

This work aims to compare, in the entropy–complexity $\mathcal{H} \times \mathcal{C}$ plane, the differential behaviour of cardiac electrical activity during the occlusion/reperfusion of the left anterior descending (LAD) coronary versus the occlusion/reperfusion of the right coronary artery (RCA). For this purpose, we have used a PTCA procedure as a model of acute reversible myocardial ischaemia.

It is hypothesised that the $\mathcal{H} \times \mathcal{C}$ plane location during the occlusion and reperfusion of the LAD and RCA arteries will exhibit differential behaviour concerning the morphological changes observed in the ECG signal, depending on which artery is occluded. According to our literature review, this study is novel in applying the $\mathcal{H} \times \mathcal{C}$ plane to assess the differential dynamics of acute ischaemia.

2. Materials and Methods

2.1. ECG Database

The present study employed the STAFF III database [5,6], which comprises ECG recordings obtained from patients undergoing elective PTCA procedures in one of their major coronary arteries. Eight leads, v1–v6, I and II, were acquired with a sampling rate and amplitude resolution of 1 KHz and 0.6 μV, respectively, using equipment manufactured by Siemens-Elena AB (Goeteborg, Sweden). Also, orthogonal XYZ leads were calculated by the Kors transform [7]. From the entire corpus of data, only those ECG signals that corresponded to the occlusion on the RCA and LAD were selected for this study. Subsequently, the ECG signals from patients were meticulously examined by an experienced cardiologist to identify those patients who did not exhibit any additional indications of cardiac pathology, i.e., myocardial infarction, bundle branch block, ventricular pre-excitation, and secondary repolarisation abnormalities. Consequently, only normal sinus rhythm was deemed suitable for the subsequent analysis.

Because of this, a population of 24 patients was included, 15 males and 9 females, ages 61.3 $\pm$ 24.4 y.o. Of ths population, LAD coronary occlusion artery was studied in 9 patients and RCA occlusion in 15 patients.

A data form indicating the anatomic site and the exact times of inflation and deflation of the balloon was completed. The mean inflation duration was 4 min 6 s with a standard deviation of 90 s. The electrocardiographic recording time after deflation ranged from 0 min to 9 min with a mean value of 3.7 min of recording. Two ECG recordings were evaluated for each patient. The initial recording was a control ECG, obtained continuously for five minutes in a supine position before the PTCA procedure. This was carried out in clinically stable conditions, within a time interval of a maximum of one hour in the room and/or catheterisation laboratory. The electrodes were maintained on the patients between both ECG recordings with marked positions, thus enabling precise comparisons of the ECG variables.

Later, a continuous ECG was recorded during the PTCA procedure, which includes the occlusion/reperfusion period. The occlusion phase is initiated by balloon inflation and concludes with balloon deflation. The reperfusion or recovery phase commences at the point of balloon deflation and concludes at the termination of the procedure. The local investigational review board approved the study, and informed consent was obtained from each subject before enrolment [6].

2.2. Dataset Pre-Processing

The ECG recordings from lead II and the orthogonal X, Y, and Z leads were preprocessed before applying the algorithm proposed in the present work. First, a notch filter was implemented to minimise power line interference, i.e., a 2nd-order Butterworth filter at 60 Hz. A cubic spline interpolation filter was then applied to attenuate the respiratory artefacts in the ECG baseline and the drift caused by variations in the skin electrode impedance [8]. Afterwards, the module vector of the ECG was calculated for each signal as

$$M = \sqrt{X^2 + Y^2 + Z^2} \tag{1}$$

For each subject, all the R-waves on lead II and their corresponding QRS endpoints were identified using a wavelet-based ECG delineation algorithm [9]. A window of $K = 512$ ms centred on each R-wave was used to cover the entire cardiac beat. After that, a QRS template was constructed through the median of the QRS-complexes detected. We have used the median instead of the mean value to avoid ectopic or noisy beats. The template was compared with each QRS-complex and the cross-correlation coefficient was obtained; if this value was greater than 75%, the corresponding QRS-complex was jitters-corrected; otherwise, it was discarded. Finally, for each of the $n = 1 \ldots N$ beats accepted and centred on lead II, the same windows of 512 ms were defined for the rest of the ECG leads to cover the corresponding ECG beats.

The Root Mean Square (RMS) voltage level was calculated in a segment of 20 msbefore the Q-point of each beat. All registers exhibiting an RMS noise level exceeding 20 µV were excluded from further analysis.

2.3. *Entropic and Complexity Quantifiers*

Using quantifiers from information theory allows us to know different signal characteristics such as its dynamics, structure, fluctuations, and changes. This is achieved by defining a probability distribution of the original signal. If this probability distribution is adequate, it will provide us with useful information. In the same way, we can make a comparative analysis between two signals by finding the distance between their probability distributions.

2.3.1. Shannon Entropy

The Shannon entropy quantifies the uncertainty within a probability distribution [10]. It is defined as

$$\mathcal{S}[\mathbf{p}] = -\sum_{j=1}^{L} p_j \, ln(p_j) \tag{2}$$

where $\mathbf{p} = (p_1, p_2, \quad, p_L)$ is a probability distribution. The normalised entropy is defined from Equation (2) as

$$\mathcal{H}[\mathbf{p}] = \frac{\mathcal{S}[\mathbf{p}]}{\mathcal{S}_{\max}} \tag{3}$$

where $\mathcal{S}_{max} = ln(L)$ and corresponds to the case the probability distribution is uniform for L i.e., $p_j = 1/L$ [10].

2.3.2. Jensen–Shannon Divergence

Given two probability distributions $\mathbf{p} = (p_1, p_2, \cdots, p_L)$ and $\mathbf{q} = (q_1, q_2, \cdots, q_L)$ associated for the same set of events, the relative entropy is a measure of the distance from $\mathbf{p}$ to $\mathbf{q}$ and is defined as

$$\mathcal{K}[\mathbf{p}, \mathbf{q}] = -\sum_{j=1}^{L} p_j \, log\left(\frac{q_j}{p_j}\right) \tag{4}$$

It is also known as the Kullback–Leibler divergence [11] and provides a measure for comparing the distribution $\mathbf{q}$ to $\mathbf{p}$. However, it is not a true distance since it is not symmetric. The Jensen–Shannon Divergence (JSD) was generalised by J. Lin in 1991 [12] as an alternative to the Kullback–Leibler divergence. It is defined as

$$\mathcal{D}_{JS}[\mathbf{p}, \mathbf{q}] = \frac{1}{2}\left[\mathcal{K}\left[\mathbf{p}, \frac{\mathbf{p}+\mathbf{q}}{2}\right] + \mathcal{K}\left[\mathbf{q}, \frac{\mathbf{p}+\mathbf{q}}{2}\right]\right] \tag{5}$$

In terms of the Shannon Entropy, the above expression can be rewritten in the following form:

$$\mathcal{D}_{JS}[\mathbf{p}, \mathbf{q}] = \mathcal{S}\left[\frac{\mathbf{p}+\mathbf{q}}{2}\right] - \frac{\mathcal{S}[\mathbf{p}]}{2} - \frac{\mathcal{S}[\mathbf{q}]}{2} \tag{6}$$

The JSD is a symmetrised version of the Kullback–Leibler divergence, is non-negative, and vanishes only if $\mathbf{p} = \mathbf{q}$, see [10]. Its square root was shown to be a proper metric for probability distributions satisfying the triangle inequality [13].

2.3.3. Statistical Complexity

Statistical Complexity measures are used to analyse the degree of structure present in a process. Statistical Complexity $\mathcal{C}[P]$ is defined as

$$\mathcal{C}[\mathbf{p}] = \mathcal{H}[\mathbf{p}]\ \mathcal{Q}[\mathbf{p}] \tag{7}$$

It represents the product of the normalised entropy, $\mathcal{H}[\mathbf{p}]$ times the disequilibrium, $\mathcal{Q}[\mathbf{p}]$ which quantifies the distance from $\mathbf{p}$ to $\mathbf{p}_e$, the uniform distribution,

$$\mathcal{Q}[\mathbf{p}] = \mathcal{Q}_0\ \mathcal{D}[\mathbf{p}, \mathbf{p}_e] \qquad 0 \leq \mathcal{Q} \leq 1 \tag{8}$$

In the above equation, $\mathcal{Q}_0$ is a normalisation constant, equal to the inverse of the maximum possible distance $\mathcal{D}[\mathbf{p}, \mathbf{p}_e]$. This distance can be provided by the JSD Equation (6), replacing $\mathbf{q}$ with $\mathbf{p}_e$.

Note that in Equation (8) $\mathcal{Q}[\mathbf{p}] = 0$, if $\mathbf{p} \equiv \mathbf{p}_e$, it will increase to the unit, as long as $\mathbf{p}$ becomes more dissimilar to $\mathbf{p}_e$. This would reflect the system's architecture [10], being different from zero if 'more likely' states exist among accessible ones.

We remark that complexity measures the degree of correlational structures and it is not a trivial function of entropy, in the sense that, for a given value of $\mathcal{H}[\mathbf{p}]$, there is a range of possible values for $\mathcal{C}[\mathbf{p}]$ between a minimum and a maximum value ($\mathcal{C}_{min}$ and $\mathcal{C}_{max}$) [10].

2.3.4. Probability Distribution Function and Permutation Entropy

It is important to point out that using quantifiers based on time-frequency methods was introduced by Rosso et al. [10]. Properly determining the underlying probability distribution function (PDF) associated with the given dynamical system or time series is essential for evaluating the information quantifiers, such as the entropy and the statistical complexity presented in the previous section.

For selecting the probability space, we apply the Bandt and Pompe methodology [14], which introduces a symbolic encoding scheme based on the ordinal relationships between neighbouring values in a data sequence. The Bandt–Pompe technique incorporates time causality into PDF construction, providing a more accurate description of the dynamics under study [13]. This approach is based on analysing ordinal patterns within a time series, which requires two key parameters: the embedding dimension D, which determines the length of the subsequences associated with permutations, and the time delay $\tau \in \mathbb{N}$, which measures the distance between consecutive observations within each subsequence. Bandt and Pompe [14] recommended using the embedding dimensions $4 \leq D \leq 6$ and $\tau = 1$. However, other values of τ may offer additional insights, particularly when this parameter is linked to the intrinsic time scales of the system being analysed.

It is important to point out that this methodology, introduced in [14] and further detailed in [15,16] through various applications, applies to any time series with minimal assumptions. Specifically, for k = D, the probability for $x_t < x_{t+k}$ should not depend on t. In addition, the length of the series K must be much larger than $D!$ to obtain reliable results.

For each time step $k = 1, \ldots, K-(D-1)\tau$, the sequence $I_k = (x_k, x_{k+\tau}, \ldots, x_{k+(D-1)\tau})$ is considered and one of the $D!$ possible permutations of order D is assigned. The procedure can be better illustrated with a simple example; let us assume that we start with the time series $\{1, 2, 5, 4, 3, 5, \ldots\}$, and we set the embedding dimension D = 3. The state space is divided into 3 partitions. The permutations are (321), (312), (231), (213), (132), and (123). If we take $\tau = 1$, we are interested in ordinal patterns of order D, (x_k, x_{k+1}, x_{k+2}).

The first 3-dimensional vector is $(1,2,5)$ and corresponds to type 6. The second 3-dimensional vector is $(2,5,4)$ and is associated with type 5; the third, $(5,4,3)$, is associated with type 1, and so on. For all the D possible orderings (permutations) π_i, the probability distribution $P = \{p(\pi_i)\}$ is defined by

$$p(\pi_i) = \frac{\#\{k \mid k \leq T - D,\, I_k \text{ has type } \pi_i\}}{K - (D-1)} \tag{9}$$

As the above formula shows, the amplitude values of I_k are not considered. Only its sequential ordering is considered. Assessing the frequency of ordinal patterns captures the degree of disorder in the time series. It provides a quantitative measure of the unpredictability or randomness in the behaviour of a time series. In this sense, Permutation Entropy, Equation (3), using this PDF, is a robust indicator of the system's dynamic: higher values correspond to greater randomness, while lower values indicate more regular or deterministic behaviour.

2.4. Implementation

2.4.1. Probability Distribution Functions

For each patient, both time series (control and PTCA) of the module vector of the ECG were calculated using Equation (1). We calculate PDFs using the permutation entropy described above. The parameters chosen for the studies carried out in this work are D = 4 and τ = 2. The state space is divided into 4 partitions and 24 mutually exclusive permutation symbols are considered. The associated PDFs are evaluated and compared.

2.4.2. Entropy, Distances and Complexity Quantifiers

From the beat information in the database, quantifiers were averaged for each patient every 30-s window during occlusion and every 15-s window during recovery. With the PDFs generated using Bandt and Pompe methodology, normalised Shannon entropy average values are computed at each time for the occlusion, recovery, and control time series. To analyse the dynamical behaviour of the occlusion and recovery against the control, distances between the PDFs are calculated with the JSD, Equation (6).

In the second stage, the disequilibrium values, i.e., the distance between the PDFs of the occlusion and recovery time series and the uniform PDF are calculated by setting $\mathbf{q} = \mathbf{p}_e$ in Equation (6). Then, the complexity values are obtained from Equation (7).

Once the entropy and complexity values for each time are obtained under control and PTCA conditions, we analyse their position in the $\mathcal{H} \times \mathcal{C}$ plane. By examining their location in this plane and considering the position of different stochastic processes with the f^{-k} power spectrum, known as k-noises [15], we aim to characterise the behaviour of the time series. The specific values of k considered in this study are $0 \leq k \leq 3$.

2.5. Statistical Analysis

To determine the statistical significance of the Jensen–Shannon Divergence between the PTCA procedure and the control situation, the D'Agostino–Pearson normality test is applied. If the analysed data follow a normal distribution, a parametric two-tailed Student's t-test is used. Otherwise, a non-parametric two-sided Mann–Whitney U test is used instead.

When the p-value was <0.05, the differences were considered statistically significant.

3. Results

As representative examples, and for a comparison of the effect of artery occlusion, we present in Figures 1 and 2 the module vector of the ECGs of two patients during control conditions and ischaemia induced by PTCA, corresponding to LAD (Figure 1) and RCA (Figure 2). For both figures, panel (a) corresponds to the control condition, while panel (b) corresponds to PTCA, respectively. They show how the intervention affects the ECG signals and their corresponding PDFs.

As mentioned in Section 2.1 and according to the database, we have the following for each patient: a control period, some ECG beats during PTCA occlusion, and another set of beats during PTCA reperfusion. Each beat is analysed using the Bandt and Pompe method to obtain the corresponding PDF, and Figures 1 and 2 on the right show the frequencies of the different patterns. The first and last patterns correspond to the increasing sequence 1234 and the decreasing sequence 4321, respectively. It can be observed that these patterns have the highest frequencies, and also that the PDFs exhibit general symmetry relative to the middle pattern. The other patterns, corresponding to different orders in the sequence, have lower probabilities, less than 0.1. As shown in these figures, the probability distributions during the control and PTCA are distinct (for both LAD and RCA). In particular, the effect of the procedure is clearly illustrated by comparing the first and last beats of occlusion (Figures 1b and 2b). In both cases, for LAD and RCA, the last beat shows more intervals of decreasing behaviour, and consequently, the last pattern, 4321, has a higher frequency than the first beat.

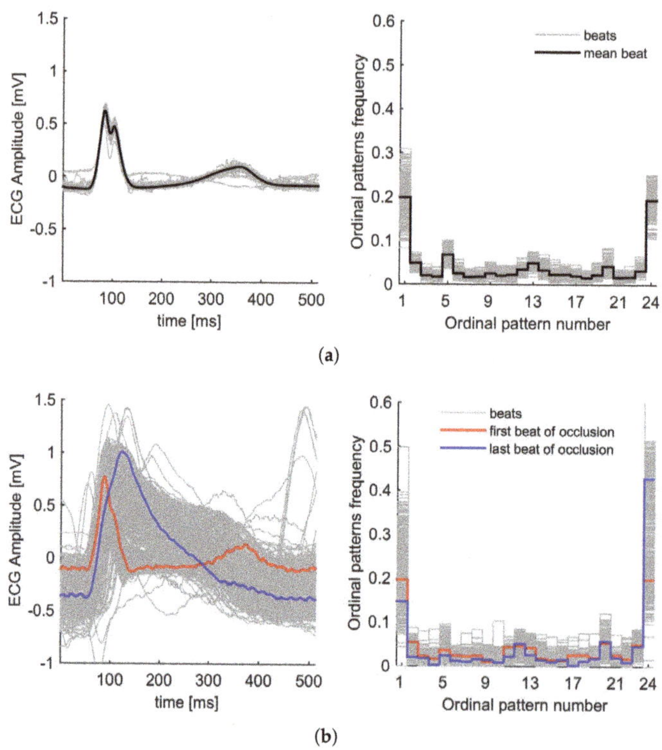

Figure 1. (**a**) ECG beats (**left**) and PDFs (**right**) of a particular patient during control conditions. (**b**) ECG beats (**left**) and PDFs (**right**) of the same patient during PTCA for LAD. Inflation time 5.2 min. The Bandt and Pompe PDFs are calculated with $D = 4$ and $\tau = 2$.

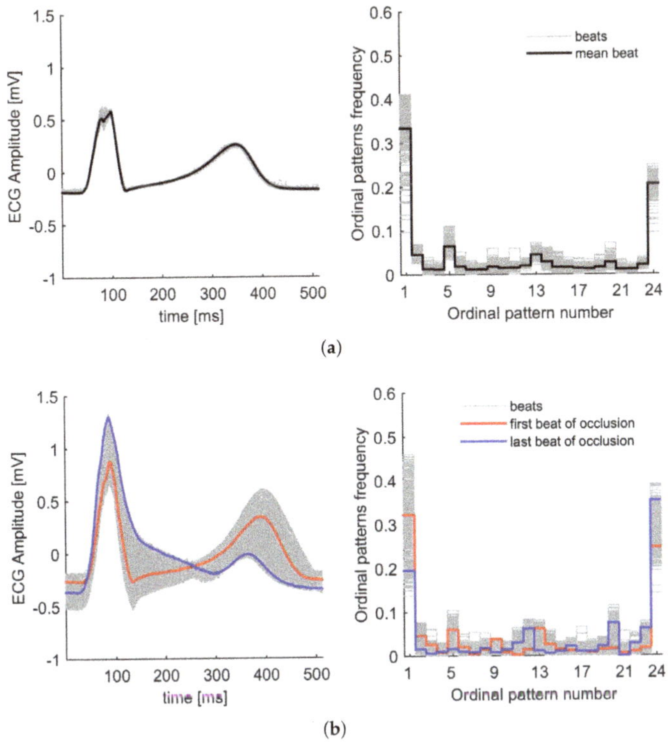

Figure 2. (**a**) ECG beats (**left**) and PDFs (**right**) of a particular patient during control conditions. (**b**) ECG beats (**left**) and PDFs (**right**) of the same patient during PTCA for RCA. Inflation time 5 min. The Bandt and Pompe PDFs are calculated with $D = 4$ and $\tau = 2$.

The Jensen–Shannon Divergence, Equation (6), is used to measure the differences between the occlusion, reperfusion, and control PDFs relative to the Mean Control PDF, calculated as the average PDF of all control beats for that patient. The JSD is calculated for each beat relative to the Mean Control PDF. To compare the JSD between control and occlusion and between control and reperfusion, the control JSD is divided into sections. The average JSD for these segments is then calculated. For example, if a patient has 4 min of occlusion, we have eight sections for the occlusion period and eight for the control period. Finally, statistical analysis is performed for each section across all patients, comparing the occlusion vs. control and reperfusion vs. control. These results are presented in Figure 3.

Figure 4a shows the localisation in the $\mathcal{H} \times \mathcal{C}$ plane of entropy and complexity values from the beginning to the end of the occlusion during the PTCA procedure, for both LAD and RCA and during control conditions. As mentioned, the corresponding complexity values are bounded by the evaluated $\mathcal{C}_{min}$ and $\mathcal{C}_{max}$ curves.

Additionally, we indicate the location of k-noises, corresponding to stochastic processes with power spectrum f^{-k} for $k = 2, 2.5$, and 3.

Figure 4b provides a zoom to examine the evolution of entropy and complexity as the occlusion progresses. Similarly, Figure 5a,b corresponds to the recovery stage.

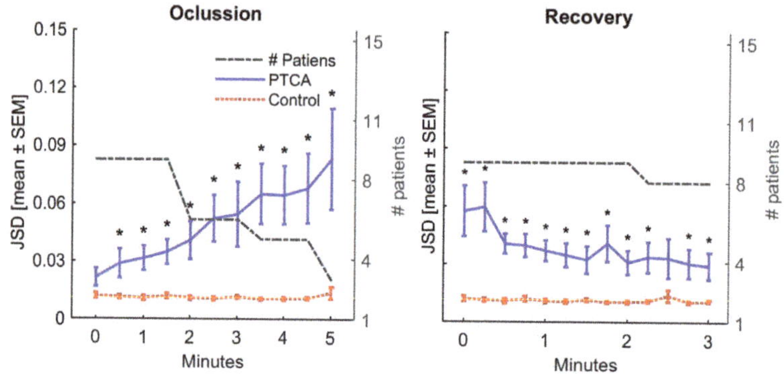

(**a**)

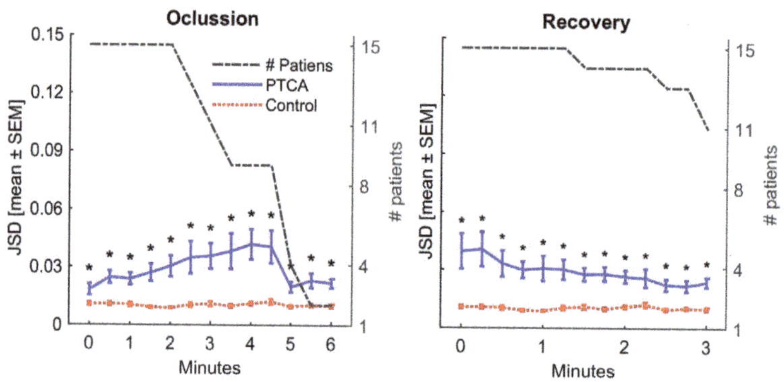

(**b**)

Figure 3. JSD distances between Bandt and Pompe PDFs as functions of occlusion time (**left**) and recovery time (**right**): (**a**) LAD and (**b**) RCA. The blue line represents the distance between PTCA and Mean Control PDFs indicating with * for p-value < 0.05. The red line shows the distance (low and almost constant) between the control PDF and the Mean Control PDF. The number of patients at each time is indicated on the right side of the y-axis.

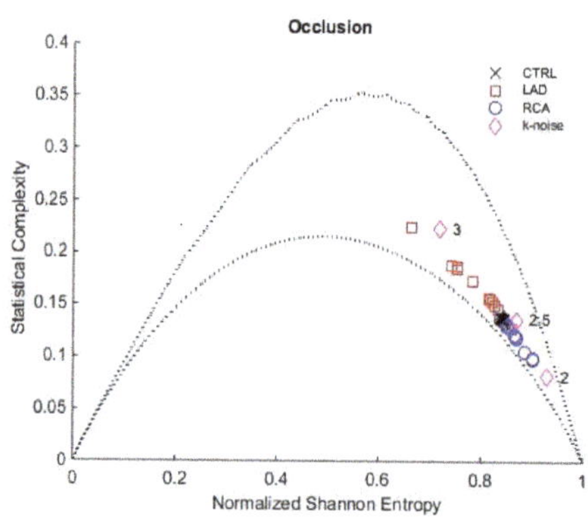

(**a**)

Figure 4. *Cont.*

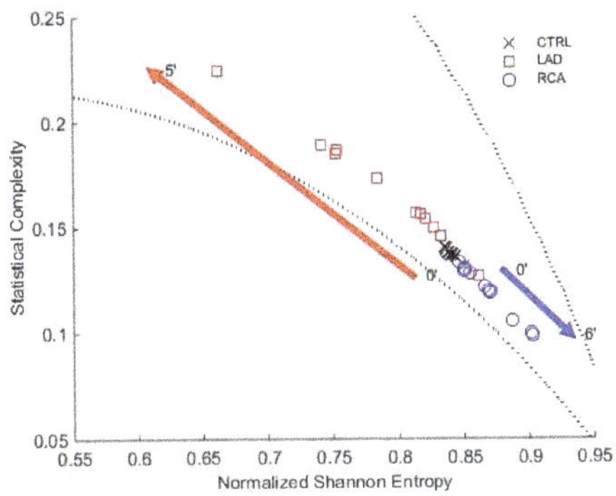

(b)

Figure 4. $\mathcal{H} \times \mathcal{C}$plane of the whole population. (**a**) Representation of the entropy and complexity values during LAD artery occlusion (red squares) and RCA occlusion (blue circles). The black crosses correspond to the control group values (CTRL). (**b**) The red arrow indicates the trajectory from minute 0 to minute 5 for LAD, while the blue arrow shows the trajectory from minute 0 to minute 6 for RCA. The dotted lines correspond to $\mathcal{C}_{min}$ and $\mathcal{C}_{max}$ evaluated for $D = 4$.

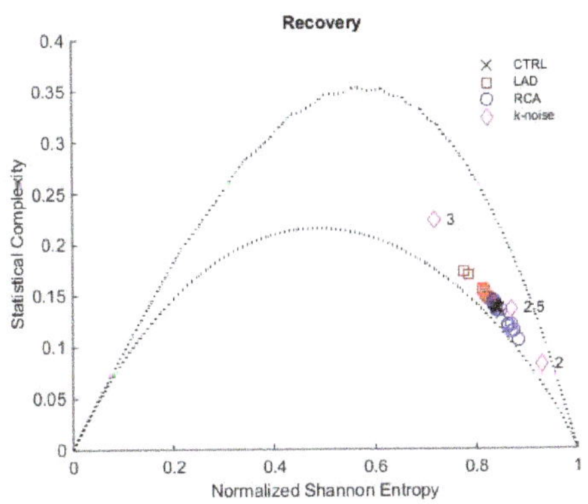

(a)

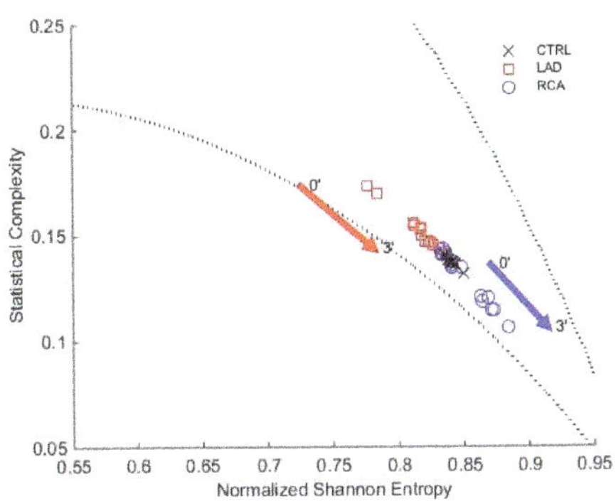

(b)

Figure 5. $\mathcal{H} \times \mathcal{C}$plane of the whole population. (**a**) Representation of the entropy and complexity values during recovery: LAD (red squares) and RCA (blue circles). The black crosses correspond to the control group values (CTRL). (**b**) Trajectory during recovery from minute 0 to minute 3. The red arrow shows the LAD trajectory, while the blue arrow shows the RCA trajectory. The dotted lines correspond to $\mathcal{C}_{min}$ and $\mathcal{C}_{max}$ evaluated for $D = 4$.

4. Discussion

Quantifiers such as entropy and complexity are significant in analysing biological time series. These metrics help us to identify irregularities, reveal hidden structures, and allow a more complete understanding of complex biological systems.

This study aims to describe how the spatio-temporal alterations of the electrocardiographic signal observed during acute ischaemia are reflected in the $\mathcal{H} \times \mathcal{C}$ plane.

To study these phenomena, we analysed a model of acute myocardial ischaemia using a PTCA procedure, which mimics the abrupt onset of symptoms associated with total thrombotic coronary occlusion [17].

We have decided to use the orthogonal XYZ approach to vectorcardiography rather than the 12-lead ECG because several investigations [18] have previously shown that the identification of ischaemia using spatio-temporal information is more sensitive and specific using the module vector of the ECG.

Figure 1 illustrates the ECG signals and the PDFs for a selected patient in the dataset under analysis. This depicts the control condition (Figure 1a) and the condition during LAD coronary artery occlusion (Figure 1b). The ECG beats show a notable degree of overlap and low dispersion during the five-minute control recording (Figure 1a(left)) and also have comparable PDFs (Figure 1a(right)). Similarly, Figure 2 presents the ECG signals and PDFs for another patient in the dataset. Figure 2a shows the control condition (exhibiting characteristics analogous to Figure 1a(left and right), while Figure 2b illustrates the RCA occlusion.

We can make two observations regarding the stability of the controls. First, when the Jensen–Shannon Divergence was computed, we observed that the control's stability was maintained in all the data under analysis (see Figure 3). Second, it is important to note that in the $\mathcal{H} \times \mathcal{C}$ plane (Figures 4 and 5), the control values are distributed within a small rectangle $\mathcal{H}$ (0.84 to 0.86) $\times$ $\mathcal{C}$ (0.13 to 0.14), showing minimal variation in entropy and complexity. We refer to this area as the control region, which is close to k-noise 2.5.

LAD and RCA occlusion exhibited distinct behaviours when comparing both PTCA procedures, evidenced by changes in the position of the $\mathcal{H} \times \mathcal{C}$ informational plane. To measure these changes, we calculate the relative differences $\delta\mathcal{H}$ and $\delta\mathcal{C}$ by subtracting the initial value from the final value and then dividing the result by the initial value.

4.1. Acute Ischaemia in LAD Coronary Artery

LAD occlusion begins near the control region and follows a trajectory during the five minutes of PTCA, where entropy decreases from 0.87 to 0.76 ($\downarrow$ $\delta\mathcal{H}$ 12%), while complexity increases from 0.12 to 0.18 ($\uparrow$ $\delta\mathcal{C}$ 50%). Regarding k-noises, with the control zone close to k = 2.5, this indicates an increase in the degree of correlation, approaching the localisation of k-noise for k = 3. As shown in Figure 3, the JSD distance to Mean Control PDF increases during the occlusion, exhibiting an almost monotonic behaviour and reaching a final value that is four times the initial.

In cardiac practice, it has been observed that significant ST-segment elevations are present in approximately 70–80% of patients with LAD occlusion. Moreover, during LAD occlusion the ST-T complex rises, transforming and deforming the ECG signal into a waveform very similar to an action potential, i.e., with greater temporal regularity with a single phase (see Figure 1b(left)). We are in the presence of a system exhibiting intermediate dynamics ($\downarrow$ $\mathcal{H}$ and $\uparrow$ $\mathcal{C}$). This phenomenon can be characterised as a mixture of order and randomness, with a tendency towards a region of intermediate entropy and complexity.

4.2. Acute Ischaemia in RCA

In contrast, during RCA occlusion, in the initial two and a half minutes of PTCA, both $\mathcal{H}$ and $\mathcal{C}$ oscillate within the reduced range of the control. This is followed by a trajectory that shows increased entropy (from 0.84 to 0.90, $\uparrow \delta\mathcal{H}$ 7%) and decreased complexity (from 0.14 to 0.10, $\downarrow \delta\mathcal{C}$ 28%), demonstrating a stronger stochastic behaviour and approaching k-noise for k = 2. As seen in Figure 3, in this case, the JSD distance strictly increases until minute 4.5, but in a lesser proportion compared to LAD; RCA only reaches double its initial value almost at the end of the period. Subsequently, it decreases back to its initial value.

Conversely to LAD occlusion, ST segment elevation is less common in patients with RCA occlusion. It occurs in about 30–40% of cases (see Figure 2b(left)). In addition, ST depression is more common in RCA occlusions, although less pronounced than in LAD. Also, absent ST changes are more likely to occur with RCA occlusions. In this context, we have observed a tendency towards a completely random system, which is located in regions of high entropy but low complexity ($\uparrow \mathcal{H}$ and $\downarrow \mathcal{C}$). Increasing permutation entropy may indicate a transition to a more irregular and complex signal. This may be due to an increased variability of ST-T complex changes, as we have recently described, during RCA occlusion.

4.3. Recovery After PTCA

During recovery, the stochastic behaviour of both LAD and RCA increases: LAD entropy rises from 0.78 to 0.84 ($\uparrow$ $\delta\mathcal{H}$ 7.7%), and complexity decreases from 0.17 to 0.14 ($\downarrow$ $\delta\mathcal{C}$ 18%), while RCA entropy rises from 0.84 to 0.87 ($\uparrow$ $\delta\mathcal{H}$ 3.5%) and complexity decreases from 0.14 to 0.12 ($\downarrow$ $\delta\mathcal{C}$ 14%). As shown in Figure 3, the JSD to Mean Control PDF is a decreasing function for both LAD and RCA. For RCA, it takes lower values than LAD, is monotonically decreasing throughout the interval, and ends at a value higher than the initial one.

4.4. General Comments

In LAD occlusion, $\downarrow \mathcal{H}$ and $\uparrow \mathcal{C}$, suggesting that the dynamics of ECG change are more regular, i.e., the behaviour is more predictable with a higher degree of complexity of the system. In contrast, during RCA occlusion, we observed that $\uparrow \mathcal{H}$ and $\downarrow \mathcal{C}$; this could be due to the ECG signal becoming more variable, i.e., less predictable with a lower degree of organisation. Moreover, for both, LAD and RCA, $\uparrow \mathcal{H}$ and $\downarrow \mathcal{C}$ in the reperfusion phase. Additionally, it is notable that the variation in $\mathcal{C}$ is approximately four times the variation in $\mathcal{H}$, except in the case of LAD recovery, where the variation in $\mathcal{C}$ is just slightly more than double the variation in $\mathcal{H}$.

4.5. Comparison of Art

A comparison of the present work with the existing literature shows that, whereas others have employed wavelet entropy, we have utilised permutation entropy to analyse acute ischaemia. It is not possible to make a direct comparison between the two approaches, although they can be considered complementary. Permutation entropy is capable of identifying regularity or randomness in a time series, which is useful for detecting patterns of behaviour or changes in signal dynamics. In contrast, wavelet entropy focuses on how energy is distributed across different scales in the signal. Moreover, Lemire et al. [1] projected the XYZ leads into one using the Karhunen–Loève transform. As we have done in this paper, this would be similar to analysing a single representative signal. Then, they applied the wavelet transform to the resulting signal and calculated the entropy at each scale.

5. Study Limitations

During the balloon occlusion of the coronary artery in the PTCA procedure, the ST segment behaves similarly to that observed during spontaneous myocardial ischaemia [17]. However, the magnitude of ST segment changes, expressed as a percentage of the QRS amplitude, and tends to be smaller. Additionally, the mean number of ECG leads exhibiting ST-segment elevation and those displaying reciprocal ST segment depression is generally lower during PTCA than during acute ischaemia.

An expert cardiologist evaluated the ECG signals from patients to identify those with no other signs of cardiac disease. Many patients with heart disease were excluded from the study, reducing the number of signals. Including more angioplasty databases could improve the study and overcome limitations related to variability and other ischaemic regions.

6. Conclusions

This work is an important and novel contribution to the spatio-temporal analysis of ECG dynamics during acute myocardial ischaemia. The estimation of PDFs for both the procedure and control series, the use of quantifiers such as entropy and complexity, and their positioning in the $\mathcal{H} \times \mathcal{C}$ causal plane enable differentiation of the behaviours for LAD and RCA occlusion and recovery. In summary, this study provides insights that had not been previously explored.

Author Contributions: Conceptualisation, E.R.V., V.V., O.A.R. and P.D.A.; methodology, E.R.V., V.V., O.A.R. and P.D.A.; software, E.R.V. and V.V.; validation E.R.V., V.V. and P.D.A.; formal analysis E.R.V., V.V., O.A.R. and P.D.A.; investigation E.R.V., V.V. and P.D.A.; resources E.R.V., V.V. and P.D.A.; data curation, E.R.V. and P.D.A.; writing—original draft preparation V.V. and P.D.A.; writing—review and editing, V.V. and P.D.A.; visualisation, E.R.V., V.V., O.A.R. and P.D.A.; supervision V.V. and P.D.A.; project administration, V.V. and P.D.A.; funding acquisition, V.V. and P.D.A. All authors have read and agreed to the published version of the manuscript.

Funding: This research was funded by Universidad de Buenos Aires (UBACyT 20020130100485BA); Agencia (MINCyT PICT 2145 2016) and by Consejo Nacional de Investigaciones Científicas y Técnicas (PIP 2014-2016 112-20130100552CO), Argentina; and by Universidad Nacional de la Plata under project I263, Argentina.

Institutional Review Board Statement: The STAFF III database was acquired during 1995–1996 at Charleston Area Medical Center (WV, USA) where single prolonged balloon inflation had been introduced to achieve optimal results of PTCA procedures, replacing the typical series of brief inflations. The lead investigator Dr. Stafford Warren designed the study protocol together with Dr. Galen Wagner at Duke University Medical Center (Durham, NC, USA); Dr. Michael Ringborn (Blekinge Hospital, Karlskrona, Sweden) was responsible for data acquisition [19].

Informed Consent Statement: Informed consent was obtained from all subjects involved in the study [6].

Data Availability Statement: The STAFF III database is available at https://physionet.org/content/staffiii/1.0.0/ [5], accessed on 27 November 2024.

Conflicts of Interest: The authors declare no conflicts of interest.

Abbreviations

The following abbreviations are used in this manuscript:

ECG	Electrocardiogram
LAD	Left anterior descending
RCA	Right coronary artery
PTCA	Percutaneous transluminal coronary angioplasty procedure
HRV	Heart rate variability
JSD	Jensen–Shannon Divergence
PDF	Probability distribution function

References

1. Lemire, D.; Pharand, C.; Rajaonah, J.; Dube, B.; Blanc, A.R.L. Wavelet time entropy, T wave morphology and myocardial ischaemia. *IEEE Trans. Biomed. Eng.* **2000**, *47*, 967–970. [CrossRef] [PubMed]
2. Farahabadi, E.; Farahabadi, A.; Rabbani, H.; Dehnavi, A.M.; Mahjoob, M.P. An entropy-based method for ischaemia diagnosis using ECG signal in wavelet domain. In Proceedings of the IEEE 10th International Conference on Signal Processing Proceedings, Beijing, China, 24–28 October 2010 ; pp. 195–198.
3. Rabbani, H.; Mahjoob, M.P.; Farahabadi, E.; Farahabadi, A.; Dehnavi, A.M. ischaemia detection by electrocardiogram in wavelet domain using entropy measure. *J. Res. Med. Sci.* **2011**, *16*, 1473 . [PubMed]
4. Calderón-Juárez, M.; Cruz-Vega, I.B.; González-Gómez, G.H.; Lerma, C. Nonlinear Dynamics of Heart Rate Variability after Acutely Induced Myocardial Ischaemia by Percutaneous Transluminal Coronary Angioplasty. *Entropy* **2023**, *25*, 469–482. [CrossRef] [PubMed]
5. Goldberger, A.; Amaral, L.; Glass, L.; Hausdorff, J.; Ivanov, P.C.; Mark, R.; Stanley, H.E. PhysioBank, PhysioToolkit, and PhysioNet: Components of a new research resource for complex physiologic signals. *Circulation [Online]* **2000**, *101*, e215–e220. [CrossRef] [PubMed]
6. Martínez, J.P.; Pahlm, O.; Ringborn, M.; Warren, S.; Laguna, P.; Sörnmo, L. The STAFF III database: ECGs recorded during acutely induced myocardial ischemia. In Proceedings of the 2017 Computing in Cardiology (CinC), Rennes, France, 24–27 September 2017; Volume 44.
7. Kors, J.A.; Van Herpen, G.; Sittig, A.C.; Van Bemmel, J.H. Reconstruction of the Frank vectorcardiogram from standard electrocardiographic leads: Diagnostic comparison of different methods. *Eur. Heart J.* **1990**, *11*, 1083–1092. [CrossRef] [PubMed]
8. Meyer, C.R.; Keiser, H. Electrocardiogram baseline noise estimation and removal using cubic spline and state-space computation techniques. *Comp. Biomed. Res.* **1977**, *10*, 459–470. [CrossRef] [PubMed]
9. Mendieta, J. Algoritmo para el Delineado de Señales Electrocardiográficas en un Modelo Animal Empleando Técnicas Avanzadas de Procesamiento de Señales. Master's Thesis, Facultad de Ingeniería de la Universidad de Buenos Aires, Buenos Aires, Argentina, 2012.
10. Kowalski, A.; Martin, M.; Plastino, A.; Rosso, O.; Casas, M. Distances in probability space and the statistical complexity setup. *Entropy* **2011**, *13*, 1055–1075. [CrossRef]
11. Kullback, S.; Leibler, R.A. On information and sufficiency. *Ann. Math. Stat.* **1951**, *22*, 79–86. [CrossRef]
12. Lin, J. Divergence Measures based on the Shannon Entropy. *IEEE Trans. Inf. Theory* **1991**, *37*, 145–151. [CrossRef]
13. Kowalski, A.M.; Rossignoli, R.D.; Curado, E.M. (Eds.) *Concepts and Recent Advances in Generalized Information Measures and Statistics*; Bentham Science Publishers: Sharjah, United Arab Emirates, 2013.
14. Bandt, C.; Pompe, B. Permutation Entropy: A Natural Complexity Measure for Time Series. *Phys. Rev. Lett.* **2002**, *88*, 174102. [CrossRef] [PubMed]
15. Rosso, O.A.; Larrondo, H.A.; Martin, M.T.; Plastino, A.; Fuentes, M.A. Distinguishing Noise from Chaos. *Phys. Rev. Lett.* **2007**, *99*, 154102. [CrossRef] [PubMed]
16. Henry, M.; Judge, G. Permutation Entropy and Information Recovery in Nonlinear Dynamic Economic Time Series. *Econometrics* **2019**, *7*, 10. [CrossRef]
17. Badir, B.F.; LeBlanc, A.R.; Nasmith, J.B.; Palisaitis, D.; Dubo, B.; Nadeau, R. Continuous ST-Segment Monitoring During Coronary Angioplasty Using Orthogonal ECG Leads. *J. Electrocardiol.* **1997**, *30*, 175–187. [CrossRef] [PubMed]
18. Pérez Riera, A.R.; Uchida, A.H.; Ferreira Filho, C.E.; Meneghini, A.; Ferreira, C.; Schapacknik, E.; Dubner, S.; Moffa, P. Significance of vectorcardiogram in the cardiological diagnosis of the 21st century. *Clin. Cardiol.* **2007**, *30*, 319–323. [CrossRef] [PubMed]
19. Warren, S.G.; Wagner, G.S. STAFF studies of the first 5 min of percutaneous coronary angioplasty balloon occlusion in man. *J. Electrocardiol.* **2014**, *47*, 402–407. [CrossRef] [PubMed]

entropy

MDPI

Article

Statistical Complexity Analysis of Sleep Stages

Cristina D. Duarte [1,*], Marianela Pacheco [1,2], Francisco R. Iaconis [1], Osvaldo A. Rosso [3], Gustavo Gasaneo [1] and Claudio A. Delrieux [2]

[1] Departamento de Física, Instituto de Física del Sur, Universidad Nacional del Sur-Consejo Nacional de Investigaciones Científicas y Técnicas (CONICET), Bahía Blanca 8000, Argentina; mpacheco@icic-conicet.gob.ar (M.P.); francisco.iaconis@uns.edu.ar (F.R.I.); ggasaneo@uns.edu.ar (G.G.)
[2] Departamento de Ingeniería Eléctrica y Computadoras, Instituto de Ciencias e Ingeniería de la Computación, Universidad Nacional del Sur-Consejo Nacional de Investigaciones Científicas y Técnicas (CONICET), Bahía Blanca 8000, Argentina; cad@uns.edu.ar
[3] Instituto de Física, Universidade Federal de Alagoas UFAL, Maceió 57072-900, Brazil; oarosso@conicet.gov.ar
* Correspondence: cristina.duarte@uns.edu.ar

Abstract: Studying sleep stages is crucial for understanding sleep architecture, which can help identify various health conditions, including insomnia, sleep apnea, and neurodegenerative diseases, allowing for better diagnosis and treatment interventions. In this paper, we explore the effectiveness of generalized weighted permutation entropy (GWPE) in distinguishing between different sleep stages from EEG signals. Using classification algorithms, we evaluate feature sets derived from both standard permutation entropy (PE) and GWPE to determine which set performs better in classifying sleep stages, demonstrating that GWPE significantly enhances sleep stage differentiation, particularly in identifying the transition between N1 and REM sleep. The results highlight the potential of GWPE as a valuable tool for understanding sleep neurophysiology and improving the diagnosis of sleep disorders.

Keywords: permutation entropy; statistical complexity; generalized weighted permutation entropy; sleep stages

1. Sleep Stage Interpretation and Analysis

Sleep and waking states are periodic biobehavioral features that are characterized by changes in brain electrical activity, manifested as alterations in consciousness, reduced sensory responsiveness, decreased muscle tone, and relative inactivity [1]. This process has an important adaptive function in the homeostasis of mammals, guaranteeing their survival [2]. In humans, alterations in the quality, quantity, and patterns of sleep (i.e., sleep disorders) negatively impact the individual's overall health and functionality [1,3–5]. Sleep disorders are extremely common in the general population [1,5].

Normal human sleep consists of two primary states, rapid eye movement (REM) and non-REM, each with distinct stages and characteristics. Furthermore, non-REM sleep includes several stages, ranging from light to deep sleep, while REM sleep is associated with vivid dreaming and brain activity patterns similar to those of wakefulness. These sleep stages undergo age-related changes and are influenced by factors such as sleep–wake history, circadian timing, and pharmacological agents, all of which impact the distribution and quality of REM and non-REM sleep [6]. Accurately assessing these sleep stages is essential for understanding various health conditions, including insomnia, sleep apnea, and neurodegenerative diseases such as Parkinson's and Alzheimer's [7].

For example, disruptions in deep sleep have been linked to cognitive decline [8,9], while abnormalities in REM sleep may indicate certain mood disorders [10,11] or post-traumatic stress disorder [12,13]. This analysis requires robust methods to discriminate sleep stages from physiological signals. The electrical signals of the brain recorded by EEG are the most used technique to understand the microstructure of sleep waves [14]. By analyzing these signals, we can identify the different stages of sleep—from light sleep (N1, N2) to deep sleep (N3) and REM stages—and assess their duration and how a person transitions between them. This allows for the accurate identification of sleep disorders, which can lead to more effective diagnostic and treatment interventions [15].

The EEG signal reflects the electrical activity of the brain, which forms a chaotic, non-linear temporal signal whose features has been studied from the standpoint of statistical complexity, with efforts aimed at developing automatic and precise methods for classifying sleep stages. Among the earliest studies to apply statistical complexity analysis in sleep stages is that presented in [16]. This paper discusses how permutation entropy (PE) can be applied to analyze EEG sleep signals in artifact-free segments of 30 s, providing measures that help identifying different sleep stages and detecting anomalies, such as those related to neurological conditions. The study shows that PE is useful for characterizing sleep EEG and offers a computationally efficient tool to measure the complexity of brain activity during different sleep stages, contributing to a better understanding of sleep physiology and potential disorders. In [14], the authors explore the statistical complexity of EEG signals during sleep, particularly focusing on the Cyclic Alternating Pattern (CAP), which is a key feature of the sleep microstructure. The study applies complexity-based features like fractal dimension and sample entropy to evaluate these activations and compares them to non-activation periods in NREM sleep. Their findings suggest that different subtypes of CAP activations (A1, A2, and A3) have distinct levels of EEG complexity, with A3 showing the highest complexity across all sleep stages. The ability of ordinal patterns to detect minute features in EEG signals is exposed in [17]. A modified PE assessment is here understood as a measure of distance from white noise, with the support of a specific statistical hypotheses test in which critical significance values are provided. That paper presents the first evidence of automated analysis that goes beyond expert human assessment since the method is able to detect delta wave patterns and graphic features which are only visible on a time scale of several seconds. At the same time, the "distance from white noise" measure serves as an indicator of sleep depth, in which the awake state is treated as the nearest to white noise.

Building on these advancements, the statistical complexity derived from Jensen–Shannon divergence has emerged as a robust alternative for capturing the intricate interplay between randomness and order in EEG signals. Unlike measures such as PE or sample entropy that primarily quantify disorder, this statistical complexity excels in identifying intermediate levels of organization within the signal. This sensitivity allows it to distinguish not only between wakefulness and sleep but also among different sleep stages with greater precision [18]. In comparison, approximate entropy (ApEn) is effective in measuring signal regularity and has been used for detecting anomalies such as epileptic seizures. However, it is sensitive to parameter selection and suffers from issues such as a bias toward shorter data lengths, which can lead to inconsistent results across datasets [19].

Lempel–Ziv complexity (LZC), on the other hand, offers a straightforward algorithm that assesses the randomness of finite sequences by the number of steps required to generate them, making it particularly suitable for real-time applications like depth-of-anesthesia monitoring [20]. However, LZC often fails to account for intermediate levels of complexity, as it primarily evaluates disorder without distinguishing structured chaotic dynamics from noise [21]. The ability of statistical complexity to integrate measures of entropy and disequilibrium positions it as a more comprehensive tool. Unlike ApEn and LZC, statistical complexity not only distinguishes structured chaotic dynamics from random patterns but also quantifies the nuanced transitions between these extremes, providing deeper insights into the non-linear dynamics of EEG signals [18].

The use of machine learning methods in classifying sleep stages is also proposed in [22], specifically to score sleep stages using single-channel electrooculogram (EOG) signals. These signals are analyzed using several features (discrete wavelet transform, spectral entropy, moment-based measures, composite multiscale dispersion entropy, and autoregressive coefficients). The discriminative power of these features is assessed through ANOVA tests, and feature reduction is applied to decrease model complexity and retain the most discriminative features. Finally, sleep stage classification is performed using Random Under-Sampling Boosting (RUSBoost), random forest (RF), and support vector machines (SVMs). On a similar setting, in [23], the authors combine entropy features (including fuzzy measure entropy, fuzzy entropy, and sample entropy), with a multi-class SVM to classify sleep stages. The method analyzes 30 s segments of Fpz-Cz and Pz-Oz EEG signals, together with horizontal EOG signals from the Sleep Telemetry Study dataset [24].

Given the widespread success of ordinal patterns and statistical complexity-based signal analysis, there has been a recent influx of alternative techniques that may outperform the more general methods. In particular, in [25], the authors propose ensemble improved PE and multiscale ensemble improved PE for taking into account both permutation relations and amplitude information. The proposed methods show better discriminating power and robustness, specifically in the analysis of EEG signals. In our study, we pursue a comparable objective: to analyze and classify sleep stages from polysomnographic data by training a classifier using novel statistical complexity measures as features, specifically generalized weighted permutation entropy (GWPE).

While deep learning methods excel at automating feature extraction, their black-box nature and high resource demands can make them less suitable for EEG analysis, where interpretability and performance with minimal data are often priorities. Our results outperform traditional analysis based on ordinal patterns, achieving accuracies comparable to those obtained with deep learning [26]. The advantage of using feature-engineered white-box models in this context (as opposed to data-driven deep learning models) is that handcrafted features are more interpretable in neurophysiological terms, providing clearer insights into the underlying neural mechanisms, which is crucial in fields like neuroscience and clinical diagnosis in which model explainability is required. Additionally, feature-engineered models are usually less computationally intensive, can work effectively with smaller datasets, and are much easier to re-train, making them a more practical choice in scenarios where labeled EEG data are varying or limited.

2. Materials and Methods

2.1. Database Description

The dataset used is obtained from Bob Kemp's publicly available database on Physionet, linked to the Sleep Telemetry Study [24] already mentioned, which consists of a set of polysomnographic signal recordings. That study investigated the effects of temazepam on sleep in 22 Caucasian men and women, ages 18 to 79, 7 men (mean age 35.71) and 15 women (mean age 42.26). The subjects had mild difficulties falling asleep, were not taking any other medications, and were healthy subjects. This dataset is made up of a total of 44 recordings in the hospital, corresponding to two nights, that is, two recordings per subject. The polysomnographic recordings lasted nine hours, including the location of the Fpz-Cz and Pz-Oz electrodes, and were performed after temazepam ingestion or after placebo ingestion. For the present work, the corresponding records are included for subjects between 18 and 40 years old after taking the placebo. The dataset is made up of 12 polysomnographic records of 12 subjects (total mean age for men and women 25.83, standard deviation: 6.55, sample rate 100 hz). More specifically, the study is composed of 8 women (mean age women 27.87, standard deviation: 7.20) and 4 men (mean age men 21.75, standard deviation: 1.71). We use only the Fpz-Cz channel of the EEG signal. These recordings include expert annotations of sleep stages, labeled in 30-s intervals, which the experts term *epochs*, each consisting of 3000 data points.

2.2. Generalized Probabilities of Ordinal Patterns

Permutation entropy (PE) [27] is a robust, non-linear time series analysis method that quantifies the complexity and randomness of a system, having gained widespread adoption due to its simplicity, computational efficiency, and ability to capture dynamic patterns in diverse fields such as neuroscience, finance, and engineering. The speed and robustness that characterize PE make it particularly useful for handling large datasets without requiring extensive preprocessing or parameter tuning [17].

The computation of PE involves transforming the time series into a sequence of symbols known as ordinal patterns (OPs). These OPs are characterized by two key parameters: the length of the pattern, D, and the time delay τ, which denotes the interval between the samples in the signal used to define the patterns. Given a time series x_t where $t = 1, \cdots, T$, overlapping segments s of length D are taken with a τ interval between consecutive samples. This results in a total of $T-(D-1)$ segments. Within each segment, the D samples are labeled with ordinal numbers $1, 2, \cdots, D$ according to their ordinal relationships with other values in the same segment. Then, each segment is labeled with an OP, which is one permutation π among the $D!$ possible permutations of D ordinal values. In a given signal, the relative frequency (or empirical probability) of each OP π_i is then

$$p(\pi_i) = \frac{\sum_{s,\pi_s=\pi_i} 1}{\sum_s 1}, \tag{1}$$

where s represents each of the sequences, and π_s is the corresponding OP of s. The probability distribution function (PDF) of the $D!$ OPs is then $P_\pi = \{p(\pi_1), \ldots, p(\pi_{D!})\}$. From this PDF, the PE of the signal can be computed using Shannon entropy:

$$S = -\sum_{i}^{D!} p(\pi_i) log(p(\pi_i). \tag{2}$$

It can be shown that S takes its minimum value of 0 when there is total certainty about the series, i.e., a given permutation has probability one. The maximum value corresponds to the case where uncertainty is maximized ($log(D!)$), which corresponds to a uniform probability distribution $P_U = \{p_i = 1/M, i = 1, \ldots, M\}$, where all states have the same occurrence probability.

As mentioned above, alternative complexity measures, such as multiscale entropy or fractal dimension, lack the capability to jointly account for randomness and chaotic patterns, which is crucial in nonstationary signals like EEG that exhibit both stochastic and deterministic behaviors. Statistical complexity addresses this gap by integrating entropy with disequilibrium, defined as the product of normalized Shannon entropy H and disequilibrium Q:

$$C = Q[P_\pi, P_u] \cdot H[P_\pi], \tag{3}$$

where normalized Shannon entropy H is defined as S/S_{max}, disequilibrium Q is defined in terms of the Jensen–Shannon divergence $Q = Q_0 \cdot D[P_\pi, P_u]$, and the divergence D provides a measure of the similarity between two probability densities. Here, the PDF is compared with the uniform distribution:

$$D[P_\pi, P_u] = S\left(\frac{P_\pi + P_u}{2}\right) - \left(\frac{S[P_\pi] + S[P_u]}{2}\right) \tag{4}$$

The lowest complexity value occurs when the system is either entirely random or entirely ordered. When the system is in a state of maximal disorder, there is no structural information, and the complexity is low. When the system is in a highly ordered state, it is fully predictable, meaning there is no uncertainty in the pattern distribution, which also results in low complexity. Entropy and complexity can be articulated together in the complexity–entropy causality plane. In [28], it is shown how to simultaneously quantify both the information content and structural complexity of a time series, distinguishing between stochastic noise and deterministic chaotic behavior, leading to many applications in analyzing and characterizing time series and signals from physiology, physics, and many other fields.

While PE has been effective in various applications, the resulting assessments may have limitations when the local amplitude of the time series is very uneven since smaller fluctuations should exert less influence in the overall analysis. To this avail, in [29], the authors propose weighted PE (WPE), which assigns more significance to OPs with higher amplitude fluctuations. This is accomplished by incorporating the local variance of each OP as a weighing factor of its contribution to the overall empirical probabilities. This modification allows WPE to capture both the ordinal structure and amplitude information of a time series, making it more suitable for applications where amplitude plays a key role. Furthermore, in [30], the authors propose the generalized weighted PE (GWPE), which also considers an entropic index q that allows to discern between the effects of small or large fluctuations in the overall complexity of the signal. The associated probability distribution is then

$$p(\pi_i, q) = \frac{\sum_{s,\pi_s=\pi_i} w^{q/2}}{\sum_s w_s^{q/2}}. \tag{5}$$

GWPE coincides with WPE for $q = 2$, and with the original PE for $q = 0$, making the notation and parameters compatible with other nonextensive measures (e.g., Tsallis entropy) or nonstationary fluctuation analysis (e.g., multifractal spectrum). As with other nonextensive/nonstationary models, entropic indices are used to generalize the concept of entropy to systems with long-range interactions or memory effects, fractal structures, or complex correlations, providing a more accurate description of their statistical behavior and properties. In our particular context, the relevance of the use of GWPE lays in how different ordinal patterns are weighted, given that negative q values enhance the contribution of small fluctuations in the OP distribution, while positive values enhance the contribution of large fluctuations.

In this study, we are focused on determining how entropic indices may enhance the ability of a PE-based polysomnographic signal analysis. As already mentioned, this approach enables a more flexible and detailed evaluation of the disorder and structure of the time series. A likely interpretation of Equation (5) is that now PE is a function of the entropic index and thus signals can be assessed according to this function. In particular, instead of evaluating this PE function for signals, or characterizing them in the complexity–entropy causality plane as in [30], our idea is to use the generalized probabilities $p(\pi_i, q)$

of the OPs as features for classification, and study the ability of the resulting classifier, as compared to the standard PE, to distinguish the different sleep stages as described in the next subsection.

2.3. Data Analysis

For each subject, the EEG signal is segmented according to the labeled epochs, with each epoch corresponding to a specific sleep stage. The numbers of epochs across all subjects are N1 = 885, N2 = 5281, N3 = 1995, R = 2599, and W = 803, for a total of 11,563 epochs. Given the uneven distribution of epochs, we downsample the N2 class by randomly dropping half of the epochs, and we perform a data augmentation procedure with N1 and W (waking state), creating new epochs in the middle of each pair of consecutive epochs belonging to the same class (i.e., starting at the midpoint of the first epoch and ending at the second epoch). Furthermore, for each sleep cycle, we remove the two boundary epochs marking the transition from one sleep phase to the next. The final number of epochs in each state is N1 = 978, N2 = 2247, N3 = 1643, R = 2444, and W = 1017. The total number of epochs across all states is 8329.

In calculating the generalized probabilities $p(\pi_i, q)$ for each subject and epoch, we use a time delay of $\tau = 1$, and a segment length D = 4. These parameter values are chosen based on [18], which demonstrate their effectiveness in characterizing sleep stages. The index q is varied in integer values in the $[-10,10]$ interval, for a total of 21 q values. Using these values, the generalized probabilities of each of the $D!$ ordinal patterns in each epoch are computed, giving, in total, 504 features (24×21) for each epoch. Thus, the shape of the dataset for training the classifier is 8329×505, the former being the rows corresponding to each epoch, and the latter the columns, including the 504 features plus the corresponding sleep stage label.

With this dataset, several classification models are trained, including random forest, support vector machines, and XGBoost. In each case, the dataset is split into 85% for training and 15% for testing. The training data are further divided using 5-fold cross validation, hyperparameter tuning procedures are performed, as well as feature selection for evaluating the effectiveness of the different features in distinguishing sleep stages W, N1, N2, N3, and REM. The optimal hyperparameters for each model are determined using RandomizedSearch implemented through the scikit-learn library [31], with 75 random permutations of parameter values. The performance of the best hyperparameters is evaluated based on accuracy. The above procedures are applied to the full set of features (corresponding to the generalized probabilities) and with only the 24 features corresponding to standard PE ($q = 0$). In addition, the PE and complexity of the 21 q values are assessed.

3. Results

3.1. Classifier and Feature Set Comparison for Sleep Stage Classification

To illustrate the advantages of the generalized probabilities of the OPs in improving sleep stage classification, we train three algorithms to automatically classify the five stages. Here, we present the results only for XGBoost, which demonstrates better performance as compared to other widely used machine learning models like random random forests and support vector machines. The results of all the methods are presented in Appendices A.1 and A.2 for comparison. We evaluate the classification performance with different feature sets to identify which features best separate the sleep stages. Each algorithm is computed in the following order:

- Permutation entropy (PE) and complexity (C);
- Generalized weighted permutation entropy (GWPE) and complexity (GWPEC);

- The probability distribution function of the ordinal patterns (PDF);
- The generalized weighted probability distribution of the ordinal patterns (GWPDF).

The normalized confusion matrices for the classification of sleep stages using XGBoost, trained with PE and C, as well as GWPE and GWC, are presented in Figure 1. Similarly, Figure 2 shows the confusion matrices for the classification using the probability distribution of the ordinal patterns (PDF) and generalized weighted probability distribution (GWPDF). All confusion matrices are calculated using the classifier optimized through cross-validation. Furthermore, the matrices are generated after performing feature importance analysis on both the generalized weighted permutation entropy and complexity, as well as on the generalized weighted probability distribution of ordinal patterns. Confusion matrices for the other classifiers are available in Appendix A.1.

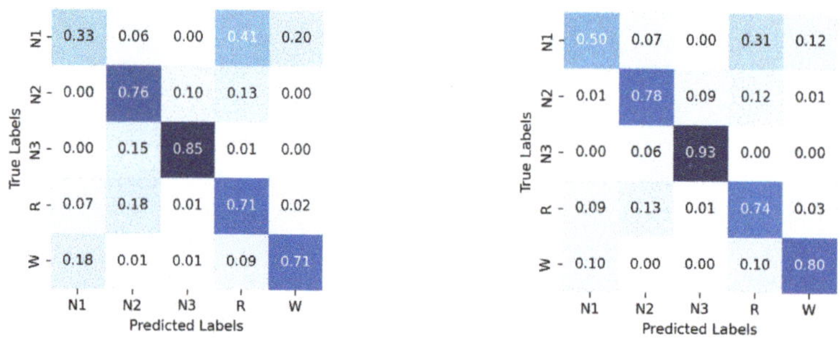

(a) (b)

Figure 1. Normalized confusion matrix (%) for XGBoost algorithm classifier, with PE and C on the left and GWPE and GWPEC on the right as features. (**a**) XGBoost algorithm; PE and C as features, (**b**) XGBoost algorithm; GWPE and GWPEC as features.

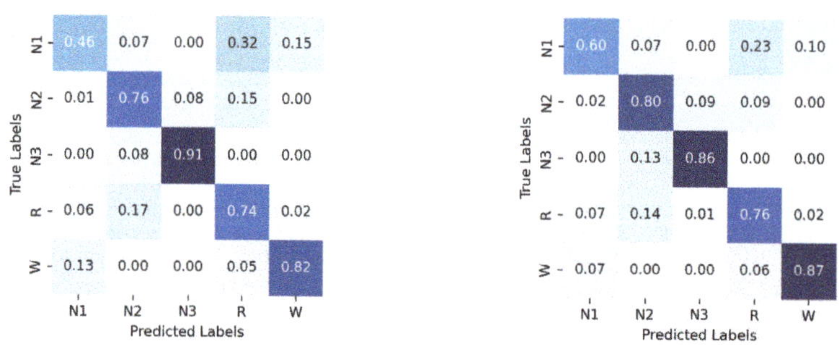

(a) (b)

Figure 2. Normalized confusion matrix (%) for XGBoost algorithm classifier, with PDF on the left and GWPDF on the right as features. (**a**) XGBoost algorithm; PDF as features, (**b**) XGBoost algorithm; GWPDF as features.

Table 1 summarizes the mean accuracy and standard deviation obtained from cross validation for XGBoost. Detailed performance metrics for random forest and support vector machine can be found in Appendix A.2. Feature importance analysis is conducted to identify the most relevant features that contribute to the model's predictive performance. The results using GWPDF as features, as well as the results for GWPE and GWPEC as features, are presented in Table 2.

The importance of each feature is calculated based on the total accuracy gain of the feature when it is used in training. In Table 2 are presented, for both features sets, the features, the corresponding index q used in the calculation, and their relative importance. Only the top 10 features are displayed, ordered based on their impact on the classification accuracy. Using OP, the top 10 features contribute to 35% of the cumulative importance, whereas for GWPE, these same top 10 features contribute to 64% of the cumulative importance.

Table 1. Mean accuracy (Acc.) and standard deviation (SD) for the XGBoost classifier.

		Acc	SD
Features	PE and C	0.71	0.01
	GWPE and GWPEC	0.77	0.01
	PDF	0.74	0.01
	GWPDF	0.80	0.01

Table 2. Top ten features ranked by relative importance for XGBoost classifiers. The classifiers are trained with GWPDF (left) and GWPE and GWPEC (right). For both sets, the first column lists the respective feature, the second column presents the entropic index q used to calculate the feature, and the third column shows the relative importance of each feature.

OPs			GWPE and GWPEC		
Features	q	**Relative Importance**	**Features**	q	**Relative Importance**
$p_{[3210]}$	0	0.0868	H	1	0.1655
$p_{[0123]}$	1	0.0716	H	0	0.1028
$p_{[0123]}$	−1	0.0355	C	3	0.0874
$p_{[1032]}$	2	0.0290	C	1	0.4366
$p_{[1032]}$	3	0.0286	H	−1	0.0595
$p_{[3210]}$	1	0.0285	C	0	0.0416
$p_{[0123]}$	0	0.0210	C	−1	0.0300
$p_{[3102]}$	2	0.0206	H	2	0.0277
$p_{[0213]}$	2	0.0204	C	2	0.0214
$p_{[0213]}$	3	0.0124	C	7	0.0208

3.2. Behavior of PE, C vs. q and POs vs. q

To illustrate how the entropic index q exerts influence in differentiating between sleep stages, we compute the median values of PE and C for all subjects across each sleep stage, as a function of q parameter. The exploration of the q parameter is shown in Figure 3. This sheds light on how different values of q affect the separation of sleep stages based on generalized permutation entropy and complexity. In both cases, PE and C, for $q < -1$, the medians across all stages remain closely clustered. Starting from $q = -1$, the median values begin to diverge. This behavior is consistent with the feature selection performed in the cases of RF and XGBoost classifiers. In these cases, it is shown that entropy and complexity with positive exponents have the highest relative importance.

Additionally, we plot the median values of the probability distributions for the patterns [0123] and [3210]. We choose these patterns because they exhibit the highest relative importance in the XGBoost classification. We observe once again that the separation between the stages begins at $q \geq -1$.

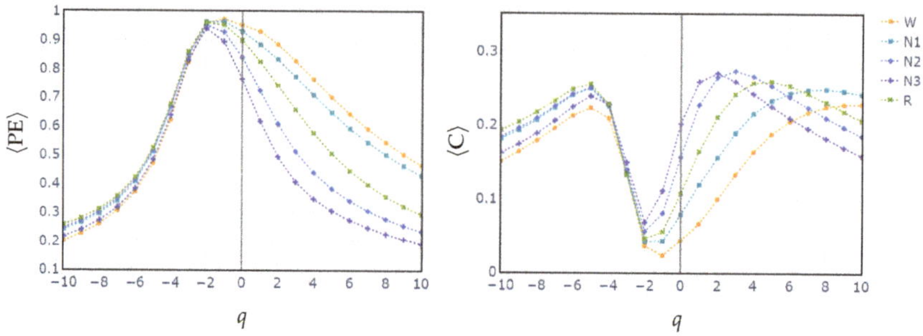

Figure 3. Median of values of generalized permutation entropy (**left**) and complexity (**right**) for q across all sleep stages.

4. Discussion

The results show improvements in median accuracy when moving from standard PE and C to GWPE and GWPEC across all classifiers (see Table A1 in the Appendix A.2). The use of PDF as a feature set improves the classification as compared with the standard PE but does not fully capture the signal's complexity.

GWPE allows for more accurate discrimination between different sleep phases, including the subtle transitions between stages such as N1 and REM. This enhanced capability is due to the fact that GWPE not only considers the order of patterns in the EEG signal as standard PE does but also weights these patterns according to their magnitude, allowing us to highlight the contributions of both small and large fluctuations. Specifically, negative values of q in GWPE increase the contribution of small fluctuations, while positive values increase the contribution of large fluctuations. This weighted approach provides a more detailed characterization of brain dynamics. In this context, this measure reflects the dynamic organization of neural activity during sleep.

As the individual progresses into deeper levels of sleep, cortical brain waves become increasingly synchronized. This synchronization increases the signal amplitude while reducing frequency, which significantly alters the probabilities of the ordinal patterns, particularly increasing the likelihood of monotonic patterns [0123] and [3210], common in deterministic and periodic signals. This feature explains the higher accuracies achieved for the N3 stage (as reflected in the confusion matrices across all feature sets) since this stage mainly contains low-frequency, high-amplitude delta waves, in which these monotonic patterns prevail.

For N2, the classification accuracy is generally high across the three algorithms using PE and C as features. However, adding the entropic index q further enhances the performance in both GWPE and GWPDF feature sets (see Figures 1, 2, A1 and A2). Stage N2 is characterized by distinctive K-complexes (brief, large-amplitude spikes) and sleep spindles (12–16 Hz oscillatory waves). Considering signal amplitude likely improves the classifiers' ability to detect these patterns by highlighting specific amplitude and frequency occurrences, thereby reducing the risk of misinterpretation as noise, which may occur when using only PE and C.

In contrast, classifying N1 sleep remains challenging. During wakefulness, the EEG generally shows lower complexity due to regular alpha or beta waves. As the signal transitions into stage N1, there is an intermediate increase in complexity as alpha waves decrease and theta waves emerge. Thus, using PE as features, particularly distinguishing N1 from wakefulness and REM, obtains low classification accuracy due to overlapping EEG signal features. However, the use of generalized probabilities of ordinal patterns, weighted by the entropic index q, significantly enhances the classifier's ability to distinguish N1 from other stages, leading to GWPE values that are distinct from W and REM. In REM sleep, the EEG signal is highly complex and disordered, with a mixture of theta and beta waves and increased signal variability, resulting in higher GWPE values (see Figure 3). The confusion matrices demonstrate a significant enhancement in N1 classification accuracy, increasing from 33% (Figure 1a) to 60% (Figure 2b) with the use of generalized probabilities. This improvement can be observed in two ways. First, as shown in Figure 3, N1 becomes more distinguishable from W and R at $q \geq -1$ in PE and C. Second, as shown in Figure 4, the probability of weighted patterns captures finer distinctions between N1 and the other sleep stages. This suggests that incorporating both the variance of the EEG signal and the occurrence probabilities of each pattern may allow for deeper capture of the microstructure within sleep stages. Specifically, at values of $q \geq -1$, there is a marked improvement in stage discrimination, suggesting that large fluctuations in EEG signals play a critical role in enhancing classification accuracy. This can be seen in Figures 3 and 4, where distinctions among all stages are apparent.

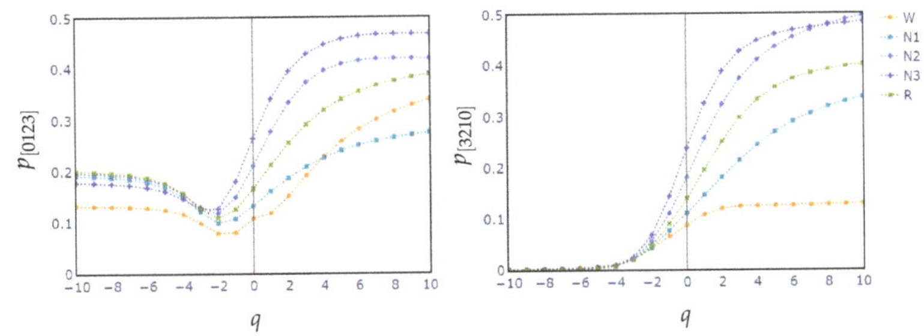

Figure 4. Median probabilities of the monotonic patterns [0123] and [3210] across all sleep stages for varying entropic index q.

The fact that XGBoost achieves the best performance can be attributed to an uneven distribution of the OP relative frequencies of each sleep stage in the feature space, which makes a boosting strategy more effective than the SVM geometric method or the RF bagging strategy. Moreover, an analysis of feature informativeness reveals a contrasting distribution in predictive power across the two feature sets. In the plain PE feature set, the ten most informative features capture only 35% of the total predictive performance gain of the classifiers, indicating a relatively dispersed distribution of informativeness across features. In contrast, in the GWPE feature set, the top ten features account for 65% of the total gain, suggesting a more concentrated distribution, where fewer features contribute a much larger amount of information relevant to the classification accuracy. As shown in Table 2, the entropic indices that achieve the highest relative importance in classifying sleep stages are those greater than or equal to -1. This applies to both the probability of ordinal patterns and the complexity and entropy measures (PE and C) derived from these probabilities. This observation points to differences in feature redundancy and offers insights into the complexity of feature selection and model optimization.

5. Conclusions

In this study, three classification algorithms (RF, SVM, and XGBoost) were evaluated to assess their performance with different feature sets, specifically examining the improvement in classification accuracy using GWPE. Our results indicate that these measures provide a more comprehensive understanding of the complexity of EEG signals compared to traditional permutation entropy and complexity. The use of the entropic index q was shown to be crucial for more effectively capturing the complexity of EEG signals, as it allows for more precise modulation of ordinal probabilities based on EEG fluctuations. This approach provides a more refined characterization of the dynamic patterns of brain activity. It is especially useful for distinguishing between sleep stages, such as N1 and REM, where brain activity is often similar. The use of generalized ordinal probabilities weighted by q contributed to improving the distinction between these phases, especially in values of $q \geq -1$, suggesting that wide fluctuations in the EEG signal are a key aspect in differentiating them.

The ability to adjust ordinal probabilities to capture the non-linear dynamics and complexity of the EEG has potential applications both in the clinical diagnosis of sleep disorders and in the study of other complex signals and time series. In future work, it would be necessary to explore in depth the effects of different values of the entropic index q in other populations and conditions such as in subjects with specific sleep disorders. On the other hand, it would be valuable to investigate whether the integration of other physiological signals, in addition to EEG, could further complement and improve the classification results, creating more robust and accurate diagnostic tools in the field of sleep disorders.

Author Contributions: Conceptualization, C.D.D., M.P., F.R.I., G.G. and C.A.D.; methodology, C.D.D. and C.A.D.; software, F.R.I. and C.D.D.; validation, C.A.D., G.G. and O.A.R.; formal analysis, O.A.R.; writing—original draft preparation, C.D.D. and M.P.; writing—review and editing, C.A.D., G.G., F.R.I. and O.A.R.; supervision, C.A.D. and G.G.; funding acquisition, C.A.D. and G.G. All authors have read and agreed to the published version of the manuscript.

Funding: This research was supported by the doctoral and postdoctoral fellowships awarded by the National Research and Technology Council of Argentina (CONICET) and was partially funded by the Science and Technology Office of the Universidad Nacional del Sur under grants 24/K093 and 24/F078, CONICET PIP KE3 11220200102879CO and by Agencia Nacional de Promocion Cientifica y Tecnológica PICT-2020-SERIE A-02450.

Data Availability Statement: The data supporting the reported results were obtained from publicly available datasets. The EEG data analyzed in this study are available at PhysioNet (https://physionet.org) under the Sleep-EDF Database Expanded.

Conflicts of Interest: The authors declare no conflicts of interest.

Appendix A

Appendix A.1. Confusion Matrix for Random Forest and Support Vector Machines

This appendix presents the confusion matrices corresponding to the classification algorithms random forest (RF) and support vector machine (SVM). The matrices are calculated using four different feature sets, allowing for the evaluation of each algorithm's performance in classifying sleep stages.

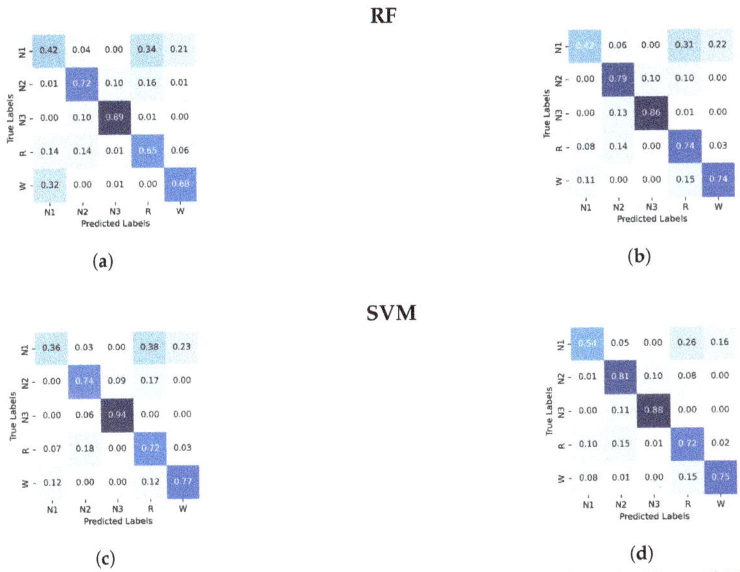

Figure A1. Normalized confusion matrix (%) for RF and SVM algorithms classifier, with PE and C on the left and, GWPE and GWPEC on the right as features. (**a**) RF algorithm; PE and C as features, (**b**) RF algorithm; GWPE and GWPEC as features, (**c**) SVM algorithm; PE and C as features, (**d**) SVM algorithm; GWPE and GWPEC as features.

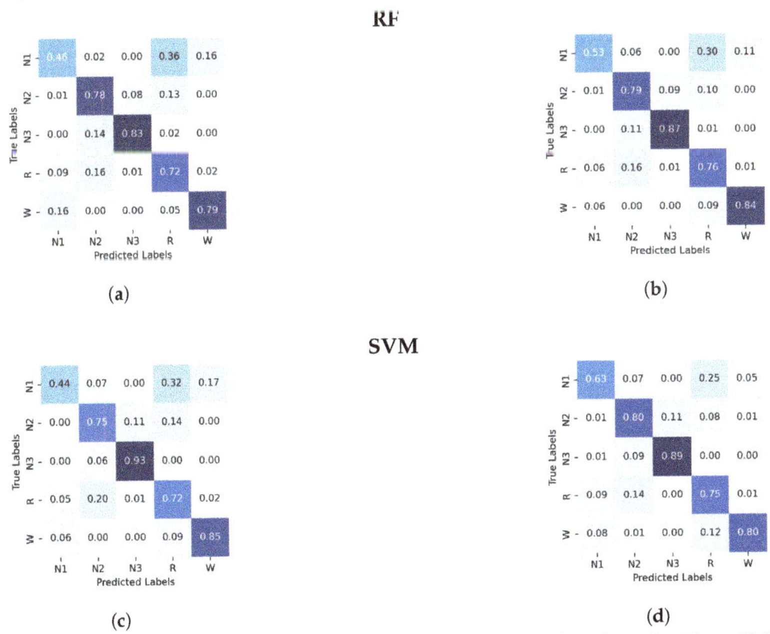

Figure A2. Normalized confusion matrix (%) for the RF and SVM algorithms classifier, with PDF on the left and GWPDF on the right as features. (**a**) RF algorithm; PDF as features, (**b**) RF algorithm; GWPDF as features, (**c**) SVM algorithm; PDF as features, (**d**) SVM algorithm; GWPDF as features.

Appendix A.2. Performance of Random Forest and Support Vector Machine

Here, we present a table summarizing the performance of the random forest and support vector machine algorithms. The table includes the mean accuracy and standard deviation (SD) for the different feature sets, highlighting their effectiveness in classifying sleep stages.

Table A1. Mean accuracy (Acc.) and standard deviation (SD) for the classifiers: random forest (RF) and support vector machine (SVM).

		RF		SVM	
		Acc	**SD**	**Acc**	**SD**
Features	PE and C	0.71	0.01	0.72	0.01
	GWPE and GWPEC	0.76	0.01	0.77	0.01
	PDF	0.74	0.01	0.74	0.01
	GWPDF	0.79	0.01	0.78	0.01

References

1. Gottesman, R.F.; Lutsey, P.L.; Benveniste, H.; Brown, D.L.; Full, K.M.; Lee, J.M.; Osorio, R.S.; Pase, M.P.; Redeker, N.S.; Redline, S.; et al. Impact of Sleep Disorders and Disturbed Sleep on Brain Health: A Scientific Statement From the American Heart Association. *Stroke* **2024**, *55*, e61–e76. [CrossRef] [PubMed]
2. Rial, R.V.; Akaârir, M.; Canellas, F.; Barceló, P.; Rubiño, J.A.; Martín-Reina, A.; Gamundí, A.; Nicolau, M.C. Mammalian NREM and REM sleep: Why, when and how. *Neurosci. Biobehav. Rev.* **2023**, *146*, 105041.
3. Kumar, V.M. Sleep and sleep disorders. *Indian J. Chest Dis. Allied Sci.* **2008**, *50*, 129.
4. Palagini, L.; Geoffroy, P.A.; Miniati, M.; Perugi, G.; Biggio, G.; Marazziti, D.; Riemann, D. Insomnia, sleep loss, and circadian sleep disturbances in mood disorders: A pathway toward neurodegeneration and neuroprogression? A theoretical review. *CNS Spectrums* **2022**, *27*, 298–308. [CrossRef]
5. Panossian, L.A.; Avidan, A.Y. Review of Sleep Disorders. *Med. Clin. N. Am.* **2009**, *93*, 407–425; Common Neurologic Disorders; [CrossRef]
6. Patel, A.K.; Reddy, V.; Shumway, K.R.; Araujo, J.F. Physiology, sleep stages. In *StatPearls [Internet]*; StatPearls Publishing: Treasure Island, FL, USA, 2024.
7. Malhotra, R.K. Neurodegenerative Disorders and Sleep. *Sleep Med. Clin.* **2022**, *17*, 307–314. [CrossRef]
8. Gilley, R.R. The Role of Sleep in Cognitive Function: The Value of a Good Night's Rest. *Clin. EEG Neurosci.* **2023**, *54*, 12–20. [CrossRef] [PubMed]
9. Porter, V.R.; Buxton, W.G.; Avidan, A.Y. Sleep, Cognition and Dementia. *Curr. Psychiatry Rep.* **2015**, *17*, 97. [CrossRef]
10. Yan, T.; Qiu, Y.; Yu, X.; Yang, L. Glymphatic Dysfunction: A Bridge Between Sleep Disturbance and Mood Disorders. *Front. Psychiatry* **2021**, *12*, 658340. [CrossRef]
11. Pearson, O.; Uglik-Marucha, N.; Miskowiak, K.W.; Cairney, S.A.; Rosenzweig, I.; Young, A.H.; Stokes, P.R. The relationship between sleep disturbance and cognitive impairment in mood disorders: A systematic review. *J. Affect. Disord.* **2023**, *327*, 207–216. [CrossRef]
12. de Boer, M.; Nijdam, M.J.; Jongedijk, R.A.; Bangel, K.A.; Olff, M.; Hofman, W.F.; Talamini, L.M. The spectral fingerprint of sleep problems in post-traumatic stress disorder. *Sleep* **2019**, *43*, zsz269. [CrossRef]
13. Barone, D.A. Dream enactment behavior—A real nightmare: A review of post-traumatic stress disorder, REM sleep behavior disorder, and trauma-associated sleep disorder. *J. Clin. Sleep Med.* **2020**, *16*, 1943–1948.
14. Chouvarda, I.; Rosso, V.; Mendez, M.O.; Bianchi, A.M.; Parrino, L.; Grassi, A.; Terzano, M.; Cerutti, S. Assessment of the EEG complexity during activations from sleep. *Comput. Methods Programs Biomed.* **2011**, *104*, e16–e28. [CrossRef]
15. Kryger, M.H. *Atlas of Clinical Sleep Medicine: Expert Consult-Online*; Elsevier Health Sciences: Amsterdam, The Netherlands, 2022.
16. Nicolaou, N.; Georgiou, J. The Use of Permutation Entropy to Characterize Sleep Electroencephalograms. *Clin. Eeg Neurosci.* **2011**, *42*, 24–28. [CrossRef]
17. Bandt, C. A new kind of permutation entropy used to classify sleep stages from invisible EEG microstructure. *Entropy* **2017**, *19*, 197. [CrossRef]
18. Mateos, D.M.; Gómez-Ramírez, J.; Rosso, O.A. Using time causal quantifiers to characterize sleep stages. *Chaos Solitons Fractals* **2021**, *146*, 110798.

19. Abásolo, D.; James, C.; Hornero, R. Non-linear Analysis of Intracranial Electroencephalogram Recordings with Approximate Entropy and Lempel-Ziv Complexity for Epileptic Seizure Detection. In Proceedings of the 2007 29th Annual International Conference of the IEEE Engineering in Medicine and Biology Society, Lyon, France, 22–26 August 2007; IEEE Engineering in Medicine and Biology Society; pp. 1953–1956. [CrossRef]
20. Zhang, X.; Roy, R.; Jensen, E. EEG complexity as a measure of depth of anesthesia for patients. *IEEE Trans. Biomed. Eng.* **2002**, *48*, 1424–1433. [CrossRef]
21. Hu, J.; Gao, J.; Principe, J. Analysis of Biomedical Signals by the Lempel-Ziv Complexity: The Effect of Finite Data Size. *IEEE Trans. Biomed. Eng.* **2007**, *53*, 2606–2609. [CrossRef]
22. Rahman, M.M.; Bhuiyan, M.; Hassan, A. Sleep stage classification using single-channel EOG. *Comput. Biol. Med.* **2018**, *102*, 211–220. [CrossRef]
23. Zhang, Z.; Wei, S.; Zhu, G.; Liu, F.; Li, Y.; Dong, X.L.; Liu, C.; Liu, F. Efficient sleep classification based on entropy features and a support vector machine classifier. *Physiol. Meas.* **2018**, *39*, 115005. [CrossRef]
24. Kemp, B.; Zwinderman, A.; Tuk, B.; Kamphuisen, H.; Oberyé, J. Sleep-Edf Database Expanded. *Physionet Org*. 2018. Available online: https://physionet.org/content/sleep-edfx/1.0.0/ (accessed on 22 November 2024).
25. Chen, Z.; Ma, X.; Fu, J.; Li, Y. Ensemble Improved Permutation Entropy: A New Approach for Time Series Analysis. *Entropy* **2023**, *25*, 1175. [CrossRef]
26. Loh, H.W.; Ooi, C.P.; Vicnesh, J.; Oh, S.L.; Faust, O.; Gertych, A.; Acharya, U.R. Automated detection of sleep stages using deep learning techniques: A systematic review of the last decade (2010–2020). *Appl. Sci.* **2020**, *10*, 8963. [CrossRef]
27. Bandt, C.; Pompe, B. Permutation entropy: A natural complexity measure for time series. *Phys. Rev. Lett.* **2002**, *88*, 174102. [CrossRef]
28. Rosso, O.A.; Larrondo, H.; Martin, M.T.; Plastino, A.; Fuentes, M.A. Distinguishing noise from chaos. *Phys. Rev. Lett.* **2007**, *99*, 154102. [CrossRef]
29. Fadlallah, B.; Chen, B.; Keil, A.; Principe, J. Weighted-permutation entropy: A complexity measure for time series incorporating amplitude information. *Phys. Rev. Stat. Nonlinear Soft Matter Phys.* **2013**, *87*, 022911. [CrossRef]
30. Stosic, D.; Stosic, D.; Stosic, T.; Stosic, B. Generalized weighted permutation entropy. *Chaos Interdiscip. J. Nonlinear Sci.* **2022**, *32*, 103105. [CrossRef] [PubMed]
31. Pedregosa, F.; Varoquaux, G.; Gramfort, A.; Michel, V.; Thirion, B.; Grisel, O.; Blondel, M.; Prettenhofer, P.; Weiss, R.; Dubourg, V.; et al. Scikit-learn: Machine Learning in Python. *J. Mach. Learn. Res.* **2012**, *12*, 2825–2830.

entropy

MDPI

Article

Evaluating Methods for Detrending Time Series Using Ordinal Patterns, with an Application to Air Transport Delays

Felipe Olivares, F. Javier Marín-Rodríguez , Kishor Acharya and Massimiliano Zanin *

Instituto de Física Interdisciplinar y Sistemas Complejos (CSIC-UIB), Campus UIB, 07122 Palma, Spain; felipe@ifisc.uib-csic.es (F.O.); fco.javier.marin.rdz@gmail.com (F.J.M.-R.); kishor@ifisc.uib-csic.es (K.A.)
* Correspondence: massimiliano.zanin@gmail.com; Tel.: +34-971-259629

Abstract: Functional networks have become a standard tool for the analysis of complex systems, allowing the unveiling of their internal connectivity structure while only requiring the observation of the system's constituent dynamics. To obtain reliable results, one (often overlooked) prerequisite involves the stationarity of an analyzed time series, without which spurious functional connections may emerge. Here, we show how ordinal patterns and metrics derived from them can be used to assess the effectiveness of detrending methods. We apply this approach to data representing the evolution of delays in major European and US airports, and to synthetic versions of the same, obtaining operational conclusions about how these propagate in the two systems.

Keywords: time series; stationarity; functional complex networks; ordinal patterns; causality

1. Introduction

In the past few decades, functional complex networks have emerged as a powerful tool for the analysis of the dynamics of real-world systems. Many of these are complex, in the sense of being composed of multiple elements interacting in a non-linear way, such that the global dynamics cannot be inferred by the behavior of the individual constituents [1]. At the same time, these interactions are seldom observable; hence, the problem is how to describe the structure created by leveraging only limited information. The solution often entails reconstructing complex networks [2], where nodes represent the individual elements of the system and links between pairs of nodes are established whenever their dynamics are shown to be connected. In other words, the dynamics of the elements are assumed to be a function of these connections, and detecting the latter from the former becomes an inverse problem. This approach has been applied in multiple scientific fields, spanning from neuroscience [3–5] and genetics [6–8] to climate [9,10]. It has more recently been used to unveil the patterns behind the propagation of delays in air transport [11–13], a problem impacting the cost-efficiency and safety of the system, and with an important negative impact on the environment [14,15]. In general terms, the analysis is conducted by extracting time series representing the evolution of delays at each airport and then applying a statistical test evaluating the presence of correlations or causalities between pairs of time series. The results of these tests can also be represented as a complex network, allowing more detailed topological analyses [2,16,17].

One of the fundamental prerequisites for obtaining representative functional networks is the stationarity of the time series used in their reconstruction. From a statistical perspective, a time series is said to be stationary when its properties do not have an explicit dependence on time and thus has no predictable patterns—e.g., there must be no trends or seasonalities. Please note that any measure averaged over a non-stationary time series will not be accurate and that standard tests (including the t- and F-tests) and models (like the Auto Regression Moving Average, ARMA) may yield incorrect results. Intuitively, time series representing the evolution of delays at a given airport are expected to be highly non-stationary; delays are usually correlated with the traffic volume, such that they will be higher at noon and lower at night and early morning. This creates regular oscillations that can be misinterpreted as correlations, or even causalities, between pairs of time series.

To illustrate this latter point, the top left panel of Figure 1 reports the fraction of pairs of airports for which metrics detect a flow of information for airports in Europe and the US. Details about the data and the four metrics are presented in Section 2 below. Results indicate that the propagation of delays forms an all-to-all complete network. While not impossible, this result has been obtained using a raw time series of delays and may thus be the consequence of their nonstationarity. Interestingly, when applying two classical stationarity tests, i.e., the Kwiatkowski–Phillips–Schmidt–Shin (KPSS) [18] and the Augmented Dickey–Fuller [19] unit root tests, only the former detects the presence of regular trends. Additionally, adding a small amount of observational noise confounds both tests, while the four metrics still yield a large fraction of causality links—see top right panel of Figure 1.

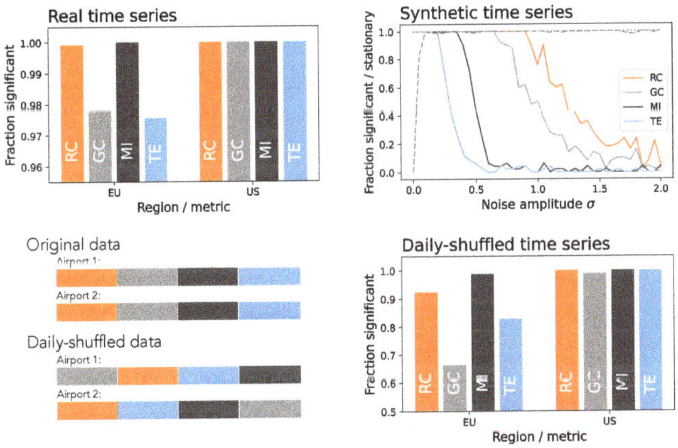

Figure 1. Nonstationarity and its consequences in the estimation of functional connectivity. (**Top left**) Fraction of pairs of time series for which functional connectivity is detected, for Europe (left bars) and the US (right bars), and four different connectivity metrics. See Section 2 for details on the data and methods. (**Top right**) Fraction of times a functional link is detected between the time series of Paris Charles de Gaulle and London Heathrow as a function of the amplitude of a Gaussian additive noise of zero mean and standard deviation σ. Please note that the amplitude is normalized, such that $\sigma = 1$ corresponds to the standard deviation of the time series. The dashed and dotted gray lines correspond to the fraction of times the time series are detected as stationary by, respectively, KPSS [18] and Augmented Dickey–Fuller [19] tests. (**Bottom left**) Graphical representation of the creation of daily shuffled surrogate time series; each color represents the data of one day. (**Bottom right**) Fraction of pairs of time series for which functional connectivity is detected when daily shuffled surrogates are used. The meaning of bars and colors is the same as in the top left panel. In all cases, RC: Rank Correlation; GC: Granger Causality; MI: Mutual Information; TE: Transfer Entropy.

In order to further validate the role of stationarity, we create surrogate time series by maintaining the same values observed within each day but randomly shuffling the days in each time series—see the bottom left panel of Figure 1 for a graphical representation. In these surrogate series, values at the same position will correspond to delays at the same time but on different days; hence, any detected relationship must be due to the trends within them and not to true propagation instances. As reported in the bottom right panel, many relationships are still detected. In short, ways to detrend these time series and to evaluate the final performance are imperative, in order to obtain reliable functional networks.

Looking at the results of Figure 1, one may be tempted to use these daily shuffled time series as a benchmark—i.e., if the number of detected links is low, it may be concluded that the detrending method is acceptable. This is, nevertheless, not a universal solution, as it depends both on the metric used to assess relationships and on the data. Again, referring to the bottom right panel of Figure 1, it can be appreciated that the Granger Causality (GC, light gray bars) yields substantially different results to the Mutual Information (MI, dark gray bars), even for the same data; and results are always worse for the US, probably due to the greater length of the time series. Consequently, when using GC on the European data, one may be misled to believe that the data are (at least partially) already detrended; but this would result in major biases when considering other metrics or datasets. In other words, obtaining few functional links with a given metric does not guarantee that the underlying time series are truly stationary; false connections may be found in different conditions.

In this contribution, we tackle the problem from an alternative perspective. Given a set of detrending methods, we ask whether independent metrics can be used to assess the number of temporal patterns in the detrended time series. The underlying hypothesis is that whenever these metrics detect no such patterns, the detrending process is effective, and hence, no spurious functional connections should be detected. Conversely, if patterns remain, the time series could not be assumed to have been effectively detrended, and the obtained functional connectivities should, therefore, not be trusted. Among the many possibilities available in the literature, we tackle this problem using ordinal patterns, i.e., a symbolic representation of time series values focusing on the relative amplitude of neighboring data points [20,21]. These patterns then allow the description of the serial dependence structure present in the studied time series in a simple way while providing advantages that include easy interpretability and low computational cost [22,23]. Two versions of the same are compared—respectively, a traditional (see Section 2.2.2) and a continuous one (Section 2.2.3), the latter being able to better incorporate amplitude information [24]. As previously hinted, we test this hypothesis using data representing the evolution of delays at major European and US airports (see Section 2.4) and synthetic versions of the same, with the aim of describing propagation patterns between them. The results presented in Section 3 depict a complex scenario in which detrending methods commonly used in other contexts here underperform. Conversely, when knowledge about the underlying system is included, highly stationary time series can be obtained, which can be used to extract interesting conclusions about the delay propagation process—as discussed in Section 4. In short, we here demonstrate that ordinal patterns-based metrics can provide a model-free methodology to evaluate and validate detrending algorithms.

2. Materials and Methods

In this section, we are going to introduce the methods and data considered in this work. Following the structure of the analysis, these will include the algorithms for detrending the data (see Section 2.1), the metrics to evaluate the correctness of the detrending process (Section 2.2), and the procedure for reconstructing functional connectivity (Section 2.3). We finally introduce the real air transport data on which the methodology will be tested in Section 2.4.

2.1. Detrending Methods

In order to remove the trends present in the considered time series, we here use some standard methods, both representing standard statistical approaches and developed using the knowledge of the problem at hand—specifically, the fact that delays have daily and weekly trends [25].

- Identity (*Ident*). As a reference, in the following analyses, we include the results corresponding to not performing any detrending on the time series.
- Delta (*Delta*). Simple detrending process based on evaluating the distance between the value at a given hour and the expected value observed for the same weekday and at the same hour: $\Delta x(t) = x(t) - \langle x(t + 24 \cdot 7 \cdot k) \rangle$, with $k \in \mathbb{Z}$ and $\langle \cdot \rangle$ representing the average.
- Independent Component Analysis (*ICA*). The time series are detrended by subtracting the three main components detected throughout all the airports by an Independent Component Analysis [26]. While computationally more costly than other solutions, it presents the advantage of being able to detect trends with variable periods, provided they are shared by multiple airports.
- Second Derivative (*SecD*). The second difference of the time series, i.e., $x''(t) = x(t) + x(t-2) - 2x(t-1)$. This approach is customary in the literature when no information about the nature of the underlying periodic trends is available.
- Z-Score by day (*ZScore24*). Detrending based on a Z-Score, defined as: $z(t) = [x(t) - \langle \tilde{x}(t) \rangle] / \sigma_{\tilde{x}(t)}$. $\tilde{x}(t)$ represents the set of values observed on different days at the same hour, i.e., $x(t + 7 \cdot k)$ with $k \in \mathbb{Z}$. In turn, $\langle \cdot \rangle$ and $\sigma_{\cdot}$ represent, respectively, the average and the standard deviation operators. The Z-Score encodes how much the observed value deviates from the expectation, in this case from the delay observed at the same hour on other days; but, as a difference with respect to the Delta approach, it takes into account the variability of the data.
- Z-Score by week (*ZScore724*). Same as the ZScore24, but taking as reference the delays observed at the same hour on the same day of the week. $\tilde{x}(t)$ is thus here defined as $x(t + 24 \cdot 7 \cdot k)$ with $k \in \mathbb{Z}$. ZScore724 should, therefore, also detrend with respect to weekly patterns, e.g., weekdays vs. weekends.

2.2. Evaluating the Detrending Process

As outlined in the introduction, the quantity of information contained in the time series is here assessed through two metrics, the JSD and the COP, both based on the application of the concept of ordinal patterns. In what follows, we first introduce the latter one and then define the two metrics.

2.2.1. Ordinal Patterns in Time Series

Bandt and Pompe introduced an encoding scheme that maps a raw time series onto a corresponding sequence of symbols called ordinal patterns [27]. This is simple, model-free, and with strong resilience to noise [28–30]. Several applications pioneered by Prof. Osvaldo Rosso have confirmed the success of this symbolization technique in the evaluation of information quantifiers [31–33]. The existence of forbidden [28,34] and missing ordinal patterns [35,36] have been applied to different areas from finance [36,37], semiconductor lasers [38], atmospheric turbulence [36], chaotic optoelectronic systems [39], and hydrology [36]. Furthermore, ordinal patterns have been used to quantify persistence [40], symmetry [40], irreversibility [41,42], and serial dependences [22,23] in time series.

Given a time series $X(t) = \{x_t; t = 1, \ldots, M\}$, sub-windows (or segments) of it, composed of D values, can be mapped into vectors. Specifically, D consecutive ($\tau = 1$)

or non-consecutive ($\tau > 1$) values starting at time t can be transformed into a vector $(x_t, x_{t+\tau}, \ldots, x_{t+(D-1)\tau})$ of dimension D. Each element of the previous vector is then replaced by its relative ranking, from zero for the smallest value up to $D-1$ for the largest one. Next, this vector is itself encoded as an ordinal pattern, representing the permutation π_i of $0, 1, \ldots, D-1$ describing the relative amplitude (strength) of each element in the considered sub-window. Finally, all ordinal patterns, calculated for all possible sub-windows, are synthesized into a probability distribution $\#(\pi_i)$, usually normalized by the total number of ordinal patterns $M-(D-1)\tau$. Please note that the condition $M \gg D!$ must be satisfied in order to obtain reliable statistics [27].

To illustrate how we build the ordinal pattern probability distribution in this study, let us consider an example comprising a synthetic hourly mean delay sequence over n days, as shown in the left panel of Figure 2. For a pattern length $D = 3$, we extract a set of ordinal patterns for each day, representing the ordinal information of the first D hours. After normalizing by n, a probability distribution is obtained. By repeating this procedure with sliding windows of length D and overlap of $D-1$, we can generate a set of 24 distributions representing the ordinal probabilities of each hour of the n days. Please note that for the 24th hour, we consider the 2 h of the following day. We follow the recipe suggested in [27] for breaking ties by adding a small amount of Gaussian noise (zero mean). This latter ingredient is important due to the inactivity periods, for which the hourly mean delay equals zero for several consecutive hours.

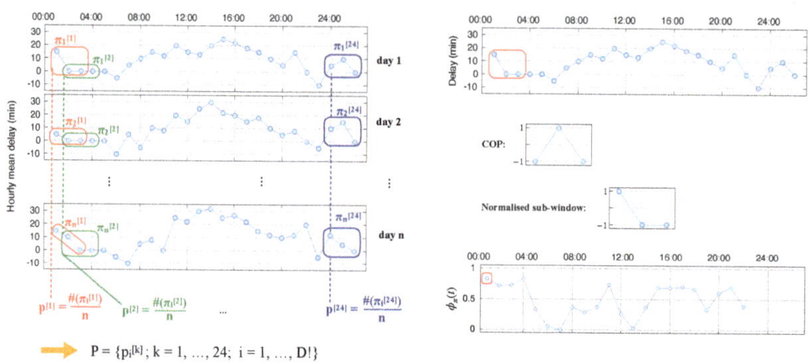

Figure 2. Graphical representations of the two methods for evaluating the detrending process. (**Left**) Illustration of building the ordinal pattern distribution over n days of hourly mean delay sequences. Red and green boxes indicate the data used to calculate the first two distributions. (**Right**) Calculation of the ϕ_π for a time series of delays using the COP methodology. See main text for details.

2.2.2. Jensen–Shannon Divergence of Ordinal Patterns

The permutation Jensen–Shannon Distance (JSD in short) of ordinal patterns was recently introduced as a versatile metric for assessing the degree of similarity between the symbolic ordinal sequence statistics of two time series [43]. This metric relies on evaluating the similarity between the ordinal pattern probability distributions $P = \{p_1, \ldots, p_N\}$ and $Q = \{q_1, \ldots, q_N\}$ associated with the two time series under analysis, using the Jensen–Shannon divergence [44], providing a quantitative measure of their statistical resemblance,

$$D_{JS}(P, Q) = S((P+Q)/2) - S(P)/2 - S(Q)/2, \quad (1)$$

where $S(P) = \sum_{i=1}^{N} p_i \ln p_i$, is the classical Shannon entropy. The JSD is hence obtained by calculating the square root of Equation (1), and its normalized version reads

$$\mathrm{JSD}(P,Q) = \sqrt{\frac{D_{JS}(P,Q)}{\ln 2}}. \tag{2}$$

Higher values of the JSD suggest greater dissimilarity between the symbolic representations of two time series, while lower values indicate closer similarity. Intuitively, signals originating from the same underlying dynamics are expected to yield small ordinal distance values approaching zero but not exactly zero due to finite-size effects [43]. Additionally, it allows for easy hypothesis testing regarding the dynamical nature of any given time series by calculating its JSD relative to reference time series generated in alignment with a null model or surrogate sequences [43,45].

In the present analysis, we calculate the JSD between each of the 24 ordinal pattern distributions (one per hour, as illustrated in Figure 2) and their shuffled surrogates. These randomized sequences are constrained realizations that satisfy the null hypothesis, i.e., stationary and absence of temporal correlations across days. Moreover, in such a way, finite-size and amplitude distribution effects are taken into account. Finally, to obtain one value for each airport we average the JSD over the 24 h.

2.2.3. Continuous Ordinal Patterns

Continuous Ordinal Patterns [24] (COPs) are a recently proposed modification of the original idea of ordinal patterns previously described, in which the output, instead of being discrete (i.e., a single ordinal pattern), is continuous [24]. Specifically, given a COP and a sub-window of the time series, this approach entails calculating the distance between both. When compared to classical ordinal patterns, this yields the advantage of naturally including the (local) amplitude information of the time series under analysis. On the other hand, the pattern moves from being an output of the analysis to being an input of the same. A brief discussion of the COP methodology is reported here for the sake of completeness; the interested reader can find additional details in [24].

We start by defining a continuous ordinal pattern π as a set of D values $\pi = (\pi_0, \pi_1, \ldots, \pi_{D-1})$ normalized in the range $[-1,1]$. D is here the embedding dimension, and it has the same meaning as the embedding dimension of ordinal patterns. We further denote by s^* the sub-window of length D of the original time series X that is currently under analysis; note that this segment is also normalized in the range $[-1,1]$. Given both π and s^*, we can then define a distance ϕ_π assessing how well the former represents the evolution of the data in the latter:

$$\phi_\pi = \frac{1}{2D}\sum_{i=1}^{D} d_i = \frac{1}{2D}\sum_{i=1}^{D} |\pi_i - s_i^*|. \tag{3}$$

Here, the subscript i denotes the ith element. Please note that $\phi_\pi = 0$ implies that s^* and π are exactly equal, or, in other words, that the pattern π is a perfect representation of the dynamics within the sub-window s^*. On the other hand, values of ϕ_π close to one imply that s^* and π have substantially different dynamics. $1 - \phi_\pi$ is then a metric assessing how well the pattern π represents the dynamics of the time series, and this can further be averaged over all possible segments of length D of the analyzed time series to quantify how important π is to understand its dynamics. Hence, the larger the value of $1 - \phi_\pi$ (and the smaller is ϕ_π), the more the pattern π is present in the data.

Please note that, up to this point, the COP π has been given as an input of the analysis; it is then necessary to define a way of finding the best π for the problem at hand. In this study, we resort to a simple strategy previously proposed in Refs. [24,46], and based on testing a large set (here, 250) of random patterns π. For each test, the distance between the values ϕ_π obtained in the original time series and those in a randomly shuffled version of the same is estimated through a Kolmogorov–Smirnov two-sample test. Finally, the π yielding the largest difference, and thus the pattern according to which the time series under analysis is farther away from a random sequence, is retained.

In synthesis, given a single time series X, the full process involves: (i) finding a COP of dimension D for which the distance between X and a randomly shuffled version of it is maximal by testing multiple random COPs; (ii) calculate the median of ϕ_π using such COP; finally, (iii) the maximum ϕ_π obtained across the multiple COPs (in what follows denoted as COP for simplicity) is understood as an index for the presence of some non-trivial structures in the time series. The whole process is illustrated in the right panel of Figure 2, representing, from top to bottom: the original time series X; an arbitrary COP of size $D = 3$; the extraction and normalization of the first sub-window, also of length $D = 3$, and corresponding to the red box; and finally, the calculation of the complete ϕ_π, in which the first value corresponds to the previous sub-window.

2.2.4. Metric Normalization

In the previous sections, we have defined two ways of detecting the presence of patterns in time series and, hence, whether they can be considered random. Understanding these metrics entails an additional challenge: while they measure a distance from randomness, a completely random time series would also give positive distances due to finite-size effects. In order to solve this, both the JSD and the COP calculated from a given time series are here normalized according to the average values obtained by the same metrics in a large ensemble of surrogate (randomly shuffled) time series. In other words, the original time series are randomly shuffled 50 times; the JSD and the COP are calculated for each one of these surrogates and averaged, respectively, obtaining $\langle JSD_{rnd} \rangle$ and the $\langle COP_{rnd} \rangle$; finally, normalized versions of the metrics are defined as $JSD^* = JSD / \langle JSD_{rnd} \rangle$ and $COP^* = COP / \langle COP_{rnd} \rangle$. Those normalized values will have an expected value of 1 in the case of random time series and greater than 1 whenever non-random structures are present.

To further validate the obtained results, we have also considered an alternative normalization method based on using both the average and the standard deviation of the values obtained in the surrogate time series to define a Z-Score; in what follows, these two will be, respectively, denoted as JSD^Z and COP^Z. This provides a more precise quantification of the distance from randomness, especially in the case of short time series, for which the variability both in the ordinal pattern frequencies and in the COP is higher. The length of the time series considered in this study (see Section 2.4) is large enough to not impact the results, and the relationship between JSD^* and COP^* and their Z-Score counterparts is almost linear. Still, for the sake of generalizability, both normalizations will be evaluated below.

2.3. *Assessing Functional Connectivity*

Once the time series have been detrended, the next step in the analysis involves the detection of functional connectivity between pairs of them or, in the context of the present study, the detection of instances of delay propagation. Thus, given two time series X and Y

for which we want to detect connectivity $X \to Y$, the analysis is conducted by applying the following four functional tests:

- Rank Correlation (RC). Spearman's Rank Correlation between the two analyzed time series, calculated over shifted time series $x(t), y(t+\lambda)$, with $\lambda \in \{0, 1, \ldots, 5\}$, to account for the time required by delays to propagate. The λ yielding the lower p-value is the one selected.
- Granger Causality (GC). The GC [47] is one of the best-known exponents of *predictive causality* [48] and assesses whether the inclusion of information about the driving element X helps predict the future dynamics of the driven element Y. As originally proposed, an autoregressive-moving-average (ARMA) model is used for the prediction. Two variants are constructed, forecasting Y by, respectively, introducing or not data about X's past. Finally, the two models' residuals are compared through an F-test, yielding a p-value indicating whether the presence of information about X is relevant—and, hence, whether a causality relationship is present.
- Mutual Information (MI). MI is an information-theoretic measure that captures the shared amount of information between any two random variables. The Shannon information for X and Y, respectively denoted as $H(X)$ and $H(Y)$, represent the corresponding amount of potential information or the degree of uncertainty [49]. MI quantifies how much of the uncertainty in Y is reduced or explained after knowing the full information of X, i.e.,

$$I(X:Y) = \sum_{y}\sum_{x} p(x,y) \log\left(\frac{p(x,y)}{p(x)p(y)}\right), \tag{4}$$

  where $p(x,y)$ represents the joint probability distribution, while $p(x)$ and $p(y)$ denote the marginal probability distributions. MI is particularly interesting for investigating the relationship between two variables because, in contrast to cross-correlation, it is sensitive to non-linear dependencies. In this work, MI is estimated using the Kozachenko–Stögbauer–Grassberger (KSG) method [50], which is well suited for continuous random variables following non-parametric (or unknown) distributions.
- Transfer Entropy (TE). TE is also an information-theoretic measure that captures the amount of directional information flow from a source variable (e.g., X) to a target variable (e.g., Y) [51]. It is the measure of the amount of information contained in the past states of a source process (i.e., X^-) about the future state of the target process (i.e., Y) given that the past states of the target (i.e., Y^-) are known:

$$TE_{X\to Y} = \sum p(y_n, y_n^-, x_n^-) \log\left(\frac{p(y_n|x_n^-, y_n^-)}{p(y_n|y_n^-)}\right). \tag{5}$$

  TE is a model-free directional measure that captures the casual relationship in Wiener sense between two variables [52]. The KSG algorithm developed for MI has been adopted to compute the TE [53], and the same adaptation has been used in this work.

Please note that the result of applying these four tests is a p-value—in the case of the MI and the TE, this is obtained by comparing the measured information to what is obtained in a set of 10^3 surrogate (randomly shuffled) time series. As a final step, the presence of a propagation link between two airports is accepted whenever such p-value is below a significance threshold $\alpha = 0.01$.

2.4. Data on Airport Dynamics

In order to validate the analysis here proposed on a real-world scenario, we consider two complementary data sets describing the hourly evolution of delays in the top-50 airports in Europe and the US. Information has, respectively, been obtained from EUROCONTROL's R&D Data Archive, a public repository of European historical flights made available for research purposes and freely accessible at https://www.eurocontrol.int/dashboard/rnd-data-archive, (accessed on 17 January 2024); and from the Reporting Carrier On-Time Performance database of the Bureau of Transportation Statistics, U.S. Department of Transportation, freely accessible at https://www.transtats.bts.gov, (accessed on 17 January 2024). Please note that both data sets have different temporal scopes due to limitations at source: while data are limited for the EU to four months (i.e., March, June, September, and December) of five years (2015–2019), the US dataset includes all months for the same five years. A list of the airports included in the study, alongside some basic statistics on operations, is reported in Tables A1 and A2. From these raw data, arrival delay time series have been extracted, calculated as the difference between the actual and scheduled landing times of each flight, and averaged at each destination airport and each hour of the day.

It is interesting to note that the two considered data sets are highly heterogeneous. Firstly, as already discussed, the European one does not cover a continuous span of time. Secondly, flights and their associated delays are reported differently [54]: in the US case, flights correspond to those operated by certified US air carriers accounting for at least one percent of domestic scheduled passenger revenues (as opposed to all flights of the European case), and their associated delays are as reported by the airline (as opposed to by a central organization, i.e., EUROCONTROL in the European case).

3. Results

We start the analysis of the results by evaluating the fraction of functional links detected in the daily shuffled surrogate time series as a function of the two evaluation metrics and under different procedures for detrending. Please note that, due to the use of surrogate time series, no functional connectivity can be present; hence, those links that are detected represent how much the functional metrics are misled by the residual nonstationarity of the time series—this fraction is therefore called "confusion" in what follows.

The top left panel of Figure 3 reports the evolution of the confusion as a function of the JSD^*, in a way that will be common to all panels in subsequent figures, e.g., Figures 3 and 4. Specifically, each point represents the result obtained using a combination of detrending strategies and functional connectivity metrics; the former is indicated by the shape of the marker, and the latter by its color—see legends in the bottom part. Please note that, according to the initial hypothesis of this work, the more a time series is detrended, the lower the quantity of information in them detected by the JSD^*, and, conversely, the lower should be the fraction of times that the functional connectivity metrics become confused by the nonstationarity. In other words, one would expect a positive correlation between the two, or at least no points corresponding to a low JSD^* and large confusion. As can be appreciated in Figure 3, this is not the case, with many points being located in the top left quadrant. In other words, even when the permutation pattern-based metric identifies no clear structures in the data (note the minimum $JSD^* \approx 1.5$), these still have residual nonstationarity that results in spurious causality relations.

The reason for this negative result is easily identifiable by recalling that permutation patterns are designed to assess the presence of temporal structures in the data at the cost of disregarding the amplitude of the same. To illustrate, the top right panel of Figure 3 reports the same results when the original time series of delays are daily normalized using a Z-Score; in other words, the segment of 24 h corresponding to one day and airport is transformed to have an average of zero and a standard deviation of one. The same results as a function of JSD^Z, divided according to European and US airports, are reported in the bottom panels of the same figure. An S-shaped curve readily emerged, suggesting that most of the seasonalities are lost using this second normalization process. However, it is worth noting that this assessment is conducted by relying on the detected functional connectivity, i.e., a posteriori.

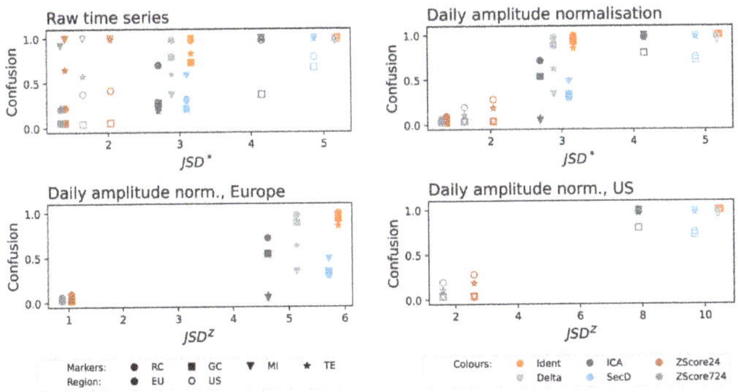

Figure 3. Confusion, i.e., fraction of times a functional relationship is detected between two airports when data are daily shuffled, as a function of the JSD^* of the detrended time series. (**Top left and right**) panels respectively correspond to the original time series, and thereto daily normalized in amplitude using a Z-Score—see main text for details. Bottom panels further report the results as a function of the JSD^Z for the daily normalized time series, distinguishing between Europe (**bottom left**) and the US (**bottom right**). Shape and colors of the markers respectively indicate the functional metric and the detrending procedure; solid and empty markers further correspond to EU and US, see bottom legends.

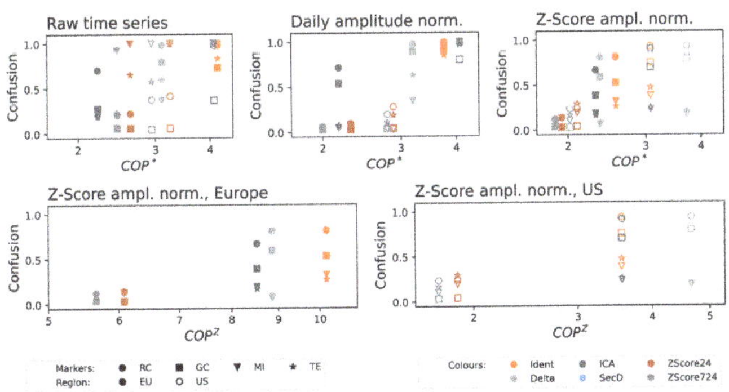

Figure 4. Confusion, i.e., fraction of times a functional relationship is detected between two airports when data are daily shuffled, as a function of the COP^* of the detrended time series. **Top** panels respectively correspond to the original time series, and daily normalized in amplitude using a Z-Score, and normalized across airports—see main text for details. **Bottom** panels further report the results as a function of the COP^Z for the Z-Score normalized time series, distinguishing between Europe (bottom left) and the US (bottom right). Shape and colors of the markers respectively indicate the functional metric and the detrending procedure; solid and empty markers further correspond to EU and US, see bottom legends. Please note that points corresponding to the SecD (light blue) are not visible due to this procedure yielding values of $COP^* > 8$ and $COP^Z > 15$.

A solution to this problem may come from the use of a modified version of the permutation pattern paradigm, taking into account the local amplitude structure, i.e., the concept of COP. Figure 4 then reports the results obtained as a function of COP^*, in a way similar to Figure 3. Specifically, the top left panel reports the confusion as a function of the COP^* when networks are reconstructed with the raw time series; the value of COP^* remains high, suggesting the presence of residual patterns. The same metric drops in the two following panels. Specifically, the top middle one corresponds to the daily amplitude normalization, i.e., the same as the right panel of Figure 3. Next, the top right panel corresponds to a normalization in which the value of the delay observed at one airport at one given hour is normalized, using a Z-Score, against the set of delays in all other airports at the same time. Please note that this second normalization accounts for the amplitude of delays across the whole system, as opposed to one single airport. As can be seen, in both cases, the value of the COP^*, at least for the Z-Score detrending methods (brown and gray markers), drops close to or below 2.0; the residual number of functional relationships is also significantly reduced. Results for the latter normalization as a function of COP^Z, divided according to European and US airports, are reported in the bottom panels of the same figure.

In order to confirm the validity of these results, we tested the previous insights using a synthetic dataset constructed by following the expected behavior of delays. We specifically started by calculating the average evolution of the hourly delays at London Heathrow across September 2019; see the solid black line in the left panel of Figure 5. Starting from this, a synthetic time series was generated by adding a noise, drawn from a distribution $\mathcal{N}(0.0, 420.0)$, that repeated every seven days and an additional noise drawn from a distribution $\mathcal{N}(0.0, 140.0)$. In other words, the time series of an airport is given by a repeated daily pattern, modulated by a weekly noise of $\approx 1/3$ of the total signal amplitude, and with an additional random component of amplitude $\approx 1/3$ of the previous one. A graphical representation of seven days is included in the left panel of Figure 5; see the thin colored lines. Ten time series, representing the dynamics of ten airports across 30 days, are finally created, each one including a random time shift—to simulate, e.g., the fact that some airports may have their peak time at different hours of the day. Please note that these time series are highly non-stationary (as, indeed, the main component repeats every day) and that, therefore, a functional connection is detected between them if not properly pre-processed. When the same detrending processes are applied to these time series, the results are as expected, see right panel of Figure 5; specifically, the Z-Score$_{724}$ and the Z-Score$_{24}$ are the two best approaches, in agreement with the way data were synthesized. Additionally, the low values of confusion confirm that no false positives are generated.

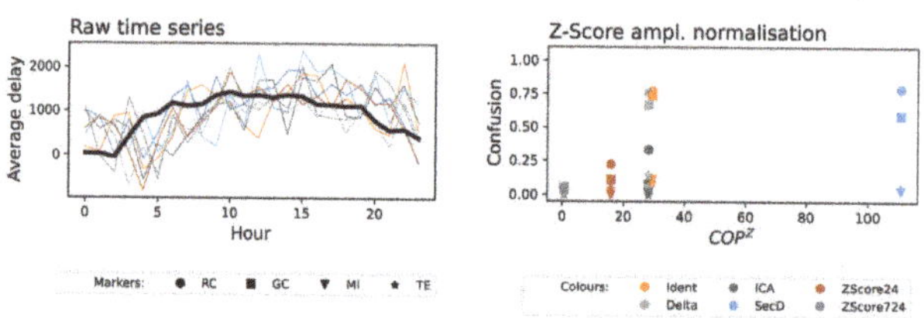

Figure 5. Analysis of synthetic data. The **left** panel reports a representation of the global average delay profile (thick black line) across the 24 h of the day and of one realization of the delay for the seven days of the week (thin colored lines). The **right** panel reports the confusion as a function of the COP^Z on the synthetic time series. See the main text for details on the generation process. Shape and colors of the markers respectively indicate the functional metric and the detrending procedure.

Given the previous results, one may be tempted to conclude that the solution is clear: the two variants of the Z-Score detrending, in conjunction with the Z-Score amplitude normalization, yield time series with low enough trends. This is, nevertheless, not the full picture, and two additional and complementary aspects must be analyzed: is a $COP^* < 2$ low enough? As discussed in Section 2.2.4, this value is equivalent to a Z-Score of ≈ 6, and hence to the presence of structures that are highly statistically significant. Conversely, does this detrending/normalization preserve enough information about the real functional relationships?

In order to answer these two questions, we calculated the number of functional links obtained when analyzing the real data, i.e., without any daily shuffling, for the four functional metrics, and using the combination of Z-Score$_{724}$ detrending and Z-Score amplitude normalization. Results, reported in the left and central panel of Figure 6, indicate that a large number of functional links are still detected. The case of the US (central panel) is interesting in that a high functional link density is obtained, much higher than in the EU case, with MI even yielding a fully connected network—which may indicate that not all trends are actually deleted. This nevertheless seems to be partly due to the much larger quantity of data available for the US; as described in Section 2.4, these include all months of the five considered years, as opposed to four months per year of the European dataset. When the time series are pruned to match the other dataset's days, the link density observed for the US substantially drops—see black bars in the central panel of Figure 6. This is, of course, to be expected, as the lower the quantity of available data, the more difficult it is to detect functional relationships, and hence, the lower the obtained link density—see also the right panel of the same figure.

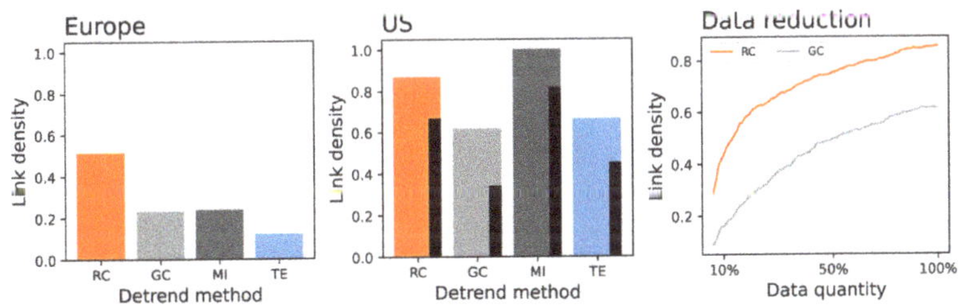

Figure 6. Real functional connectivity. (**Left and center**) Fraction of pairs of time series for which functional connectivity is detected, for Europe (**left panel**) and the US (**central panel**), and four different connectivity metrics. In the latter case, black bars report the link density obtained when US time series are pruned to match European ones. (**Right**) Evolution of the detected link density in the US system as a function of the percentage of data used in the evaluation. Original time series have been processed using a combination of Z-Score$_{724}$ detrending and Z-Score amplitude normalization.

4. Discussion and Conclusions

This contribution tackled two complementary questions: whether ordinal patterns can be used to evaluate the stationarity of time series for then using these to assess the suitability of standard detrending methods in the context of air transport delay data. The discussion of the results must, therefore, be addressed from this two-fold perspective.

On the one hand, ordinal patterns seem to be a good quantifier of the quantity of temporal structure in the data and, hence, their nonstationarity. These work better than classical statistical tests [18,19] and may be the solution to some known problems they pose [55]. As frequently discussed in the literature, the original version of this metric disregards the amplitude of the time series and, therefore, cannot detect non-stationarities that are amplitude-based—a problem well-known in the literature [56,57]. This can be improved by resorting to modified versions of the same that include local or global amplitude information. We have here considered Continuous Ordinal Patterns—see Figure 4; the reader should nevertheless take into account that other alternatives exist, including weighted permutation patterns [58], generalized patterns [59], or slope entropy [60]. In synthesis, given a detrending method, this can easily be evaluated using an ordinal-based approach, and this conclusion is of general applicability beyond the specific dataset considered here.

On the other hand, we have evaluated a suite of detrending methods commonly used to reconstruct functional networks representing the propagation of delays in air transport; and what obtained depicts a complex scenario. First, a detrending method alone is not enough to obtain stationary data—see left panels of Figures 3 and 4. This can be explained by taking into account the presence of confounding factors that may affect the observed values, both within the same day and across multiple airports of the system. To illustrate, suppose a day with adverse weather throughout the continent. Delays will be higher than expected, both in a given airport for multiple hours (possibly the whole day) and across all the airports of the affected region. A detrend method alone cannot account for this, and its use would result in many functional links that do not necessarily correspond to true propagations. This problem is easily solved by incorporating an additional normalization of the values, either through time (right panel of Figure 3 and central panel of Figure 4) or through airports (right panel of Figure 4). In other words, the problem at hand requires a multi-variate style detrending, something not common in the literature [61]. It is worth noting that, to the best of our knowledge, the proposed normalizations have hitherto never been applied in this context.

On a negative note, we also observed that the use of this combination of detrending/normalization methods is not enough to ensure a complete filtering of the time series. Specifically, the COP^*, while lower than in the raw time series, is still quite high—see Figures 4 and 7; and this results in a link density > 0.1 for daily shuffled time series, and suspiciously high connectivity for the raw data—see Figure 6. This is more prominent in the case of the US. Beyond the availability of more data, we hypothesize that this is a consequence of the larger geographical dispersion of this air transport system and of the presence of multiple time zones, which may hinder the Z-Score normalization across airports. Thus, when considering the problem of air transport delay propagations, the conclusion that can here be drawn is that better (possibly tailored) methods for data detrending are needed.

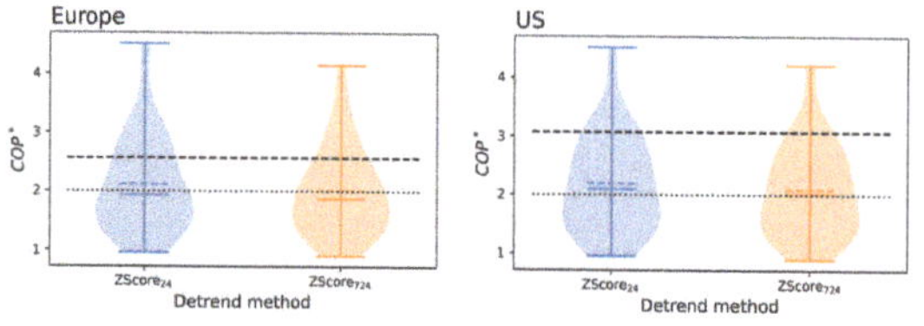

Figure 7. Distribution of the COP^* of time series for all airports in Europe (**left panel**) and US (**right panel**), when these are detrended using Z-Score$_{24}$ (left graph, blue) and Z-Score$_{724}$ (right graph, orange) methods, and with a Z-Score amplitude normalization. Within each distribution, the middle horizontal lines indicate the median (solid line) and the mean (dashed line). The dashed black horizontal lines indicate the median value observed in the raw data, i.e., before any detrending/normalization. Finally, the dotted lines correspond to a $COP^* = 2$ and are included as a visual reference.

It is worth noting that some detrending approaches, which are common in the literature, yield results that are actually worse than performing no pre-processing—see, for instance, the cases of SecD in Figure 3 and ICA in Figure 4, respectively yielding higher JSD^* and COP^*. This may be due to underlying hypotheses that are not fulfilled by the system under analysis. To illustrate, ICA expects components in the signals to be synchronized across all of them [26]; yet, rush hours at airports may be different, especially when these are located in different time zones.

As a final point, some interesting operational conclusions can be drawn from Figure 6. Even when pruning the time series to include the same number of days, higher link densities are obtained for the US; in addition, this difference is especially notable for MI and TE. This indicates that the US air transportation system is more prone to propagate delays, possibly in a more non-linear and complex way. Please note that this may be caused by multiple factors. To illustrate, the EU strongly relies on strategic Air Traffic Flow Management (ATFM) regulations, which aim at avoiding airborne holdings, while the US equivalent Ground Delay Program (GDP) works on shorter time scales [62]. Additionally, delays are reported in slightly different ways, as these are calculated in Europe using the last-filed flight plans and thus already contain strategic ATFM adjustments [54]. Lastly, even the origin of the data may have an impact: while in Europe, delays are calculated by a central authority, in the US, they are reported by individual airlines, which may follow different standards for their estimation.

In conclusion, the detrending of real-world time series has been shown to be a complex problem in which the idiosyncrasies of the studied system have a non-negligible impact on the validity of the process. Ordinal patterns-based metrics can nevertheless be used to validate the detrending process by quantifying the remaining amount of temporal structures.

Author Contributions: Conceptualization, M.Z.; methodology, F.O., F.J.M.-R., K.A. and M.Z.; formal analysis, F.O., F.J.M.-R., K.A. and M.Z.; writing—original draft preparation, F.O., F.J.M.-R., K.A. and M.Z.; writing—review and editing, F.O., F.J.M.-R., K.A. and M.Z. All authors have read and agreed to the published version of the manuscript.

Funding: This project has received funding from the European Research Council (ERC) under the European Union's Horizon 2020 research and innovation program (grant agreement No. 851255). This work was partially supported by the María de Maeztu project CEX2021-001164-M, funded by the MICIU/AEI/10.13039/501100011033.

Institutional Review Board Statement: Not applicable.

Data Availability Statement: No new data were created or analyzed in this study. The original data analyzed in the study are openly available at https://www.eurocontrol.int/dashboard/rnd-data-archive (accessed on 13 July 2023) and https://www.transtats.bts.gov (accessed on 24 February 2024).

Conflicts of Interest: The authors declare no conflicts of interest.

Appendix A

Table A1. Information about European airports. As well as ICAO codes and official names, the table includes the average number of hourly departures and arrivals using the operations available in the dataset.

Rank	Code	Name	Departures	Arrivals
1	EGLL	Heathrow Airport	26.39	26.41
2	LFPG	Charles de Gaulle Airport	26.97	26.91
3	EHAM	Amsterdam Airport Schiphol	27.37	27.31
4	EDDF	Frankfurt am Main Airport	26.92	26.89
5	LEMD	Adolfo Suárez Madrid-Barajas Airport	21.93	21.90
6	LEBL	Josep Tarradellas Barcelona-El Prat Airport	17.77	17.76
7	EDDM	Munich Airport	22.13	22.14
8	EGKK	Gatwick Airport	15.83	15.79
9	LIRF	Leonardo da Vinci-Fiumicino Airport	17.17	17.17
10	LFPO	Orly Airport	12.96	12.94
11	EIDW	Dublin Airport	12.24	12.22
12	LSZH	Zurich Airport	14.15	14.11
13	EKCH	Copenhagen Airport	14.62	14.64

Table A1. *Cont.*

Rank	Code	Name	Departures	Arrivals
14	LEPA	Palma de Mallorca Airport	10.97	10.99
15	LPPT	Lisbon Airport	11.00	11.01
16	ENGM	Oslo Airport, Gardermoen	13.75	13.75
17	EGCC	Manchester Airport	10.84	10.82
18	EGSS	London Stansted Airport	10.01	10.00
19	LOWW	Vienna International Airport	13.91	13.89
20	ESSA	Stockholm Arlanda Airport	13.31	13.31
21	EBBR	Brussels Airport	12.54	12.51
22	LIMC	Malpensa Airport	10.49	10.49
23	EDDL	Düsseldorf Airport	11.96	11.94
24	LGAV	Athens International Airport	10.62	10.63
25	EDDT	Berlin Tegel Airport	10.18	10.17
26	LEMG	Málaga Airport	6.82	6.82
27	EPWA	Warsaw Chopin Airport	9.21	9.21
28	LSGG	Geneva Airport	9.63	9.63
29	EDDH	Hamburg Airport	8.16	8.16
30	LKPR	Václav Havel Airport Prague	7.80	7.81
31	EGGW	Luton Airport	6.84	6.85
32	LHBP	Budapest Ferenc Liszt International Airport	5.78	5.77
33	EGPH	Edinburgh Airport	6.87	6.86
34	LEAL	Alicante-Elche Airport	4.94	4.94
35	LFMN	Nice Côte d'Azur Airport	7.22	7.23
36	LROP	Henri Coanda International Airport	6.20	6.19
37	EDDK	Cologne Bonn Airport	7.27	7.26
38	LIME	Orio al Serio International Airport	4.69	4.68
39	UKBB	Boryspil International Airport	4.88	4.89
40	EGBB	Birmingham Airport	6.01	6.00
41	LPPR	Porto Airport	4.73	4.73
42	EDDS	Stuttgart Airport	6.45	6.45
43	LIPZ	Venice Marco Polo Airport	4.99	4.99
44	LFLL	Lyon-Saint-Exupéry Airport	6.30	6.30
45	LICC	Catania-Fontanarossa Airport	3.66	3.66
46	LIRN	Naples Airport	3.85	3.85
47	EGPF	Glasgow Airport	4.86	4.84
48	LFBO	Toulouse-Blagnac Airport	5.14	5.11
49	LFML	Marseille Provence Airport	5.21	5.20
50	LIML	Linate Airport	5.88	5.88

Table A2. Information about European airports. As well as IATA codes and official names, the table includes the average number of hourly departures and arrivals using the operations available in the dataset.

Rank	Code	Name	Title 3	Rank
1	ATL	Hartsfield-Jackson Atlanta International Airport	41.63	41.61
2	DEN	Denver International Airport	25.41	25.38
3	DFW	Dallas Fort Worth International Airport	25.42	25.35
4	LAX	Los Angeles International Airport	23.76	23.77
5	ORD	Chicago O'Hare International Airport	31.42	31.37
6	PHX	Phoenix Sky Harbor International Airport	17.81	17.79
7	MSP	Minneapolis-Saint Paul International Airport	14.95	14.95

Table A2. *Cont.*

Rank	Code	Name	Title 3	Rank
8	CLT	Charlotte Douglas International Airport	16.17	16.14
9	SEA	Seattle-Tacoma International Airport	14.88	14.88
10	SFO	San Francisco International Airport	18.77	18.77
11	JFK	John F. Kennedy International Airport	11.69	11.68
12	IAH	George Bush Intercontinental Airport	17.04	17.00
13	MCO	Orlando International Airport	14.51	14.51
14	EWR	Newark Liberty International Airport	13.48	13.46
15	LAS	Harry Reid International Airport	17.10	17.12
16	FLL	Fort Lauderdale-Hollywood International Airport	9.87	9.87
17	BOS	General Edward Lawrence Logan International Airport	14.34	14.36
18	DTW	Detroit Metropolitan Wayne County Airport	14.45	14.47
19	MIA	Miami International Airport	8.31	8.31
20	LGA	LaGuardia Airport	12.84	12.80
21	IAD	Washington Dulles International Airport	5.37	5.37
22	BWI	Baltimore/Washington International Thurgood Marshall Airport	10.96	10.96
23	PHL	Philadelphia International Airport	9.53	9.53
24	SAN	San Diego International Airport	9.28	9.28
25	MDW	Chicago Midway International Airport	9.45	9.43
26	SLC	Salt Lake City International Airport	11.91	11.93
27	DCA	Ronald Reagan Washington National Airport	10.32	10.32
28	TPA	Tampa International Airport	7.89	7.90
29	PDX	Portland International Airport	6.58	6.59
30	STL	St. Louis Lambert International Airport	6.47	6.48
31	BNA	Nashville International Airport	6.78	6.79
32	AUS	Austin-Bergstrom International Airport	6.04	6.05
33	HNL	Daniel K. Inouye International Airport	5.44	5.44
34	SJC	San José International Airport	5.50	5.50
35	MCI	Kansas City International Airport	5.24	5.25
36	DAL	Dallas Love Field	7.65	7.63
37	SMF	Sacramento International Airport	4.97	4.98
38	MSY	Louis Armstrong New Orleans International Airport	5.40	5.40
39	SNA	John Wayne Airport	4.56	4.57
40	RDU	Raleigh-Durham International Airport	4.88	4.89
41	RSW	Southwest Florida International Airport	3.42	3.42
42	PIT	Pittsburgh International Airport	3.78	3.79
43	HOU	William P. Hobby Airport	6.17	6.16
44	IND	Indianapolis International Airport	3.75	3.76
45	SAT	San Antonio International Airport	3.87	3.88
46	SJU	Luis Mu noz Marín International Airport	2.84	2.85
47	CLE	Cleveland Hopkins International Airport	4.36	4.37
48	OAK	Oakland International Airport	5.54	5.53
49	CVG	Cincinnati/Northern Kentucky International Airport	3.10	3.11
50	CMH	John Glenn Columbus International Airport	3.44	3.45

References

1. Anderson, P.W. More Is Different: Broken symmetry and the nature of the hierarchical structure of science. *Science* **1972**, *177*, 393–396. [CrossRef] [PubMed]
2. Strogatz, S.H. Exploring complex networks. *Nature* **2001**, *410*, 268–276. [CrossRef] [PubMed]
3. Bullmore, E.; Sporns, O. Complex brain networks: Graph theoretical analysis of structural and functional systems. *Nat. Rev. Neurosci.* **2009**, *10*, 186–198. [CrossRef] [PubMed]
4. Sporns, O. Structure and function of complex brain networks. *Dialogues Clin. Neurosci.* **2013**, *15*, 247–262. [CrossRef]
5. Park, H.J.; Friston, K. Structural and functional brain networks: From connections to cognition. *Science* **2013**, *342*, 1238411. [CrossRef]

6. Lee, I.; Date, S.V.; Adai, A.T.; Marcotte, E.M. A probabilistic functional network of yeast genes. *Science* **2004**, *306*, 1555–1558. [CrossRef]
7. Lezon, T.R.; Banavar, J.R.; Cieplak, M.; Maritan, A.; Fedoroff, N.V. Using the principle of entropy maximization to infer genetic interaction networks from gene expression patterns. *Proc. Natl. Acad. Sci. USA* **2006**, *103*, 19033–19038. [CrossRef]
8. Boone, C.; Bussey, H.; Andrews, B.J. Exploring genetic interactions and networks with yeast. *Nat. Rev. Genet.* **2007**, *8*, 437–449. [CrossRef]
9. Tsonis, A.A.; Roebber, P.J. The architecture of the climate network. *Phys. A Stat. Mech. Its Appl.* **2004**, *333*, 497–504. [CrossRef]
10. Donges, J.F.; Zou, Y.; Marwan, N.; Kurths, J. The backbone of the climate network. *Europhys. Lett.* **2009**, *87*, 48007. [CrossRef]
11. Zanin, M.; Belkoura, S.; Zhu, Y. Network analysis of Chinese air transport delay propagation. *Chin. J. Aeronaut.* **2017**, *30*, 491–499. [CrossRef]
12. Pastorino, L.; Zanin, M. Air delay propagation patterns in Europe from 2015 to 2018: An information processing perspective. *J. Phys. Complex.* **2021**, *3*, 015001. [CrossRef]
13. Guo, Z.; Hao, M.; Yu, B.; Yao, B. Detecting delay propagation in regional air transport systems using convergent cross mapping and complex network theory. *Transp. Res. Part E Logist. Transp. Rev.* **2022**, *157*, 102585. [CrossRef]
14. Carlier, S.; De Lépinay, I.; Hustache, J.C.; Jelinek, F. Environmental impact of air traffic flow management delays. In Proceedings of the 7th USA/Europe air traffic management research and development seminar (ATM2007), Barcelona, Spain, 2–5 July 2007; Volume 2, p. 16.
15. Peterson, E.B.; Neels, K.; Barczi, N.; Graham, T. The economic cost of airline flight delay. *J. Transp. Econ. Policy (JTEP)* **2013**, *47*, 107–121.
16. Boccaletti, S.; Latora, V.; Moreno, Y.; Chavez, M.; Hwang, D.U. Complex networks: Structure and dynamics. *Phys. Rep.* **2006**, *424*, 175–308. [CrossRef]
17. Costa, L.d.F.; Rodrigues, F.A.; Travieso, G.; Villas Boas, P.R. Characterization of complex networks: A survey of measurements. *Adv. Phys.* **2007**, *56*, 167–242. [CrossRef]
18. Kwiatkowski, D.; Phillips, P.C.; Schmidt, P.; Shin, Y. Testing the null hypothesis of stationarity against the alternative of a unit root: How sure are we that economic time series have a unit root? *J. Econom.* **1992**, *54*, 159–178. [CrossRef]
19. Dickey, D.A.; Fuller, W.A. Distribution of the estimators for autoregressive time series with a unit root. *J. Am. Stat. Assoc.* **1979**, *74*, 427–431.
20. Zanin, M.; Olivares, F. Ordinal patterns-based methodologies for distinguishing chaos from noise in discrete time series. *Commun. Phys.* **2021**, *4*, 190. [CrossRef]
21. Leyva, I.; Martínez, J.H.; Masoller, C.; Rosso, O.A.; Zanin, M. 20 years of ordinal patterns: Perspectives and challenges. *Europhys. Lett.* **2022**, *138*, 31001. [CrossRef]
22. Bandt, C.; Shiha, F. Order patterns in time series. *J. Time Ser. Anal.* **2007**, *28*, 646–665. [CrossRef]
23. Weiß, C.H. Non-parametric tests for serial dependence in time series based on asymptotic implementations of ordinal-pattern statistics. *Chaos Interdiscip. J. Nonlinear Sci.* **2022**, *32*, 093107. [CrossRef] [PubMed]
24. Zanin, M. Continuous ordinal patterns: Creating a bridge between ordinal analysis and deep learning. *Chaos Interdiscip. J. Nonlinear Sci.* **2023**, *33*, 033114. [CrossRef] [PubMed]
25. Acharya, K.; Olivares, F.; Zanin, M. How representative are air transport functional complex networks? A quantitative validation. *Chaos Interdiscip. J. Nonlinear Sci.* **2024**, *34*, 043133. [CrossRef]
26. Lee, T.W.; Lee, T.W. *Independent Component Analysis*; Springer: Berlin/Heidelberg, Germany, 1998.
27. Bandt, C.; Pompe, B. Permutation entropy: A natural complexity measure for time series. *Phys. Rev. Lett.* **2002**, *88*, 174102. [CrossRef]
28. Amigó, J.M.; Zambrano, S.; Sanjuán, M.A. True and false forbidden patterns in deterministic and random dynamics. *Europhys. Lett.* **2007**, *79*, 50001. [CrossRef]
29. Zunino, L.; Soriano, M.C.; Fischer, I.; Rosso, O.A.; Mirasso, C.R. Permutation-information-theory approach to unveil delay dynamics from time-series analysis. *Phys. Rev. E—Stat. Nonlinear Soft Matter Phys.* **2010**, *82*, 046212. [CrossRef]
30. Rosso, O.A.; Carpi, L.C.; Saco, P.M.; Ravetti, M.G.; Plastino, A.; Larrondo, H.A. Causality and the entropy–complexity plane: Robustness and missing ordinal patterns. *Phys. A Stat. Mech. Its Appl.* **2012**, *391*, 42–55. [CrossRef]
31. Rosso, O.A.; Larrondo, H.; Martin, M.T.; Plastino, A.; Fuentes, M.A. Distinguishing noise from chaos. *Phys. Rev. Lett.* **2007**, *99*, 154102. [CrossRef]
32. Zunino, L.; Soriano, M.C.; Rosso, O.A. Distinguishing chaotic and stochastic dynamics from time series by using a multiscale symbolic approach. *Phys. Rev. E—Stat. Nonlinear Soft Matter Phys.* **2012**, *86*, 046210. [CrossRef]
33. Amigó, J.M.; Rosso, O.A. Ordinal methods: Concepts, applications, new developments, and challenges-In memory of Karsten Keller (1961–2022). *Chaos Interdiscip. J. Nonlinear Sci.* **2023**, *33*, 080401. [CrossRef] [PubMed]
34. Amigó, J.M.; Kocarev, L.; Szczepanski, J. Order patterns and chaos. *Phys. Lett. A* **2006**, *355*, 27–31. [CrossRef]

35. Carpi, L.C.; Saco, P.M.; Rosso, O.A. Missing ordinal patterns in correlated noises. *Phys. A Stat. Mech. Its Appl.* **2010**, *389*, 2020–2029. [CrossRef]
36. Olivares, F.; Zunino, L.; Pérez, D.G. Revisiting the decay of missing ordinal patterns in long-term correlated time series. *Phys. A Stat. Mech. Its Appl.* **2019**, *534*, 122100. [CrossRef]
37. Zunino, L.; Zanin, M.; Tabak, B.M.; Pérez, D.G.; Rosso, O.A. Forbidden patterns, permutation entropy and stock market inefficiency. *Phys. A Stat. Mech. Its Appl.* **2009**, *388*, 2854–2864. [CrossRef]
38. Tiana-Alsina, J.; Buldu, J.M.; Torrent, M.; García-Ojalvo, J. Quantifying stochasticity in the dynamics of delay-coupled semiconductor lasers via forbidden patterns. *Philos. Trans. R. Soc. A Math. Phys. Eng. Sci.* **2010**, *368*, 367–377. [CrossRef]
39. Olivares, F.; Zunino, L.; Soriano, M.C.; Pérez, D.G. Unraveling the decay of the number of unobserved ordinal patterns in noisy chaotic dynamics. *Phys. Rev. E* **2019**, *100*, 042215. [CrossRef]
40. Bandt, C. Statistics and contrasts of order patterns in univariate time series. *Chaos Interdiscip. J. Nonlinear Sci.* **2023**, *33*, 033124. [CrossRef]
41. Martínez, J.H.; Herrera-Diestra, J.L.; Chavez, M. Detection of time reversibility in time series by ordinal patterns analysis. *Chaos Interdiscip. J. Nonlinear Sci.* **2018**, *28*, 123111. [CrossRef]
42. Zanin, M.; Rodríguez-González, A.; Menasalvas Ruiz, E.; Papo, D. Assessing time series reversibility through permutation patterns. *Entropy* **2018**, *20*, 665. [CrossRef]
43. Zunino, L.; Olivares, F.; Ribeiro, H.V.; Rosso, O.A. Permutation Jensen-Shannon distance: A versatile and fast symbolic tool for complex time-series analysis. *Phys. Rev. E* **2022**, *105*, 045310. [CrossRef] [PubMed]
44. Lin, J. Divergence measures based on the Shannon entropy. *IEEE Trans. Inf. Theory* **1991**, *37*, 145–151. [CrossRef]
45. Olivares, F.; Zunino, L.; Zanin, M. Markov-modulated model for landing flow dynamics: An ordinal analysis validation. *Chaos Interdiscip. J. Nonlinear Sci.* **2023**, *33*, 033142. [CrossRef] [PubMed]
46. Zanin, M. Augmenting granger causality through continuous ordinal patterns. *Commun. Nonlinear Sci. Numer. Simul.* **2024**, *128*, 107606. [CrossRef]
47. Granger, C.W. Investigating causal relations by econometric models and cross-spectral methods. *Econom. J. Econom. Soc.* **1969**, *37*, 424–438. [CrossRef]
48. Diebold, F.X. *Independent Component Analysis*; Elements of Forecasting; Cengage Learning: Mason, OH, USA, 1998.
49. Shannon, C.E. A mathematical theory of communication. *Bell Syst. Tech. J.* **1948**, *27*, 623–656. [CrossRef]
50. Kraskov, A.; Stögbauer, H.; Grassberger, P. Estimating mutual information. *Phys. Rev. E* **2004**, *69*, 066138. [CrossRef]
51. Schreiber, T. Measuring Information Transfer. *Phys. Rev. Lett.* **2000**, *85*, 461–464. [CrossRef]
52. Wiener, N. The theory of prediction. In *Modern Mathematics for the Engineer*; Beckenbach, E.F., Ed.; McGraw-Hill: New York, NY, USA, 1956; pp. 125–139.
53. Gomez-Herrero, G.; Wu, W.; Rutanen, K.; Soriano, M.; Pipa, G.; Vicente, R. Assessing Coupling Dynamics from an Ensemble of Time Series. *Entropy* **2010**, *17*, 1958–1970. [CrossRef]
54. Cook, A.; Belkoura, S.; Zanin, M. ATM performance measurement in Europe, the US and China. *Chin. J. Aeronaut.* **2017**, *30*, 479–490. [CrossRef]
55. Raffalovich, L.E. Detrending time series: A cautionary note. *Sociol. Methods Res.* **1994**, *22*, 492–519. [CrossRef]
56. Azami, H.; Escudero, J. Amplitude-aware permutation entropy: Illustration in spike detection and signal segmentation. *Comput. Methods Programs Biomed.* **2016**, *128*, 40–51. [CrossRef] [PubMed]
57. Cuesta Frau, D. Permutation entropy: Influence of amplitude information on time series classification performance. *Math. Biosci. Eng.* **2019**, *16*, 6842–6857. [CrossRef] [PubMed]
58. Fadlallah, B.; Chen, B.; Keil, A.; Príncipe, J. Weighted-permutation entropy: A complexity measure for time series incorporating amplitude information. *Phys. Rev. E—Stat. Nonlinear Soft Matter Phys.* **2013**, *87*, 022911. [CrossRef]
59. Stosic, D.; Stosic, D.; Stosic, T.; Stosic, B. Generalized weighted permutation entropy. *Chaos Interdiscip. J. Nonlinear Sci.* **2022**, *32*. [CrossRef]
60. Cuesta-Frau, D. Slope entropy: A new time series complexity estimator based on both symbolic patterns and amplitude information. *Entropy* **2019**, *21*, 1167. [CrossRef]
61. Knight, M.I.; Nunes, M.; Nason, G. Modelling, detrending and decorrelation of network time series. *arXiv* **2016**, arXiv:1603.03221.
62. Eurocontrol; FAA. *Comparison of Air Traffic Management Related Operational and Economic Performance U.S.—Europe*; Technical Report; 2024. Available online: https://www.eurocontrol.int/publication/comparison-air-traffic-management-related-operational-and-economic-performance (accessed on 25 December 2024).

entropy

MDPI

Article

Complexity Analysis of Environmental Time Series

Holger Lange [1,*] and Michael Hauhs [2]

[1] Division of Forest and Forest Resources, Norwegian Institute of Bioeconomy Research (NIBIO), N-1433 Ås, Norway
[2] Faculty of Biology, Chemistry and Earth Sciences, University of Bayreuth, D-95440 Bayreuth, Germany; michael.hauhs@uni-bayreuth.de
* Correspondence: holger.lange@nibio.no; Tel.: +47-90088460

Abstract: Small, forested catchments are prototypes of terrestrial ecosystems and have been studied in several disciplines of environmental science over several decades. Time series of water and matter fluxes and nutrient concentrations from these systems exhibit a bewildering diversity of spatiotemporal patterns, indicating the intricate nature of processes acting on a large range of time scales. Nonlinear dynamics is an obvious framework to investigate catchment time series. We analyzed selected long-term data from three headwater catchments in the Bramke valley, Harz mountains, Lower Saxony in Germany at common biweekly resolution for the period 1991 to 2023. For every time series, we performed gap filling, detrending, and removal of the annual cycle using singular system analysis (SSA), and then calculated metrics based on ordinal pattern statistics: the permutation entropy, permutation complexity, and Fisher information, as well as their generalized versions (q-entropy and α-entropy). Further, the position of each variable in Tarnopolski diagrams is displayed and compared to reference stochastic processes, like fractional Brownian motion, fractional Gaussian noise, and β noise. Still another way of distinguishing deterministic chaos and structured noise, and quantifying the latter, is provided by the complexity from ordinal pattern positioned slopes (COPPS). We also constructed horizontal visibility graphs and estimated the exponent of the decay of the degree distribution. Taken together, the analyses create a characterization of the dynamics of these systems which can be scrutinized for universality, either across variables or between the three geographically very close catchments.

Keywords: time series; ordinal patterns; catchments; ecosystems; permutation entropy; permutation complexity; fisher information; Tarnopolski diagrams; horizontal visibility graphs

1. Introduction

Long-term monitoring of terrestrial ecosystems is a key activity producing insights into trends, pertinent oscillations, and system dynamics in general. It is the backbone of statements about changes in the environment on different time scales, whether these are natural phenomena (e.g., succession), related to human activities like land use change, or associated with climate change. The monitoring programs generate long-term time series, often spanning several decades, and Earth system models (ESMs) are attempting to reproduce the observations assuming a set of processes and eventually to predict them using scenarios like shared socioeconomic pathways (SSPs), or to classify them by the urgency of intervention, in the case of management-related variables.

ESMs have to assume a set of processes acting within the system and across its boundaries and need to be parametrized prior to simulations. The set of parameters needed for model calibration is often beyond what monitoring can deliver, and assumptions have to be made for the values of unobserved (or even unobservable) parameters. This is often conducted through an inverse modelling approach, where minimizing a cost function describing the data-model discrepancy is used to estimate the parameters needed. This is the recipe of most process-based approaches in the environmental sciences.

The alternative, data-driven approach does not assume given processes and requires fewer or even no parameters to be estimated. It starts with a set of observed data, often time series, and concludes on their multivariate spatiotemporal structure. This is the route we follow in this article.

We aim at a thorough characterization of this spatiotemporal structure through a set of metrics obtained from methods from nonlinear dynamics. These metrics separate deterministic from stochastic parts of the time series, elucidate the stochastic properties of them, and provide insights into their information content and complexity, thereby also indicating the efforts needed to successfully model them in a process-based way. It is reasonable to assume that reproducing rather complex data might also require complex models, although there might be exceptions. The opposite is not necessarily true: there are simple processes that generate complex data, as convincingly demonstrated by toy chaotic maps like the logistic map or the Rössler attractor, and so on. From the many approaches to investigating the complexity of time series [1], we focus mainly on those where Osvaldo Rosso had a leading role or made significant contributions [2–10], and predominantly on methods utilizing ordinal pattern statistics.

Using a set of variables across several locations allows us to investigate the classic question of whether the dynamics of a given variable (here, ion concentrations in water solutions) are universal for that variable, governed by the same processes at different locations (spatial universality), or, alternatively, the location (point of measurement) determines the dynamics (temporal universality), i.e., we observe similar dynamics within the given ecosystem, but the same variables at different locations show diverging dynamics. The modelling framework suitable in each case might be rather different. Classifying the dataset by appropriate metrics supports decisions about the most suitable modelling approach. It is, however, notoriously difficult to reproduce all of the complexity metrics with any process-based model.

Two typical modelling approaches prominent in environmental sciences, in particular in forest science and in hydrology in our case, can be described as follows: firstly, physical transport models based on a dynamic system approach (the Richards equations solved with appropriate boundary conditions [11]) and, secondly, forest growth models (e.g., yield tables [12] or growth simulators [13]) based on regional growth histories of the same species and treatment. The first one focuses on abiotic aspects of the system and keeps the acting organisms at an abstract level; the second one considers growing trees and keeps the physicochemical aspects at a rather simplified level. Of course, in between these cartoon representations there is a continuum of hybrid model classes, with agent-based models with a detailed environment description as important examples.

In this contribution, however, we are occupied with the classification of time series from a long-term ecosystem research site, intended as inputs to informed decision-making about the most suitable model classes to successfully describe the system's behavior.

2. Materials and Methods

2.1. Site Description

Hydrological catchments or watersheds covered with forests are often used as monitoring units of semi-natural landscapes, not least since these landscape units conceptually allow for a closed input–output balance for matter fluxes. Such catchments should be small to allow for a homogenous, even-aged forest stand, but they need to be sufficiently large to allow for a perennial stream. If there is only a single perennial stream, the catchment is called first-order or "headwater". The dataset of this study is derived from three first-order headwater catchments with an area between 33 and 75 ha within the Bramke valley (Harz, Germany; center coordinates 51.858° N, 10.423° E). The catchments are underlain by fissured Lower Devonian rocks, with deeply weathered soils. They are dominated by Norway spruce stands (*Picea abies* Karst. L). The observation period covers a full rotation period of the forest stand, i.e., from clearcut to clearcut. For most of this time the catchment was under monitoring of forest growth and hydrochemistry [14]. Since 1994 it also has been a level II site of the ICP-Forests monitoring program [15].

All observations included here stem from three adjacent small headwater catchments in the Harz mountains, known as "Lange Bramke" (LB), "Dicke Bramke" (DB), and "Steile Bramke" (SB) (Figure 1). At LB, Norway spruce (*Picea abies* (L.)) is practically the only species (62–70 yrs.), whereas at DB (1.1% of the area) and SB (14.4%) replanting with European beech (*Fagus sylvatica* L.) and European alder (*Alnus glutinosa* L.) near streams has occurred since 1986. The area has a long history of timber use and charcoal production due to mining, and the Lange Bramke catchment was clearcut in 1947 as part of reparation payments of Germany to Great Britain after World War II. It was then replanted starting in 1948. Environmental monitoring started the same year, first with discharge measurements (runoff rates) of the stream at LB. After the clearcut, there were concerns about soil erosion at the steep slopes. The influence of forest cover on the quantity and quality of streamwater was in focus when the Bramke catchment study was implemented in 1948 [16]. Water budgets were of major interest at the time. Starting in the 1960s, water samples were also taken for water quality assessment. Today, the long-term hydrochemical data allow a nuanced view of internal processes. Here, we use the maximal (biweekly) resolution to characterize four major dissolved ions. In addition, we consider air temperature and runoff coarse-scaled to the same resolution.

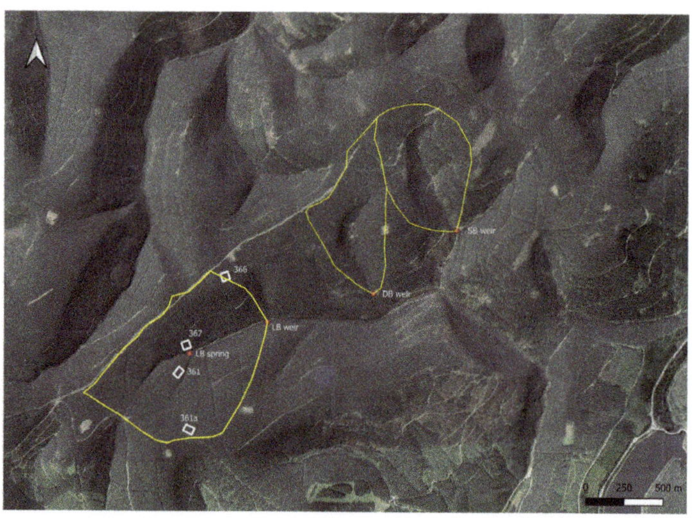

Figure 1. The three catchments with four hydrochemical sampling points: LB spring (LBQ (Lange Bramke Quelle (in German) ≙ Lange Bramke Spring)) and LB weir (LBW), and DB weir and SB weir (DBW and SBW). Rectangles are the forest inventory plots; 361, 361a, 366, and 367 are the numbers of the forest inventory plots, some of them being ICP-Forest Level II monitoring sites. Meteorological data are obtained at the clearing near and northeast of site 366.

The time series analyzed here, except temperature, are all from one of these four sampling points, and we will refer to them as DBW, LBQ, LBW, and SBW from now on.

2.2. Time Series from the Sites

Water samples from all four sampling points (cf. Figure 1) are taken at regular two weeks intervals and analyzed for chemical composition. Among the many ions determined in the chemical analysis, we selected only four: sulfate (SO_4^{2-}), nitrate (NO_3^-), chloride (Cl^-), and potassium ions (K^+). Chloride is part of the atmospheric deposition, reaching the soil dissolved in rain and in throughfall washed off from the needles; it is not actively processed by plants and does not react with the soil matrix. It is thus considered as "ecosystem-inert". This is very much contrary to sulfate with its intricate impact on soil chemistry through both bacterial reduction and inorganic sorption processes, while nitrate, as part of the plant nitrogen cycle, is also processed by soil bacteria (process of denitrification) but is not well retained in the soil matrix and can reach the groundwater. Potassium is a major plant nutrient and can be retained in the soil. Thus, these variables are important representatives for plant nutrition or are reflections of soil chemical properties, i.e., they represent different influences from either physical or biological processes, or both, on runoff hydrochemistry, and are therefore expected to exhibit different dynamics.

As these chemical concentrations in streamwater are influenced by precipitation, temperature, radiation, and other variables, we expect them to display both long-term trends and an annual cycle (seasonality). The impact of these two deterministic properties of the time series on the complexity metrics will be investigated by comparing the original time series with versions where the trend or the seasonality has been removed.

Water sampling started in the LB catchment at the end of the 1960s and in the 1980s for the other two catchments (DB and SB). However, due to irregular sampling in the beginning and some changes in the chemical–analytical methods over the decades, we extracted data from all four sampling points for a common period of 33 years, 1991 to 2023. Due to certain irregularities in the sampling intervals and some gaps contained in the records, we also decided to use the data at regular 14-day intervals, averaging all observations within each given period when more than one was obtained.

We supplemented the ion concentrations by time series for air temperature, taken from a meteorological station within a clearing near the LB catchment (Figure 1) at 2 m height above surface, and with runoff (stream discharge) from the Lange Bramke weir, which is the longest and most continuous record available. Both are obtained at daily resolution but were downsampled for our purposes to 14-day resolution.

The original dataset used in the analysis thus is a set of 18 time series (four ions at four sampling locations, plus air temperature plus runoff at one location) of length $N = 860$ values each.

Over the lifetime of the forest stand (1948–2022), two major environmental changes occurred at the Harz: an increase and decrease in the deposition of air pollutants by long-range transport (acid rain), and climate change. These phenomena are reflected in the trends for some of the variables, most prominently in the decline of sulphate as a result of substantially reduced atmospheric deposition of sulfur dioxide (SO_2). We also mention that the catchments were recently severely affected by stand-replacing bark beetle attacks. The planted species was considered as well-adapted to the site conditions, i.e., able to respond and survive the environmental conditions to be encountered over its planned rotation period of 120 years. However, the stand was practically killed (83% of the forest in the catchment) by a bark beetle epidemic after 71 years only. We expect a dramatic response of stream chemistry due to this damage; this is already seen for nitrate concentrations in the last two years reported here, 2022 and 2023.

2.3. *Data Preparation and Analysis Methods*

Many of the methods applied to our time series here require gap-free data or at least are biased or difficult to interpret when gaps in the data are present. We thus spent some efforts to generate time series at a completely regular temporal resolution with no values missing. We conceptualized each time series as additively composed of trend, periodic, and noise components, where the latter include a potentially complex mix of correlation structure. We compared the partial series, e.g., the trend component only or the original series detrended, and thus needed an operational method to decompose the series into these three components. We used a general notion of a trend as any "static" component, i.e., having no identified periods and no high-frequency noise present. Given the temporal resolution and the time span covered, the only periodic component notoriously present is the annual cycle. Among the existing methods for decomposition, we chose a fully data-adaptive flexible method which at the same time can be utilized for gap filling as well: singular system analysis [17].

2.3.1. Gap Filling, Detrending, and Deseasonalization: Singular System Analysis

Annual cycles are notorious for most of the water chemistry variables, basically induced by yearly cycles of temperature and radiation, but also crucially determined by the biological activity during the growing season and hibernation during winter. The presence of trends is due to a mixture of the growth of the forest stands—the observation period is a significant fraction of the average lifespan of a spruce stand after plantation—and nonstationary environmental conditions (atmospheric deposition of air pollutants, climate change, disruptive events). As a result, our typical time series would not pass any classical stationary tests.

Some of the 14-day intervals did not contain a single value for some of the ions, i.e., the original time series also contained some gaps. As most of the methods require gap-free data to avoid bias, one ought to fill the gaps prior to their application.

The following Figures 2–6 show gap-filled versions of the time series, representing the whole data set going into all further analysis.

Figure 2 shows runoff at the gauged weir at Lange Bramke. The LB stream is perennial but almost ceased in the very dry summers of 2003 and 2018. The temperature record (14-day minimum: -12.4 °C, 14-day maximum: 22.4 °C) has a highly significant positive trend in the observation period with a slope of +0.062 °C/year, which is, however, difficult to identify in the figure.

Our tool to deal with all three issues of gap filling, detrending, and deseasonalization is singular system analysis (SSA) in the version described in [18], i.e., a fully data-adaptive decomposition method based on the lagged covariance matrix. SSA leads to an orthogonal set of eigenmodes, ranked according to their explained variance, which are the eigenvalues of a singular value decomposition. SSA comes with one parameter, the embedding dimension or window length L. No strict rules exist to find an optimal value of L; however, $N/L \approx 2 - 10$ is recommended. Periodic components appear as pairs of eigenmodes with nearly identical eigenvalues, and single components with no detectable period smaller than L (quasi-static) are considered as a trend. The periodic components are not necessarily sinusoidal, nor is the trend necessarily monotonous.

We used the R package R-SSA [19] to decompose the time series. The SSA also allows filling in missing data with several methods; in our case, we used the "Caterpillar" algorithm [20], which requires selecting a group of SSA components to base the gap filling on. This group should contain trends and major periodic components as a minimum. In our case, we selected the components with the six highest ranks, the ones with the annual cycle (or season) and the trend always among them. Isolated single missing values were eliminated by simple linear interpolation.

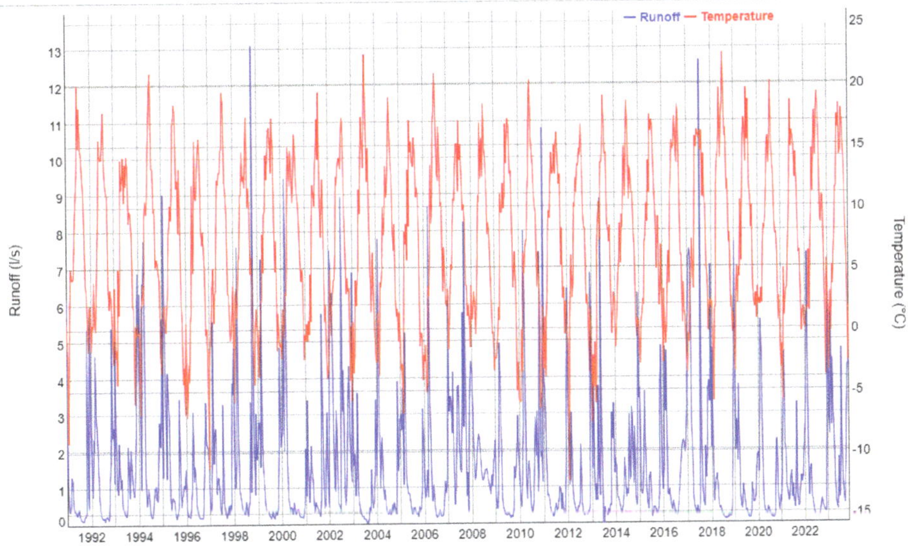

Figure 2. Runoff at the weir of the Lange Bramke catchment (LBW) (blue, left axis) and air temperature at the meteorological station within the catchment (red, right axis), values aggregated to 14 days based on daily observations, 1991–2023.

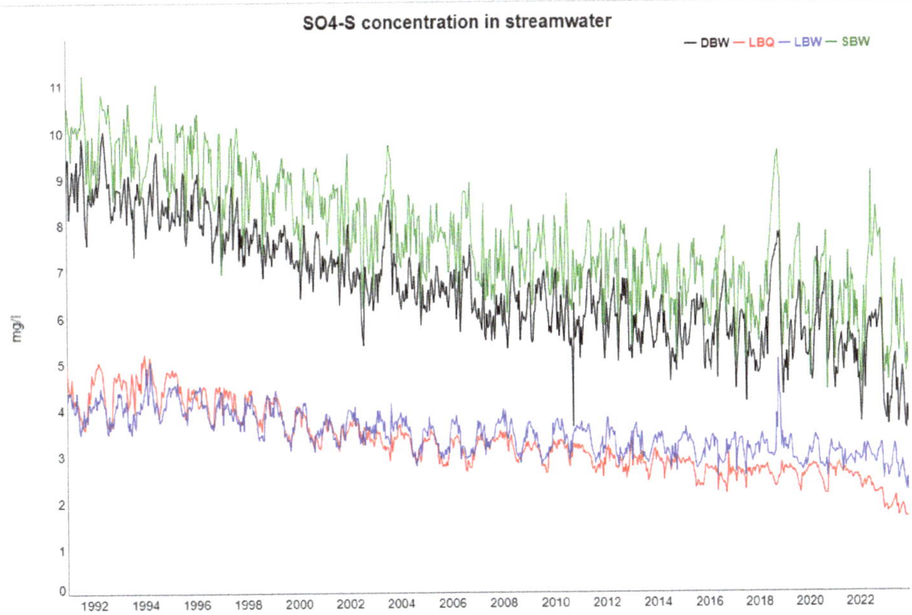

Figure 3. Sulfate concentrations at the four locations (DBW, LBQ, LBW, and SBW). The clear decreasing trend is due to recovery from acid deposition since the late 1980s.

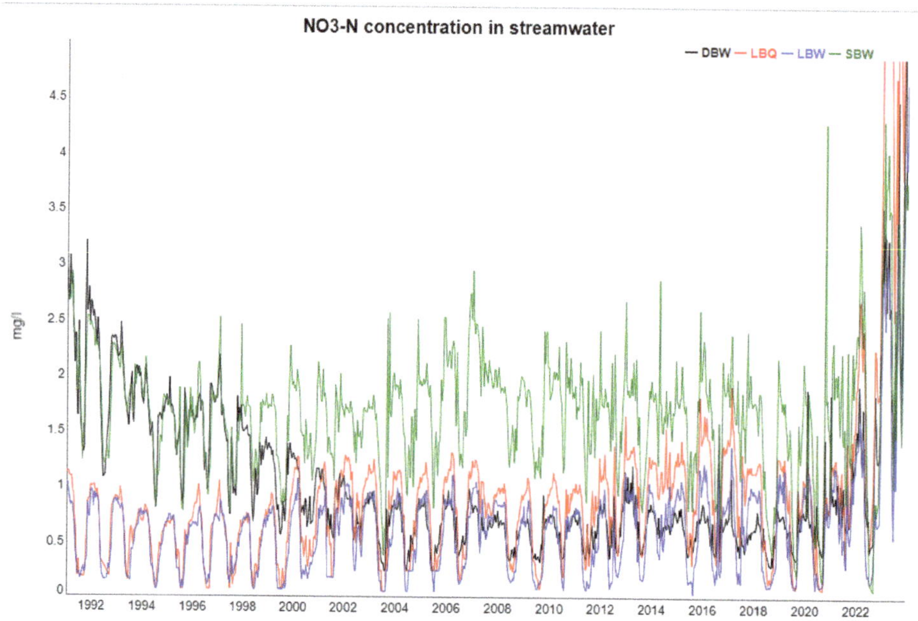

Figure 4. Nitrate concentrations in streamwater at the four locations, 1991–2023. The sharp increase since 2022, exceeding even the concentration range displayed here, is due to forest dieback induced by a bark beetle attack.

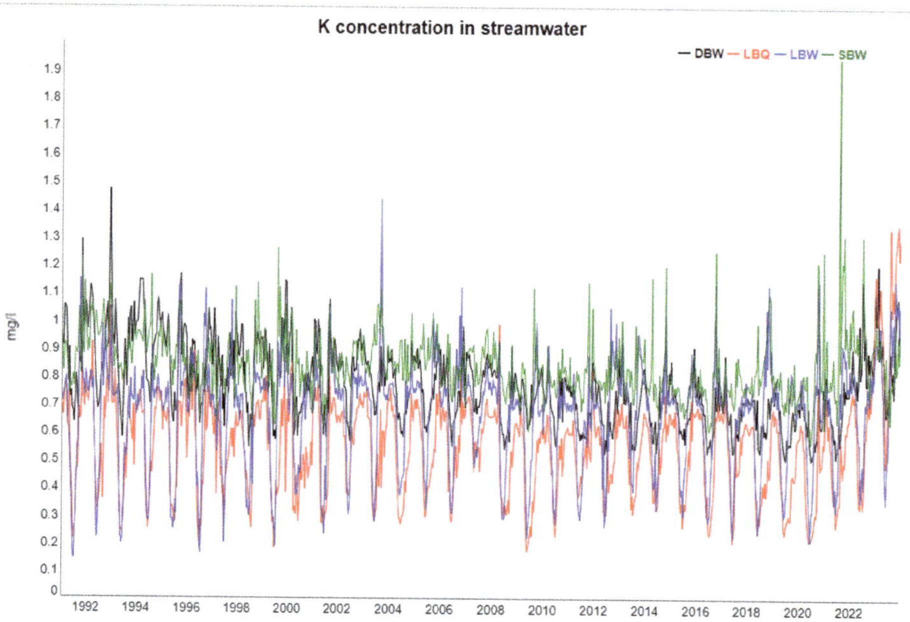

Figure 5. Potassium concentrations in streamwater at the four locations, 1991–2023.

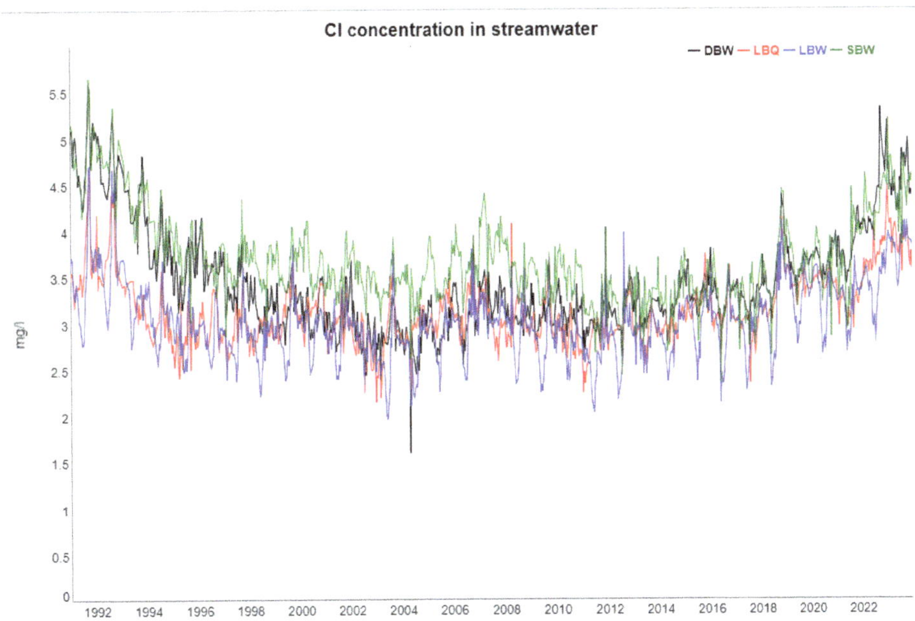

Figure 6. Similar to Figures 2–5, here for Cl concentrations.

After gap filling, the complete time series were used to identify the components containing the trend and the ones with the annual cycle. An overview of their contribution to the total variance of the time series is provided in Table 1. This left us with six versions of the respective time series: the original one, a detrended one where the trend is subtracted but the annual cycle is retained, one where the annual cycle is removed but the trend is retained, the trend alone, the annual cycle (season) alone, and the residual time series where both the trend and the annual cycle have been removed. Further analysis was conducted on all six versions and then compared.

Table 1. Percentage of explained variance relative to the total variance of the original time series for the variables investigated, based on SSA decomposition.

Variable	Process	%Variance per Location			
Temperature	Trend	0.97			
	Season	80.99			
		DBW	LBQ	LBW	SBW
Runoff	Trend	-	-	0.48	-
	Season	-	-	20.69	-
Cl	Trend	86.05	57.43	24.00	73.93
	Season	4.42	5.57	19.15	4.90
K	Trend	39.74	24.51	0.76	19.41
	Season	31.84	46.11	45.89	4.48
NO_3	Trend	72.06	74.16	45.41	20.59
	Season	7.39	11.45	30.27	32.56
SO_4	Trend	75.91	88.02	64.65	70.12
	Season	3.17	4.46	15.33	6.68

It is obvious from Table 1 that in most cases, the trend and seasonal components combined contain a large fraction of the total variance. On the other hand, there are major differences between the variables, as is also recognizable from the time series plots, e.g., for runoff and K at LBW, the trend is negligible, whereas for SO_4, the trend dominates by far over the seasonal component. NO_3 has a very strong trend, but that consists of a decline during the 1990s and an increase in the last years, so it is non-monotonous.

The percentages reported in Table 1 do not reveal whether this periodic component is synchronized between the variables, or if they show different phases but with a constant phase relation. Insight into these connections can be gained by extracting the seasonal component only (usually a pair of eigenmodes of the SSA decomposition) and then calculating the instantaneous phase using the Hilbert transformation of the time series. Time series of the difference between the two instantaneous phases were constructed and interpreted as lag times between the two annual cycles by converting it to time scales. The degree of synchronization, a measure of the stability of the phase relation between the two, can be obtained by calculating the mean resultant length [21].

The SSA decomposition is illustrated in Figure 7, using Cl at LBW as an example. For most variables, removing the seasonal component or the trend, or both, to obtain the residual does not change the visual appearance of the time series substantially. This observation indicates that the stochastic component dominates the dynamics. The extracted trend and seasonal components separately (the lowermost two time series in Figure 7) exhibit rather smooth and regular dynamics and are the deterministic part of the time series in this framework of additive SSA decomposition.

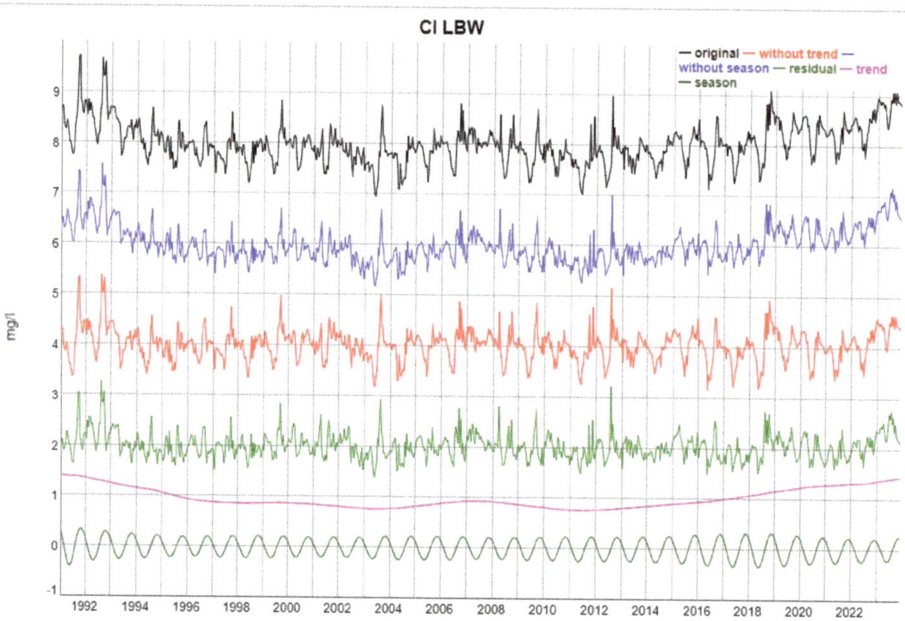

Figure 7. Example of the SSA-based time series decomposition: Cl concentrations at LBW. The individual time series have been shifted for easier identification, e.g., the original time series was increased by a constant of 5 mg/L. From top to bottom: original time series; original minus annual component; original minus trend component; residual = original − trend component − annual component; the trend component alone; and the annual component alone.

Note that the trend component generated by SSA is a nonlinear, non-monotonic function.

2.3.2. Permutation Entropy and Complexity

We used symbolic dynamics for quantifying the entropy and complexity of the time series. Following the seminal approach of [22], we constructed ordinal patterns from the real-value series. The crucial parameter to do that is the embedding dimension (aka the word length or pattern length) D. Considering that, in order to capture as much structure on different timescales as possible, one wants to maximize D, but the factorial explosion of the number of possible patterns quickly prevents statistical saturation for a given time series length N, we respected the rule of thumb that $N > 5D!$ should hold and thus fixed $D = 5$. This implies that the temporal distance between the first and the last value in any pattern is roughly two months (precisely 56 days). The regularity of the annual cycle is undetectable with these settings. Ordinal pattern analysis is strongly dependent on temporal resolutions.

From the ordinal pattern probability distributions (OPDs), we calculated the Shannon entropy, in this context also known as the permutation entropy, and the permutation complexity, also known as the MPR complexity, following the seminal work of Rosso and co-workers [8,23]. Permutation complexity is based on the permutation Jensen–Shannon divergence [24,25]:

$$JSD(p,q) = S(\frac{p+q}{2}) - \frac{1}{2}(S(p) + S(q)) \tag{1}$$

with $S(p)$ being the permutation entropy for the OPD p, and with white noise with its equidistributed OPD q as the reference process. The square root of JSD can be shown to be a proper metric [25].

The entropy–complexity plane contains regions which are unreachable for any time series; the accessible area is delimited by a lower and an upper limit curve, with their shape depending on D. It also turns out that power law noise, i.e., correlated noise whose power spectral density scales as $P(f) \sim f^{-k}$, forms a single one-dimensional curve in this plane. This curve was used as a reference to judge on the type of stochastic process we observed in our time series.

2.3.3. Fisher Information

As a third metric quantifying an information-related property of the time series, we considered the Fisher information adapted to ordinal pattern series [12]:

$$F(P) = \frac{1}{2}\sum_{i=1}^{D!-1} [\sqrt{p_{i+1}} - \sqrt{p_i}]^2 \tag{2}$$

where P denotes an ordinal pattern distribution and the p_i are the probabilities of the patterns. The Fisher information is not a unique quantity, since a numbering scheme for the patterns is required, inducing an ambiguity which is, however, insignificant for interpretation purposes, as our experience indicates. In this work, we used the coding scheme of Karsten Keller [6], and $D = 5$ as before. A two-dimensional plot of Fisher information versus permutation entropy is not known to have limit curves, i.e., every point in the square $[0, 1] \times [0, 1]$ is reachable, and the positions of the time series in this plot can be used to draw conclusions about stochasticity and to compare to reference processes.

2.3.4. Rényi and Tsallis Entropy and Complexity

Permutation entropy and complexity can be considered as special cases of a class of entropies and complexities, introduced by Rényi [26] and Tsallis [27], respectively. Either class is parametrized through a non-negative real number; within the concept of Rényi, the parameter α is used to apply power weights to the probabilities of Equation (1), either enhancing ($\alpha < 1$) or suppressing ($\alpha > 1$) rare patterns. The Rényi approach assigns power weights to the probabilities of the ordinal pattern distribution for the calculation of the entropy:

$$H_\alpha(p) = \frac{1}{(1-\alpha)ln\ D!} ln \sum_{i=1}^{D!} p_i^\alpha \tag{3}$$

In the case of the Tsallis entropy, its parameter is usually called q, a generalized logarithm log_q is introduced, and the Tsallis entropy is defined as

$$H_q(p) = \frac{1}{D!} \sum_{i=1}^{D!} p_i\ log_q \frac{1}{p_i} \tag{4}$$

Both entropies converge to the usual Shannon/Boltzmann entropy when $\alpha \to 1$ or $q \to 1$, respectively. Only the Rényi entropy shares the property of extensivity, i.e., additivity in the case of independent distributions, with the conventional entropy.

Starting from these entropy generalizations, corresponding complexities were defined, formed as the product of the respective entropies with appropriate distance measures to a reference process (white noise), and normalized with a maximum distance.

The resulting Rényi complexity [28] allows for a qualitative distinction between stochastic and deterministic–chaotic time series; for the former, in an entropy–complexity plane, varying α leads to a monotonous behavior of both entropy and complexity, which is not the case for well-studied deterministic maps. In the case of the Tsallis complexity [29], while the deterministic processes lead to open curves with two ends, the stochastic ones form closed loops when running through all q values. The area covered by the loops is related to the Hurst parameter for the stochastic series.

2.3.5. Tarnopolski Diagrams

Another method to locate our time series in the context of standard reference processes is provided by the Tarnopolski diagram [30]. Here, one plots two rather simple and parameter-free quantities of each time series against each other: the number of turning points T and the sum of squared differences of adjacent values, also known as the Abbe value $\mathcal{A}$. Reference stochastic processes like fBm or fGn build invertible functions in the $T-\mathcal{A}$ plane. Properly normalized, they are largely insensitive to the exact time series length; however, Tarnopolski found exact equations for the two reference processes, fBm and fGn, depending on the Hurst parameter at any time series length [31].

2.3.6. Horizontal Visibility Graphs

A conceptually simple geometric visualization of the correlations (in a rather general sense, not restricted to linear ones) can be articulated through the following question: sitting on a point in the time series, how far could you see in a horizontal direction before other (higher) values block your view? This is the idea behind horizontal visibility graphs (HVGs) [32], a member of the family of complex networks. The resulting network of visibility is analyzed, e.g., through its degree distribution; for some processes, it is known that the probability of finding a network node with degree k is exponentially decaying:

$$P(k) = \frac{3}{4} e^{-\lambda_{HVG} k} \tag{5}$$

which is a robust result independent of the time series distribution [33]; in the absence of autocorrelations, there is even the analytical result $\lambda_{HVG} = ln(3/2)$. From the observed degree distributions for our time series, we estimated the slope of the relation (5), i.e., λ_{HVG}, and compared it in particular to the uncorrelated case, and among the time series.

2.3.7. Complexity of Ordinal Pattern Positioned Slopes (COPPS)

Another recent approach to quantifying the complexity of time series is combining ordinal pattern statistics and network construction. It is based on the slopes present in the ordinal patterns [34]. Encoding the ranks within a pattern of embedding dimension D as integer numbers, one determines the maximum slope (differences of ranks) within the pattern and its position (ordinal pattern positioned slope, OPPS). Grouping OPPSs together with a group length s leads to a transition network of depth s. The ability of a system to generate new patterns whenever the network depth s is increased constitutes the notion of complexity in this context. Thus, the COPPS variable $\lambda_s(D)$ introduced in [34] quantifies the growth rate of the OPPS patterns when the network depth is increased by one unit. For reference processes, this complexity indicator is robust already at small to intermediate time series lengths. We compared the COPPS values obtained for our time series with a few of the reference processes.

3. Results

For each of the 18 time series, we derived six different versions from the SSA decomposition: (1) the original as obtained from field samples; (2) the one where the trend component has been removed; (3) a set where the respective annual cycles have been removed, but the trends remain; (4) the version where both the annual cycle and the trend have been removed, which we call the "residual" time series; (5) the pure annual cycles; and (6) the pure trend components. Note that the last one is nonlinear and might be complicated, although not complex in terms of our notion of complexity.

We compared the different versions of the time series for each method, focusing on the difference in dynamical structure obtained. We started, however, with a linear method to visualize the correlation between the datasets as a correlogram in Figure 8.

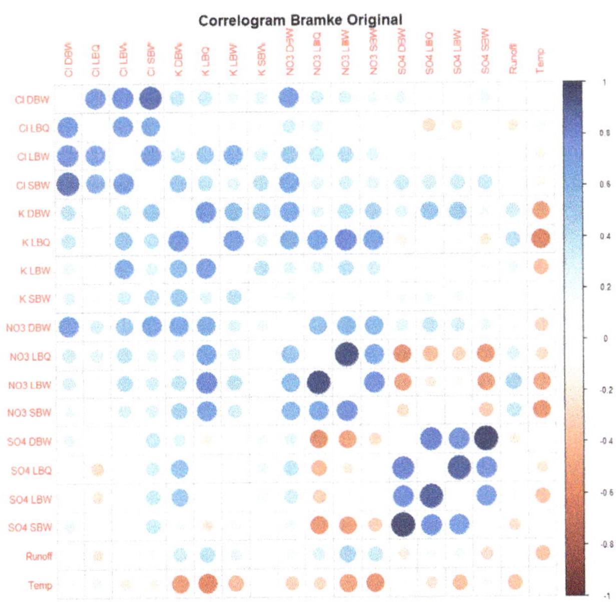

Figure 8. Correlogram of the original time series. Both the color as well as the size of the circles indicate the Pearson correlation coefficient of each pair of time series (its absolute value in the case of the size). The main diagonal (with $R^2 = 1$) is blanked out for better visualization.

3.1. Correlograms, Phase Shifts, and Jensen–Shannon Divergence of the Time Series Set

The correlogram identifies the group of Cl ions as rather closely connected; it is much less so for K, apart from the pair of K LBQ/K LBW, which are from the same stream. NO_3 is again more strongly connected, and SO_4 even more so. Since potassium (K^+) is strongly processed by plants, this might point to differences in the vegetation dominating the three catchments.

There are also some anti-correlations, e.g., between NO_3 and SO_4, but also between runoff and K, NO_3, and SO_4. Given the strong seasonality, this indicates that these ions are not in synchrony: there is a lag of several months between the different pairs.

Correlograms for other variants (trend only, annual cycle only, residuals) are shown in the Appendix (Figures A1–A3).

The presence of non-synchronous dynamics is partially confirmed by the phase shift analysis, where we calculated the instantaneous phase between the annual components of two variables, recalculated to a time lag bound between −6 months and +6 months.

In Figure 9, the time lags between NO_3 concentration from all four locations and temperature are shown. The proper time lag revolves around 6 months and flips around between +6 months and −6 months, which means the same shift for annual cycles. Nitrate in runoff is high when root uptake and temperature are low, and we used the negative temperature instead. The proper interpretation of Figure 9 is, therefore, that nitrate and temperature are half a year apart, easily explaining the negative correlation coefficient seen in the correlogram. The phase shifts are reasonably stable most of the time, although they also contain periodic components of unknown origin. In the last two years, NO_3 and temperature have decoupled from each other due to tree mortality, so the biological control on nitrate largely disappears.

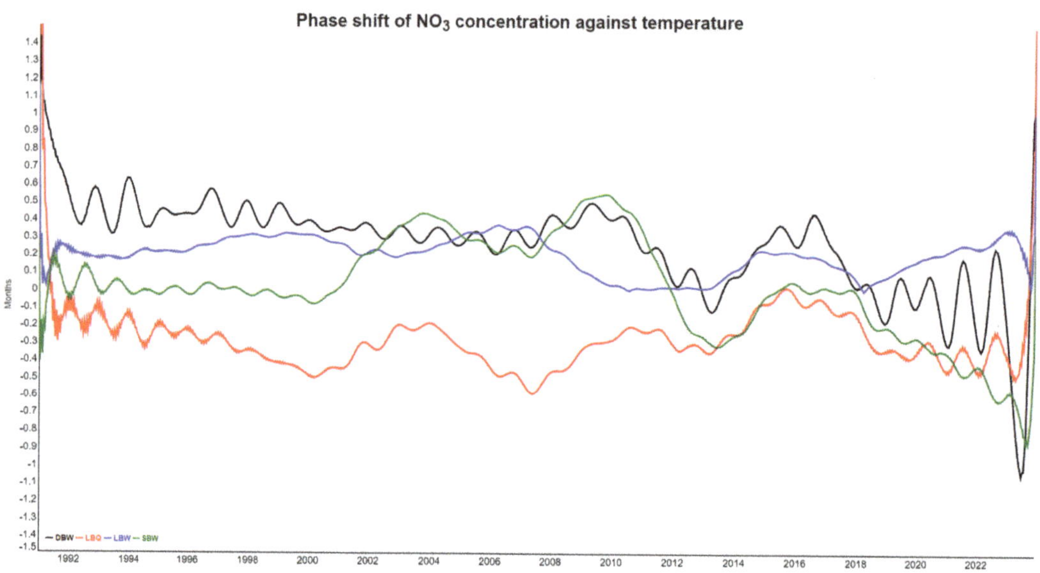

Figure 9. Phase shift between NO_3 concentrations and air temperature. For better visualization, temperature went into the instantaneous phase calculation with a reversed sign, which basically means a shift by six months.

The phase shift between SO_4 at all locations and air temperature, analogous to Figure 9, is shown in Figure A8.

Figure 10 shows the phase shift between SO_4 at SBW and NO_3 at all locations. There is no obvious coupled cycle between the two, but both ions are dominated by biological processes and also through soil interactions. For LBQ, LBW, and SBW, the sulfate signal comes mostly first, but the lag is rather small, typically less than a month. For DBW, however, NO_3 is completely out of phase with lags around 6 months, the biggest possible. Again, in the last two years, NO_3 has decoupled from the SO_4 dynamics.

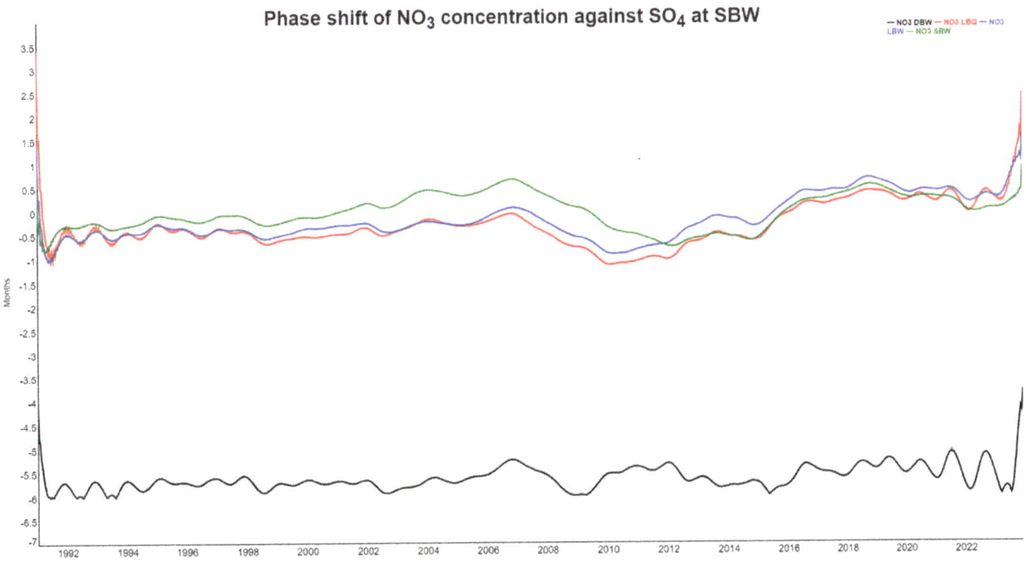

Figure 10. Phase shift between NO_3 at all locations and SO_4 at DBW. Positive values imply that SO_4 comes first in the annual cycle. Since the phase shift is cyclic (minus six months is equivalent to plus six months), we used the negative absolute value for the phase in the case of NO_3 at DBW, which would otherwise flip around frequently.

Phase shift analysis for a pair of time series can be condensed to a single number, the phase synchronization index ρ $(0 \leq \rho \leq 1)$. An overview of the phase synchronization index for all pairs of time series is provided in Figure A7.

As an alternative approach to visualizing the connections between the different variables, we calculated the Jensen–Shannon divergence of the respective ordinal distributions. As this is a distance measure (We are aware that a proper metric is only obtained when using the square root of the JSD instead. This is however not a relevant aspect for our application here) with $JSD(p,p) = 0$, and we wanted to compare to the correlogram, we used the complementary $JSD^*(p,q) = 1 - JSD(p,q)$ with $0 \leq JSD^*(p,q) \leq 1$, and values close to 1 if the two distributions were rather similar, as with the correlation coefficient.

Figure 11 displays JSD^* for the 18 time series. Note that we are comparing pairs of temporal structures, as expressed through the ordinal pattern distributions at $D = 5$. Synchronicity of pattern occurrences is not part of the comparison directly; you might shift one time series relative to the other and still get the same result. There are three pairs of variables which are very close to each other in the ordinal pattern space: NO_3 at the two locations LBQ and LBW (which is the same stream); NO_3 SBW and Runoff; and SO_4 at SBW and at DBW. Some of the pairs which are Pearson-anticorrelated appear with small values for JSD^* in Figure 11. The pair (SO_4 LBW, NO_3 DBW) which particularly showed this behavior in Figure 10 had a low but not exceptional value for JSD^*.

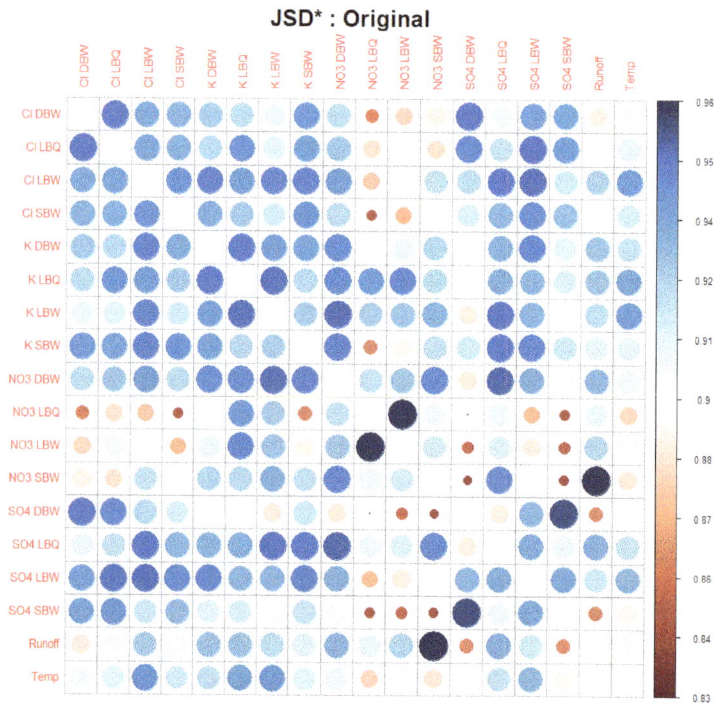

Figure 11. Complementary Jensen–Shannon divergence ($JSD^* = 1 - JSD$) for the original time series. The trivial main diagonal entries are suppressed.

3.2. Entropy–Complexity Plane

The permutation entropy (PE), permutation complexity (MPR), and Fisher information (Fis) were calculated with $D = 5$. For the PE versus MPR plane, the results are shown in Figure 12.

Most of the time series are close to, or even on, the k noise curve. The removal of the trend and/or the annual component allows them to move a bit upwards to lower entropy and higher complexity, but still along the powernoise curve. The trend and annual component alone, however, occupy an area on the left side of the maxima of the curves, with corresponding low values for the entropy. In particular, the "trend only" variant exhibits higher complexity values than the powernoise would indicate, demonstrating the less stochastic nature of the trend component. The OPDs of these time series also have a lot of missing patterns.

3.3. Entropy–Fisher Information Plane

The conclusion from Figure 12 that the original and detrended series are quite compatible with k noise cannot be drawn in the same manner for the entropy–Fisher information plane (Figure 13). Again, the trend and annual component alone are very distinct from the other variants, but for the latter, every time series shows higher Fisher information than the k noise. This indicates that the OPDs for them are more heterogeneous; in particular, some of the patterns might occur very rarely or not at all at that time series length, which is contrary to the k noise, where all patterns appear. In fact, each of our time series exhibits missing patterns in all variants, resembling deterministic parts still contained in them.

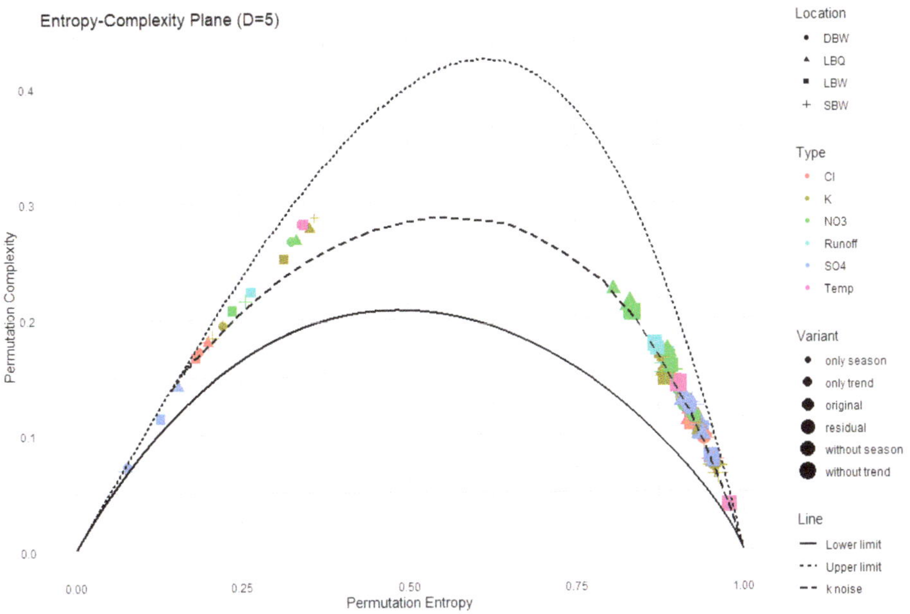

Figure 12. The entropy–complexity plane (PE, MPR) for our time series. The plot symbol indicates the location, the size of the symbol corresponds to the variant ("original" to "trend only"), and the color refers to the variable observed. The upper and lower limit curves for $D = 5$ as well as the powernoise curve are also shown.

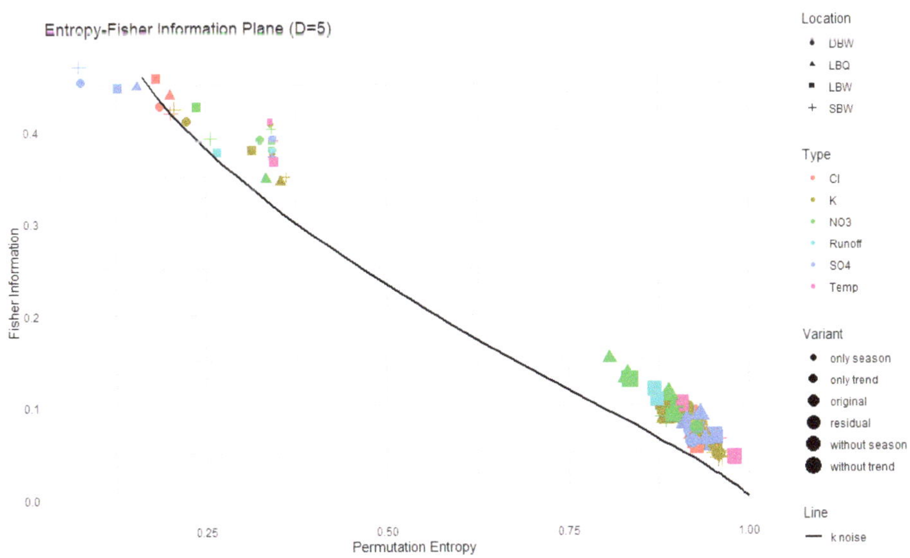

Figure 13. The Fisher information–entropy plane for our time series. For the legend, see Figure 12. The curve for k noise is plotted for reference.

3.4. *Rényi and Tsallis Entropy–Complexity Planes*

Varying the Rényi parameter α across a wide range (we chose $0.01 \leq \alpha \leq 100$) ensures that the whole range of possible enhancements or discriminations for the pattern probabilities is covered. The result is one curve per variable per treatment. This is shown in Figures 14 and 15. For the original time series (Figure 14) and the three variants where part of the signal was removed (Figures A4–A6), all curves have the common property that Rényi complexity decreases with increasing Rényi entropy. This is a behavior characteristic of correlated stochastic processes. The opposite is true for the trend and the annual component (Figure 15), as these resemble deterministic processes.

The Tsallis parameter q also was selected from the interval $0.01 \leq q \leq 100$; note that $q = 0$ is a pathological case. Figures 16 and 17 show the (H_q, C_q) planes for the original time series and for the annual components, respectively. Note that none of these curves are closed loops, contrary to the powernoise reference processes (the white noise (k = 0) curve is actually just a point at $(H_q, C_q) = (1, 0)$ independent of q). This is due to the presence of missing patterns: the authors of [29] have shown that the Tsallis curves start in our case at $(H_{0^+}, C_{0^+}) = (\frac{119-m}{119}, \frac{m(119-m)}{119^2})$ when $q \to 0^+$, where m is the number of missing patterns. The latter are notorious in deterministic time series, but also occur in stochastic processes, depending on the embedding dimension [6]. Still, the original time series curves for H_q and C_q resemble very much the loops of stochastic processes (Figure 16), whereas the annual component does not (Figure 17). They are clearly open curves. For their endpoints for $q \to \infty$, the authors of [29] provide an analytical equation, which can be checked against the value obtained for the largest q used; our choice $q = 100$ is clearly sufficiently asymptotic.

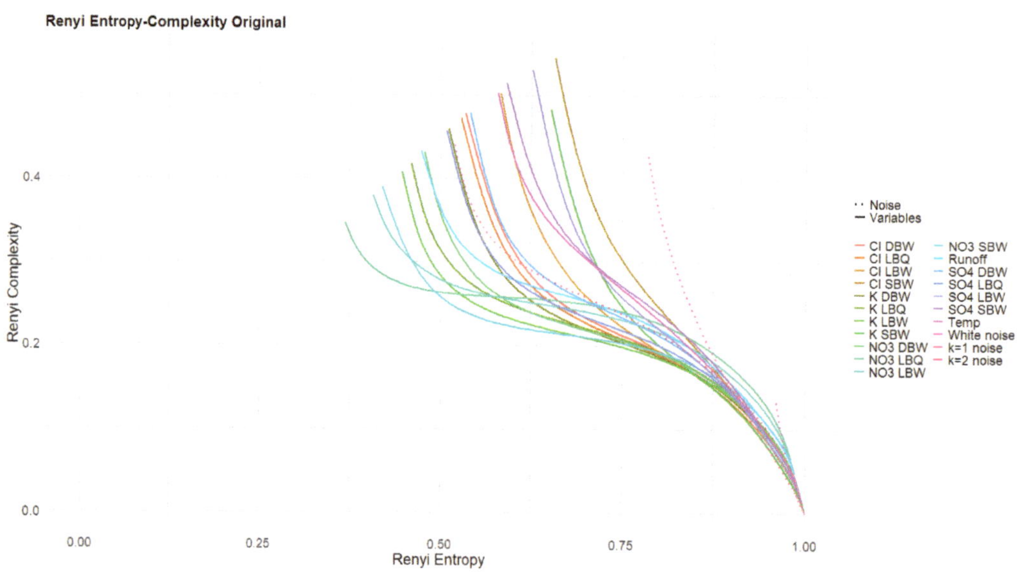

Figure 14. Rényi entropy–complexity plane for the original time series. Three powernoise processes are displayed for comparison.

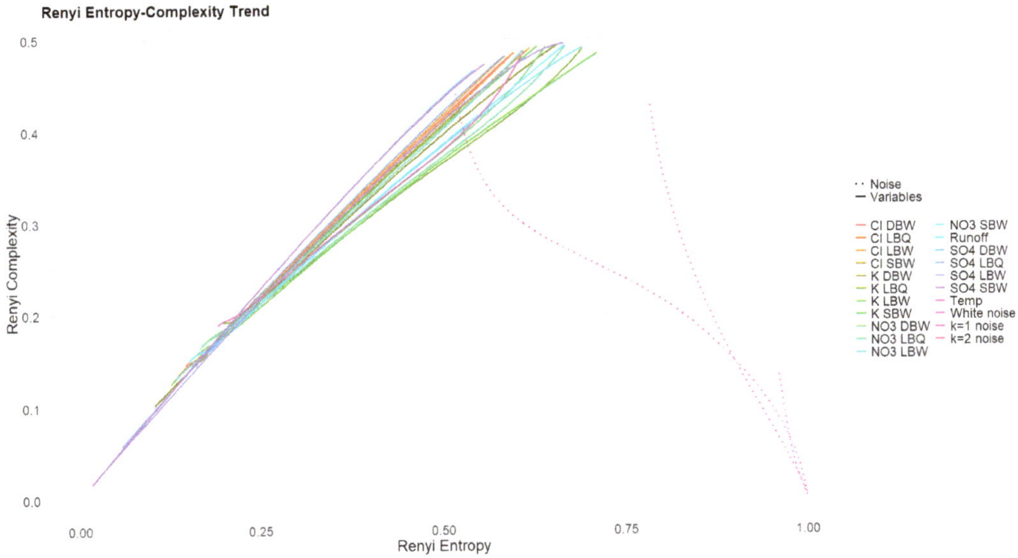

Figure 15. Rényi entropy–complexity plane for the trend components. Three powernoise processes are displayed for comparison.

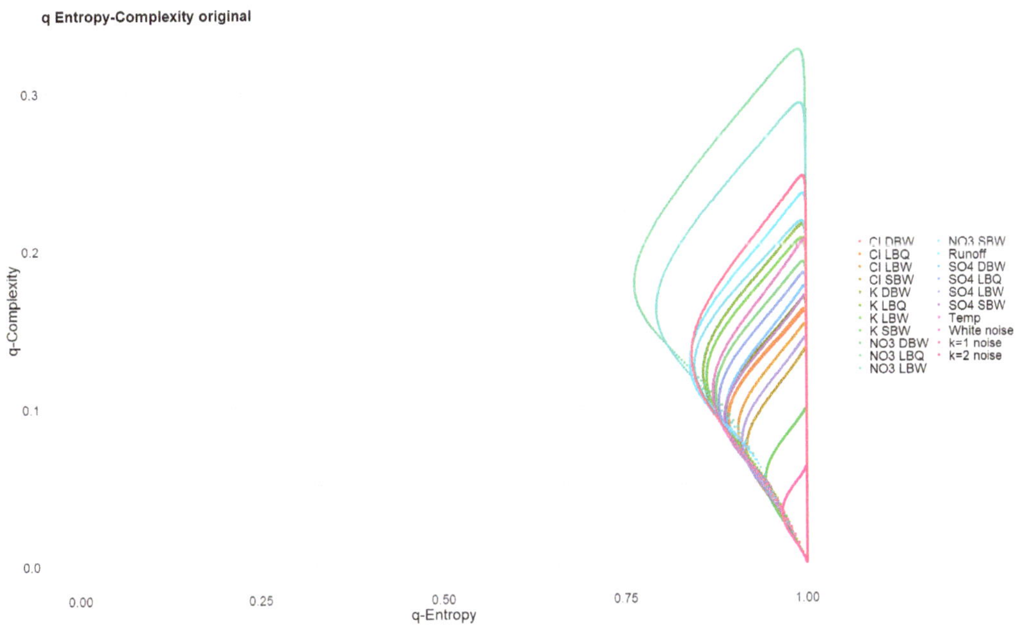

Figure 16. Tsallis q-entropy versus q-complexity for the original time series. The scaling of the entropy axis is deliberately chosen to cover the whole possible interval [0, 1]. Powernoise curves for k = 0, 1 and 2 are indistinguishably part of this set of curves.

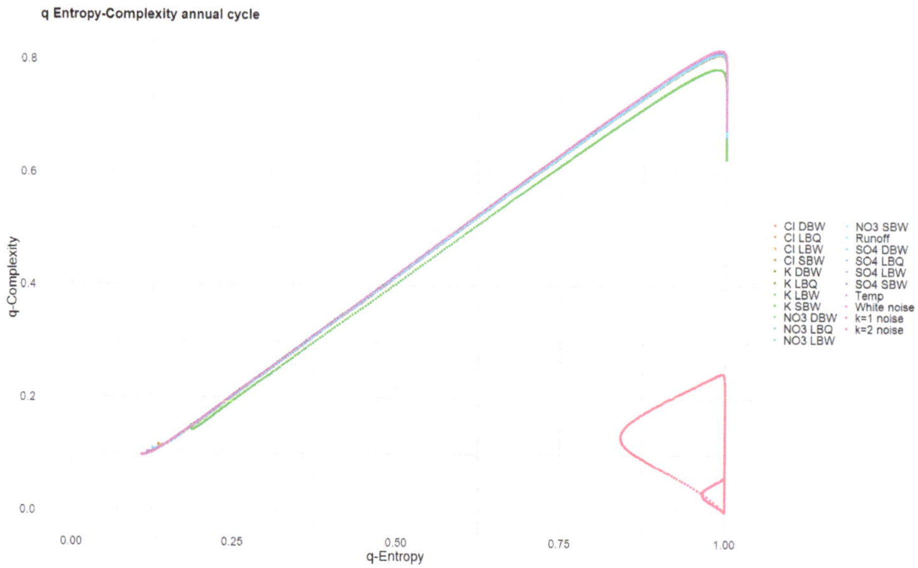

Figure 17. Same as Figure 16, but for the annual components of the time series. The powernoise curves occupy a quite different region of the plane.

3.5. Tarnopolski Diagram

The representation of a time series in the Abbe value/turning points diagram is parameter-free and does not use ordinal pattern statistics. Time series values are taken at face value without preprocessing. For some stochastic reference processes, the inventor of the method derived analytical results depending on time series length [31] and indicated regions in that plane occupied by time series models like ARMA(p, q) processes. The equation refers to the mean value of $\mathcal{A}$ and T when generating a large set of similar time series of the given length; however, the finite size effects are already quite small for our $N = 860$.

Figure 18 shows $\mathcal{A}$ and T for all time series in all variants, together with the curves for fractional Brownian motion, fractional Gaussian noise, and k-noise (the latter being a numerical result). The trend and annual components have a zero or very low $\mathcal{A}$, quite contrary to the original and the detrended variants. The latter do not fit nicely to any of the three processes, cover a large range of $\mathcal{A}$ values, and have T values between the fBm and the k noise. These points certainly do not represent a one-dimensional curve and will not fit to any simple stochastic process. In that regard, the Tarnopolski diagram gives a rather different perspective on the time series compared to the entropy–complexity diagram.

3.6. Horizontal Visibility Graph Analysis

The construction of the network of visibilities, considering every value of a time series as a node and two nodes connected with a link when the older one can "see" the later one in horizontal direction, is another nonparametric way to characterize the dynamical structure of a time series. From the networks obtained, we extracted just one quantity: the slope of the decay of the degree distribution, assuming the exponential relationship of Equation (5). This assumption is empirically justified for many time series; however, at very large degrees, it always fails since there are gaps in the degree distribution since these degrees are simply non-existent in the distribution. We therefore used a cutoff for the survival function of the degree distribution adapted to our time series length.

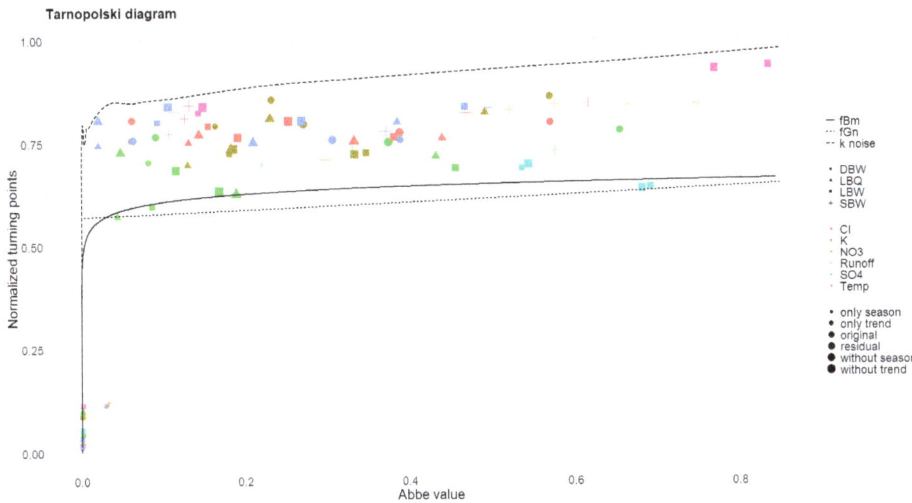

Figure 18. Tarnopolski diagram for the time series together with three different reference processes. The turning points are normalized to cover strictly the interval [0, 1]; the Abbe value extends to 1.5, which is reached by the fBM but not by our datasets.

The slopes λ_{HVG} for the exponential decay for the horizontal visibility graphs obtained from the time series in this way are shown in Figure 19. Here, the only reference process with a known theoretical value for this slope is white noise, where $\lambda_{HVG}{}^{WN} = ln(3/2)$. This is drawn as a horizontal line in the figure. For colored noise, it is not even known whether the degree distribution is of an exponential type, and even assuming it is, the spread of estimates for λ_{HVG} for relatively short time series, as is the case here, is substantial [7], so we do not show them. However, if anything, the slope values would be *larger* than for the white noise.

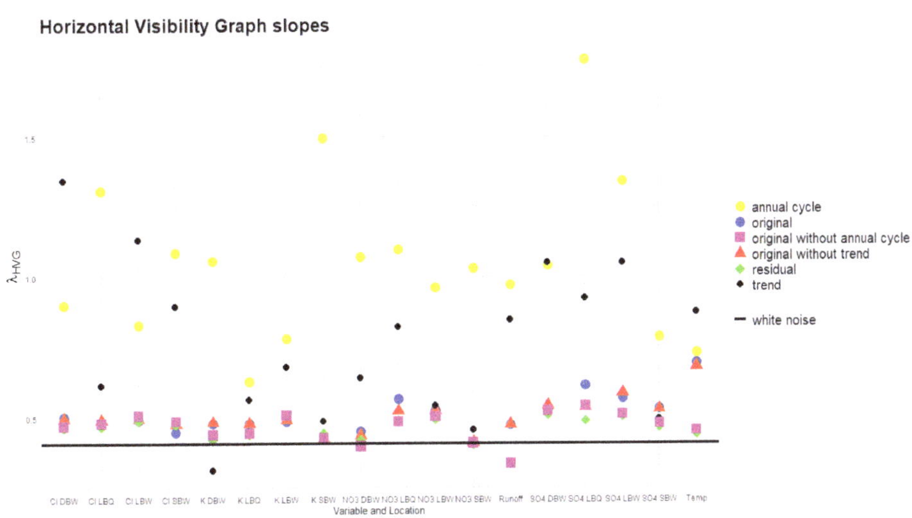

Figure 19. Slopes of the decay of the degree distribution of horizontal visibility graphs.

The HVG slope is rather insensitive to the removal of a trend component, and the original, detrended, and residual time series are by and large close to the white noise case from this perspective. However, a notable exception is temperature, and to a lesser extent also NO_3 and SO_4 at LBQ. The annual and trend components, however, extend to much higher slopes; their networks thus have a much tighter degree distribution, as is also reflected in smaller mean degrees for them. Notable outliers for the annual cycle are K at SBW (with a particular small annual amplitude, explaining only 4.5% of the total variance) and SO_4 at both LBQ and LBW.

3.7. COPPS Analysis

When applying the COPPS procedure to measured time series, the choices for the network depth s and the embedding dimension D are strongly constrained by the length of the time series, N. The number of possible groups/patterns is $\Lambda_s(N) = ((N-1)^2+1)^{s+1}$ [34]. The resulting complexity $\lambda_s(D)$ is also strongly dependent on N.

With our time series length $N = 860$, we chose $s = 1$ and $D = 4$. Figure 20 shows the resulting λ_1 (4) for all variants. These were compared to the results for Gaussian white noise, pink noise (k = 1), and red noise (k = 2), and the logistic map at fully developed chaos (r = 4), all of which were generated with the same length $N = 860$. The results for these processes are rather robust, repeatedly generating time series of the same length and calculating λ leads to rather narrow distributions.

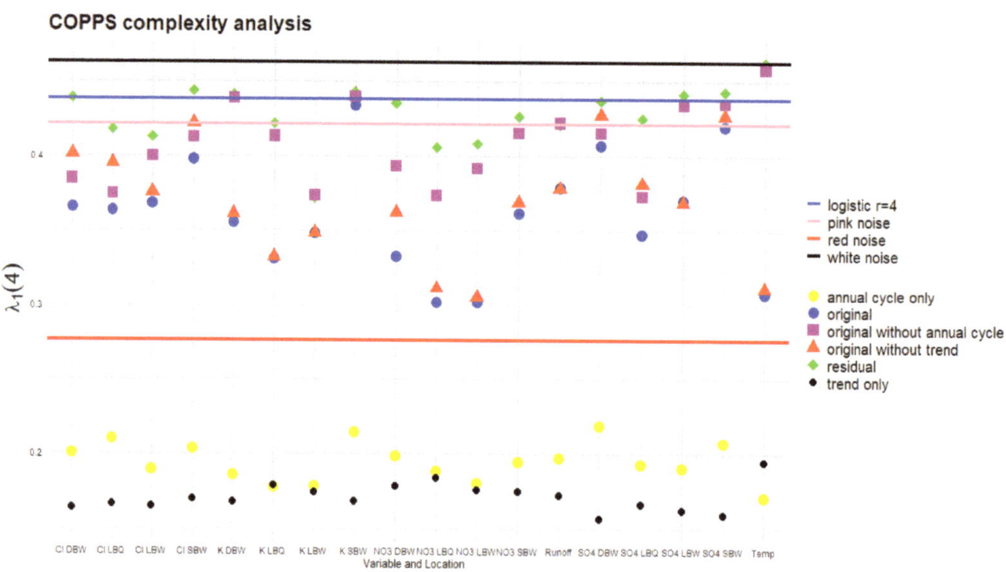

Figure 20. The complexity parameter $\lambda_{s=1}(D = 4)$ of the COPPS analysis [34]. Four reference stochastic processes are shown for comparison.

Removing the trend or the annual component always increases the complexity. The magnitude of the effect is related to the strength (explained variance) of the component, as expected. Two extreme cases are the temperature, where removing the annual cycle lets the λ parameter jump to almost the white noise case, and K LBW, where removal of the trend does not change λ at all; the trend component for this ion has an explained variance of only 0.76% (Table 1). The complexity of the residuals is the highest in all cases. The opposite is true for the λ values of the trend and annual components alone; they all are rather low, well below the reference value for red noise, with the trend λ values being generally the lowest.

It would be possible to calculate an effective powernoise–k for each time series, i.e., one where the λ value of the noise coincides with that of the observed time series. However, the relative position of the noise and the deterministic–chaotic map is not stable as a function of the time series length; at higher N, the noise curves cross the logistic map, all converging to a value of 1 for very long series, whereas the logistic map has a limit value well below that. Thus, conclusions on which "color" our time series have (which noise process they are resembling) depend on their length.

4. Discussion

The complexity analysis of the time series from the Bramke valley, and, we would claim, also for similar observations of water chemistry from other catchments, reveals that the stochastic component overwhelmingly determines the dynamics of the system. From a linear perspective, this is surprising since the trend component and the seasonal cycle explain a lot of the total variance, in a few cases more than 90% (Table 1). However, the nonlinear properties of the time series dominate the complexity metrics considered.

Permutation entropy and complexity, Fisher information, and in particular Rényi and Tsallis entropy and complexity classify the trend and seasonal cycle as deterministic signals. Although we chose a linear decomposition technique, SSA, for disentangling the deterministic from the stochastic part, the metrics show strong non-additivity. Some quantifiers are almost unaffected by removal of one or even two of the deterministic signals, thus the residual is still a rather complex signal; it is in this sense that the stochastic parts of the time series are rather strong.

4.1. Trend and Seasonality

The ecosystems at the Bramke valley are exposed to important environmental trends and disturbances. The major concern of the 1970s and 1980s, acidification, basically came to halt when flue-gas desulfurization was set into action. The consequence is a strong decline in the SO_4 concentrations in the streamwater of all four locations. Since this change is spatially large-scale, it is no surprise that the trend components for SO_4 are strongly correlated to each other (Figure A2). SO_4 at DBW and SBW is also strongly correlated to temperature ($r = 0.71$ and $r = 0.88$, respectively), whereas at LBW and LBQ, SO_4 is anticorrelated ($r = -0.77$ and $r = -0.85$, respectively) with a stable phase shift of ca. 5 months (Figure A8). The annual cycle at LB reflects the expositional difference between a dry south-facing slope and a wet north-facing slope with SO_4 peaks in late winter. At the steeper, parallel sloping catchments DB and SB, the annual variation reflects probably an SO_4 depth gradient in soil storage.

For Cl, there is no linear trend but there is a decadal structure (Figure 6) which is also simultaneously present at the four locations, thus the correlation coefficients for the trend component are also very high in the Cl group (Figure A2). The Cl found in the streams is in large part from sea salt spray, which also acts on broad spatial scales.

For NO_3 and K, the trends are more diverse between the four locations: for K, the strength of the trend component and the amplitude of the annual cycle vary a lot between locations. At SBW, which was limed in 1989 and is the only catchment with a feasible areal cover of deciduous trees, NO_3 concentrations are much higher for most of the period compared to the other three ones; and the K dynamics are quite different from all other ions, as revealed already by the correlogram.

The residuals still show long-range correlations; it is unlikely that these are induced by periodic components at longer timescales, as those were not discovered by the SSA.

4.2. Complexity

The trend and annual components show low entropy values and a MPR complexity higher than that of k noise at the same entropy level, again confirming their deterministic nature. The other variants are compatible with *k* noise, i.e., follow that curve, varying in their equivalent *k* value and thus in correlations strength. NO_3 appears to be the most complex; at the other end of the complexity spectrum, temperature residuals are almost white noise.

Contrary to pure long-term noise processes, all our time series contain missing patterns. For the stochastic variants, the number of missing patterns varies between 1 and 10 (out of 120). The most complex variable, NO_3, also has the highest number of missing patterns (Figure A9). There is still the possibility that the missing patterns would occur for longer time series with the same dynamics [35]. For the deterministic parts, this number is substantially higher than expected, around 100 (Figure A10).

The Fisher information for the stochastic variants (Figure 13) is rather clumped together, but in all cases higher than for the *k* noise.

The q- and α-entropies and complexities confirm the huge difference between the stochastic and the deterministic parts of our time series. It is difficult, however, to draw conclusions on differences within the stochastic group. As the area covered by the (almost) loops for the q-entropy–complexity relationship depends on the Hurst parameter for fBM, one could calculate an effective Hurst parameter for our time series and compare it to the one obtained with more conventional methods.

4.3. Abbe Values

The Abbe values of the Tarnopolski diagram might be used as a classifier for the dynamics of our time series, as they spread over almost the whole available range. They do not seem to make a distinction between the locations easy; the more complex variables (as judged from the MPR plot and Fisher information plot) appear at lower Abbe values. For the stochastic part, the number of turning points is generally higher and the position of our time series interpolates between fBm and *k* noise.

4.4. HVG Slopes

Among the quantifiers considered, this might be the least discriminating for the stochastic parts. Most of the λ_{HVG} values obtained are rather close to white noise, and removing trend or seasonality has a minor effect. The slopes for the deterministic parts are very different, so this basic distinction is possible also here.

4.5. COPPS

The slopes for the ordinal patterns interpolate between red and pink noise in most cases for the stochastic parts. The effect of removing trend or seasonality differs between the variables and is related to the strength of these components (explained variance percentages). The slopes increase from the original to the residual, and reach values compatible with the logistic map, e.g., for SO_4. The location SBW reaches the highest slopes for the residuals in three of the four cases. The trend and seasonal components are characterized by rather low λ values. The COPPS slope seems to be an alternative good indicator of the complexity of a time series; however, the strong dependence on time series length necessitates comparisons only between time series of the same length.

4.6. Summary

Complexity and information measures were used here as efficient tools comparing and classifying this set of environmental time series. They separate the classification task, e.g., between stochastic and deterministic parts, from the more difficult issue of reconstructing a given time series. We found similarities in dissolved ions across sites (especially Cl), but also highly site-specific behavior (K at SBW). Especially for the behavior of variables for which biotic interactions are implicated (K, NO_3, water), reference processes were difficult to identify. Some of the time series are easy to classify by these methods, but difficult to reproduce (and explain) by the respective process-based models. The suite of methods presented here is a rather stringent test environment for them.

Complexity measures have become a critical tool to compare documented environmental behavior relative to candidate references processes from various disciplines. They may thus form a building in search of a common formalization of environmental processes.

Author Contributions: Conceptualization, H.L. and M.H.; methodology, H.L.; software, H.L. and M.H.; validation, H.L. and M.H.; formal analysis, H.L.; investigation, H.L.; resources, M.H.; data curation, M.H.; writing—original draft preparation, H.L.; writing—review and editing, M.H.; visualization, H.L.; funding acquisition, H.L. All authors have read and agreed to the published version of the manuscript.

Funding: H.L. acknowledges partial financial support from the Research Council of Norway (Contract No. 342631/L10) and the EU H2020 Climb-Forest project (project number 101059888).

Institutional Review Board Statement: Not applicable.

Data Availability Statement: The original time series of the Bramke catchments as well as their decomposition are available upon requests. All calculations and analyses were done in the R programming environment. Contact the corresponding author for more information.

Acknowledgments: It is our great pleasure to thank Osvaldo Rosso for collaboration during the last 15 years, and his hospitality during the sabbatical stay of the first author of this contribution in Maceió. Osvaldo's ideas and method development were the main inspiration for this study. The data used here are collected, shared and discussed with us by Henning Meesenburg (Northwest German Forest Research Institute at Göttingen, Germany) since many years. The work here could not have been carried out without his permanent support. Katharina Funk and Britta Aufgebauer (Univ. of Bayreuth, Germany) helped organizing the data set and contributed to software development. We also would like to thank Eyebe Fouda (University of Yaoundé, Cameroon) for sharing his COPPS code with us and fruitful discussions. The complexity measures were calculated using the R package statcomp which Sebastian Sippel (now Univ. of Leipzig, Germany) developed, and the phase shift calculations used the package simsalabim which Lukas Gudmundsson (now ETH Zürich, Switzerland) developed, both in the context of their respective master theses where the two authors of this article were their supervisors.

Conflicts of Interest: The authors declare no conflicts of interest. Also, the funders had no role in the design of the study; in the collection, analyses, or interpretation of data; in the writing of the manuscript; or in the decision to publish the results.

Appendix A. Additional Figures

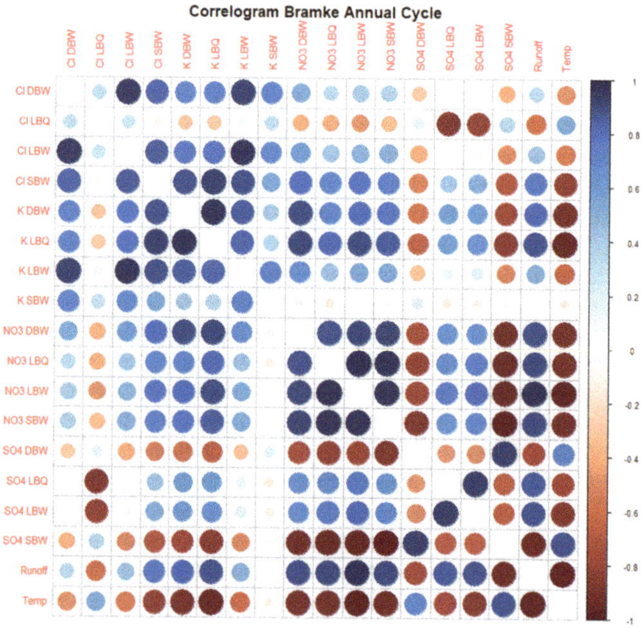

Figure A1. Correlogram for the annual cycle alone.

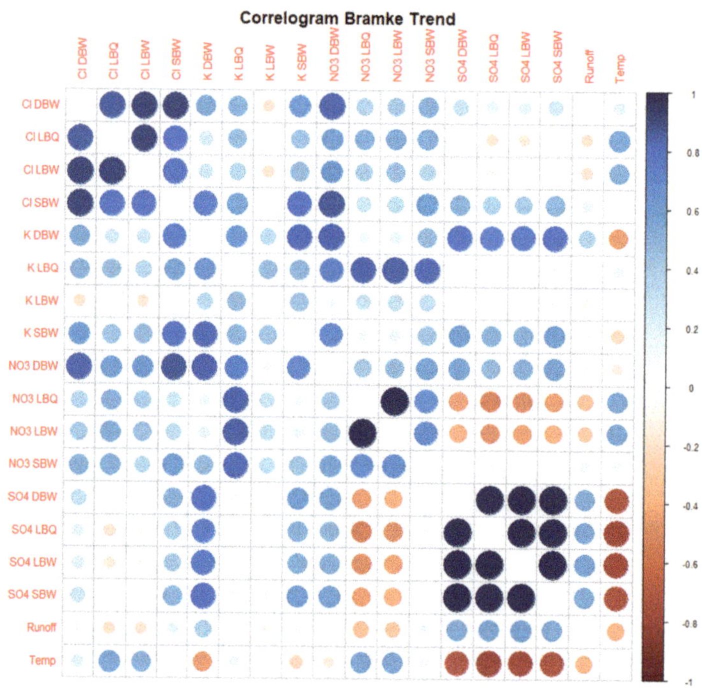

Figure A2. Correlogram for the trend component alone.

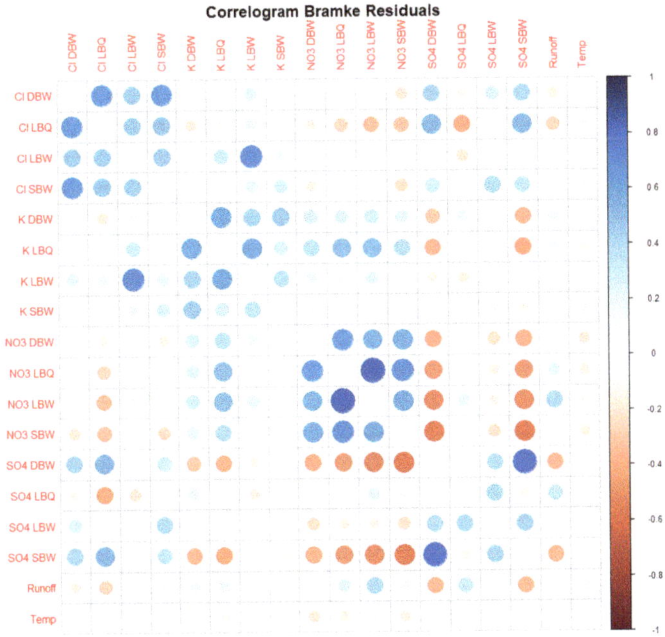

Figure A3. Correlogram for the residuals (after removing both the trend and the annual component).

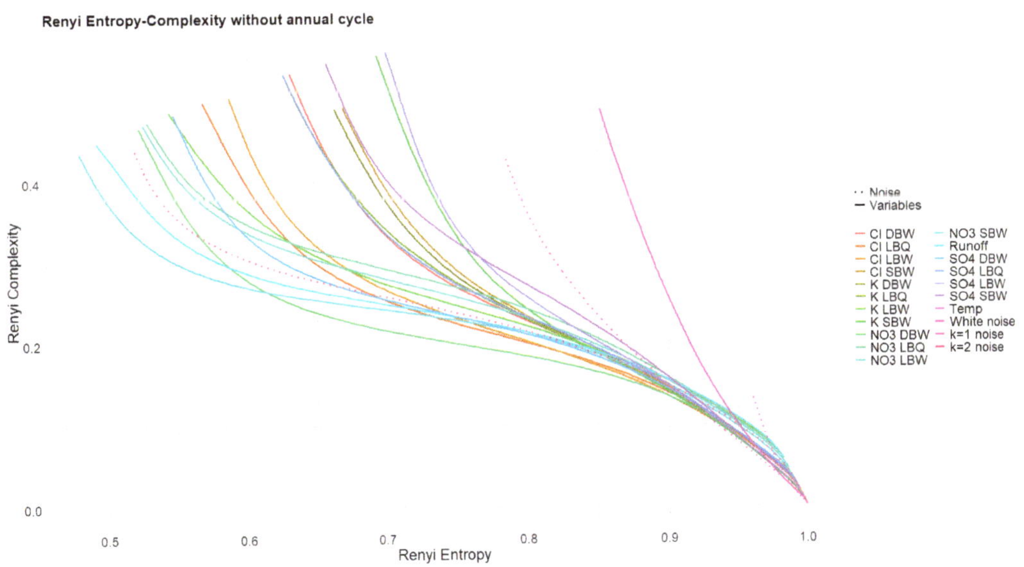

Figure A4. Rényi entropy–complexity plane for the time series where the annual cycle is removed. Similar to Figure 14.

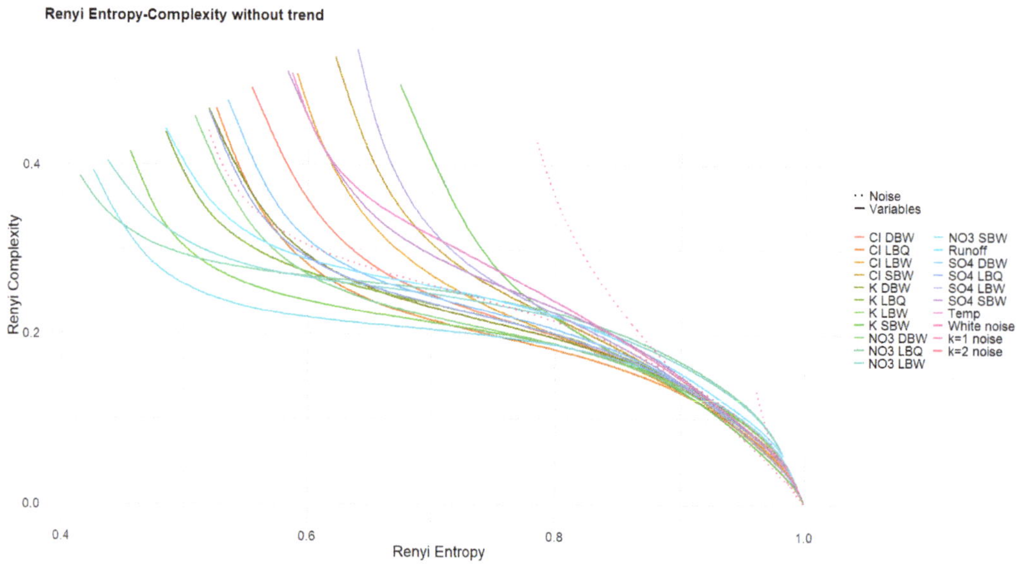

Figure A5. Rényi entropy–complexity plane for the time series with the trend removed. Similar to Figure 14.

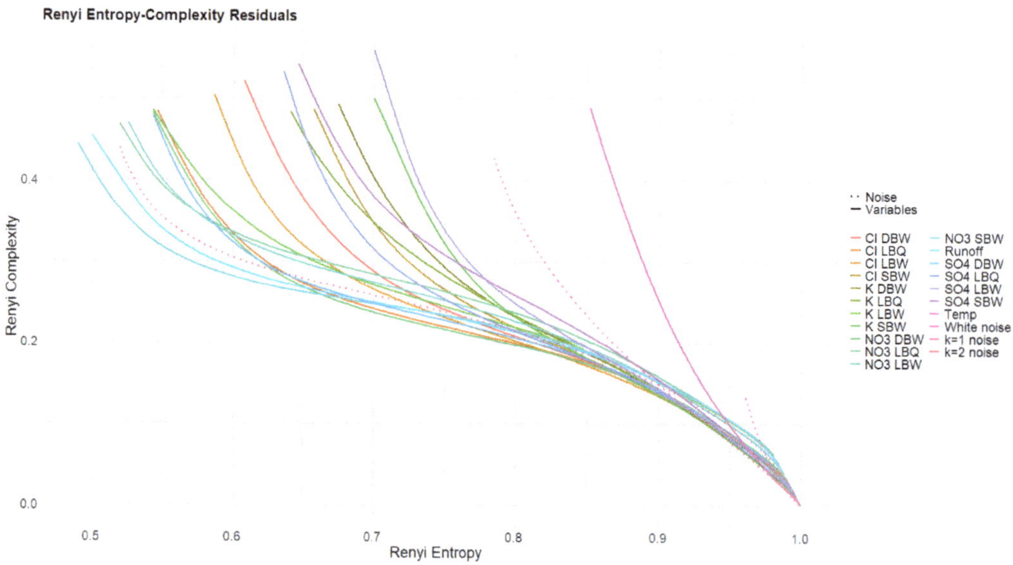

Figure A6. Rényi entropy–complexity plane for the residual time series. Similar to Figure 14.

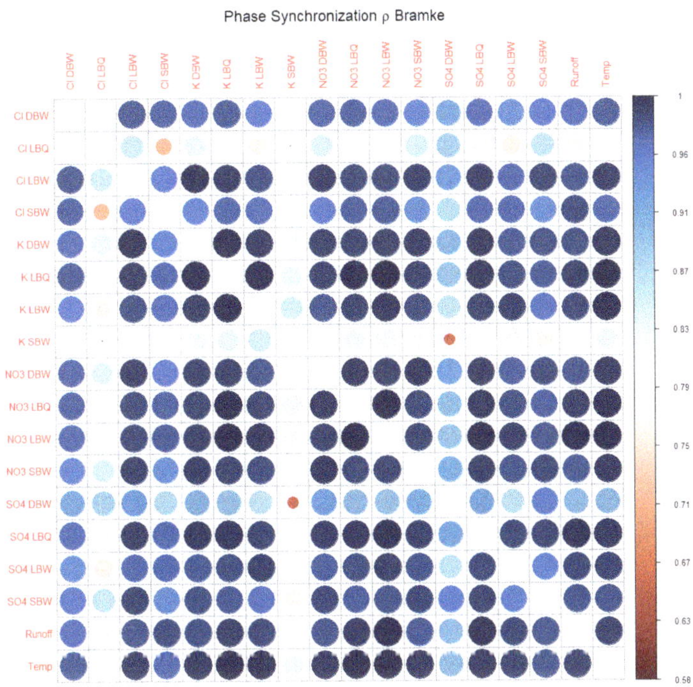

Figure A7. Phase synchronization index ρ for the annual component of the time series.

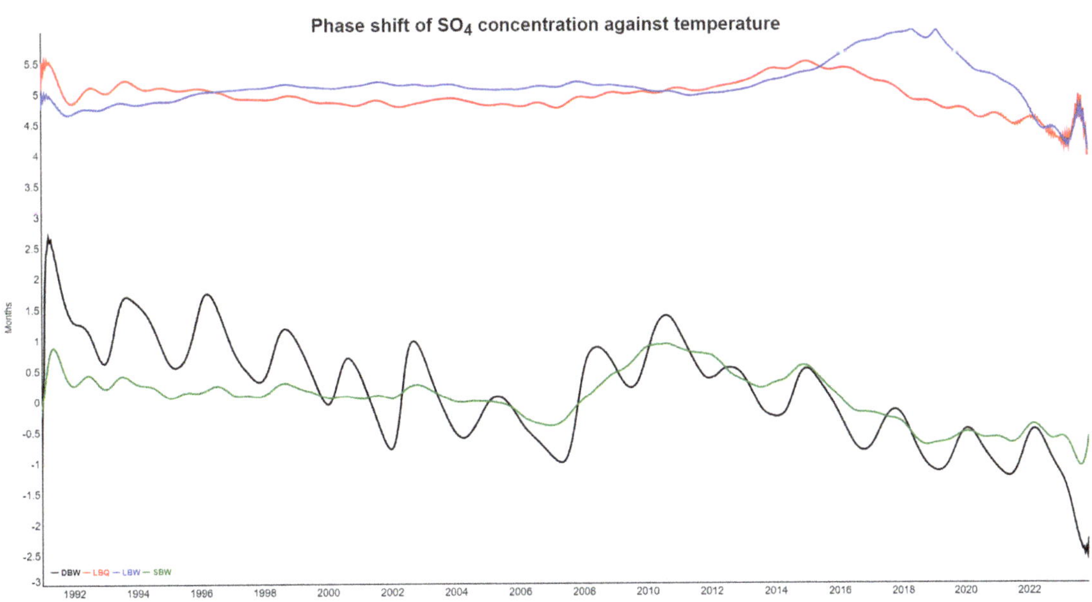

Figure A8. Phase shift time series for SO_4 at the four locations and temperature.

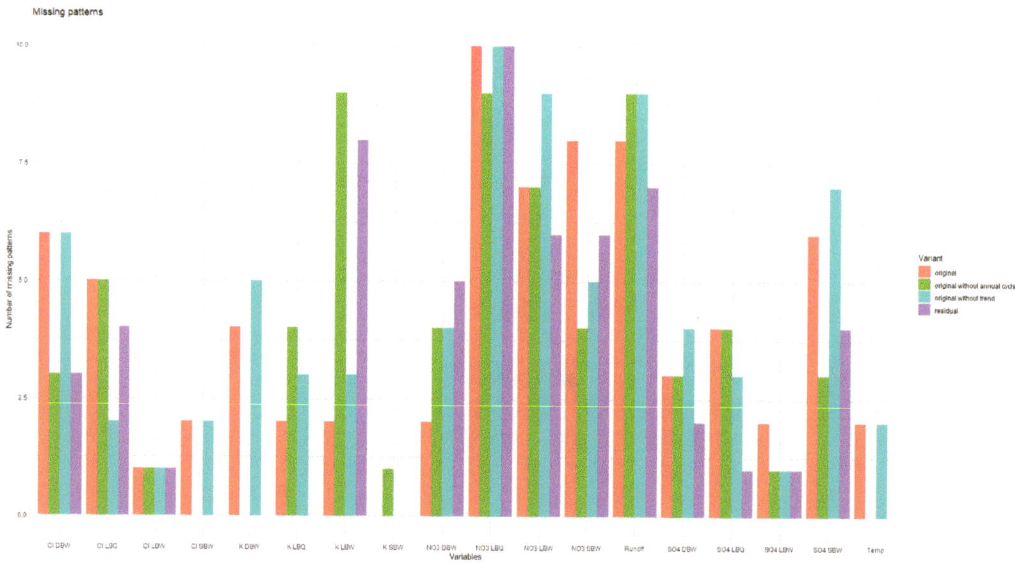

Figure A9. The number of missing patterns for the original, detrended, deseasonalized, and residual variant.

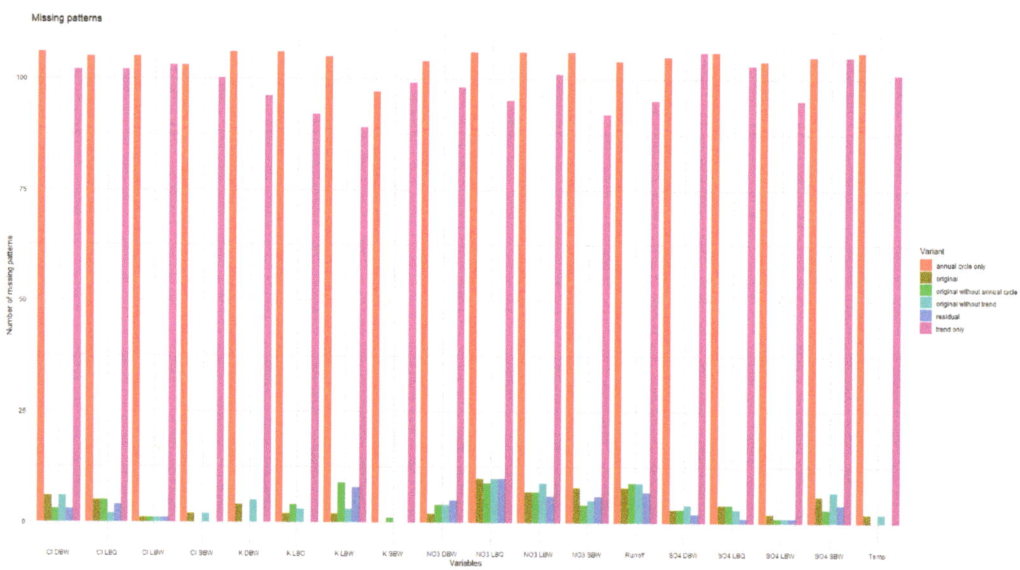

Figure A10. The number of missing patterns for all time series variants. This graph is dominated by the deterministic components (trend and annual cycle only); it is a zoom out of Figure A9.

References

1. Tang, L.; Lv, H.; Yang, F.; Yu, L. Complexity testing techniques for time series data: A comprehensive literature review. *Chaos Solitons Fractals* **2015**, *81*, 117–135. [CrossRef]
2. Fernandez Bariviera, A.; Belen Guercio, M.; Martinez, L.B.; Rosso, O.A. A permutation information theory tour through different interest rate maturities: The Libor case. *Philos. Trans. R. Soc. A-Mathematical Phys. Eng. Sci.* **2015**, *373*, 20150119. [CrossRef]

3. Zunino, L.; Olivares, F.; Bariviera, A.F.; Rosso, O.A. A simple and fast representation space for classifying complex time series. *Phys. Lett. A* **2017**, *381*, 1021–1028. [CrossRef]
4. Montani, F.; Rosso, O.A.; Matias, F.S.; Bressler, S.L.; Mirasso, C.R. A symbolic information approach to determine anticipated and delayed synchronization in neuronal circuit models. *Philos. Trans. R. Soc. A-Math. Phys. Eng. Sci.* **2015**, *373*, 20150110. [CrossRef]
5. Carpi, L.C.; Rosso, O.A.; Saco, P.M.; Ravetti, M.G. Analyzing complex networks evolution through Information Theory quantifiers. *Phys. Lett. A* **2011**, *375*, 801–804. [CrossRef]
6. Rosso, O.A.; Carpi, L.C.; Saco, P.M.; Gómez Ravetti, M.; Plastino, A.; Larrondo, H.A. Causality and the entropy-complexity plane: Robustness and missing ordinal patterns. *Phys. A Stat. Mech. Its Appl.* **2012**, *391*, 42–55.
7. Rosso, O.A.; Olivares, F.; Zunino, L.; De Micco, L.; Aquino, A.L.L.; Plastino, A.; Larrondo, H.A. Characterization of chaotic maps using the permutation Bandt-Pompe probability distribution. *Eur. Phys. J. B* **2013**, *86*, 116.
8. Rosso, O.A.; Larrondo, H.A.; Martin, M.T.; Plastino, A.; Fuentes, M.A. Distinguishing noise from chaos. *Phys. Rev. Lett.* **2007**, *99*, 154102.
9. Rosso, O.A.; De Micco, L.; Plastino, A.; Larrondo, H.A. Info-quantifiers' map-characterization revisited. *Phys. A-Stat. Mech. Its Appl.* **2010**, *389*, 4604–4612. [CrossRef]
10. Rosso, O.A.; Olivares, F.; Plastino, A. Noise versus chaos in a causal Fisher-Shannon plane. *Pap. Phys.* **2015**, *7*, 070006. [CrossRef]
11. Farthing, M.W.; Ogden, F.L. Numerical Solution of Richards' Equation: A Review of Advances and Challenges. *Soil Sci. Soc. Am. J.* **2017**, *81*, 1257–1269. [CrossRef]
12. Pretzsch, H. Forest Dynamics, Growth, and Yield. In *Forest Dynamics, Growth and Yield: From Measurement to Model*; Pretzsch, H., Ed.; Springer: Berlin/Heidelberg, Germany, 2009; pp. 1–39.
13. Cattaneo, N.; Astrup, R.; Antón-Fernández, C. PixSim: Enhancing high-resolution large-scale forest simulations. *Softw. Impacts* **2024**, *21*, 100695. [CrossRef]
14. Hauhs, M. A model of ion-transport through a forested catchment at Lange Bramke, West-Germany. *Geoderma* **1986**, *38*, 97–113. [CrossRef]
15. Forests, I. International Co Operative Programme on Assessment and Monitoring of Air Pollution Effects on Forests. Available online: http://icp-forests.net/ (accessed on 4 March 2025).
16. Keller, R. Wald und Wasserhaushalt: Die Bedeutung neuer Versuche im Harz. *Erdkunde* **1953**, *7*, 52–57.
17. Broomhead, D.S.; King, G.P. On the qualitative analysis of experimental dynamical systems. In *Nonlinear Phenomena and Chaos*; Sarkar, S., Ed.; Adam Hilger: Bristol, UK, 1986; pp. 113–144.
18. Golyandina, N.; Korobeynikov, A. Basic Singular Spectrum Analysis and forecasting with R. *Comput. Stat. Data Anal.* **2014**, *71*, 934–954. [CrossRef]
19. Korobeynikov, A. *Rssa: A Collection of Methods for Singular Spectrum Analysis*; CRAN project 2024. Available online: https://cran.r-project.org/ (accessed on 4 March 2025).
20. Golyandina, N.; Osipov, E. The "Caterpillar" SSA method for analysis of time series with missing values. *J. Stat. Plan. Inference* **2007**, *137*, 2642–2653.
21. Palus, M.; Novotna, D. Common oscillatory modes in geomagnetic activity, NAO index and surface air temperature records. *J. Atmos. Sol.-Terr. Phys.* **2007**, *69*, 2405–2415.
22. Bandt, C.; Pompe, B. Permutation Entropy: A Natural Complexity Measure for Time Series. *Phys. Rev. Lett.* **2002**, *88*, 174102.
23. Martin, M.T.; Plastino, A.; Rosso, O.A. Generalized statistical complexity measures: Geometrical and analytical properties. *Phys. A Stat. Mech. Its Appl.* **2006**, *369*, 439–462.
24. Zunino, L.; Olivares, F.; Ribeiro, H.V.; Rosso, O.A. Permutation Jensen-Shannon distance: A versatile and fast symbolic tool for complex time-series analysis. *Phys. Rev. E* **2022**, *105*, 045310. [CrossRef]
25. Zunino, L.; Porte, X.; Soriano, M.C. Identifying Ordinal Similarities at Different Temporal Scales. *Entropy* **2025**, *26*, 1016. [CrossRef]
26. Rényi, A. On Measures of Entropy and Information. In *Berkeley Symposium on Mathematical Statistics and Probability*; University of California Press: Berkeley, CA, USA, 1961; Volume 1961, pp. 547–561.
27. Tsallis, C. Possible generalization of Boltzmann-Gibbs statistics. *J. Stat. Phys.* **1988**, *52*, 479–487. [CrossRef]
28. Jauregui, M.; Zunino, L.; Lenzi, E.K.; Mendes, R.S.; Ribeiro, H.V. Characterization of time series via Rényi complexity–entropy curves. *Phys. A Stat. Mech. Its Appl.* **2018**, *498*, 74–85. [CrossRef]
29. Ribeiro, H.V.; Jauregui, M.; Zunino, L.; Lenzi, E.K. Characterizing time series via complexity-entropy curves. *Phys. Rev. E* **2017**, *95*, 062106. [CrossRef]
30. Tarnopolski, M. On the relationship between the Hurst exponent, the ratio of the mean square successive difference to the variance, and the number of turning points. *Phys. A Stat. Mech. Its Appl.* **2016**, *461*, 662–673. [CrossRef]
31. Tarnopolski, M. Analytical representation of Gaussian processes in the A-T plane. *Phys. Rev. E* **2019**, *100*, 062144. [CrossRef]

32. Luque, B.; Lacasa, L.; Ballesteros, F.; Luque, J. Horizontal visibility graphs: Exact results for random time series. *Phys. Rev. E* **2009**, *80*, 046103. [CrossRef]
33. Lange, H.; Sippel, S.; Rosso, O.A. Nonlinear dynamics of river runoff elucidated by horizontal visibility graphs. *Chaos* **2018**, *28*, 075520. [CrossRef]
34. Eyebe Fouda, J.S.A.; Koepf, W.; Marwan, N.; Kurths, J.; Penzel, T. Complexity from ordinal pattern positioned slopes (COPPS). *Chaos Solitons Fractals* **2024**, *181*, 114708. [CrossRef]
35. Amigó, J.M.; Zambrano, S.; Sanjuán, M.A.F. True and false forbidden patterns in deterministic and random dynamics. *Europhys. Lett.* **2007**, *79*, 50001. [CrossRef]

entropy

MDPI

Article

Information Theory Quantifiers in Cryptocurrency Time Series Analysis

Micaela Suriano [1,2,*], Leonidas Facundo Caram [2], Cesar Caiafa [3], Hernán Daniel Merlino [4] and Osvaldo Anibal Rosso [5,6]

1 Departamento de Hidráulica, Facultad de Ingeniería, Universidad de Buenos Aires, Av. Las Heras 2214, Buenos Aires C1127AAR, Argentina
2 Laboratorio de Redes y Sistemas Móviles, Departamento de Electrónica, Facultad de Ingeniería, Universidad de Buenos Aires, Buenos Aires C1063ACV, Argentina; fcaram@fi.uba.ar
3 Instituto Argentino de Radioastronomía-CCT La Plata, CONICET/CIC-PBA/UNLP, Camino Gral. Belgrano Km 40, Berazategui B1894XAB, Argentina; ccaiafa@gmail.com
4 Grupo IngenIA, Facultad de Ingeniería, Universidad de Buenos Aires, Buenos Aires C1063ACV, Argentina; hmerlino@fi.uba.ar
5 Instituto de Física (IFLP), Universidad Nacional de La Plata, CONICET, La Plata B1900AJJ, Argentina; oarosso@gmail.com
6 Instituto de Física, Universidade Federal de Alagoas (UFAL), Maceió 57072-970, Brazil
* Correspondence: msuriano@fi.uba.ar

Abstract: This paper investigates the temporal evolution of cryptocurrency time series using information measures such as complexity, entropy, and Fisher information. The main objective is to differentiate between various levels of randomness and chaos. The methodology was applied to 176 daily closing price time series of different cryptocurrencies, from October 2015 to October 2024, with more than 30 days of data and not completely null. Complexity–entropy causality plane (CECP) analysis reveals that daily cryptocurrency series with lengths of two years or less exhibit chaotic behavior, while those longer than two years display stochastic behavior. Most longer series resemble colored noise, with the parameter k varying between 0 and 2. Additionally, Natural Language Processing (NLP) analysis identified the most relevant terms in each white paper, facilitating a clustering method that resulted in four distinct clusters. However, no significant characteristics were found across these clusters in terms of the dynamics of the time series. This finding challenges the assumption that project narratives dictate market behavior. For this reason, investment recommendations should prioritize real-time informational metrics over whitepaper content.

Keywords: permutation entropy; statistical complexity; cryptocurrency

1. Introduction

The relationship between complex systems and cryptocurrencies has been a topic that has captured the interest of the scientific community specializing in cryptocurrencies in recent years. This interest was further amplified by the COVID-19 pandemic, as [1] found an increase in volatility during this period. This is hypothetically due to the time that more people were dedicated to trading cryptocurrencies, especially Bitcoin, as more individuals engaged in trading during lockdowns.

This issue was addressed through the analysis of entropy by [2], concluding that the pandemic altered the volatility inherent in cryptocurrency markets. A study by [3] analyzed the increase in entropy in periods of turbulence, such as during the COVID-19 pandemic when the entire market showed increased volatility, rather than isolated cryptocurrencies. Paradoxically [4] observed volatility consolidation in mature cryptocurrencies, suggesting market maturation phases where entropy metrics require multi-scale approaches.

In [5], the authors employed the Fisher information measure alongside permutation entropy to assess the evolving informational efficiency and price disorder dynamics of five major cryptocurrencies across pre- and intra-pandemic periods, offering critical insights into their maturity and predictability for liquidity risk diversification strategies.

Studies have identified persistent chaos patterns in these markets, proposing alternative methods for volatility analysis [6]. Incorporating sentiment analysis improves volatility analysis in chaotic markets [7]. The unpredictability of cryptocurrency markets reflects a chaotic system with high entropy [8].

The authors in [9] examined how entropy, as a variable in the analysis of network decentralization, contributes to market volatility. The intrinsic volatility of cryptocurrency markets renders traditional volatility evaluation methods inadequate [10]. Research on market risk using high-frequency entropy concluded that this method is a good predictor of Bitcoin's value risk [11]. Entropy is also applied as a measure of uncertainty in portfolio management [12], showing that portfolio diversification is a reasonable practice for reducing return uncertainty. The complexity–entropy causality plane (CEPC) analysis method has been used to distinguish stages of stock market development [13] and to analyze algorithmic behavior in cryptocurrencies [14], serving as a precedent for complexity-based approaches in this field.

Previous studies have highlighted the utility of clustering in cryptocurrency analysis, such as the work of [15], which identified temporal efficiency patterns using permutation entropy and statistical complexity. Our approach extends this paradigm by applying clustering techniques to whitepaper analysis, offering a complementary perspective focused on project fundamentals rather than market metrics.

This paper analyzes the temporal evolution of different cryptocurrency time series, utilizing information quantifiers to map the complexity–entropy causality plane (CECP) and describe system dynamics. By combining global statistical complexity and local Fisher information measures, we classify clusters based on whitepaper attributes, which group projects according to their declared rules and governance structures. These clusters are then evaluated using information theoretic metrics to identify links between project frameworks and statistical market behavior.

This paper is structured as follows:

Materials and Methods: description of applied methods, equations, data sources, and time series characteristics;

Results and Discussion: interpretation of outcomes and analysis;

Conclusions: summary of key findings.

2. Materials and Methods

2.1. Ordinal Patterns

Cryptocurrencies, like many other economic phenomena, produce time series observations. To describe the nature of the processes, it is necessary to determine the appropriate probability density function associated with the analyzed time series. For this analysis, the probability distribution function $\boldsymbol{P}$ can be determined using the method of ordinal patterns developed by Bandt and Pompe (2002) [16]. This method replaces the values that appear in the time series with their corresponding range sequence. This procedure considers the temporal causality within the dynamics of the process.

For a given time series $\{x_t: t = 1, \ldots, N\}$, the delay time τ ($\tau \in \mathbb{N}$), an embedding dimension $D \geq 2$ ($D \in \mathbb{N}$), and the D-ordinal pattern are defined by

$$s \to \left\{X_{s-(D-1)\tau}, X_{s-(D-2)\tau}, \ldots, X_{s-\tau}, X_s\right\} \tag{1}$$

A D-dimensional vector is assigned for every time instant s, and it is determined from the evaluation of the time series in the $s-(D-1)\,\tau, \ldots, s-\tau, s$ instants. The D-ordinal patterns related to the instant s are referred to as the permutation $\pi = \{r_0, r_1, \ldots, r_{D-1}\}$ of $\{0, 1, \ldots, D-1\}$ described by

$$x_{s-r_0\tau} \geq x_{s-r_1\tau} \geq \ldots \geq x_{s-r_{D-2}\tau} \geq x_{s-r_{D-1}\tau}. \tag{2}$$

To find a unique result, $r_i < r_{i-1}$ if $x_{s-r_i\tau} < x_{s-r_{i-1}\tau}$ is considered and equal consecutive values are not usual. The vector defined by Equation (2) becomes the unique symbol π.

Figure 1 illustrates the determination of the permutations π_i for $D = 3$ with a time lag $\tau = 1$. At the top of the figure the possible ordinal patterns π_i for $D = 3$ are presented, while the analysis of a cryptocurrency time series with $\tau = 1$ is displayed at the bottom. For each possible permutation π_i derived from the order $D!$, the associated relative frequencies can be calculated by counting how many times each sequence appears in the series and dividing that number by the total number of sequences. This process yields the probability distribution of the ordinal patterns for the given time series.

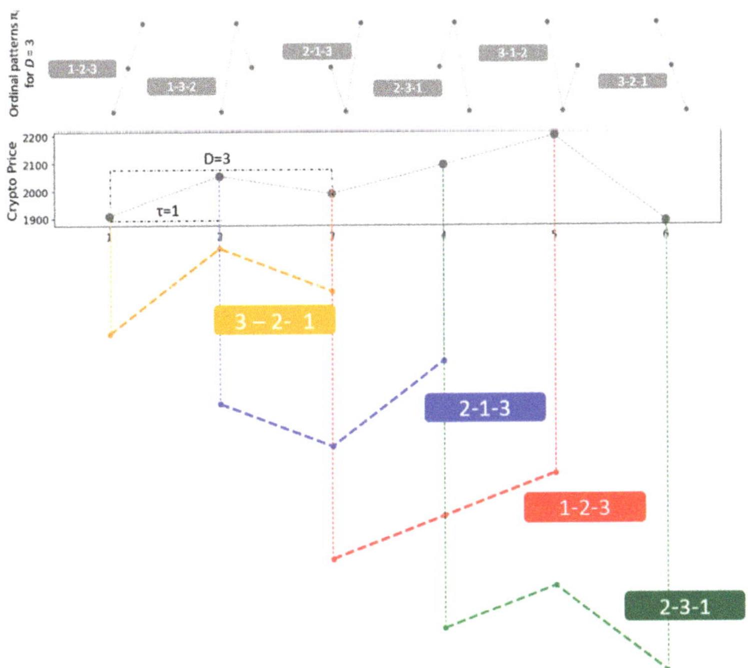

Figure 1. An illustration of identifying ordinal patterns to a cryptocurrency time series (dimension $D = 3$ and an embedding delay $\tau = 1$). At the top, the resultant ordinal patterns π_i for $D = 3$ are presented. At the bottom, the methodology applied to a cryptocurrency time series with $\tau = 1$ is found. The pattern is obtained by replacing the original values with their corresponding rankings, (adapted from [17,18]).

The selection of the value of D should correspond to the minimum sampling frequency needed to capture all the information about the signal's time structure [19]. To differentiate between deterministic and stochastic dynamics, it is recommended that $N \gg D!$, where N denotes the length of the time series [20]. Bandt and Pompe [16] proposed $3 \leq D \leq 7$ and $\tau = 1$. The embedding delay $\tau = 1$ captured the daily temporal structure of the cryptocurrency time series.

2.2. Complexity–Entropy Causality Plane (CECP)

The Shannon logarithmic information measure $S[\boldsymbol{P}]$ is considered a measure of the uncertainty associated with a phenomenon described by $\boldsymbol{P}$, with $\boldsymbol{P}$ being a given arbitrary probability distribution $\boldsymbol{P} = \{p_i: i = 1, \ldots, N\}$, and determined by

$$S[\boldsymbol{P}] = -\sum_{i=1}^{N} p_i ln(p_i) \tag{3}$$

On the one hand, when $S[\boldsymbol{P}] = 0$, it is possible to predict with certainty which of the possible scenarios i, associated with probabilities p_i, will occur. On the other hand, when $\boldsymbol{P}$ is the uniform distribution, our ignorance is maximum. A measure of statistical complexity [21] capable of detecting key details of the dynamics is defined through the product

$$C_{JS}[\boldsymbol{P}] = Q_J[\boldsymbol{P}, \boldsymbol{P_e}] \cdot H_S[\boldsymbol{P}] \tag{4}$$

and the generalized Shannon entropy [22] is

$$H_S = \frac{S[\boldsymbol{P}]}{S_{mx}} \tag{5}$$

with $S_{máx} = S[\boldsymbol{P_e}] = ln(N)$, $0 \leq H_s \leq 1$ and $\boldsymbol{P_e} = 1/N, \ldots, 1/N$ the uniform distribution.

The disequilibrium is defined in terms of the Jensen–Shannon divergence:

$$Q_j[\boldsymbol{P}, \boldsymbol{P_e}] = Q_0 J[\boldsymbol{P}, \boldsymbol{P_e}] \tag{6}$$

with Q_0 being a normalization constant equal to the inverse of the maximum possible value of $J[\boldsymbol{P},\boldsymbol{P_e}]$ and the Jensen–Shannon divergence:

$$J[\boldsymbol{P}, \boldsymbol{P_e}] = S\left[\frac{\boldsymbol{P} + \boldsymbol{P_e}}{2}\right] - \frac{S[\boldsymbol{P}]}{2} - \frac{S[\boldsymbol{P_e}]}{2} \tag{7}$$

The complexity–entropy causality plane (CECP) represents the plot of the permutation statistical complexity C_{JS} versus the generalized Shannon entropy H_S, including the bounds that determine the admissible region which depends solely on the embedding dimension D. The maximum and minimum envelope complexity can be calculated as a function of the entropy [23]. Figure 2 provides a schematic illustration of the CECP. The maximum and minimum boundaries are calculated for $D = 5$ (denoted $C_{máx}$ and $C_{mín}$, respectively) along with an approximate classification of chaotic and stochastic zones.

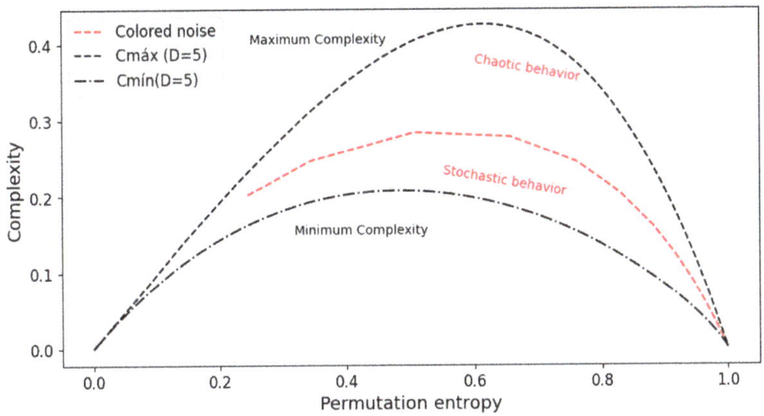

Figure 2. Illustrative representation of the complexity–entropy causality plane (CECP) [13]. In the dashed line are represented the maximum and minimum boundaries for $D = 5$, as well in red the colored noise with the approximate identification of chaotic and stochastic zones [18].

2.3. Fisher's Information Measure (FIM)

The Fisher–Shannon plane is another graphic representation used to identify the characteristics of the time series dynamics based on the probability distribution $\boldsymbol{P}$. This approach can reveal the informational characteristics of the planar position [24]. The Fisher's information measure (FIM) [25] represents a measure of the ability to estimate the amount of information that can be extracted from a set of measurements [26] and F serves as a measure of the gradient content of the distribution $f(x)$, calculated by

$$F[f] = \int_{\Delta} \frac{1}{f(x)} \left[\frac{df(x)}{dx}\right]^2 dx = 4 \int_{\Delta} \left[\frac{d\Psi(x)}{dx}\right]^2 \quad (8)$$

For a discrete environment, the best-behaved expression to use [27] is the discrete normalized FIM, as obtained by

$$F[\boldsymbol{P}] = F_0 \sum_{i=1}^{N-1} \left[(p_{i+1})^{\frac{1}{2}} - (p_i)^{\frac{1}{2}}\right]^2 \quad (9)$$

and the normalization constant F_0 is obtained by

$$F_0 = \begin{Bmatrix} 1\ if\ p_{i*} = 1\ for\ i^* = 1\ or\ i^* = N\ and\ p_i = 0\ \forall i \neq i^* \\ 0\ otherwise \end{Bmatrix} \quad (10)$$

Then, the Fischer–Shannon causality plane Hs x F can be calculated, where Hs is the generalized Shannon entropy in Equation (3).

2.4. Clustering

Cryptocurrencies can be described based on their white papers, which are technical documents that outline the project, its value proposition, the underlying technology, and other relevant details. A white paper is similar to a business plan and informs investors and users about the project.

By applying Natural Language Processing (NLP) analysis, it is possible to identify the most relevant terms in each whitepaper, allowing us to apply a clustering method that results in four clusters. Consequently, each token is assigned four membership coefficients indicating the degree of similarity with each cluster.

To identify clusters, the Latent Dirichlet Allocation (LDA) model was used. LDA is a probabilistic model that attempts to generate clusters based on the similarities between documents [28]. The basic concept behind this clustering is that similar documents are grouped into the same cluster, implying that cryptocurrencies with comparable characteristics should exhibit similar behavior.

2.5. Cryptocurrency Time Series

This paper applied the methods described to 176 daily closing price time series of different cryptocurrencies. The data were obtained from CoinMarketCap (coinmarketcap.com) (accessed on 15 October 2024). The data selection was based on the length of the series, spanning from October 2015 to October 2024, requiring series of more than 30 days in length and not completely null. A second classification was conducted to compare the behavior of series with a length of two years or less to those of greater size, as detailed in Tables A1 and A2, respectively. In [29], the authors suggest exercising caution when applying permutation-based entropies to short high-frequency variability series characterized by a low signal-to-noise ratios. However, daily closing price time series are analyzed in this study, which smooths out intraday fluctuations and, therefore, is not considered to be significantly affected by high-frequency noise.

The algorithms used for analysis were obtained from the Python3.11.12 library ordpy 1.2.0 [30] to apply the methodology to the data. The color noise series was generated with the library colorednoise 2.2.0 (by Felix Patzel), which generates Gaussian distributed noise with a power law spectrum based on the algorithm [31]. Plots are created using the Matplotlib 3.10.0 [32], seaborn 0.13.2 [33], and GeoPandas 1.0.1 [34] Python libraries.

3. Results and Discussion

The results were obtained by applying the methodology to the daily closing price time series of different cryptocurrencies, as detailed in Appendix A. The data sets were divided into two groups: those with lengths of two years or less and those with lengths longer than two years. Bandt and Pompe [16] proposed $3 \leq D \leq 7$ and $\tau = 1$. In [35] it was found that an embedding dimension $3 \leq D \leq 5$, as well as a window length of about 10^3 sampling points, is favorable. Moreover, increasing the value of D results in a greater incorporation of temporal causality into the embedding vectors [14]. In this case, the embedding delay is considered to be $\tau = 1$ in order to capture the daily temporal structure of the cryptocurrency time series. The embedding dimension is set to $D = 5$ to effectively capture the variability observed within a typical working week in cryptocurrency markets. This choice enables the construction of embedding vectors that represent five consecutive daily observations, aligning with the standard trading cycle and enabling the model to incorporate temporal dependencies and weekly behavioral patterns specific to cryptocurrencies. Since the lengths of the series are variable, considering an average length $Na = 1950$, $D = 5$ satisfied the requirement that $Na \gg D!$ [20].

The methodology was applied to dynamic stochastic series of k-noises (noise with a power spectrum frequency dependence characterized by $f^{(-k)}$), where k ranged from 0.00 to 3.50 in increments of 0.25, with 10 random simulations conducted for each k value.

The complexity–entropy causality planes (CECPs) using Shannon entropy and the Fischer–Shannon plane for the daily cryptocurrency time series are illustrated in Figures 3 and 4, respectively.

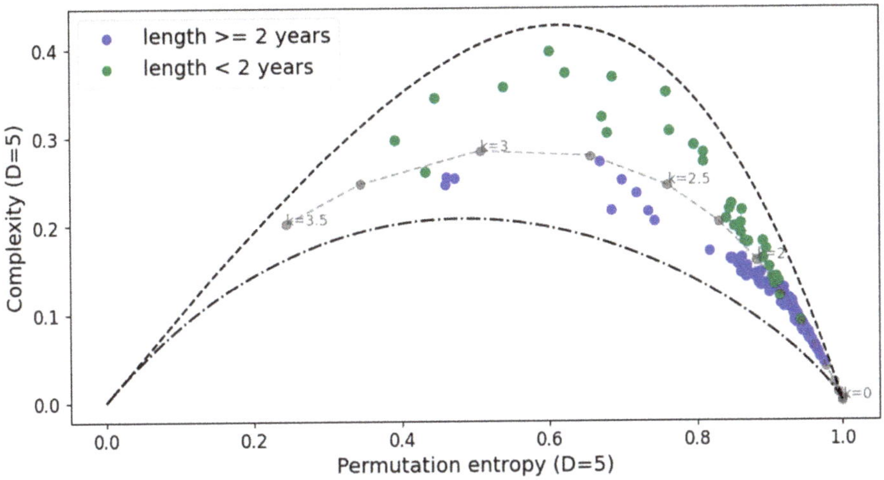

Figure 3. Complexity–entropy causality plane analyses for all the cryptocurrency series and noise with power spectrum frequency-varying parameter k (gray dashed line). In color blue the time series with lengths of more than two years and in green those with equal two years or less. The parameters used are $D = 5$ and $\tau = 1$.

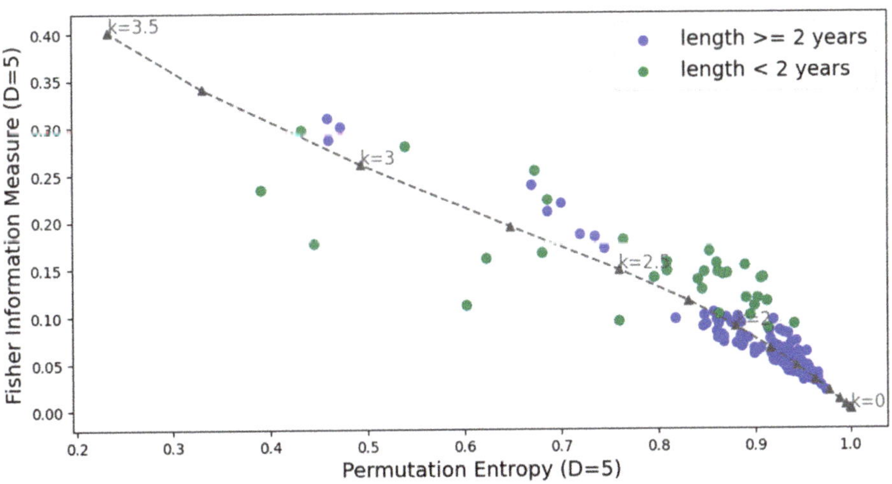

Figure 4. Fisher information measures and permutation entropy planes for all the cryptocurrency series analyzed and noise with power spectrum frequency-varying parameter k (gray dashed line). In color blue the time series with a length of more than two years and in green those with equal two years or less. The parameters used are $D = 5$ and $\tau = 1$.

The CECP analyses showed that, in most cases, daily cryptocurrency time series with lengths of two years or less exhibit chaotic behavior, while those longer than two years display stochastic behavior. For longer duration series, it can be observed that colored noise represents an upper bound on the CECP.

As shown in Figure 5, in all instances, the complexity, entropy, and Fisher parameters exhibit significantly greater variability in the short series compared to the long series. The long series demonstrate more stable parameter values, with a higher median entropy, lower complexity, and lower Fisher values when compared to the short series.

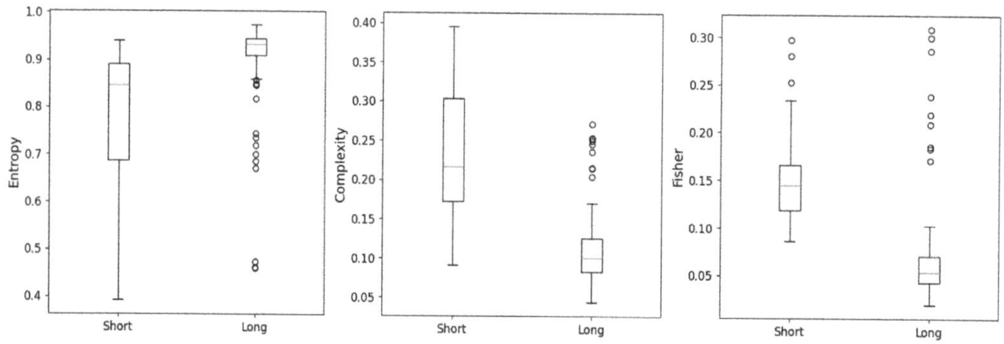

Figure 5. Box plots to compare short (equal or less than two years) and long (more than two years) cryptocurrency time series in terms of entropy, complexity and Fisher parameters. The parameters used are $D = 5$ and $\tau = 1$.

Subsequently, a clustering analysis was conducted for the subgroup of cryptocurrencies identified based on the characteristics outlined in the white papers across four topics. After applying the Latent Dirichlet Allocation (LDA) algorithm, the resulting clusters are as follows:

Topic 1 has 51 cryptocurrency time series: 'ADC', 'ANC', 'ARG', 'BITS', 'BLOCK', 'BSD', 'BTB', 'CANN', 'CLAM','CSC', 'CURE', 'DEM', 'EMD', 'ETH', 'FLT', 'FRC', 'GRC', 'IFC', 'IXC', 'LOG', 'NAV', 'NET', 'NOBL', 'NTRN', 'NVC', 'OMNI', 'ORB', 'PINK', 'POP', 'PTC', 'RBT', 'RDD', 'RED', 'SLG', 'SMLY', 'THC', 'TRK', 'TRUST', 'TTC', 'USNBT', 'UTC', 'VIA', 'XCN', 'XDN', 'XLM', 'XMR', 'XPD', 'XPM', 'XQN', 'XST', and 'XVG'.

Topic 2 has 45 cryptocurrency time series: 'ARI', 'BBR', 'BLU', 'BTC', 'BTS', 'CLOAK', 'CRW', 'CRYPT', 'DIME', 'DMD', 'DP', 'DTC', 'EFL', 'EMC2', 'FJC', 'FTC', 'GLD', 'GRS', 'HBN', 'LDOGE', 'LOG', 'MAX', 'MEC', 'NOTE', 'NYC', 'POT', 'PPC', 'PXC', 'RBY', 'SKC', 'SLR', 'SOON', 'STV', 'SXC', 'TGC', 'TIPS', 'UBQ', 'UNIT', 'UNO', 'VTC', 'XBC', 'XCO', 'XMY', 'XPY', and 'XWC'.

Topic 3 has 13 cryptocurrency time series: 'AC', 'BSTY', 'DGB', 'DGC', 'GAME', 'GRN', 'PHO', 'PLNC', 'TES', 'TROLL', 'VRC', 'WDC', and 'XCP'.

Topic 4 has 30 cryptocurrency time series: '42', 'ACOIN', 'AIB', 'ANC', 'AUR', 'BAY', 'BCN', 'BLC', 'BTA', 'BTCD', 'CASH', 'DON', 'DOPE', 'EMC', 'ENRG', 'FAIR', 'FLO', 'GP', 'IOC', 'KOBO', 'MINT', 'MONA', 'NXS', 'NXT', 'OK', 'PND', 'SAK', 'SUPER', 'SYS', and 'VIA'.

Figure 6 presents the CECP analyses for clustering classification. It can be observed that Topics 1 and 2 exhibit similar behaviors. Most series are concentrated near the noise curve between $k = 0$ and $k = 2$, displaying chaotic behavior for lower entropy values. For Topic 2, the majority of the series are close to the noise zone region between $k = 0$ and $k = 2.5$, with some values bordering the curve for lower entropy values. Finally, for Topic 3, the values appear near the noise curve between $k = 0$ and $k = 2.5$, but it should be noted that the sample size is smaller than in the other cases, with only 13 cases analyzed.

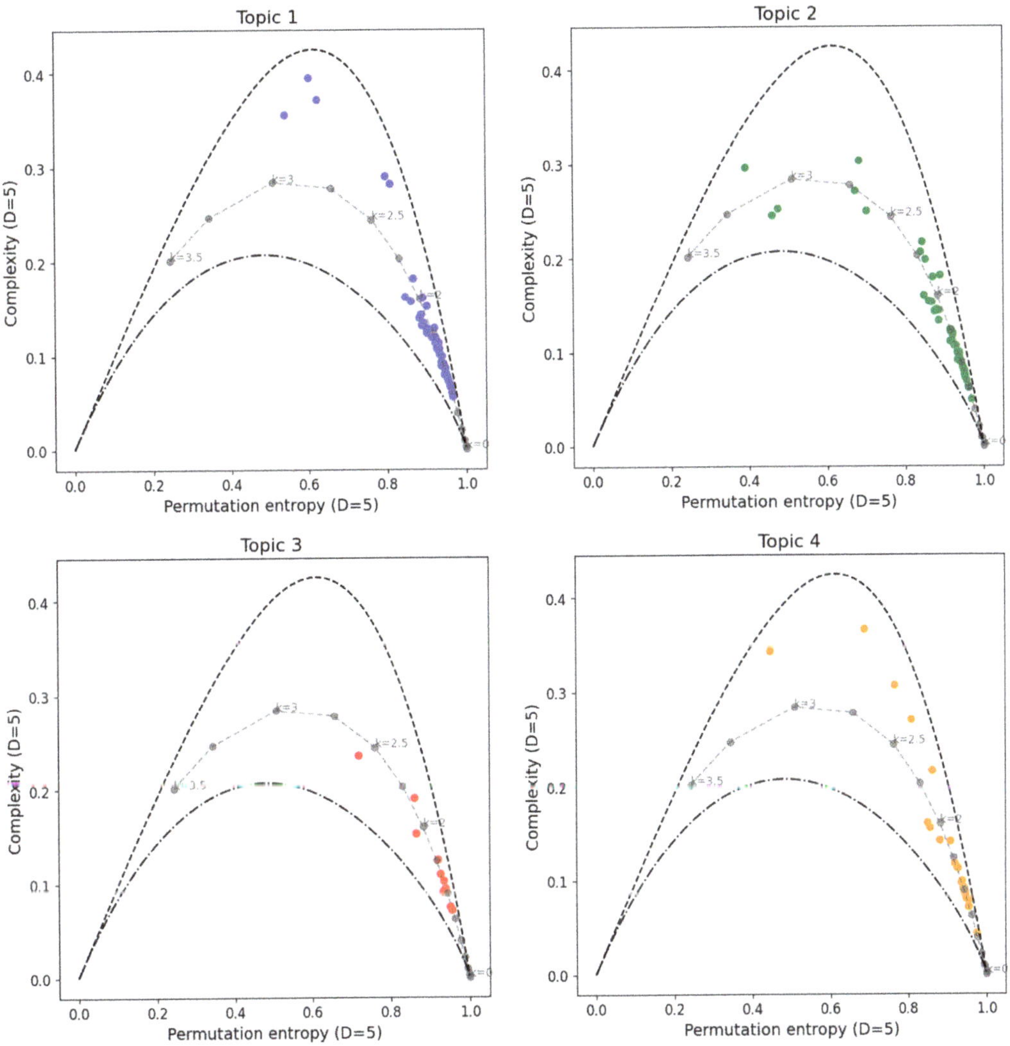

Figure 6. Complexity–entropy causality plane analyses for the cryptocurrency series divided in clusters and noise with power spectrum frequency-varying parameter k (gray dashed line). Topic 1 in color blue, Topic 2 in color green, Topic 3 in color red and Topic 4 in color orange. The parameters used are $D = 5$ and $\tau = 1$.

The Fisher information measure vs. permutation entropy plane is shown in Figure 7, comparing the different topics identified in this study. No significant differences in behavior are observed between topics. In all cases, most series are concentrated near the noise curve between $k = 0$ and $k = 2.5$, with greater dispersion for higher k values.

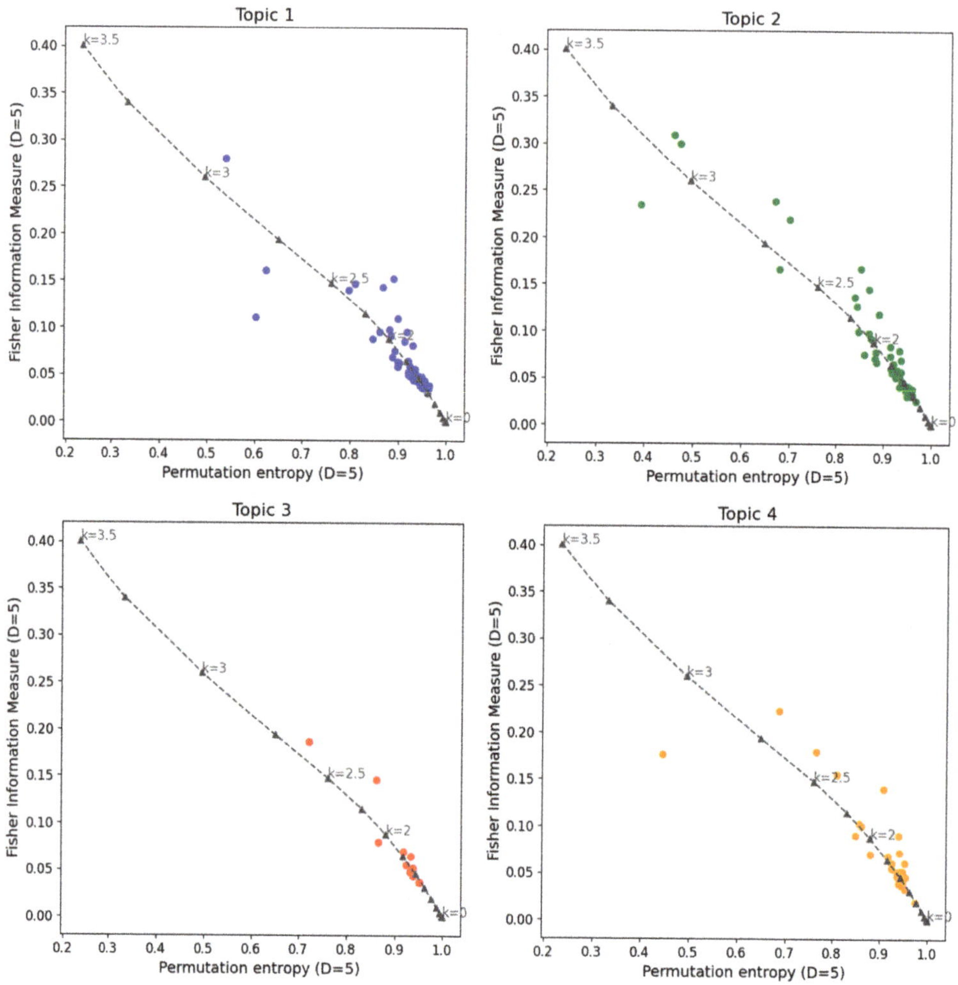

Figure 7. Fisher information measures and permutation entropy planes for all the cryptocurrency series analyzed and noise with power spectrum frequency-varying parameter k (gray dashed line). Topic 1 in color blue, Topic 2 in color green, Topic 3 in color red, and Topic 4 in color orange. The parameters used are $D = 5$ and $\tau = 1$.

In Figure 8, the box plots compare the different topics in terms of parameters such as entropy, complexity, and Fisher information. The graphs show that no significant differences exist in the mean values.

The clustering technique successfully identified four distinct groups based on the main characteristics in their white papers. However, these differences were not significant when evaluated using information measures such as entropy, complexity, and Fisher information. Despite the variations in the white papers, the time series dynamics remained similar. This result reveals the limited predictive power of white papers, suggesting that declared project rules do not directly govern observed market efficiency or complexity.

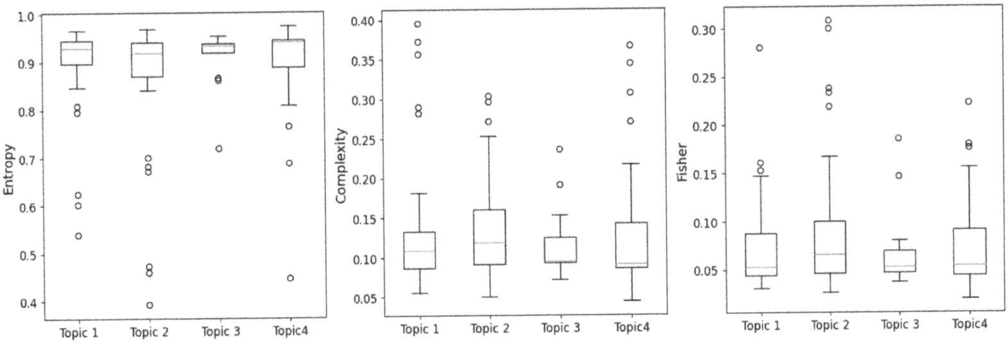

Figure 8. Boxplots to compare the different topics clustering of the cryptocurrency time series in terms of entropy, complexity and Fisher parameters. The parameters used are $D = 5$ and $\tau = 1$.

4. Conclusions

The temporal evolution of cryptocurrency time series was analyzed using information measures. The dynamics of these series were described in terms of complexity, entropy, and Fisher information. The main objective was to distinguish between different levels of randomness and chaos.

The representation in the complexity-entropy causality plane (CECP) indicates that daily cryptocurrency series with a length of two years or less exhibit chaotic behavior, while series longer than two years are associated with stochastic behavior. Most of the longer series display characteristics similar to colored noise, with the parameter k varying between 0 and 2. The CECP framework could help investors distinguish between speculative assets (chaotic phases) and mature markets (stochastic phases), informing portfolio diversification strategies.

By applying Natural Language Processing (NLP) analysis, the most relevant terms in each whitepaper were identified, enabling the use of a clustering method that resulted in four distinct clusters. However, no significant characteristics were found in terms of information measures, challenging the assumption that project narratives dictate market behavior. For this reason, an investment recommendation should prioritize real-time informational metrics over white paper content.

Future research should investigate other variables, including technological innovations, investor sentiment metrics, and macro-financial drivers (regulatory shifts and cross-asset correlations), that may be influencing the behavior of time series in cryptocurrencies.

Understanding these dynamics is critical for constructing diversified portfolios that mitigate risks associated with chaotic market phases while capitalizing on efficiency regimes. For instance, identifying assets transitioning from chaos to stochasticity—via complexity-entropy metrics—can inform strategic asset allocation, balancing high-volatility cryptocurrencies with stable, entropy-resilient ones.

Author Contributions: All the authors made significant contributions to this work. Conceptualization, M.S., H.D.M. and L.F.C.; methodology, M.S., H.D.M., L.F.C., C.C. and O.A.R.; supervision, L.F.C. and O.A.R.; validation, C.C. and O.A.R.; visualization, M.S.; writing—original draft, M.S.; writing—review and editing, H.D.M., C.C., L.F.C. and O.A.R. All authors have read and agreed to the published version of the manuscript.

Funding: We acknowledge the financial support by the Universidad de Buenos Aires (UBACYT-20020220400162BA).

Institutional Review Board Statement: Not applicable.

Data Availability Statement: The data used to support the findings from this study are included within the article. The processed data are available from the corresponding author upon request.

Acknowledgments: We would like to thank the Academic Editor and the two anonymous reviewers for their constructive criticism and comments, which greatly contributed to enhancing the quality of the manuscript.

Conflicts of Interest: The authors declare no conflicts of interest.

Appendix A

In this appendix, the information from the cryptocurrency time series is shown. Table A1 shows the series with lengths of two years or less and Table A2 shows those of greater length.

Table A1. Information from the cryptocurrency time series with two years or less length.

Crypto	Length	Start	End
ARI	927	20/12/2021	3/7/2024
BAY	250	1/11/2021	8/7/2022
BCF	298	25/5/2022	18/3/2023
BTCD	227	19/11/2023	2/7/2024
BTCS	301	19/12/2023	14/10/2024
CBX	1075	5/11/2021	14/10/2024
CRYPT	563	10/10/2021	25/4/2023
DON	992	28/8/2020	28/5/2023
DP	342	22/11/2021	31/10/2022
FCN	88	7/10/2021	3/1/2022
FLO	317	10/3/2021	20/1/2022
FLT	204	25/3/2024	14/10/2024
GLD	573	14/12/2021	9/7/2023
GP	45	1/12/2021	14/1/2022
HYPER	487	27/4/2021	11/9/2022
NET	58	8/3/2023	4/5/2023
NOBL	182	16/4/2024	14/10/2024
PND	136	28/3/2022	10/8/2022
PTC	95	12/7/2024	14/10/2024
RPC	450	24/1/2022	30/8/2023
SAK	373	8/9/2022	15/9/2023
SLG	720	19/7/2022	7/7/2024
SLR	434	18/11/2021	29/1/2023
SOON	505	14/10/2022	1/3/2024
SRC	71	5/8/2024	14/10/2024
TES	430	9/8/2023	14/10/2024
TGC	99	4/1/2022	12/4/2022
TIT	353	23/6/2023	9/6/2024
TRI	376	15/12/2020	27/12/2021
TRK	57	30/11/2021	2/2/2022
TRUST	755	31/7/2020	24/8/2022
TTC	712	29/7/2022	9/7/2024
UNB	1036	14/12/2021	14/10/2024
UTC	408	22/3/2022	5/6/2023

Table A2. Information from the cryptocurrency time series with lengths of more than two years.

Crypto	Length	Start	End
42	2532	9/11/2017	14/10/2024
AC	1410	13/10/2020	25/8/2024
ACOIN	2527	9/11/2017	14/10/2024
ADC	2532	9/11/2017	14/10/2024
AIB	1998	9/11/2017	13/6/2023
ANC	2532	9/11/2017	14/10/2024
ARG	1162	10/8/2021	14/10/2024
AUR	2532	9/11/2017	14/10/2024
BBR	1656	27/11/2017	9/6/2022
BCN	2531	9/11/2017	14/10/2024
BITB	2532	9/11/2017	14/10/2024
BITS	2518	9/11/2017	14/10/2024
BLC	1537	9/11/2017	23/1/2022
BLK	2532	9/11/2017	14/10/2024
BLOCK	2532	9/11/2017	14/10/2024
BLU	2532	9/11/2017	14/10/2024
BSD	2284	9/11/2017	9/2/2024
BSTY	2532	9/11/2017	14/10/2024
BTA	2532	9/11/2017	14/10/2024
BTB	2105	10/1/2019	14/10/2024
BTC	3285	18/10/2015	14/10/2024
BTS	2532	9/11/2017	14/10/2024
C2	2526	9/11/2017	14/10/2024
CANN	2532	9/11/2017	14/10/2024
CASH	1637	22/4/2020	14/10/2024
CCN	2055	1/3/2019	14/10/2024
CLAM	2532	9/11/2017	14/10/2024
CLOAK	2532	9/11/2017	14/10/2024
CRW	2532	9/11/2017	14/10/2024
CSC	2500	9/11/2017	14/10/2024
CURE	2532	9/11/2017	14/10/2024
DASH	2532	9/11/2017	14/10/2024
DEM	2461	9/11/2017	14/10/2024
DGB	2532	9/11/2017	14/10/2024
DGC	3285	18/10/2015	14/10/2024
DIME	2532	9/11/2017	14/10/2024
DMD	2532	9/11/2017	14/10/2024
DOGE	2532	9/11/2017	14/10/2024
DOPE	2532	9/11/2017	14/10/2024
DTC	2237	21/2/2018	14/4/2024
ECC	1762	9/11/2017	5/9/2022
EFL	2120	9/11/2017	14/10/2024
EMC	2532	9/11/2017	14/10/2024
EMC2	2222	9/11/2017	12/12/2023
EMD	1573	9/11/2017	10/3/2022
ENRG	1601	9/11/2017	30/6/2022
ETH	2532	9/11/2017	14/10/2024
EXCL	1871	9/11/2017	23/12/2022
FAIR	1154	9/11/2017	10/5/2022
FJC	1689	9/11/2017	8/7/2022
FRC	3181	18/10/2015	14/10/2024
FST	1775	8/4/2019	15/2/2024
FTC	3285	18/10/2015	14/10/2024
GAME	2532	9/11/2017	14/10/2024
GCN	2104	9/11/2017	16/1/2024

Table A2. *Cont.*

Crypto	Length	Start	End
GRC	2284	9/11/2017	9/2/2024
GRN	1922	21/5/2018	24/8/2023
GRS	2532	9/11/2017	14/10/2024
HBN	1821	9/11/2017	14/10/2024
IFC	2012	9/11/2017	14/10/2024
IOC	1673	9/11/2017	8/6/2022
IXC	3208	18/10/2015	14/10/2024
KOBO	2524	9/11/2017	14/10/2024
LDOGE	996	17/4/2020	14/10/2024
LOG	1795	9/11/2017	14/10/2024
LTC	3285	18/10/2015	14/10/2024
MAX	2531	9/11/2017	14/10/2024
MEC	2533	18/10/2015	2/11/2023
MINT	2284	9/11/2017	9/2/2024
MNC	1301	21/2/2019	13/9/2022
MONA	2532	9/11/2017	14/10/2024
MUE	1674	9/11/2017	9/6/2022
NAV	2532	9/11/2017	14/10/2024
NLG	1839	9/11/2017	21/11/2022
NMC	3285	18/10/2015	14/10/2024
NOTE	2312	9/11/2017	3/5/2024
NTRN	2532	9/11/2017	14/10/2024
NVC	3002	18/10/2015	14/10/2024
NXS	2333	9/11/2017	14/10/2024
NXT	2532	9/11/2017	14/10/2024
NYAN	1206	1/9/2018	14/2/2023
NYC	2522	9/11/2017	14/10/2024
OK	2532	9/11/2017	14/10/2024
OMNI	2532	9/11/2017	14/10/2024
ORB	2051	9/11/2017	24/6/2023
PHO	1516	9/11/2017	23/1/2022
PIGGY	1097	14/10/2021	14/10/2024
PINK	2284	9/11/2017	9/2/2024
PLNC	2496	9/11/2017	14/10/2024
POP	1466	10/10/2020	14/10/2024
POT	2529	9/11/2017	14/10/2024
PPC	3285	18/10/2015	14/10/2024
PXC	3071	18/10/2015	14/10/2024
PXI	2532	9/11/2017	14/10/2024
QRK	2530	9/11/2017	14/10/2024
RBT	2532	9/11/2017	14/10/2024
RBY	2532	9/11/2017	14/10/2024
RDD	2532	9/11/2017	14/10/2024
RED	2337	23/5/2018	14/10/2024
SKC	1356	4/5/2020	19/1/2024
SMC	1667	9/11/2017	2/6/2022
SMLY	2275	9/11/2017	9/2/2024
SONG	2532	9/11/2017	14/10/2024
SPHR	1673	9/11/2017	9/6/2022
SPR	2532	9/11/2017	14/10/2024
START	2532	9/11/2017	14/10/2024
STV	1306	19/3/2021	14/10/2024
SUPER	1765	9/11/2017	12/9/2022
SXC	1199	4/7/2021	14/10/2024
SYS	2532	9/11/2017	14/10/2024

Table A2. *Cont.*

Crypto	Length	Start	End
TAG	2520	9/11/2017	14/10/2024
THC	2532	9/11/2017	14/10/2024
TIPS	2532	9/11/2017	14/10/2024
TRC	3037	18/10/2015	9/2/2024
TROLL	2491	9/11/2017	14/10/2024
UBQ	2225	9/11/2017	12/12/2023
UFO	2188	9/11/2017	14/10/2024
UNIT	2532	9/11/2017	14/10/2024
UNO	2532	9/11/2017	14/10/2024
USNBT	1650	9/11/2017	16/5/2022
VIA	2532	9/11/2017	14/10/2024
VRC	1669	9/11/2017	4/6/2022
VTC	2532	9/11/2017	14/10/2024
WBB	2521	9/11/2017	14/10/2024
WDC	2759	18/10/2015	14/10/2024
XBC	2532	9/11/2017	14/10/2024
XCN	2532	9/11/2017	14/10/2024
XCO	2393	9/11/2017	28/5/2024
XCP	2532	9/11/2017	14/10/2024
XDN	2532	9/11/2017	14/10/2024
XEM	2532	9/11/2017	14/10/2024
XLM	2532	9/11/2017	14/10/2024
XMR	2532	9/11/2017	14/10/2024
XMY	2225	9/11/2017	12/12/2023
XPD	2532	9/11/2017	14/10/2024
XPM	2532	9/11/2017	14/10/2024
XPY	2517	9/11/2017	14/10/2024
XQN	2526	9/11/2017	14/10/2024
XRP	2532	9/11/2017	14/10/2024
XST	2349	9/11/2017	14/10/2024
XVG	2532	9/11/2017	14/10/2024
XWC	2532	9/11/2017	14/10/2024
ZET	2532	9/11/2017	14/10/2024

References

1. Drożdż, S.; Kwapień, J.; Oświęcimka, P.; Stanisz, T.; Wątorek, M. Complexity in Economic and Social Systems: Cryptocurrency Market at around COVID-19. *Entropy* **2020**, *22*, 1043. [CrossRef]
2. Kim, K.; Lee, M. The Impact of the COVID-19 Pandemic on the Unpredictable Dynamics of the Cryptocurrency Market. *Entropy* **2021**, *23*, 1234. [CrossRef] [PubMed]
3. García-Medina, A.; Hernández, C.J.B. Network Analysis of Multivariate Transfer Entropy of Cryptocurrencies in Times of Turbulence. *Entropy* **2020**, *22*, 760. [CrossRef]
4. Kwapień, J.; Wątorek, M.; Drożdż, S. Cryptocurrency Market Consolidation in 2020–2021. *Entropy* **2021**, *23*, 1674. [CrossRef] [PubMed]
5. Fernandes, L.H.S.; Bouri, E.; Silva, J.W.L.; Bejan, L.; de Araujo, F.H.A. The resilience of cryptocurrency market efficiency to COVID-19 shock. *Physica A* **2022**, *607*, 128218. [CrossRef] [PubMed]
6. Dehouche, N. Revisiting the volatility of bitcoin with approximate entropy. *Cogent Econ. Financ.* **2022**, *10*, 2013588. [CrossRef]
7. Valencia, F.; Gómez-Espinosa, A.; Valdés-Aguirre, B. Price Movement Prediction of Cryptocurrencies Using Sentiment Analysis and Machine Learning. *Entropy* **2019**, *21*, 589. [CrossRef]
8. Stosic, D.; Stosic, D.; Ludermir, T.B.; Stosic, T. Exploring disorder and complexity in the cryptocurrency space. *Phys. A Stat. Mech. Its Appl.* **2019**, *525*, 548–556. [CrossRef]
9. Liu, F.; Fan, H.-Y.; Qi, J.-Y. Blockchain Technology, Cryptocurrency: Entropy-Based Perspective. *Entropy* **2022**, *24*, 557. [CrossRef]

10. Puoti, F.; Pittorino, F.; Roveri, M. Quantifying Cryptocurrency Unpredictability: A Comprehensive Study of Complexity and Forecasting. *arXiv* **2025**, arXiv:2502.09079.
11. Pele, D.T.; Mazurencu-Marinescu-Pele, M. Using High-Frequency Entropy to Forecast Bitcoin's Daily Value at Risk. *Entropy* **2019**, *21*, 102. [CrossRef] [PubMed]
12. Rodriguez-Rodriguez, N.; Miramontes, O. Shannon Entropy: An Econophysical Approach to Cryptocurrency Portfolios. *Entropy* **2022**, *24*, 1583. [CrossRef]
13. Zunino, L.; Zanin, M.; Tabak, B.M.; Pérez, D.G.; Rosso, O.A. Complexity-entropy causality plane: A useful approach to quantify the stock market inefficiency. *Phys. A Stat. Mech. Its Appl.* **2010**, *389*, 1891–1901. [CrossRef]
14. Bariviera, A.F.; Zunino, L.; Rosso, O.A. An analysis of high-frequency cryptocurrencies prices dynamics using permutation-information-theory quantifiers. *Chaos* **2018**, *28*, 075511. [CrossRef] [PubMed]
15. Sigaki, H.Y.D.; Perc, M.; Ribeiro, H.V. Clustering patterns in efficiency and the coming-of-age of the cryptocurrency market. *Sci. Rep.* **2019**, *9*, 1440. [CrossRef]
16. Bandt, C.; Pompe, B. Permutation Entropy: A Natural Complexity Measure for Time Series. *Phys. Rev. Lett.* **2002**, *88*, 174102. [CrossRef]
17. Zanin, M.; Olivares, F. Ordinal patterns-based methodologies for distinguishing chaos from noise in discrete time series. *Commun. Phys.* **2021**, *4*, 190. [CrossRef]
18. Suriano, M.; Caram, L.F.; Rosso, O.A. Daily streamflow of Argentine Rivers Analysis Using Information Theory Quantifiers. *Entropy* **2024**, *26*, 56. [CrossRef]
19. Serinaldi, F.; Zunino, L.; Rosso, O.A. Complexity–entropy analysis of daily stream flow time series in the continental United States. *Stoch. Environ. Res. Risk Assess.* **2014**, *28*, 1685–1708. [CrossRef]
20. Rosso, O.A.; Larrondo, H.A.; Martín, M.T.; Plastino, A.; Fuentes, M.A. Distinguishing Noise from Chaos. *Phys. Rev. Lett.* **2007**, *99*, 154102. [CrossRef]
21. Stosic, T.; Telesca, L.; Vicente de Souza Ferreira, D.; Stosic, B. Investigating anthropically induced effects in streamflow dynamics by using permutation entropy and statistical complexity analysis: A case study. *J. Hydrol.* **2016**, *540*, 1136–1145. [CrossRef]
22. Shannon, C.E. A Mathematical Theory of Communication. *Bell System Technol.* **1948**, *27*, 379–423. [CrossRef]
23. Calbet, X.; López-Ruiz, R. Tendency towards maximum complexity in a nonequilibrium isolated system. *Phys. Rev. E Stat. Nonlinear Soft Matter Phys.* **2001**, *63*, 066116. [CrossRef]
24. Rosso, O.A.; Olivares, F.; Plastino, A. Noise versus chaos in a causal Fisher-Shannon plane. *Pap. Phys.* **2015**, *7*, 1852–4249. [CrossRef]
25. Fisher, R.A. On the mathematical foundations of theoretical statistics. *Philos. Trans. R. Soc. Lond. Ser. A* **1922**, *222*, 309–368. [CrossRef]
26. Frieden, B. *Science from Fisher Information: A Unification*; Cambridge University Press: Cambridge, UK, 2004. [CrossRef]
27. Sánchez-Moreno, P.; Yanez, R.; Dehesa, J. Discrete densities and Fisher information. In Proceedings of the 14th International Conference on Difference Equations and Applications, Instanbul, Turkey, 21–25 July 2008; Bohner, M., Ed.; Difference Equations and, Applications. Ugur–Bahçeşehir University Publishing Company: Istanbul, Turkey, 2009; pp. 291–298.
28. Hasan, M.; Rahman, A.; Karim, M.R.; Khan, M.S.I.; Islam, M.J. Normalized approach to find optimal number of topics in Latent Dirichlet Allocation (LDA). In *International Conference on Trends in Computational and Cognitive Engineering: Proceedings of TCCE 2020*; Springer: Singapore, 2021; pp. 341–354. [CrossRef]
29. Porta, A.; Bari, V.; Marchi, A.; De Maria, B.; Castiglioni, P.; di Rienzo, M.; Guzzetti, S.; Cividjian, A.; Quintin, L. Limits of permutation-based entropies in assessing complexity of short heart period variability. *Physiol. Meas.* **2015**, *36*, 755–765. [CrossRef]
30. Pessa, A.A.; Ribeiro, H.V. Ordpy: A Python package for data analysis with permutation entropy and ordinal network methods. *Chaos* **2021**, *31*, 063110. [CrossRef] [PubMed]
31. Timmer, J.; Koenig, M. On generating power law noise. *Astron. Astrophys.* **1995**, *300*, 707–710.
32. Hunter, J.D. Matplotlib: A 2D Graphics Environment. *Comput. Sci. Eng.* **2007**, *9*, 90–95. [CrossRef]
33. Waskom, M.L. Seaborn: Statistical data visualization. *J. Open Source Softw.* **2021**, *6*, 3021. [CrossRef]
34. Jordahl, K.; Van den Bossche, J.; Fleischmann, M.; Wasserman, J.; McBride, J.; Gerard, J.; Fleischmann, M.; Tratner, J.; Perry, M.; Farmer, C.; et al. *Geopandas/Geopandas: v0.8.1 (Version v0.8.1)*; Zenodo: Geneva, Switzerland, 2020. [CrossRef]
35. Staniek, M.; Lehnertz, K. Parameter selection for Permutation Entropy Measurements. *Int. J. Bifurc. Chaos* **2007**, *17*, 3729–3733. [CrossRef]

entropy

MDPI

Article

Quantum-Like Approaches Unveil the Intrinsic Limits of Predictability in Compartmental Models

José Alejandro Rojas-Venegas [1,2,†], Pablo Gallarta-Sáenz [3,4,†], Rafael G. Hurtado [2,*], Jesús Gómez-Gardeñes [3,4,*] and David Soriano-Paños [4,5,*]

1 Departamento Administrativo Nacional de Estadística (DANE), Bogotá 111321, Colombia; jarojasv@dane.gov.co
2 Departamento de Física, Facultad de Ciencias, Universidad Nacional de Colombia, Bogotá 111321, Colombia
3 Departamento de Física de la Materia de Condensada, Universidad de Zaragoza, 50009 Zaragoza, Spain; pgallarta@unizar.es
4 GOTHAM Lab, Instituto de Biocomputación y Sistemas Complejos (BIFI), Universidad de Zaragoza, 50018 Zaragoza, Spain
5 Departament d'Enginyería Informática i Matemátiques, Universitat Rovira i Virgili, 43007 Tarragona, Spain
* Correspondence: rghurtadoh@unal.edu.co (R.G.H.); gardenes@unizar.es (J.G.-G.); sorianopanos@gmail.com (D.S.-P.)
† These authors contributed equally to this work.

Abstract: Obtaining accurate forecasts for the evolution of epidemic outbreaks from deterministic compartmental models represents a major theoretical challenge. Recently, it has been shown that these models typically exhibit trajectory degeneracy, as different sets of epidemiological parameters yield comparable predictions at early stages of the outbreak but disparate future epidemic scenarios. In this study, we use the Doi–Peliti approach and extend the classical deterministic compartmental models to a quantum-like formalism to explore whether the uncertainty of epidemic forecasts is also shaped by the stochastic nature of epidemic processes. This approach allows us to obtain a probabilistic ensemble of trajectories, revealing that epidemic uncertainty is not uniform across time, being maximal around the epidemic peak and vanishing at both early and very late stages of the outbreak. Therefore, our results show that, independently of the models' complexity, the stochasticity of contagion and recovery processes poses a natural constraint for the uncertainty of epidemic forecasts.

Keywords: epidemic dynamics; compartmental models; Doi–Peliti formalism

1. Introduction

Understanding the temporal dynamics of epidemic outbreaks is critical for pandemic management [1,2]. Classical compartmental models for disease transmission, grounded in the pioneering work of Kermack and McKendrick [3], are considered the keystones of mathematical epidemiology to quantify the public health threat posed by a novel pathogen. Indeed, different indicators, such as the effective reproduction number [4] or the expected outbreak size, have long served as hallmarks for the design of non-pharmaceutical interventions or vaccine rollouts to mitigate the social impact of infectious diseases.

The use of compartmental models for the assessment of epidemic scenarios implicitly assumes that the ultimate consequences of control policies on disease outbreaks can be predicted. However, forecasting the long-term evolution of infectious disease outbreaks still remains a major challenge [5,6], both theoretically [7–9] and from a data-driven perspective [10,11].

Focusing on real data, such challenge can be attributed to the unpredictability and intricacies of the variety of biological and social factors neglected in simple compartmental models but ultimately shaping the long-term propagation of infectious diseases [12,13].

Along this line, multimodel forecasting efforts [14,15], integrating predictions from multiple frameworks with different underlying assumptions, have been recently proposed as a solution to partially overcome the former fundamental limitation and to extend the time horizon over which accurate forecasts can be made.

Beyond external factors not included in their formulation, the intrinsic mathematical properties of simple deterministic compartmental models also pose limitations for the reliability of their long-term epidemic forecasts. Recent works have shown that the parameter identification issue [16,17], measuring whether epidemiological parameters can be retrieved when calibrating models with limited data, represents an important source of uncertainty for epidemic forecasts. This issue is exacerbated when noisy points [18] or very early stages of the outbreak [19] are considered for calibration purposes, as trajectory degeneracy ultimately makes divergent predictions compatible with these data. The latter issue hampers the accuracy of the predictions of key quantities in mathematical epidemiology such as the time and size of the epidemic peak or the duration of an epidemic wave [20].

Recognizing the limitations of deterministic models, the role of stochasticity in epidemic models has been also addressed over recent years [21–23]. For instance, stochastic Markovian models allow for capturing the uncertain course of epidemic trajectories in small size populations [24–27] and improving the calibration of epidemic frameworks to real noisy data through the use of adapted Kalman filters [28,29]. In general, accounting for the inherent stochastic dynamics in epidemic processes requires moving from a set of equations governing the time evolution of the expected number of cases to a master equation approach yielding a probabilistic ensemble of epidemic trajectories. One simple approach consists of introducing random fluctuations in the spreading dynamics to improve the compartmental differential equations [30], while more sophisticated frameworks rely on the use of quantum mechanics tools to study different dynamic systems [31,32] or the Doi–Peliti formalism to study the critical behaviour of epidemiological models using the Hamilton–Jacobi equations [33].

Despite these novel approaches, determining the influence of the inherent stochasticity of epidemic processes on forecast uncertainty remains an open problem. To fill this gap, we follow previous works and propose a quantum-like formalism to model epidemic dynamics by extending the Doi–Peliti approach [34,35] to the classical susceptible–infected–susceptible (SIS) and susceptible–infected–recovered (SIR) models. By leveraging the Doi–Peliti formalism, our paper aims at (i) unravelling hidden behaviours in classical deterministic compartmental models and (ii) showcasing how stochasticity shapes the uncertainty of epidemic outbreaks. In particular, the main contribution of this study is to reveal that the stochastic nature of epidemic processes hinders obtaining faithful long-term forecasts on the magnitude and position of the epidemic peak at early stages of an epidemic outbreak.

This article is organized as follows. We first introduce the theoretical formalism of our work in Section 2, including the basic rules to describe the system and the Master Equation. Then, in Section 3, we present the main results of our work, related to the simulation of the Master Equation, the probability of finding minor outbreaks and the stochastic determinants for the uncertainty of epidemic forecasts. Finally, we discuss the implications of our findings and future research venues in Section 4.

2. Doi–Peliti Approach to Compartmental Models

In this section, we present the theoretical background of our work, both the classical compartmental models and Doi–Peliti formalism, with equations that lead to the description of this new approach. In both cases, we consider a closed population with N individuals, thus neglecting any changes in the population size as a result of birth–death processes. Regarding the structure of contacts, we restrict to the simplest scenario and follow a mean-field assumption considering well-mixed populations.

2.1. Deterministic Equations for the SIS and SIR Models

The most usual way to tackle the modelling of SIS and SIR dynamics is to consider a set of ordinary differential equations (ODEs) governing the time evolution of the expected occupation of the different compartments. In the simplest case, the SIS model assumes that each individual can be either in the Susceptible or in the Infected state and that transitions between them correspond to contagion and recovery processes. In particular, Susceptible individuals become infectious at a β rate upon contact with Infected individuals. Moreover, Infected individuals recover at a γ rate, without acquiring any immunity against the circulating pathogen. Denoting the occupation of the Susceptible (Infected) compartment by S (I), the deterministic equations capturing the SIS dynamics are as follows:

$$\frac{d}{dt}S = -\beta S\frac{I}{N} + \gamma I, \quad \frac{d}{dt}I = \beta S\frac{I}{N} - \gamma I. \tag{1}$$

The SIR model, instead, accounts for those infections conferring immunity to the host upon recovery. This aspect is included by assuming that infectious individuals recover at a γ rate and enter into a new compartment, the Removed state R, rather than returning to the Susceptible state. From these simple rules, the differential equations of the SIR model read as follows:

$$\frac{d}{dt}S = -\beta S\frac{I}{N}, \quad \frac{d}{dt}I = \beta S\frac{I}{N} - \gamma I, \quad \frac{d}{dt}R = \gamma I. \tag{2}$$

2.2. The Doi–Peliti Master Equation

Going beyond the traditional deterministic approach, epidemic dynamics can be interpreted as stochastic birth–death processes. In these models, individuals transition (or 'die') from one compartment to 'be born' into another. In this context, Doi–Peliti formalism [34,35] takes advantage of the quantum field theory to build a Markovian Master Equation (MME). This approach requires two fundamental components: the vectors, $|\varphi\rangle$, describing the dynamical state of the system, and the creation–annihilation operators, a, $a^{\dagger}$, which create or annihilate individuals, respectively, in the compartments described in the epidemic models.

Regarding the first component, we follow a probabilistic approach considering that the state of our system $|\varphi\rangle$ exists in the space in which the elements of the basis $|\phi\rangle$ span, with each one representing a possible configuration of the model under consideration. Mathematically, we assume the following:

$$|\varphi\rangle = \sum_{\phi} P(\phi)|\phi\rangle \,. \tag{3}$$

The elements $|\phi\rangle$ of the basis depend on the compartmental model and are explained below for both the SIS and the SIR models. The second component concerns the ladder operators for each compartment, a and $a^{\dagger}$, creating or removing individuals, respectively. Assuming that $|x\rangle$ represents the element of the basis corresponding to an occupation number x of a given compartment, the previous operators are defined as follows:

$$a^{\dagger}|x\rangle = |x+1\rangle, \tag{4}$$

$$a|x\rangle = x|x-1\rangle. \tag{5}$$

Likewise, these operators follow some quantum mechanics principles: their eigenvalues allow us to write the states as $|x\rangle = \left(a^{\dagger}\right)^{x}|0\rangle$, and they satisfy the usual commutation rule $\left[a, a^{\dagger}\right] = aa^{\dagger} - a^{\dagger}a = \mathbb{1}$. Furthermore, it is also possible to define a number operator, $\hat{n} = a^{\dagger}a$, which returns the occupation number of the state $\hat{n}|x\rangle = x|x\rangle$.

With these two ingredients, one can construct the Hamiltonian-governing different dynamics in systems with many-body interactions as outlined in [33,36]. Regardless of the chosen dynamics, the Doi–Peliti approach allows us to capture the evolution of the system state, $|\varphi\rangle$, using a backward master equation (BME) [34,35], analogous to Schrödinger's equation with an imaginary time as follows:

$$\frac{d}{dt}|\varphi\rangle = \mathcal{H}|\varphi\rangle. \tag{6}$$

2.3. The Doi–Peliti Approach to the SIS Model

In an SIS dynamics with a closed population of N individuals, the number of infected individuals I provides enough information to describe the state of the system $|\varphi_{SIS}\rangle$, given the constraint $S = N - I$. Thus, the state $|\varphi_{SIS}\rangle$ can be written as a combination of all possible occupation numbers, $\{|I\rangle\}_{I=0,\ldots,N}$, forming the following basis of the system:

$$|\varphi_{SIS}\rangle = \sum_{I} P(I)|I\rangle. \tag{7}$$

In the former linear combination, the coefficients $P(I) = \langle I|\varphi_{SIS}\rangle$ measure the probability associated with each occupation number, unequivocally defining the state $|\varphi_{SIS}\rangle$ by the set $\{P(I)\}_{I=0,\ldots,N}$.

To construct the Hamiltonian governing SIS dynamics, we consider different ladder operators acting on each compartment. We consider that the operators a, $a^{\dagger}$ act on the susceptible states, whereas b, $b^{\dagger}$ act on the infectious ones. Therefore, the operator $(ab^{\dagger})$ models contagion processes, creating an infectious individual and removing a susceptible one, whereas $(ba^{\dagger})$ captures recovery processes. Consequently, the Hamiltonian of the SIS dynamics, $\mathcal{H}_{\mathcal{SIS}}$, reads as follows:

$$\mathcal{H}_{\mathcal{SIS}} = -\frac{\beta}{N}n_I\left(n_S - ab^{\dagger}\right) - \gamma\left(n_I - ba^{\dagger}\right). \tag{8}$$

2.4. The Doi–Peliti Approach to the SIR Model

The SIR model requires a basis that accounts for the occupation numbers of two of the three compartments (S, I, and R). Here, without loss of generality, we take the infected and susceptible occupation numbers to fully capture the system composition. Therefore, we define the state basis as $\{|S,\, I\rangle\}_{S,\,I=0,\ldots,N;\,S+I\leq N}$. Hence, the state $|\varphi_{SIR}\rangle$ is expressed as follows:

$$|\varphi_{SIR}\rangle = \sum_{S,I} P(S,I)|S,\, I\rangle, \tag{9}$$

with $P(S,I) = \langle S,\, I|\varphi_{SIR}\rangle$. Analogous to the SIS case, we assume that the set $\{P(S,I)\}$ defines $|\varphi_{SIR}\rangle$, simplifying the notation and the representation of the states.

To construct the SIR Hamiltonian, we must define the ladder operators c and $c^{\dagger}$ acting on the recovered compartment. In the SIR model, the transition of an infected individual moving to the recovered state $(bc^{\dagger})$ replaces the transition to the susceptible state of the SIS model. Then, the Hamiltonian $\mathcal{H}_{\mathcal{SIR}}$ is as follows:

$$\mathcal{H}_{\mathcal{SIR}} = -\frac{\beta}{N}n_I\left(n_S - ab^{\dagger}\right) - \gamma\left(n_I - bc^{\dagger}\right). \tag{10}$$

3. Results

In this section, we present the main results of our work, derived from simulations of the Master Equation discussed earlier. The results are organized into three subsections: the dynamics of both models (Section 3.1), a formal analysis and computational solution for the probability of no-outbreak results (Section 3.2), and an examination of the predictability problem using entropy measures of temporal dynamics (Section 3.3).

3.1. Dynamics of the Doi–Peliti Master Equation

Figure 1 summarizes the two different approaches of classical compartmental models, highlighting the conceptual differences between both the determistic ODEs for the SIS and SIR models and their corresponding equations based on the Doi–Peliti approach.

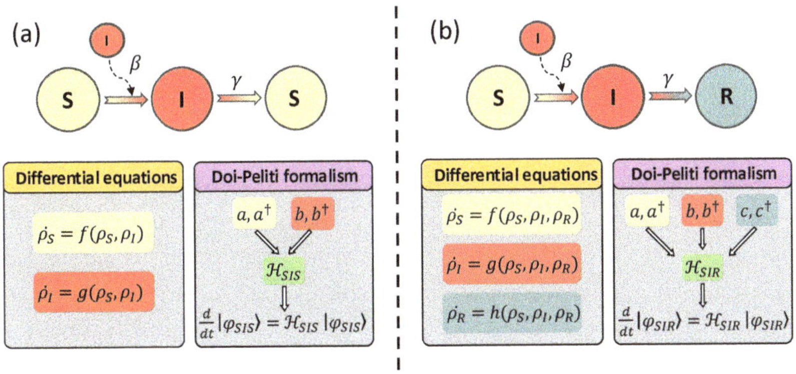

Figure 1. Comparison between theoretical approaches to compartmental models based on differential equations and the Doi–Peliti equations for the SIS (Panel (**a**)) and SIR (Panel (**b**)) models. The classical approach consists of deriving a set of deterministic ODEs governing the time evolution of the expected occupation of each compartment m, denoted by ρ_m. Conversely, the Doi–Peliti approach involves a quantum-like approach, constructing the Hamiltonian for both dynamics from the ladder operators determining the occupation of each compartment and using the time-dependent Schrödinger equation for the evolution of the dynamical state of the system.

To obtain the time evolution of epidemic outbreaks in both the SIS and the SIR model under the Doi–Peliti approach, we should use the matrix representation of the Hamiltonian operator and compute its elements $\mathcal{H}_{x,x'} = \langle x|\mathcal{H}|x'\rangle$ capturing the transitions between the possible configurations $|x\rangle$ and $|x'\rangle$ for each model. Once the Hamiltonian is defined, the analytical solution of Equation (6) can be readily obtained as follows [37]:

$$|\varphi(t)\rangle = e^{\mathcal{H}t}|\varphi(0)\rangle, \tag{11}$$

where $\exp \mathcal{H}t$ represents the propagator of each dynamics.

The computation of the system propagator for each time t might be cumbersome, especially for high-dimensional systems where the diagonalization of the Hamiltonian is computationally expensive. To overcome this limitation, we rely on the Markovian property of the master equation and consider the time evolution of the system over multiple discrete time steps of duration Δt. In each time step, the state of the system is updated as follows:

$$|\varphi(t+\Delta t)\rangle = e^{\mathcal{H}\Delta t}|\varphi(t)\rangle. \tag{12}$$

Therefore, the time evolution of the system can be obtained as the subsequent action of a single propagator $e^{\mathcal{H}\Delta t}$ on the updated state according to Equation (12), thus saving the computational time associated with the computation of the propagator. Throughout the manuscript, we assume $\Delta t = 0.1$.

As explained above, the evolution of epidemic outbreaks in the SIS model is fully characterized by monitoring the time evolution of the probabilities $P(I(t))$ of finding I individuals in the infected compartment at time t. Without any loss of information, let us instead focus on the probability of finding a fraction of population $\rho_I(t)$ in such a compartment at time t, with $\rho_I(t) = I(t)/N$. Figure 2a represents the evolution of this set of probabilities for an epidemic triggered by a single infectious individual in a population of $N_{SIS} = 1000$ individuals, characterized by $\beta = 0.6$, $\gamma = 0.1$, thus with a basic reproduction number $\mathcal{R}_0 = \beta/\gamma = 6$. We also represent the analytical solution of the deterministic ODE, Equation (1) in the same figure, governing the evolution of the SIS model (dashed line of Figure 2a).

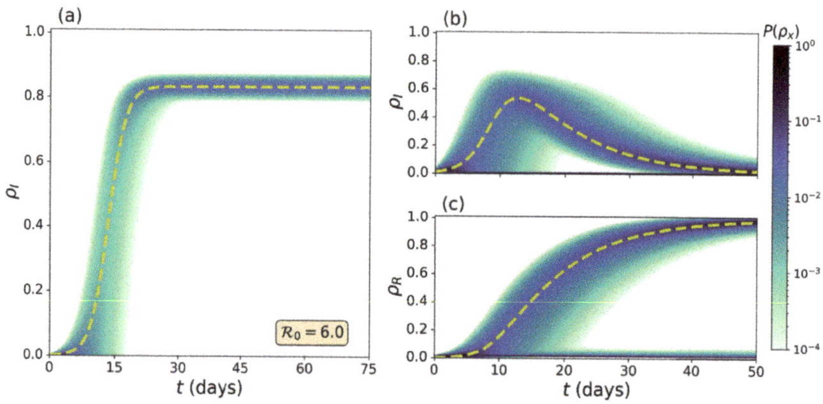

Figure 2. (**a**): Time evolution of the probability of finding a fraction ρ_I of individuals in the infected state $P(\rho_I)$ (color code) for an SIS dynamics. (**b**,**c**) The time evolution of the probability of finding a fraction ρ_m of the population in the compartment m, $P(\rho_m)$ (color code) for an SIR dynamics. The compartments shown are (**b**) the Infected compartment and (**c**) the Recovered compartment. The deterministic solutions of Equations (1) and (2) (dashed lines) are shown over the cloud probability to compare both frameworks. In panel (**a**), we consider a population of $N_{SIS} = 1000$ individuals and, in panels (**b**,**c**), a population of $N_{SIR} = 100$ individuals. In all panels, we fix the contagion rate to $\beta = 0.6$ and the recovery rate to $\gamma = 0.1$, yielding a basic reproduction number $\mathcal{R}_0 = 6$.

The comparison between both probabilistic and deterministic approaches reveals the wealth of information typically overlooked by classical deterministic models. First, epidemic uncertainty is not uniform across time, being maximal at intermediate stages. For instance, at $t = 15$, the deterministic equations predicts a widespread epidemic ($\rho_I \simeq 0.4$), whereas the probabilistic ensemble of trajectories also shows a significant probability of finding a small epidemic outbreak $\rho_I \simeq 0$. The latter shows how the classical indicator, i.e., the expected fraction of population in the infected state, might not be a representative quantity to capture the stochastic transient dynamics of epidemic outbreaks. This behaviour cannot be reproduced by the deterministic model. Remarkably, this uncertainty shrinks around the deterministic value at later stages $t > 30$, showing the robustness of classical deterministic approaches in determining the metastable epidemic state of the system. Note, however, that the steady state of the stochastic system is always $\rho_I^\infty = 0$. This occurs because the absence of infected individuals represents an absorbing state with a probability of occupation always increasing over time as a result of stochastic fluctuations destabilizing the metastable epidemic state.

For the SIR model, we characterize the evolution of epidemic outbreaks by monitoring the occupation of both the infected and the recovered compartments. As the state of the system is described by $\{P(S, I)\}$ (see Section 2.4), we should compute the marginal probabilities $\{P(I)(t)\}$ and $\{P(R)(t)\}$ as follows:

$$P(I) = \sum_S P(S, I), \tag{13}$$

$$P(R) = \sum_{S,I|S+I=N-R} P(S, I). \tag{14}$$

Figure 2b,c represent the time evolution of the infected and recovered compartments, respectively, considering a population of $N_{SIR} = 100$ individuals due to computational memory limits, including the deterministic solution of Equation (2) with dashed lines. In these plots, we can observe that the Doi–Peliti approach reproduces the wave-like behaviour of the epidemic outbreaks under the SIR model. As in the case of the SIS model, we observe that epidemic uncertainty is not uniform across time and that the deterministic trajectories capture the time evolution of the expected value of the probability distributions yielded by the Doi–Peliti approach. Additionally, we observe a bimodal probability density function of the recovered individuals. While the region closer to the deterministic equations captures major epidemic outbreaks, the probabilistic cloud with negligible recovered population is a consequence of the stochastic effects driving the system to the absorbing state before any epidemic is observed in the population.

Our results reveal that epidemic uncertainty is not uniform in a single epidemic trajectory for both the SIS and SIR models. To fully characterize the impact of stochasticity on epidemic dynamics, we now analyze the uncertainty of the order parameters of both models as a function of the basic reproduction number $\mathcal{R}_0$ of the disease, as shown in Figure 3. For the SIS model, Figure 3a represents the prevalence of the disease in the metastable epidemic state, at $t = 1000$ days, with a deterministic value that is given by $\rho_I^{SIS,met} = 1 - \mathcal{R}_0^{-1}$. Conversely, for the SIR model, we analyze in Figure 3b the attack rate of the disease ρ_R^∞ defined as the fraction of recovered individuals at equilibrium. The deterministic value for this parameter is obtained by solving the implicit equation $\rho_R^\infty = 1 - s_0 e^{-R_0(\rho_R^\infty - r_0)}$, where s_0 and r_0 represent the initial proportion of susceptible and recovered individuals in the population. In both cases, the Doi–Peliti framework proves that there is a high probability of observing the order parameters predicted by the deterministic ODEs. Notably, as previously stated, the SIR model reveals a notable feature: a non-negligible probability of having a minor outbreak in the stationary state, even for values of $\mathcal{R}_0 > 1.0$, as shown in Figure 3b. We further explore this phenomenon in Section 3.2.

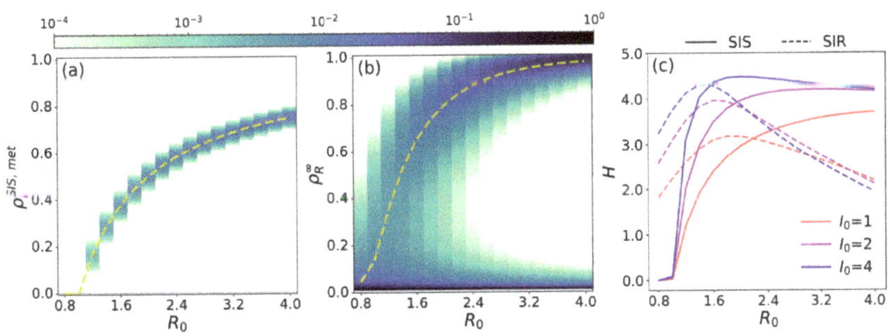

Figure 3. (**a**,**b**): Evolution of the cloud probability of the order parameters as R_0 evolves, using the prevalence of the metastable epidemic state computed at $t = 1000$ days $\rho_I^{SIS,met}$ in the SIS model (Panel (**a**)) and the attack rate ρ_R^∞ in the SIR model (Panel (**b**)). The deterministic solution for the order parameters of both models is also shown in both panels with dashed lines. (**c**) The evolution of the entropy as a function of $\mathcal{R}_0$ for the SIS (solid lines) and SIR (dashed lines) and three different initial conditions, varying the number of initially infected individuals I_0 (color code). For the SIS model, we consider a population of $N_{SIS} = 1000$ individuals and, for the SIR, a population of $N_{SIR} = 100$ individuals. In all panels, we fix the recovery rate to $\gamma = 0.1$ and modify the infection rate β, such that $\mathcal{R}_0 \in [0.8, 4.0]$.

Qualitatively, Figure 3 reveals that the uncertainty of the predictions for the order parameter is not uniform but instead varies as a function of the basic reproduction number of the disease $\mathcal{R}_0$. To quantify such behaviour, we compute the entropy H of the marginal probability distributions for the occupation of a single compartment X in each epidemic model, denoted by $P(X)$ as follows:

$$H = -\sum_X P(X) \log P(X), \tag{15}$$

where X stands for the infected (recovered) compartment I (R) in the case of the SIS (SIR) model.

Figure 3c represents the evolution of entropy for both models as a function of $\mathcal{R}_0$ and the number of individuals initially infected I_0. First, we observed that entropy does not follow a monotonic dependence on $\mathcal{R}_0$ for the SIR model. Instead, entropy reaches a maximum value and then drops as $\mathcal{R}_0$ increases. This behaviour is driven by the presence of a no-outbreak probability, which significantly influences the system when $\mathcal{R}_0 \sim 1.0$, leading to a greater uncertainty. As $\mathcal{R}_0$ grows further, the probability of a minor outbreak diminishes, and the cloud probability around $\rho_R^\infty \sim 0$ shrinks. Analogously, for the SIS model, there is a high probability of falling into the absorbing state in the vicinity of $\mathcal{R}_0 \simeq 1$, which starts decreasing with $\mathcal{R}_0$, favoring the occupation of the epidemic metastable state to the detriment of the absorbing state. This phenomenon is reflected by the increase in entropy observed for the SIS model with $\mathcal{R}_0$, which might give rise to a non-monotonic behaviour when the occupation of the metastable epidemic state becomes dominant over the absorbing one. Note that the entropy for the SIS model is analyzed at $t = 1000$ as in the steady state $H_{SIS}^\infty = 0$, provided the absorbing state is the single equilibrium of the dynamics.

3.2. The Probability of No-Outbreak in the Doi–Peliti Formalism

The interplay between the stochastic nature of compartmental models and the existence of absorbing states has been long studied in the mathematical epidemiology field. In 1955, Whittle [38] derived the probability to find a minor outbreak caused by a pathogen with a given reproduction number $\mathcal{R}_0$. Using the theory of branching processes, the probability of extinction of an outbreak given a number of I_0 individuals, $\pi_{NO}(I_0)$, is fulfilled as follows:

$$\pi_{NO}(I_0) = \frac{\beta S_0 I_0/N}{\beta S_0 I_0/N + \gamma I_0}\pi_{NO}(I_0+1) + \frac{\gamma I_0}{\beta S_0 I_0/N + \gamma I_0}\pi_{NO}(I_0-1)\,. \tag{16}$$

Note that the two terms in the previous equation can be related to $\mathcal{H}_{SIR}$ in the Doi–Peliti approach. In particular, $\beta S_0 I_0/N$ corresponds to the rate of transition $\mathcal{H}_{|S_0,I_0\rangle,|S_0-1,I_0+1\rangle}$, whereas the term γI_0 corresponds to $\mathcal{H}_{|S_0,I_0\rangle,|S_0,I_0-1\rangle}$.

Considering a small number of infectious individuals, we can assume $S_0 \simeq N$, turning Equation (16) into the following:

$$\pi_{NO}(I_0) = \frac{\mathcal{R}_0}{\mathcal{R}_0+1}\pi_{NO}(I_0+1) + \frac{1}{\mathcal{R}_0+1}\pi_{NO}(I_0-1) \tag{17}$$

For the simplest case $I_0 = 1$, and bearing in mind that $\pi_{NO}(0) = 1$, since no outbreak can take place without the initial infectious individuals, the equation is reduced to a quadratic form with the following roots:

$$\frac{\mathcal{R}_0}{\mathcal{R}_0+1}\pi_{NO}(1)^2 - \pi_{NO}(1) + \frac{1}{\mathcal{R}_0+1} = 0 \longrightarrow \pi_{NO}(1) = \begin{cases} 1 & \mathcal{R}_0 < 1 \\ \dfrac{1}{\mathcal{R}_0} & \mathcal{R}_0 > 1 \end{cases} \tag{18}$$

Finally, introducing a tree-like assumption [22] in the presence of multiple initially infected individuals, i.e., $\pi_2 = (\pi_1)^2$, we can generalize the former equation as follows:

$$\pi_{NO}(I_0) = \begin{cases} 1 & \mathcal{R}_0 < 1 \\ \left(\frac{1}{\mathcal{R}_0}\right)^{I_0} & \mathcal{R}_0 > 1 \end{cases} \tag{19}$$

To validate this theoretical expression, we perform agent-based simulations relying on the τ-leap algorithm. In particular, we consider a population of $N = 10^5$ individuals and compute the probability of not observing an outbreak by varying the initial number of infected individuals I_0 and the basic reproduction number of the pathogen $\mathcal{R}_0$. The probability of no outbreak $\pi_{NO}(I_0)$ is computed as the fraction of simulations giving rise to minor outbreaks with little impact on the population. To compute such quantity, we classify an epidemic trajectory as a minor outbreak when less than 20% of the expected attack rate of the major outbreak, R_∞^{det}, has been infected throughout the dynamics. Mathematically, we assume that the attack rate R_∞^{minor} of minor outbreaks fulfils $R_\infty^{minor} \leq 0.2R_\infty^{det} + I_0$. Since no major outbreak can be reached when $R_0 < 1$, we assume $R_\infty^{minor} \leq 0.2N + I_0$ in this case. Figure 4 shows that the theoretical expression fairly captures the results from the stochastic simulations. Note that the choice of R^∞_{minor} is somehow arbitrary as there is not a unique way of defining the attack rate corresponding to a minor outbreak in the SIR model as a function of $\mathcal{R}_0$. Nonetheless, we have checked that the results remain consistent under other choices for R^∞_{minor}.

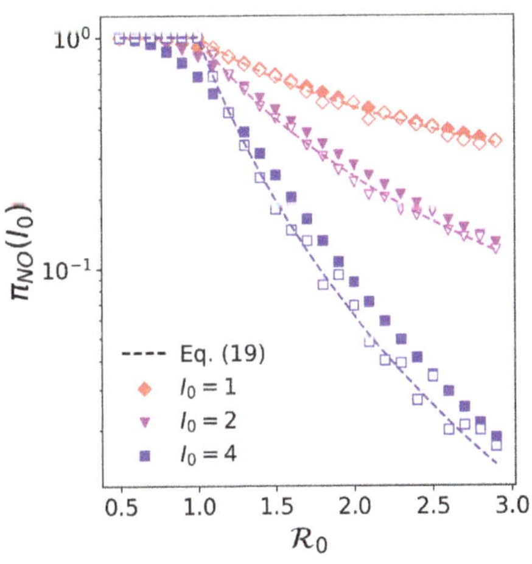

Figure 4. Probability of generating minor epidemic outbreaks as a function of the basic reproduction number $\mathcal{R}_0$ and the number of individuals initially infected I_0 (symbols and color code). Dashed lines represent theoretical estimations obtained from Equation (19). Filled dots represent integration in the region of low attack rates (see text for details) of the probability density function for recovered individuals obtained through the Doi–Peliti equations. Empty dots represent the results from agent-based simulations using the τ-leap algorithm. We simulate $n = 1000$ epidemic trajectories, considering that a minor outbreak is characterized by an attack rate $R_\infty^{minor} < 0.20N + I_0$ if $R_0 < 1$ and $R_\infty^{minor} < 0.05R_\infty + I_0$ if $R_0 > 1$.

Despite the former agreement, performing agent-based simulations comes with a high computational cost. Obtaining significant results requires considering very large populations to avoid finite size effects, and many outbreaks should be simulated to attain enough statistics to compare them with the theoretical predictions. In contrast, the Doi–Peliti equations can be readily leveraged to compute the probability of minor outbreaks. To do so, we must restrict ourselves to the region of low attack rates and integrate the probability density function of the recovered individuals once the steady state has been reached, fixing the upper bound of integration to the values of R_{∞}^{minor} considered in the agent-based simulations. Figure 4 confirms that the Doi–Peliti equations allow for characterizing the probability of observing minor outbreaks in the population without the need to perform any agent-based simulations.

3.3. The Predictability Problem of the SIR Model

Apart from capturing the probability of minor outbreaks, the Doi–Peliti equations can be leveraged to quantify how the inherent stochasticity of epidemic processes shapes the uncertainty of forecasts during an epidemic outbreak. This uncertainty does not come from the model complexity or trajectory degeneracy in the space of parameters but instead is a consequence of the existence of an underlying probabilistic ensemble of trajectories which can be generated with fixed epidemiological parameters and initial conditions.

To tackle this problem, we generate a synthetic trajectory with the deterministic equations of the SIR model and assume that these data represent the actual time evolution of an epidemic outbreak with $\mathcal{R}_0 = 6.0$. To address how epidemic uncertainty changes over time, we run Equation (12), assuming that the initial conditions correspond to the epidemic state of the system across different time points of the epidemic trajectory. Note that, for each time step, starting the epidemic outbreak from an epidemic point resembles the effect of measurements in quantum mechanics (epidemic data), which change the probabilistic state of the system to a well-defined one.

Figure 5 represents the time evolution of the probability density functions for the density of infected individuals assuming four different initial times: $t_0 = 0$ (Panel a), $t_0 = t_{peak}/2$ (Panel b), $t_0 = t_{peak}$ (Panel c), and $t_0 = 20$ (Panel d). Several key insights can be drawn from this figure. First, the width of the probability cloud decreases as the initial condition time increases, indicating a reduction in the uncertainty present in the system. Moreover, the Doi–Peliti formalism reveals vastly different epidemic impacts around the epidemic peak, which shows how deterministic approaches overlook many possible epidemic scenarios [39]. Additionally, in Panels b–d, the probability of no outbreak disappears due to the large initial conditions.

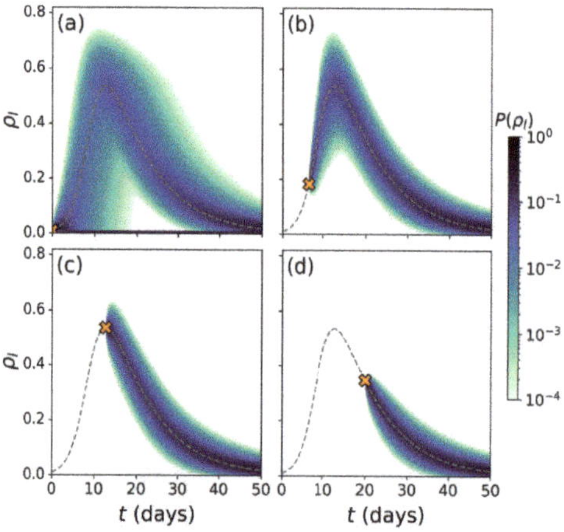

Figure 5. Time evolution of the probability of finding a fraction ρ_I of individuals in the infected compartment I, $P(\rho_I)$ (color code), for the SIR dynamics of a pathogen with $\mathcal{R}_0 = 6$ propagating across a population of $N = 100$ individuals. The initial conditions to run the Doi–Peliti equations (cross symbol) are set according to the values of the deterministic epidemic trajectories at different stages t_0 of the outbreak: (**a**) $t_0 = 0$, (**b**) $t_0 = t_{peak}/2$, (**c**) $t_0 = t_{peak}$ and (**d**) $t_0 = 20$.

To further quantify the time evolution of the predictability of the outbreak, we consider different initial time points and compute the entropy of the generated probability distributions of infected individuals at the epidemic peak, $H_{inf}(t_{peak})$. The latter reads as follows:

$$H_{inf}(t_{peak}) = -\sum_{I} P\Big(I(t_{peak})\Big) \log P\Big(I(t_{peak})\Big) \tag{20}$$

For the sake of generality, we perform the former analysis by considering several epidemic outbreaks characterized by different $\mathcal{R}_0$ values. For a fair comparison between epidemic scenarios, we consider the time points $\tilde{t_0}$ in terms of the relative difference between the time where forecasts are made and the time of the epidemic peak. In particular, we define $\Delta\tilde{t}_0 = \Big(t_0 - t_{peak}\Big)/t_{peak}$ with $\Delta\tilde{t}_0 \in [-1, 0]$.

Figure 6a represents the time evolution of the epidemic uncertainty of the peak $H_{inf}(t_{peak})$ as a function of the time taken for forecasting purposes $\Delta\tilde{t}_0$. There, we observe that the entropy H_{inf} initially rises around $\Delta\tilde{t} \approx -1$, reaches a maximum, and then drops to zero at $\Delta\tilde{t} = 0$. The initial rise in entropy may seem counterintuitive at first sight, but it is linked to the no-outbreak probability. When I_0 is small ($\Delta\tilde{t}_0 \approx -1$), there is a non-negligible probability of no outbreak ($P_0 \equiv \pi_{NO}(I_0) > 0$) present throughout the time series (Figure 5a). As $\Delta\tilde{t}_0$ grows, P_0 decreases, leading to a less defined state (inset of Figure 6a). This loss of information about the state is reflected in the slight rise in the entropy. Once the initial conditions discard minor outbreaks, the uncertainty of forecasts decreases as they are made closer to the epidemic peak, thus replicating the observed phenomena in classical deterministic models [20].

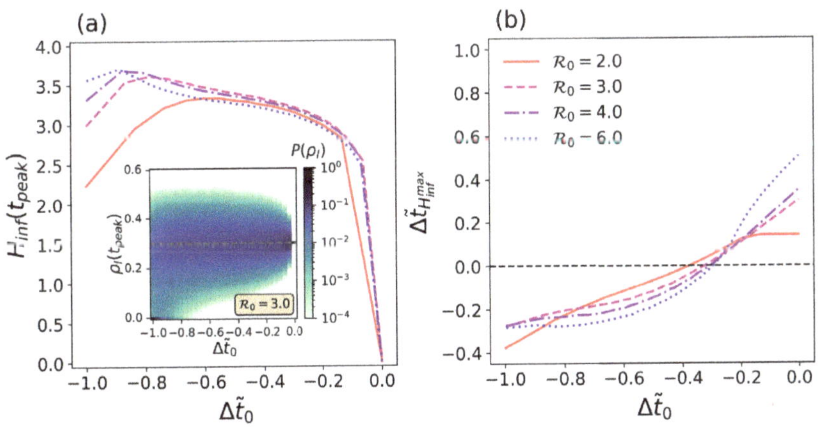

Figure 6. (**a**): Entropy of the marginal distribution of infected individuals at the epidemic peak $H_{inf}(t_{peak})$ as a function of the time from which forecasts are made $\Delta\tilde{t}_0$ and the reproduction number of the pathogen $\mathcal{R}_0$ (color code). The inset panel shows the time evolution of the marginal probability distribution for the fraction of population in the infected state at the epidemic peak $P(\rho_I(t_{peak}))$ (color code). (**b**) Relative position of the highest entropy observed in the epidemic trajectory $\Delta\tilde{t}_{H^{max}_{inf}}$ as a function of $\Delta\tilde{t}$ and the reproduction number of the pathogen $\mathcal{R}_0$ (color code). In all panels, time is measured in relative units to the position of the epidemic peak (see text for further details).

In Figure 6b, we represent

$$\Delta\tilde{t}_{H^{max}_{inf}} = \frac{\Big(t_{H^{max}_{inf}} - t_{peak}\Big)}{t_{peak}} \tag{21}$$

against $\Delta\tilde{t}_0$, measuring the relative position of the maximum entropy found in the epidemic trajectory, $t_{H_{inf}^{max}}$, compared to the time at which the peak of infected individuals occurs. $\Delta\tilde{t}_{H_{inf}^{max}} < 0$ indicates that the maximum of entropy occurs before the peak of contagions, and $\Delta\tilde{t}_{H_{inf}^{max}} > 0$ implies that the uncertainty is higher at later stages. At early times of the outbreak, we find $\Delta\tilde{t}_{H_{inf}^{max}} < 0$, indicating that maximum uncertainty precedes the infection peak. Regardless of the basic reproduction number $\mathcal{R}_0$, the latter position is delayed as forecasts are performed at later stages. Interestingly, such delay is not linear with the forecasting time, highlighting a complex interplay between the underlying stochastic dynamics and the probabilistic output of epidemic processes in determining the uncertainty associated with epidemic forecasts.

4. Discussion

Compartmental models are widely used to characterize mathematically epidemic outbreaks, obtain short-term forecasts on their evolution, and design control policies to mitigate their impact on society [40,41]. These models usually rely on deterministic approaches based on ODEs to produce the expected time evolution of the number of cases in the population. Therefore, most of the epidemic trajectories obtained do not account for the stochastic nature of the underlying epidemiological processes driving the onset of infectious diseases. To overcome this limitation, in this work, we have developed a quantum-like approach, based on a Hamiltonian formulation of both the SIS and SIR dynamics through Doi–Peliti equations. Our approach provides a probabilistic description of the ensemble of possible epidemic trajectories yielded by the stochasticity of both contagion and recovery processes.

The analysis of the probabilistic cloud of infections reveals interesting phenomena which cannot be observed through the lens of deterministic models. For the SIR model, we have first shown how such clouds typically present two disjoint areas with high density of trajectories, corresponding to the propagation of major and minor outbreaks in the population. Indeed, our results showed that the Doi–Peliti equations yield a fair estimation of probability that a given pathogen generates a minor outbreak without the need of performing computationally expensive agent-based simulations.

Focusing on major outbreaks, our results show that the uncertainty of epidemic trajectories is not uniform across time, being maximal around the peak of contagions. This finding poses theoretical constraints to the accuracy of long-term forecasts on the position and magnitude of the epidemic peak. Indeed, for several pathogens with different infectiousness, we show how the forecasts uncertainty around the epidemic peak is only reduced when those are made considerably close to its position. Therefore, our results prove that the reliability of epidemic forecasts is not only limited by the intrinsic complexity of compartmental models [42,43] but also by the stochasticity of the epidemiological processes determining the onset of pathogens in the population.

Regarding the limitations of the present work, one significant challenge is the amount of memory needed to store the propagators in the equations. These propagators are typically sparse as transitions between many epidemic states are not allowed given the definition of the ladder operators. Nonetheless, memory demands can become a limiting factor to analyze epidemic outbreaks in large size populations as the size of the propagators scales with the number of individuals N. From a practical point of view, both the compartmental model describing the course of the disease and the assumptions on the structure of contacts should be improved to use our formalism in a real epidemic scenario [44]. Factors such as political interventions, mobility patterns, demographic information, or spatial structure must be incorporated to account for more complex behaviours within the model [45]. However, yet improving the realism of the model, such refinements would also enlarge the propagators of the system considerably, as more information would be required to capture the evolution of any possible microstate in the system.

In summary, our study highlights that the mathematical characterization of epidemic dynamics through deterministic ODEs misses very rich phenomena arising from the stochasticity of epidemic outbreaks. We believe our theoretical framework provides a solid ground for the future development of more complex models, leveraging advanced probabilistic models to refine our understanding of epidemic phenomena reported in agent-based simulations. For instance, the extension of this framework to networked populations could improve our understanding of the Griffiths phases [46,47] appearing close to the epidemic threshold in complex networks. Likewise, the Doi–Peliti equations on metapopulations could serve as a benchmark to characterize the so-called invasion threshold [48,49] of pathogens driven by epidemic mobility without the need for agent-based simulations. This ongoing development is promising for enhancing the predictive capabilities and improving our responses to future outbreaks, contributing to better public health outcomes and more accurate interventions.

Author Contributions: Conceptualization, J.A.R.-V. and R.G.H.; methodology, J.A.R.-V., R.G.H., J.G.-G. and D.S.-P.; software, J.A.R.-V. and P.G.-S.; validation, J.A.R.-V. and P.G.-S.; writing—original draft preparation, J.A.R.-V. and P.G.-S.; writing—review and editing, R.G.H., J.G.-G. and D.S.-P. All authors have read and agreed to the published version of the manuscript.

Funding: P.G.S. and J.G.G. acknowledge financial support from the Departamento de Industria e Innovación del Gobierno de Aragón y Fondo Social Europeo (FENOL group grant E36-23R) and from Ministerio de Ciencia e Innovación through project PID2020-113582GB-I00/AEI/10.13039/501100011033. P.G.S. acknowledges financial support from the European Union-NextGenerationEU and Servicio Público de Empleo Estatal through Programa Investigo 024-67. D.S.P. acknowledges financial support through grants JDC2022-048339-I and PID2021-128005NB-C21 funded by the European Union "NextGenerationEU"/PRTR" and MCIN/AEI/10.13039/501100011033. J.A.R.-V. and R.G.H. acknowledge financial support from Universidad Nacional de Colombia, Centro de Excelencia en Computación Científica, through project HERMES 52625.

Data Availability Statement: The original contributions presented in the study are included in the article, further inquiries can be directed to the corresponding author/s.

Conflicts of Interest: The authors declare no conflicts of interest.

References

1. Morris, D.H.; Rossine, F.W.; Plotkin, J.B.; Levin, S.A. Optimal, near-optimal, and robust epidemic control. *Commun. Phys.* **2021**, *4*, 78. [CrossRef]
2. Castioni, P.; Gòmez, S.; Granell, C.; Arenas, A. Rebound in epidemic control: How misaligned vaccination timing amplifies infection peaks. *arXiv* **2024**, arXiv:2405.17189.
3. Kermack, W.; Mckendrick, A.G. A contribution to the mathematical theory of epidemics. *Proc. R. Soc. Lond. Ser. A Contain. Pap. Math. Phys. Character* **1927**, *115*, 700–721. [CrossRef]
4. Diekmann, O.; Heesterbeek, J.; Metz, J. On the definition and the computation of the basic reproduction ratio R 0 in models for infectious diseases in heterogeneous populations. *J. Math. Biol.* **1990**, *28*, 365–382. [CrossRef]
5. Desai, A.N.; Kraemer, M.U.; Bhatia, S.; Cori, A.; Nouvellet, P.; Herringer, M.; Cohn, E.L.; Carrion, M.; Brownstein, J.S.; Madoff, L.C.; et al. Real-time epidemic forecasting: Challenges and opportunities. *Health Secur.* **2019**, *17*, 268–275. [CrossRef]
6. Moran, K.R.; Fairchild, G.; Generous, N.; Hickmann, K.; Osthus, D.; Priedhorsky, R.; Hyman, J.; Valle, S.Y.D. Epidemic Forecasting is Messier Than Weather Forecasting: The Role of Human Behavior and Internet Data Streams in Epidemic Forecast. *J. Infect. Dis.* **2016**, *214*, S404–S408. [CrossRef]
7. Krapivsky, P.; Redner, S. Epidemic forecast follies. *npj Complex.* **2024**, *1*, 7. [CrossRef]
8. Penn, M.J.; Laydon, D.J.; Penn, J.; Whittaker, C.; Morgenstern, C.; Ratmann, O.; Mishra, S.; Pakkanen, M.S.; Donnelly, C.A.; Bhatt, S. Intrinsic randomness in epidemic modelling beyond statistical uncertainty. *Commun. Phys.* **2023**, *6*, 146. [CrossRef]
9. Colizza, V.; Barrat, A.; Barthélemy, M.; Vespignani, A. Predictability and epidemic pathways in global outbreaks of infectious diseases: The SARS case study. *BMC Med.* **2007**, *5*, 34. [CrossRef]
10. Datilo, P.M.; Ismail, Z.; Dare, J. A review of epidemic forecasting using artificial neural networks. *Epidemiol. Health Syst. J.* **2019**, *6*, 132–143.
11. Scarpino, S.V.; Petri, G. On the predictability of infectious disease outbreaks. *Nat. Commun.* **2019**, *10*, 898. [CrossRef] [PubMed]

12. Bavel, J.J.V.; Baicker, K.; Boggio, P.S.; Capraro, V.; Cichocka, A.; Cikara, M.; Crockett, M.J.; Crum, A.J.; Douglas, K.M.; Druckman, J.N.; et al. Using social and behavioural science to support COVID-19 pandemic response. *Nat. Hum. Behav.* **2020**, *4*, 460–471. [CrossRef] [PubMed]
13. Markov, P.V.; Ghafari, M.; Beer, M.; Lythgoe, K.; Simmonds, P.; Stilianakis, N.I.; Katzourakis, A. The evolution of SARS-CoV-2. *Nat. Rev. Microbiol.* **2023**, *21*, 361–379. [CrossRef] [PubMed]
14. Shea, K.; Borchering, R.K.; Probert, W.J.; Howerton, E.; Bogich, T.L.; Li, S.L.; van Panhuis, W.G.; Viboud, C.; Aguás, R.; Belov, A.A.; et al. Multiple models for outbreak decision support in the face of uncertainty. *Proc. Natl. Acad. Sci. USA* **2023**, *120*, e2207537120. [CrossRef]
15. Sherratt, K.; Gruson, H.; Johnson, H.; Niehus, R.; Prasse, B.; Sandmann, F.; Deuschel, J.; Wolffram, D.; Abbott, S.; Ullrich, A.; et al. Predictive performance of multi-model ensemble forecasts of COVID-19 across European nations. *eLife* **2023**, *12*, e81916. [CrossRef]
16. Gallo, L.; Frasca, M.; Latora, V.; Russo, G. Lack of practical identifiability may hamper reliable predictions in COVID-19 epidemic models. *Sci. Adv.* **2022**, *8*, 5234. [CrossRef]
17. Gutenkunst, R.N.; Waterfall, J.J.; Casey, F.P.; Brown, K.S.; Myers, C.R.; Sethna, J.P. Universally sloppy parameter sensitivities in systems biology models. *PLoS Comput. Biol.* **2007**, *3*, e189. [CrossRef]
18. Melikechi, O.; Young, A.L.; Tang, T.; Bowman, T.; Dunson, D.; Johndrow, J. Limits of epidemic prediction using SIR models. *J. Math. Biol.* **2022**, *85*. [CrossRef]
19. Case, B.K.; Young, J.G.; Hébert-Dufresne, L. Accurately summarizing an outbreak using epidemiological models takes time. *R. Soc. Open Sci.* **2023**, *10*, 230634. [CrossRef]
20. Castro, M.; Ares, S.; Cuesta, J.A.; Manrubia, S. The turning point and end of an expanding epidemic cannot be precisely forecast. *Proc. Natl. Acad. Sci. USA* **2020**, *117*, 26190–26196. [CrossRef]
21. Greenwood, P.E.; Gordillo, L.F. Stochastic Epidemic Modeling. In *Mathematical and Statistical Estimation Approaches in Epidemiology*; Springer: Dordrecht, The Netherlands, 2009; pp. 31–52. [CrossRef]
22. Allen, L.J. A primer on stochastic epidemic models: Formulation, numerical simulation, and analysis. *Infect. Dis. Model.* **2017**, *2*, 128–142. [CrossRef] [PubMed]
23. Champredon, D.; Li, M.; Bolker, B.M.; Dushoff, J. Two approaches to forecast Ebola synthetic epidemics. *Epidemics* **2018**, *22*, 36–42. [CrossRef] [PubMed]
24. Artalejo, J.R.; Economou, A.; Lopez-Herrero, M.J. Stochastic epidemic models revisited: Analysis of some continuous performance measures. *J. Biol. Dyn.* **2012**, *6*, 189–211. [CrossRef] [PubMed]
25. Artalejo, J.; Economou, A.; Lopez-Herrero, M. The stochastic SEIR model before extinction: Computational approaches. *Appl. Math. Comput.* **2015**, *265*, 1026–1043. [CrossRef]
26. Papageorgiou, V.E.; Tsaklidis, G. A stochastic SIRD model with imperfect immunity for the evaluation of epidemics. *Appl. Math. Model.* **2023**, *124*, 768–790. [CrossRef]
27. Papageorgiou, V.E. Novel stochastic descriptors of a Markovian SIRD model for the assessment of the severity behind epidemic outbreaks. *J. Frankl. Inst.* **2024**, *361*, 107022. [CrossRef]
28. Papageorgiou, V.E.; Tsaklidis, G. An improved epidemiological-unscented Kalman filter (hybrid SEIHCRDV-UKF) model for the prediction of COVID-19. Application on real-time data. *Chaos Solitons Fractals* **2023**, *166*, 112914. [CrossRef]
29. Sebbagh, A.; Kechida, S. EKF-SIRD model algorithm for predicting the coronavirus (COVID-19) spreading dynamics. *Sci. Rep.* **2022**, *12*, 13415. [CrossRef]
30. Nakamura, G.M.; Martinez, A.S. Hamiltonian dynamics of the SIS epidemic model with stochastic fluctuations. *Sci. Rep.* **2019**, *9*, 15841. [CrossRef]
31. Merbis, W.; Domenico, M.D. Emergent information dynamics in many-body interconnected systems. *Phys. Rev. E* **2023**, *108*, 014312. [CrossRef]
32. Merbis, W.; de Domenico, M. Complex information dynamics of epidemic spreading in low-dimensional networks. *arXiv* **2023**, arXiv:2305.05429.
33. Visco, G.V.; Artime, O.; Nauta, J.; Scagliarini, T.; Domenico, M.D. Effective one-dimension reduction of multi-compartment complex systems dynamics. *arXiv* **2024**, arXiv:2404.11366.
34. Doi, M. Second quantization representation for classical many-particle system. *J. Phys. A Math. Gen.* **1976**, *9*, 1465. [CrossRef]
35. Peliti, L. Path integral approach to birth-death processes on a lattice. *J. Phys.* **1985**, *46*, 1469–1483. [CrossRef]
36. Dodd, P.J.; Ferguson, N.M. A many-body field theory approach to stochastic models in population biology. *PLoS ONE* **2009**, *4*, e6855. [CrossRef]
37. Dirac, P.A.M. *The Principles of Quantum Mechanics*; Clarendon Press: Oxford, UK, 1930.
38. Whittle, P. The outcome of a stochastic epidemic—A note on Bailey's paper. *Biometrika* **1955**, *42*, 116–122. [CrossRef]
39. House, T.; Ross, J.V.; Sirl, D. How big is an outbreak likely to be? Methods for epidemic final-size calculation. *Proc. R. Soc. A Math. Phys. Eng. Sci.* **2013**, *469*, 20120436. [CrossRef]
40. Arenas, A.; Gómez-Gardeñes, J.; Granell, C.; Soriano-Paños, D. Epidemic spreading: Tailored models for COVID-19. *Europhys. News* **2020**, *51*, 38–40. [CrossRef]
41. Tolles, J.; Luong, T. Modeling epidemics with compartmental models. *JAMA* **2020**, *323*, 2515–2516. [CrossRef]

42. Myasnikova, E.; Spirov, A. Relative sensitivity analysis of the predictive properties of sloppy models. *J. Bioinform. Comput. Biol.* **2018**, *16*, 1840008. [CrossRef]
43. Rosenkrantz, D.J.; Vullikanti, A.; Ravi, S.; Stearns, R.E.; Levin, S.; Poor, H.V.; Marathe, M.V. Fundamental limitations on efficiently forecasting certain epidemic measures in network models. *Proc. Natl. Acad. Sci. USA* **2022**, *119*, e2109228119. [CrossRef] [PubMed]
44. Sudhakar, T.; Bhansali, A.; Walkington, J.; Puelz, D. The disutility of compartmental model forecasts during the COVID-19 pandemic. *Front. Epidemiol.* **2024**, *4*, 1389617. [CrossRef] [PubMed]
45. Estrada, E. COVID-19 and SARS-CoV-2. Modeling the present, looking at the future. *Phys. Rep.* **2020**, *869*, 1–51. [CrossRef] [PubMed]
46. Cota, W.; Ferreira, S.C.; Ódor, G. Griffiths effects of the susceptible-infected-susceptible epidemic model on random power-law networks. *Phys. Rev. E* **2016**, *93*, 032322. [CrossRef]
47. Cota, W.; Ódor, G.; Ferreira, S.C. Griffiths phases in infinite-dimensional, non-hierarchical modular networks. *Sci. Rep.* **2018**, *8*, 9144. [CrossRef]
48. Balcan, D.; Vespignani, A. Invasion threshold in structured populations with recurrent mobility patterns. *J. Theor. Biol.* **2012**, *293*, 87–100. [CrossRef]
49. Colizza, V.; Vespignani, A. Invasion threshold in heterogeneous metapopulation networks. *Phys. Rev. Lett.* **2007**, *99*, 148701. [CrossRef]

entropy

MDPI

Article

Effects of Multiplicative Noise in Bistable Dynamical Systems

Sara C. Quintanilha Valente [1], Rodrigo da Costa Lima Bruni [1], Zochil González Arenas [2,*] and Daniel G. Barci [3,*]

1 PPG-CompMat, Universidade do Estado do Rio de Janeiro, Rua São Francisco Xavier 524, Rio de Janeiro 20550-013, RJ, Brazil; saravalenteq@gmail.com (S.C.Q.V.); bruni.r.c.l@gmail.com (R.d.C.L.B.)
2 Departamento de Matemática Aplicada, Universidade do Estado do Rio de Janeiro, Rua São Francisco Xavier 524, Rio de Janeiro 20550-013, RJ, Brazil
3 Departamento de Física Teórica, Universidade do Estado do Rio de Janeiro, Rua São Francisco Xavier 524, Rio de Janeiro 20550-013, RJ, Brazil
* Correspondence: zochil@ime.uerj.br (Z.G.A.); daniel.barci@fis.uerj.br (D.G.B.)

Abstract: This study explores the escape dynamics of bistable systems influenced by multiplicative noise, extending the classical Kramers rate formula to scenarios involving state-dependent diffusion in asymmetric potentials. Using a generalized stochastic calculus framework, we derive an analytical expression for the escape rate and corroborate it with numerical simulations. The results highlight the critical role of the equilibrium potential $U_{\rm eq}(x)$, which incorporates noise intensity, stochastic prescription, and diffusion properties. We show how asymmetries and stochastic calculus prescriptions influence transition rates and equilibrium configurations. Using path integral techniques and weak noise approximations, we analyze the interplay between noise and potential asymmetry, uncovering phenomena such as barrier suppression and metastable state decay. The agreement between numerical and analytical results underscores the robustness of the proposed framework. This work provides a comprehensive foundation for studying noise-induced transitions in stochastic systems, offering insights into a broad range of applications in physics, chemistry, and biology.

Keywords: stochastic dynamic; Langevin equations; multiplicative noise; decay rates; bistable systems

1. Introduction

Dynamical systems often exhibit multiple local equilibrium configurations separated by potential barriers. The primary mechanism for the decay of these states is thermal activation over a potential barrier. This phenomenon has been widely investigated due to its significance in diverse fields such as chemistry, physics, and biology [1].

A simple representation for thermal activation considers a classical particle in a bistable potential, $U(x)$, undergoing stochastic dynamics governed by a Langevin equation with additive white noise. A key quantity in this framework is the escape rate, which characterizes the particle's transition from a local minimum of the potential.

Kramers's pioneering work [2] provides an elegant formula for the escape rate in symmetric potentials:

$$r_{\rm add} = \frac{\sqrt{\omega_{\rm min}|\omega_{\rm max}|}}{2\pi} e^{-\frac{\Delta U}{\sigma^2}}, \tag{1}$$

where $\Delta U = U(x_{\rm max}) - U(x_{\rm min})$ is the barrier height, with $x_{\rm max}$ and $x_{\rm min}$ denoting the positions of the maximum and the minimum of the potential, respectively. The noise intensity, represented as $\sigma^2 \sim k_B T$, where $k_B T$ is the thermal energy, quantifies the strength of the stochastic fluctuations in the system, and $\omega_{\min} = U''(x_{\min})/m$ and $\omega_{\max} = U''(x_{\max})/m$ are frequencies associated with a parabolic approximation of the potential energy. $U''(x_{\min})$ and $U''(x_{\max})$ are the local curvatures of the potential at the minimum and the maximum and m is the mass of the particle. This result holds in the weak noise regime ($\sigma^2 \ll \Delta U$).

Recently, we extended Kramers's formula, given by Equation (1), to systems with multiplicative noise [3]. In such systems, the stochastic dynamics are driven by a state-dependent diffusion function, $g(x)$. This state-dependent nature add further challenges and require a careful treatment using stochastic calculus, where the choice of interpretation, such as those by Itô, Stratonovich, or others, can influence the resulting dynamics [4–6]. Understanding Kramers's rate in the context of multiplicative noise is not merely of theoretical interest but has practical implications. A significant application is stochastic resonance, a phenomenon in which noise amplifies the response of a nonlinear system to weak periodic signals [7].

In addition, evidence of heterogeneous diffusion has been observed in very different systems such as gene expression processes [8], microparticles in confined geometries [9,10], and colloids [11,12]. From a theoretical perspective, stochastic systems with multiplicative noise, i.e., with heterogeneous diffusion, have been investigated using a variety of approaches. These range from numerical simulations of Langevin equations [13–16] to mean-field approximations of Fokker–Planck equations [17] and further to path integral techniques [18–21], which are particularly well suited for exploring symmetries, conserved quantities, and fluctuation theorems [22].

While prior studies of Kramers's escape rate and its generalizations have primarily addressed symmetric potentials, real-world systems often exhibit asymmetric potential landscapes. Asymmetries introduce new complexities, such as unequal barrier heights and different curvatures at the minima, which significantly impact escape dynamics. This requires further generalization of classic escape rate formulas.

In this work, we derive an explicit analytic formula for the escape rate in stochastic systems with multiplicative noise under asymmetric potentials. We use a broad stochastic framework to define the Langevin equation, which includes the most common interpretations of stochastic integration. Our analysis highlights how asymmetry modifies the escape rate and its dependence on the stochastic prescription. The main result of this paper is the explicit analytic expression for the decay rate

$$r = \frac{g(x_m)^2}{2}\left(\frac{\sqrt{\tilde{\omega}_a|\tilde{\omega}_m|}}{2\pi}e^{-\frac{\Delta U^a_{\rm eq}}{\sigma^2}} + \frac{\sqrt{\tilde{\omega}_b|\tilde{\omega}_m|}}{2\pi}e^{-\frac{\Delta U^b_{\rm eq}}{\sigma^2}}\right) \tag{2}$$

where $\Delta U_{\rm eq}$ is the height of the equilibrium potential barrier related to each of the minima $x = a$ or $x = b$, $\tilde{\omega}$s are the local curvatures of the equilibrium potential at the minima or the maximum of the potential, and $g(x_m)$ is the diffusion function evaluated at the local maximum of $U_{\rm eq}$. In this case, the square of the diffusion function is playing the role of the inverse of the mass, since the actual frequency is proportional to $g^2\tilde{\omega}$. The main feature of this expression is that it depends neither on the details of the equilibrium potential nor on those of the diffusion function. Instead, it relies solely on the local properties of these quantities, such as the height of the barrier and the local curvatures at the minima and maximum. On the other hand, all the information about the specific stochastic prescription defining the multiplicative noise process is encapsulated in the equilibrium potential. This potential incorporates not only the original bare potential but also information on the multiplicative noise fluctuations. Moreover, we observe that both minima, the global and the metastable one, contribute to the decay rate, independently of the initial conditions of the process.

We calculate the escape rate using the conditional probabilities of state transitions over time. The computation employs path integral techniques, utilizing an instanton–anti-instanton expansion valid in the weak noise limit and long times. Our approach closely follows the methods in [3,23]. Additionally, we perform extensive numerical simulations of the Langevin equation for specific potential and diffusion functions. These results enable us to evaluate the validity and limitations of our approximations.

The structure of this paper is as follows. Section 2 present the theoretical framework for analyzing asymmetric potentials in the presence of additive and multiplicative noise. Section 3 reviews the path integral approach to stochastic processes, emphasizing its application to escape rate calculations. In Section 4, we derive the weak noise approximations for escape rates in asymmetric potentials, presenting the main result of this paper given by Equation (2). Numerical simulations and comparisons with analytic results are presented in Section 5. Finally, we provide a summary of our findings and discuss potential applications in Section 6.

2. A Simple Representation of State-Dependent Diffusion

Consider a one-dimensional conservative system characterized by a potential energy function, $U(x)$, that has a double-well structure. The system interacts with a thermal bath described by a state-dependent diffusion function, $g(x)$. The overdamped Langevin equation is given by

$$\frac{dx(t)}{dt} \stackrel{\alpha}{=} f(x(t)) + g(x(t))\eta(t), \tag{3}$$

where $\eta(t)$ is a Gaussian white noise, thus satisfying

$$\langle \eta(t) \rangle = 0 \ , \ \ \langle \eta(t)\eta(t') \rangle = \sigma^2 \delta(t-t') \, . \tag{4}$$

The drift force and the diffusion function are represented by $f(x)$ and $g(x)$, respectively, which are arbitrary smooth functions of x, and σ measures the noise intensity. The symbol α over the equal sign indicates that the Langevin equation is interpreted in the generalized Stratonovich [24] convention or α-prescription [25]. In this prescription, α is defined as a continuous parameter, $0 \leq \alpha \leq 1$, and each of its values corresponds with a different discretization rule for the stochastic differential equation. $\alpha = 0$ corresponds to the Itô interpretation, while $\alpha = 1/2$ corresponds to the Stratonovich one. Additionally, $\alpha = 1$ is known as the Hänggi–Klimontovich (or kinetic) interpretation [6,26]. It is always good to emphasize that each stochastic prescription leads to a different stochastic evolution of Equation (3) and determines specific rules of differential calculus. Thus, the discretization prescription is an inherent component of the model.

To achieve thermodynamic equilibrium, the drift force $f(x)$ must be connected to the potential $U(x)$ and the dissipation function $g(x)$ via a generalized Einstein relation [19], in such a way that

$$f(x) = -\frac{1}{2} g^2(x) \frac{dU(x)}{dx}. \tag{5}$$

Under this condition, the Langevin Equation (3) can be rewritten as

$$\frac{dx}{dt} \stackrel{\alpha}{=} -\frac{1}{2} g^2(x) \frac{dU(x)}{dx} + g(x)\eta(t). \tag{6}$$

The equilibrium probability distribution over long times takes the form

$$P_{\rm eq}(x) = \mathcal{N}\, e^{-\frac{1}{\sigma^2}U_{\rm eq}(x)}, \tag{7}$$

where $\mathcal{N}$ is a normalization constant, and the equilibrium potential $U_{\rm eq}(x)$, obtained by solving the related stationary Fokker–Planck equation [19], is represented by

$$U_{\rm eq}(x) = U(x) + (1-\alpha)\sigma^2 \ln\Big(\mu g^2(x)\Big), \tag{8}$$

where μ is an arbitrary scale parameter. Naturally, no physical observable can depend on this parameter, as it merely represents a constant shift in the potential energy reference. Notice that the equilibrium distribution depends on the original potential as well as on the diffusion function $g(x)$ and the stochastic prescription α. Notably, $\alpha = 1$ (the kinetic interpretation) is the unique stochastic prescription which results in the Boltzmann distribution, $U_{\rm eq}(x) = U(x)$.

There is a relation between stochastic differential equations defined in different prescriptions. Sometimes, it could be useful to represent the same stochastic process in a different prescription. In this context, the process given by Equation (3) in the α prescription can be formulated by using an alternative differential equation defined in another discretization (say, in the stochastic prescription β) through the following equation:

$$\frac{dx(t)}{dt} \stackrel{\beta}{=} F_\beta(x(t)) + g(x(t))\eta(t), \tag{9}$$

with

$$F_\beta(x) = f(x) + (\alpha-\beta)\sigma^2 g(x) g'(x)\,, \tag{10}$$

where primes, $(\;)'$, means derivative with respect to x. A thorough proof is provided in the appendix of ref. [3]. This property is very useful because, depending on the calculation techniques, some prescriptions are more convenient or easier to be implemented. For instance, for analytic calculations, the Stratonovich prescription $\beta = 1/2$ is simpler because the calculus rules, such as the chain rule or integration by parts, turn out to be the usual ones. Alternatively, the Itô interpretation ($\beta = 0$) is the unique prescription which guarantees the non-anticipating property for stochastic calculus [4] and makes easier to perform numerical simulations.

Although the techniques and results presented here are general, we illustrate the equilibrium potential $U_{\rm eq}(x)$ using a simple bistable model,

$$U(x) = \frac{1}{4}x^4 - \frac{1}{2}x^2 + px, \tag{11}$$

in which the term px breaks the parity symmetry. For the diffusion function, we choose

$$g(x) = 1 + \lambda x^2, \tag{12}$$

where λ quantifies the multiplicative nature of the noise. For $\lambda = 0$, the noise is additive. To outline the general characteristics of the equilibrium potential, we express Equations (11) and (12) in arbitrary units. A discussion about dimensions will follow the presentation of our main result, Equation (35).

The potential $U(x)$ exhibits two minima (degenerated for $p = 0$) and a local maximum. The multiplicative noise introduces significant effects in the equilibrium potential. In Figure 1, the equilibrium potential $U_{\rm eq}(x)$, defined by Equation (8), is illustrated for the simplest symmetric model described by Equations (11) and (12) (with $p = 0$), considering different values of the parameters σ and α. Figure 1a shows the equilibrium potential for $\sigma^2 = 0.095$ under various stochastic prescriptions, $\alpha = 0, 1/2, 1$. For $\alpha = 1$, we find $U_{\text{eq}}(x) = U(x)$, with the minima fixed at $x_{\text{min}} = \pm 1$. In contrast, under the Stratonovich and Itô prescriptions, the minima shift toward the origin. In Figure 1b, the curves are computed using the Itô prescription for different noise intensities, $\sigma^2 = 0.055, 0.12, 0.25$. In this case, as the noise increases, the minima move closer to zero; for the largest value $\sigma^2 = 0.25$, the equilibrium potential exhibits a single global minimum at $x_{\text{min}} = 0$. As noted in ref. [3], this behavior resembles a second-order phase transition induced by spontaneous symmetry breaking. This is in contrast to the asymmetric case shown in Figure 2, where we fix $p = 0.05$. In this case, there is a metastable local minimum and a global minimum separated by a barrier. Similarly to the symmetric case, in Figure 2a, we show how the minima approach as the prescription goes from $\alpha = 1$ to $\alpha = 0$, as well as the differences in the height of the barrier for the stochastic prescriptions. This aspect clearly impacts the validity range of the Kramers rate. In Figure 2b, we show the equilibrium potential in the Itô prescription for different values of noise. It is possible to see that with increasing noise, the barrier height decreases and the metastable state disappears, in a similar way to a spinodal decomposition mechanism. Therefore, the effects of multiplicative noise on the equilibrium properties of the systems are nontrivial, in the symmetric as well as in the asymmetric case.

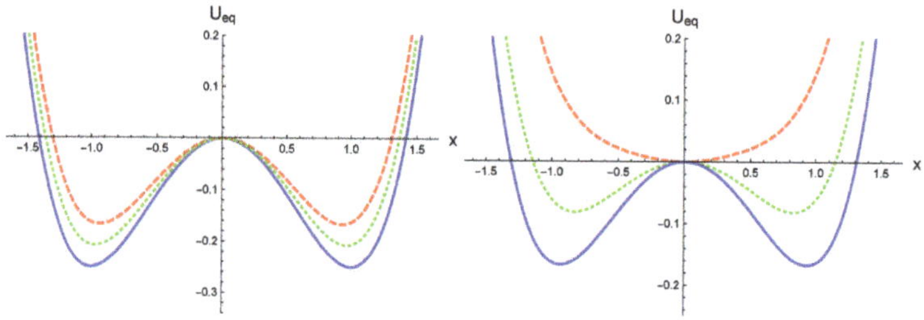

(a) (b)

Figure 1. The equilibrium potential $U_{\text{eq}}(x)$, as defined in Equation (8) with $p = 0$, is shown. In panel (**a**), $\sigma^2 = 0.095$ is fixed. The solid line corresponds to the anti-Itô prescription ($\alpha = 1$), the dotted line represents the Stratonovich prescription ($\alpha = 1/2$), and the dashed line depicts the Itô interpretation ($\alpha = 0$). In panel (**b**), all curves are calculated using the Itô interpretation. The solid line corresponds to $\sigma^2 = 0.055$, the dotted line to $\sigma^2 = 0.12$, and the dashed line to $\sigma^2 = 0.25$. In both panels, we set $\lambda = 1.2$.

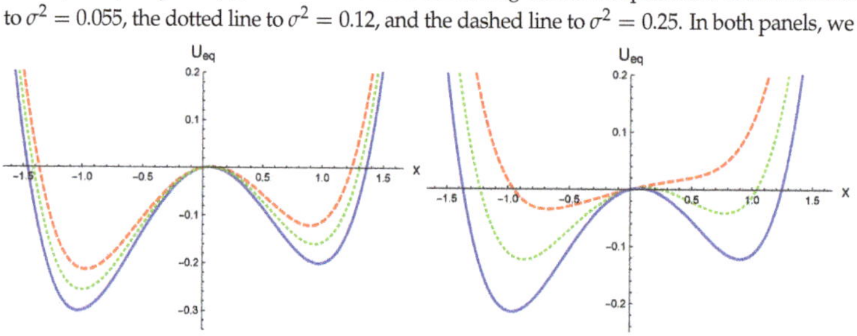

(a) (b)

Figure 2. The equilibrium potential $U_{\text{eq}}(x)$, given by Equation (8), with $p \neq 0$. In panel (**a**), $\sigma = 0.5$ is fixed. The solid line corresponds to the anti-Itô prescription ($\alpha = 1$), the dotted line represents the Stratonovich prescription ($\alpha = 1/2$), and the dashed line depicts the Itô interpretation ($\alpha = 0$). In panel (**b**), all curves are calculated using the Itô interpretation. The solid line corresponds to $\sigma^2 = 0.055$, the dotted line to $\sigma^2 = 0.12$, and the dashed line to $\sigma^2 = 0.2$. In both panels, we set $\lambda = 1.2$ and $p = 0.05$.

3. Path Integral Representation of Langevin Dynamics with Multiplicative White Noise

We present here the formalism used to compute conditional probabilities. This section is not entirely novel; however, it is necessary to introduce the notation and to make a self-contained presentation. We briefly outline the method and refer the interested reader to the cited references for technical details.

Conditional probabilities are expressed by the means of the path integral representation of the generating functional of stochastic correlations. In order to build the path integral, it is convenient to work in the Stratonovich prescription. Thus, using Equation (9), with $\beta = 1/2$, the stochastic process given by Equation (6), defined in the α-prescription sense, is described by the Langevin equation

$$\frac{dx(t)}{dt} \stackrel{S}{=} F_S(x(t)) + g(x(t))\eta(t), \tag{13}$$

with

$$F_S(x) = f(x) + \left(\frac{2\alpha - 1}{2}\right)\sigma^2 g(x)g'(x) \tag{14}$$

where, now, the stochastic differential equation is integrated in the Stratonovich sense. In this equation, α is just a parameter contained in the definition of the drift force F_S.

The transition probability $P(x_f, t_f|x_i, t_i)$ plays a fundamental role in studying any dynamical property of a stochastic process. It represents the conditional probability of the system being in the state x_f at the time t_f, given that it was in the state x_i at the time t_i. Within the path integral formalism, this probability can be expressed as [21,23,27]

$$P(x_f, t_f|x_i, t_i) = \int \mathcal{D}x \, \det{}^{-1}(g) \, e^{-\frac{1}{\sigma^2}S[x]} \,, \tag{15}$$

where the "action" $S[x]$ is given by

$$S[x] = \int_{t_i}^{t_f} dt \left\{ \frac{1}{2g^2}\left[\frac{dx}{dt} - F_S + \frac{1}{2}\sigma^2 g g'\right]^2 + \frac{\sigma^2}{2}F_S' - \frac{\sigma^4}{8}g'^2 \right\}, \tag{16}$$

with the boundary conditions $x(t_i) = x_i$ and $x(t_f) = x_f$. Equations (15) and (16) coincide with the Onsager–Mashlup representation [28] of the conditional probability for the stochastic process governed by Equation (13). Although slightly different versions of Equation (16) have been reported [21,23,27], essentially due to the covariant properties of the formalism, the differences are of the order $O(\sigma^4)$ [29] and do not affect the present computation of the transition probability, as we explain below.

Rewriting the action in an alternative form is very insightful. By expanding the squared bracket in Equation (16) and applying Equations (14), (5), and (8), followed by integration by parts, we obtain

$$S[x] = \frac{\Delta U_{\rm eq}}{2} + \int_{t_i}^{t_f} dt \; L(x, \dot{x}) \,, \tag{17}$$

where $\dot{x}$ denotes the time derivative. The first term is a state function defined by the equilibrium potential $U_{\rm eq}$, evaluated in the system's initial and final states, such that

$\Delta \mathcal{U}_{\text{eq}} = \mathcal{U}_{\text{eq}}(x_f) - \mathcal{U}_{\text{eq}}(x_i)$, where $\mathcal{U}_{\text{eq}}$ is determined by Equation (8). The Lagrangian can be expressed in the following suggestive form:

$$L = \frac{1}{2}\left(\frac{1}{g^2(x)}\right)\dot{x}^2 + V(x)\,, \tag{18}$$

where

$$V(x) = \frac{g^2}{2}\left[\left(\frac{\mathcal{U}'_{\text{eq}}}{2}\right)^2 - \sigma^2\left(\frac{\mathcal{U}''_{\text{eq}}}{2} + \frac{g'}{g}\mathcal{U}'_{\text{eq}}\right)\right] + O\left(\sigma^4\right). \tag{19}$$

The first interesting result is that, for the order σ^2, all the dependence on the parameter α is encoded in the equilibrium potential $\mathcal{U}_{\text{eq}}$ (Equation (8)), strengthening the role of this quantity not only for equilibrium properties but also for dynamic evolution.

Replacing Equation (17) in Equation (15), the conditional probability takes the form

$$P(x_f, t_f | x_i, t_i) = e^{-\frac{\Delta \mathcal{U}_{\text{eq}}}{2\sigma^2}} K(x_f, t_f | x_i, t_i) \tag{20}$$

with $K(x_f, t_f | x_i, t_i)$ represented by

$$K(x_f, t_f | x_i, t_i) = \int [\mathcal{D}x]\, e^{-\frac{1}{\sigma^2}\int_{t_i}^{t_f} dt\, L(x,\dot{x})}\,, \tag{21}$$

which is commonly known as a *propagator* in the quantum mechanics literature. Here, the initial and final conditions are considered $x_i = x(t_0)$ and $x_f = x(t_N)$. The functional integration measure takes the form

$$[\mathcal{D}x] = \mathcal{D}x\, \det{}^{-1} g = \lim_{\substack{N\to\infty \\ \Delta t\to 0}} \prod_{n=0}^{N-1} \frac{dx_n}{\sqrt{\Delta t\, g^2\left(\frac{x_n + x_{n+1}}{2}\right)}} \tag{22}$$

with $x_n = x(t_n)$.

Notably, the Equation (21) is the exact propagator for a quantum particle with a position-dependent mass, $m(x) = 1/g^2(x)$, moving in a potential, $V(x)$, written within the imaginary time path integral formalism $t \to -it$ [23]. In this context, the noise intensity σ^2 serves as the counterpart of $\hbar$ in the quantum theory.

4. Weak Noise Expansion and the Kramers Escape Rate

We analytically compute $K(x_f, t_f | x_i, t_i)$ in Equation (21) in a weak noise approximation. The propagator can be written in the saddle-point expansion with Gaussian fluctuations as

$$K(x_f, t_f | x_i, t_i) = \sum_n e^{-\frac{1}{\sigma^2} S[x_{cl}^{(n)}]} \int [\mathcal{D}\delta x_n]\, e^{-\frac{1}{2}\int dt dt'\, \delta x_n(t) S^{(2)}[x_{cl}^{(n)}](t,t') \delta x_n(t')}\,. \tag{23}$$

The functions $x_{cl}^{(n)}(t)$, with $n = 1, 2, 3, \ldots$, denote the solutions (or approximate solutions) of the equation of motion

$$\left.\frac{\delta S[x]}{\delta x(t)}\right|_{x = x_{cl}^{(n)}} = \frac{d^2 x_{cl}^{(n)}}{dt^2} - g^2 V' - \frac{g'}{g}(\dot{x}_{cl}^{(n)})^2 = 0 \tag{24}$$

with $x_{cl}^{(n)}(t_i) = x_i$ and $x_{cl}^{(n)}(t_f) = x_f$ as initial and final conditions.

The classical action in Equation (23) is

$$S[x_{cl}^{(n)}] = \int_{t_i}^{t_f} dt\; L(x_{cl}^{(n)}(t), \dot{x}_{cl}^{(n)}(t)) \tag{25}$$

while the fluctuation kernel is given by

$$S_{cl}^{(2)}(t,t') = \left.\frac{\delta^2 S[x]}{\delta x(t')\delta x(t)}\right|_{x(t)=x_{cl}^{(n)}(t)} . \tag{26}$$

In Equation (23), the functional integration measure is represented as

$$[\mathcal{D}\delta x_n] = \lim_{\substack{N\to\infty \\ \Delta t\to 0}} \prod_{j=0}^{N-1} \frac{d\delta x_j^{(n)}}{\sqrt{\Delta t\, g^2\left(\frac{x_{cl}^{(n)}(t_j)+x_{cl}^{(n)}(t_{j+1})}{2}\right)}} \tag{27}$$

and fluctuations satisfy the boundary conditions $\delta x_n(t_i) = \delta x_n(t_f) = 0$.

The main task is to compute all solutions of Equation (24) and the corresponding fluctuation integral around each one. In Figure 3, the opposite of the potential $V(x)$ is displayed, considering its general form given by Equation (19). The first observation is that $-V(x)$ has three maxima and two minima. The positions of the lateral maxima roughly align with the minima of the potential $U(x)$, with a difference of the order σ^2. The primary effect of the diffusion function is an increase in the curvature at each maximum by a factor proportional to $g^2(x_{\max}) > 1$. A key feature, important for computing conditional probabilities, is that the height differences between the peaks are of the order σ^2. Therefore, in a weak noise regime, the three maxima are quasi-degenerated.

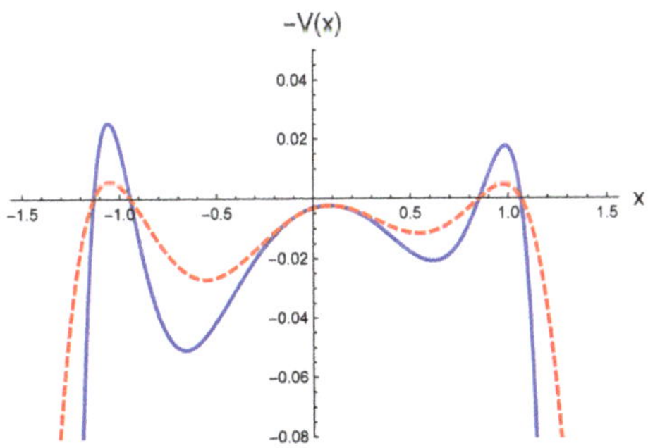

Figure 3. The opposite of the potential $V(x)$ given by Equation (19), for the Itô prescription ($\alpha = 0$), with $\sigma = 0.1$ and $p = 0.08$. The additive noise case, given by $\lambda = 0$ in the diffusion function (Equation (12)), is depicted with dashed lines, while multiplicative noise, taking $\lambda = 1$, is illustrated with a continuous line.

For simplicity, let us generically call $x = a$ the position of one of the lateral maxima, say, $x \sim -1$ in Figure 3, and $x = b$ the position of the other maximum $x \sim 1$ in the same figure. Moreover, we call $x = x_m$ the position of the central maximum ($x \sim 0$ in the figure).

We are interested in computing, for example, the probability that the system remains within one of the wells of $U(x)$, specifically around $x = a$. In this case, given that $x_i = x_f$, $\Delta U_{eq} = 0$ in Equation (20), and the propagator corresponds to the probability of the system staying in the same state. Thus, we want to compute

$$P(a, t/2|a, -t/2) = K(a, t/2|a, -t/2) . \tag{28}$$

The first task is to compute solutions of the equation of motion, Equation (24), with the conditions $x(\pm t/2) = a$. A trivial solution is simply $x_{cl} = a$, and it is possible to compute its contribution to the propagator. Details of the calculation can be found in ref. [3]. So, we obtain

$$K^{(0)}(a, t/2|a, -t/2) = \left(\frac{g_a^2 U''_{\text{eq}}(a)}{2\pi\sigma^2} \right)^{1/2} , \tag{29}$$

denoting $g(a)$ by g_a and where the superscript (0) is used to highlight the contribution of the constant solution to the propagator.

Moreover, for very long times, the dynamics of the system exhibit topological time-dependent solutions with finite action that connect the maxima of the potential. These solutions, commonly referred to as instantons and anti-instantons, play a critical role in calculating the propagator. Over extended time intervals, nontrivial contributions to the path integral arise from the well-separated superposition of instantons and anti-instantons. The methodology for summing over these configurations, known as the dilute instanton/anti-instanton gas approximation, has been extensively developed to compute tunneling amplitudes in quantum mechanics [30–32]. In the context of stochastic processes, this approach has been applied to systems driven by additive white noise [33], as well as to scenarios involving colored noise [34–37]. Stochastic systems with multiplicative noise and symmetric potentials were recently treated in ref. [3]. In this paper, we generalize the results presented in the last reference to asymmetric potentials.

In our case, the simplest instanton/anti-instanton solution that satisfies the boundary conditions is a trajectory that begins at $x = a$ at the time $-t/2$, goes to $x = x_m$ at some intermediate time, t_m, and goes back to $x = a$ at the time $t/2$. We use the schematic notation $a \leftrightarrows x_m$ to indicate this trajectory. By following standard, albeit laborious, procedures, we obtain [3]

$$K^{(1)}\left(a, \frac{t}{2} \Big| a, -\frac{t}{2}\right) = -g^2(x_m) t \, K^{(0)} \, \Gamma_a \, , \tag{30}$$

where $K^{(0)}$ represents the contribution from the constant solution, as expressed in Equation (29), and

$$\Gamma_a = \frac{\left(U''_{\text{eq}}(a)|U''_{\text{eq}}(x_m)|\right)^{1/2}}{2\pi} \exp\left\{ -\frac{U_{\text{eq}}(x_m) - U_{\text{eq}}(a)}{\sigma^2} \right\} . \tag{31}$$

As can be observed in Equation (30), at long times, the contribution to the propagator of a simple instanton/anti-instanton configuration is a linear function of time. Furthermore, the structure of the coefficient Γ_a is revealing. This is the unique term in the theory containing the equilibrium potential, U_{eq}, which, at the same time, completely encloses the information relative to the stochastic calculus. Additionally, Γ_a depends on the curvature at the maximum and the minimum of the equilibrium potential, $U''_{\text{eq}}(x_m)$ and $U''_{\text{eq}}(a)$, respectively, and on the barrier height, $U_{\text{eq}}(x_m) - U_{\text{eq}}(a)$, without any dependence on the explicit details of $U_{\text{eq}}(x)$.

The propagator also receives important contributions from other solutions of Equation (24). There are trajectories that contain two instantons and two anti-instantons, which we schematically depict as $a \leftrightarrows b$. This is the case, for example, of trajectories beginning in $x = a$ that go to $x = b$, passing through $x = x_m$, and then return to $x = a$. The contribution of these trajectories to the propagator can be computed by following the same technique as for the previous calculation of $a \leftrightarrows x_m$ trajectories [3]. Performing so, in this case, we find

$$K^{(2)}\left(a, \frac{t}{2}\Big|a, -\frac{t}{2}\right) = \frac{(g^2(x_m)t)^2}{2!} K^{(0)}\, \Gamma_a \Gamma_b\,, \tag{32}$$

where Γ_b has the same structure of Γ_a given by Equation (31), just by changing $a \to b$ in all places. As can be seen, trajectories of this kind ($a \leftrightarrows b$) produce a quadratic time contribution to the propagator, with a coefficient proportional to $\Gamma_a\Gamma_b$.

Combinations of these types of configurations are, in the long-time approximation, also quasi-solutions of Equation (24) and should be considered in the summation of the propagator of Equation (23). Typical trajectories would be

$$a \overset{p \text{ times}}{\leftrightarrows} x_m \overset{q \text{ times}}{\leftrightarrows} b \tag{33}$$

where p, q are integer numbers. The final infinite summation is a combinatory problem that can be exactly summed [3,33], obtaining

$$P(a,t/2|a,-t/2) = \frac{K^{(0)}}{\Gamma_a + \Gamma_b}\{\Gamma_b + \Gamma_a e^{-rt}\} \tag{34}$$

where the decay rate is given by $r = g(x_m)^2(\Gamma_a + \Gamma_b)/2$. Explicitly, we find, for the decay rate,

$$r = \frac{g(x_m)^2}{2}\left(\frac{\sqrt{\tilde{\omega}_a|\tilde{\omega}_m|}}{2\pi} e^{-\frac{\Delta U_{\rm eq}^a}{\sigma^2}} + \frac{\sqrt{\tilde{\omega}_b|\tilde{\omega}_m|}}{2\pi} e^{-\frac{\Delta U_{\rm eq}^b}{\sigma^2}}\right) \tag{35}$$

where $\tilde{\omega}_a = U''_{\rm eq}(a)$, $\tilde{\omega}_b = U''_{\rm eq}(b)$ and $\tilde{\omega}_m = U''_{\rm eq}(x_m)$ are the local curvatures at the minima and the maximum of the equilibrium potential. Moreover,

$$\Delta U_{\rm eq}^a = U_{\rm eq}(x_m) - U_{\rm eq}(x_a) \tag{36}$$

$$\Delta U_{\rm eq}^b = U_{\rm eq}(x_m) - U_{\rm eq}(x_b) \tag{37}$$

are the barrier height, measured from each of the asymmetric minima.

Equation (35) is the main contribution of this paper. It presents an analytic expression for the Kramers decay rate in the case of state-dependent diffusion for asymmetric potentials. The main approximations involved in our calculation are the assumption of a weak noise, $\sigma^2 \ll \Delta U_{\rm eq}^a \sim \Delta U_{\rm eq}^b$, and the expansion for long times, $rt \gg 1$.

At this point, it is useful to verify the dimensional consistency of Equation (35). Starting from the overdamped Langevin equation, Equation (3), and using the Einstein relation, Equation (5), we deduce that $[g] = L/([\sigma]\sqrt{[t]})$ and $[U] = [\sigma^2]$, where $[\cdot]$ denotes the dimensional units of a quantity and L represents a characteristic unit of length. From these considerations, it follows straightforwardly that $[r] = 1/[t]$, as expected. Interestingly, given a characteristic length scale, $[x] = L$, and the energy scale $[\sigma^2]$, the characteristic time scale is determined by the value of the diffusion function g^2. These are the only parameters that appear in the Langevin equation, Equation (3).

Equation (35) is a nontrivial generalization of previous results. For instance, for $p = 0$ in the original potential, $U_{\rm eq}(x) = U_{\rm eq}(-x)$ and, in this case, $a = -b$. In these conditions, Equation (35) reduces to the expression computed in ref. [3]. On the other hand, for $\lambda = 0$ or $g(x) = 1$, the stochastic dynamics are additive and our results converge to that of reference [33].

5. Numerical Simulations

For the purpose of validating the expression for the Kramers escape rate given by Equation (35), we performed extensive numerical simulations for the stochastic process governed by the Langevin Equation (6) by using the Euler–Maruyama scheme. This algorithm was based on an Itô interpretation of the stochastic differential equation (SDE). Therefore, given that the Langevin equation was initially considered in the general α-prescription, it had to be transformed to an Itô-defined Langevin equation by the means of Equation (10), taking $\beta = 0$. Consequently, any α-defined SDE, for $0 \leq \alpha \leq 1$, could be represented through the following Itô differential equation:

$$\frac{dx}{dt} = -\frac{1}{2}g^2(x)\frac{dU(x)}{dx} + \sigma^2\alpha g(x)g'(x) + g(x)\eta(t). \tag{38}$$

For the model described by Equations (11) and (12), the Itô stochastic differential equation can be explicitly written as

$$dx = \frac{(1+\lambda x^2)}{2}\left\{\left(1+\lambda x^2\right)\left[x\left(1-x^2\right) - p\right] + 4\lambda\sigma^2\alpha x\right\}dt + \left(1+\lambda x^2\right)dW\,, \tag{39}$$

where $W(t)$ is a standard Wiener process with $\langle W(t)\rangle = 0$ and $\langle W(t)W(t')\rangle = \sigma^2 \min(t,t')$. The white noise $\eta(t)$ is formally defined as the time derivative of the Wiener process. This relation is understood in the context of distributions, where $\eta(t)$ represents a generalized derivative [4].

In Figure 4, a typical path from Equation (39) is depicted for a particular noise realization and by making $p = 0.08$ in the expression of the potential function $U(x)$, Equation (11). With the initial condition $x(0) = 1$ fixed, the dynamics of the stochastic variable $x(t)$ become evident, as they fluctuate around the potential minima $x_{\text{min}} \sim \pm 1$, transitioning between them at seemingly irregular intervals. In this case, the global minimum of the potential is the negative one, $x_{\text{min}} \sim -1$, and its barrier is higher, further increasing the likelihood that the system remains around this minimum for longer periods.

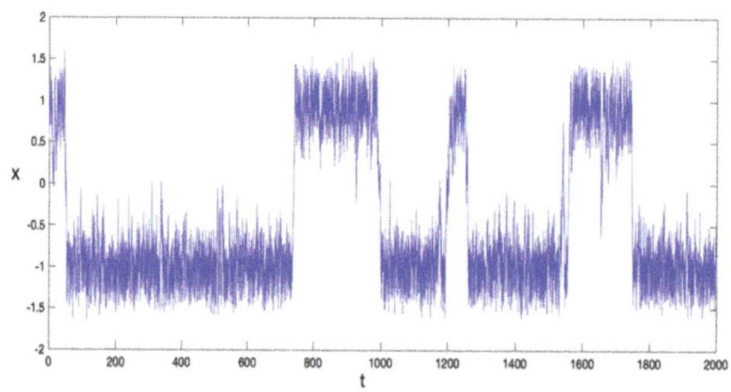

Figure 4. $x(t)$, obtained by numerically integrating Equation (39) for a specific noise realization, with the parameters $p = 0.08$; $\lambda = 0.5$; $\alpha = 1$; and $\sigma^2 = 0.085$. The time interval $0 < t < 2000$ was divided into 7×10^4 steps.

To ensure reliable statistical results, the mean value $\langle x(t)\rangle$ was computed over a significant number of distinct noise realizations. Figure 5 displays the results obtained by averaging over 8×10^4 noise configurations for various stochastic prescriptions. As expected, $\langle x(t)\rangle$ converges exponentially toward the equilibrium value

$$x_{\text{eq}} \equiv \langle x \rangle_{\text{eq}} = \int_{-\infty}^{+\infty} dx\; x\; P_{\text{eq}}(x) \tag{40}$$

with the equilibrium probability distribution $P_{\text{eq}}(x)$ represented by Equation (7).

Figure 5 also reveals that the typical decay rate varied across different stochastic prescriptions, with $r_I > r_S > r_K$, where r_I, r_S, and r_K represent the decay rates for the Itô, Stratonovich, and Kinetic prescriptions, respectively. This is aligned with the observation in Figure 2a, which shows that the height of the equilibrium potential barrier increases as α grows. This behavior is independent of the initial condition used in the SDE integration. As shown in Figure 6, simulations starting near both the metastable and global minima illustrate this point. It is possible to observe that the reached equilibrium state as well as the decay rates do not depend on initial conditions. It can also be distinguished that the equilibrium state for the Itô prescription has the smallest absolute value, while it is the largest for the anti-Itô convention. This is further represented in Figures 1a and 2a; the Itô-defined SDE always exhibits the lowest potential barrier, facilitating transitions between minima in this prescription.

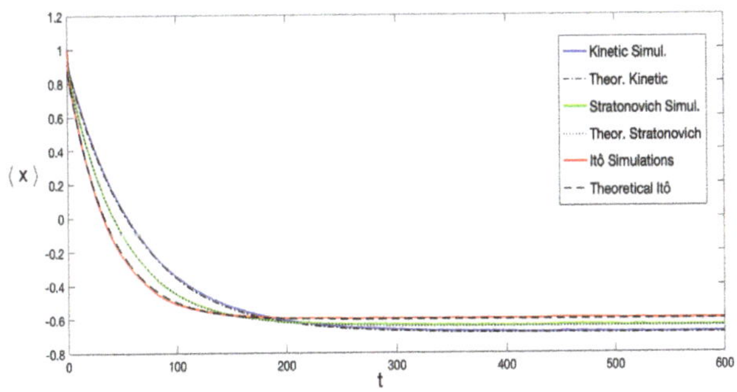

Figure 5. $\langle x(t) \rangle$ computed from Equation (39), obtained by averaging over 8×10^4 noise configurations. The initial condition $x(0) = 1$ and parameter values $\lambda = 0.5$ and $\sigma^2 = 0.085$ were fixed. The curves represent different stochastic prescriptions: $\alpha = 0$; $\alpha = 1/2$; and $\alpha = 1$. Solid lines represent the numerical simulations while the dashed, dotted, and dash–dotted lines correspond to the theoretical expression given by Equation (41), in the Itô, Stratonovich, and Kinetic prescriptions, respectively. The only fitting parameter was the amplitude $A \sim 1$.

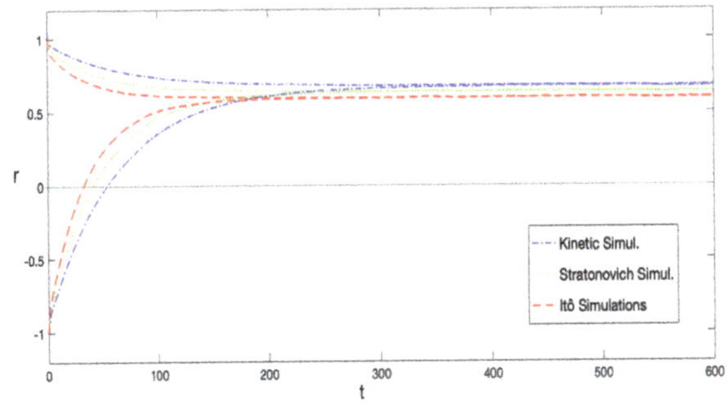

Figure 6. $\langle x(t) \rangle$ computed from Equation (39), obtained by averaging over 8×10^4 noise configurations for two different initial conditions, $x(0) = 1$ and $x(0) = -1$. Numerical simulations were performed for different stochastic prescriptions: $\alpha = 0$; $\alpha = 1/2$; and $\alpha = 1$. The parameter values $\lambda = 0.5$ and $\sigma^2 = 0.085$ were fixed. The values of r and x_{eq} were independent of the initial conditions.

By means of the asymptotic conditional probability distribution derived in Section 4, Equation (34), it was possible to obtain an analytical expression for $\langle x(t)\rangle$ in the long-time limit. It can be shown that, for $t \gg r^{-1}$,

$$\langle x(t)\rangle = A\, e^{-rt} + x_{\rm eq}, \tag{41}$$

where r is given by Equation (35) and A is a constant used as a fitting parameter. Equation (41) was used to fit the numerical simulations curves and is graphically represented in Figure 5, demonstrating excellent agreement between the theoretical result and simulations across all three stochastic prescriptions.

As an additional validation of the results, we plotted the logarithm of fluctuations around the equilibrium state,

$$\ln[\delta x(t)] = -rt + \ln A\,, \tag{42}$$

where $\ln[\delta x(t)] = \ln\left[\langle x(t)\rangle - x_{\rm eq}\right]$, and a linear least-square fitting of $\ln[\delta x(t)]$ was made. We used an arbitrary scale to split up $\ln[\delta x/A]$, since it does not affect the value of rt. Notice that the decay rate is given by the slope of the linear function and does not depend on any fitting parameter. This number, measured from the data, must be compared with the analytic expression of the decay rate given by Equation (35). Using this approach, we explored a broad range of the parameter space $\{\alpha, \sigma^2\}$.

In Figure 7, the decay rate r, represented as a function of the noise intensity σ^2, is displayed for the three stochastic prescriptions under consideration. The continuous, dashed, and dash–dotted curves were derived from Equation (35), corresponding to the Itô, Stratonovich, and kinetic (or anti-Itô) interpretations, respectively. Numerical results, obtained via the least-squares fitting of $\ln[\delta x(t)]$, are represented by diamond markers. Each diamond reflects computations based on at least 8×10^4 numerical simulations for each pair of values of α and σ^2, with $\lambda = 0.5$ and $p = 0.08$. A remarkable agreement can be perceived across nearly the entire range of noise intensity. A slight deviation occurs for higher noise values, where $\Delta U_{\rm eq}/\sigma^2 \sim 1$, as expected, since the Arrhenius approximation becomes less reliable in this regime. The first noticeable deviations in the results correspond to the Itô interpretation, as further highlighted in Figure 8.

The decay rate r as a function of the stochastic prescription α, with $0 \le \alpha \le 1$, is shown in Figure 9, for a noise range from $\sigma^2 = 0.055$ to $\sigma^2 = 0.085$. Once again, an excellent agreement between the theoretical predictions derived from Equation (35) and the results obtained through numerical simulations of the Langevin equation is observed. It is important to highlight that this analytic expression was obtained in the limit $\sigma^2 \ll \Delta U_{\rm eq} = U_{\rm eq}(x_{max}) - U_{\rm eq}(x_{min})$. Therefore, the most precise and accurate values of $r(\sigma^2)$, given by Equation (35) are expected for $\sigma^2/\Delta U_{\rm eq} \ll 1$.

For the sake of comparison between the different stochastic prescriptions, in Figure 8, $\sigma^2/\Delta U_{\rm eq}$ is displayed as a function of the noise σ^2 for the Itô (dash–dotted line), Stratonovich (dotted line), and kinetic (dashed line) conventions. The figure shows that in the kinetic prescription, this quantity is always less than unity within the studied noise range. This indicates that the noise intensity is consistently much lower than the potential barrier, resulting in an excellent agreement, as observed in Figures 7 and 9. On the other hand, for the Stratonovich and Itô prescriptions, we observe a critical value of σ^2 in which $\sigma^2/\Delta U_{\rm eq} \sim 1$. For noise intensity above this critical value, a good agreement between analytical and simulation results is no longer expected.

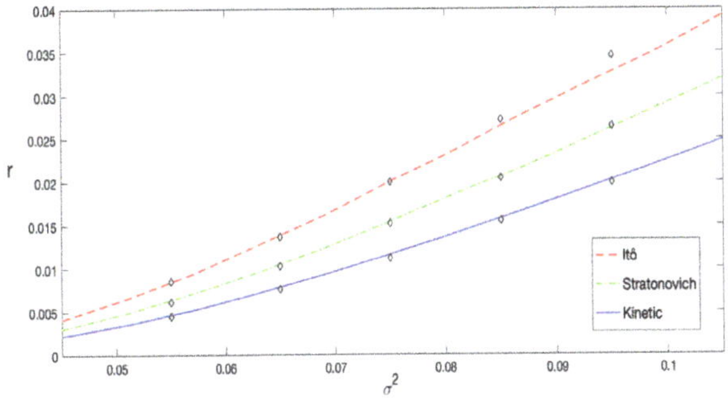

Figure 7. The decay rate r computed through Equation (35), represented as a function of the noise intensity σ^2. The solid line represents the decay rate for the Itô prescription, while the dashed and dash–dotted curves correspond to the Stratonovich and kinetic (or anti-Itô) interpretations, respectively. The points (diamonds) were obtained via the least-squares fitting of $\ln\left[\langle x(t)\rangle - x_{\text{eq}}\right]$ based on numerical simulations for each case. For all cases, the parameters were fixed at $\lambda = 0.5$ and $p = 0.08$.

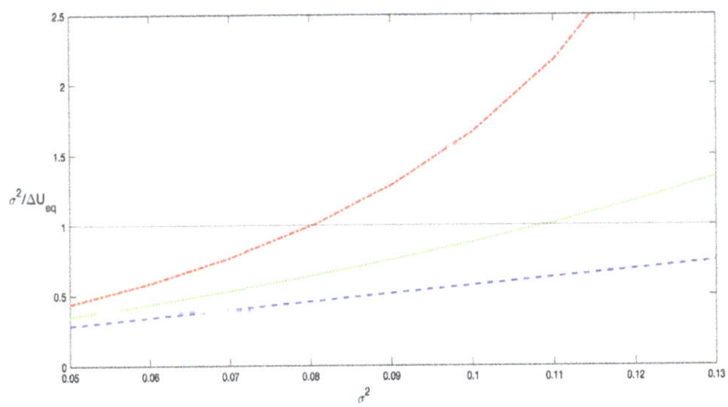

Figure 8. $\sigma^2/\Delta U_{\text{eq}}$ as a function of the noise σ^2. The dashed line represents the kinetic prescription $\alpha = 1$, the dotted line corresponds to the Stratonovich prescription, $\alpha = 1/2$, while the Itô prescription, $\alpha = 0$, is represented by the dash–dotted line. The parameters $\lambda = 0.5$ and $p = 0.08$ were fixed for all the curves.

Most of the results exhibited in Figures 7 and 9 correspond to values of σ^2 within the interval $[0.055, 0.085]$. The results are highly accurate within this range; however, as pointed before, for the Itô prescription, the value of r at $\sigma^2 = 0.085$ slightly deviates from the numerical data. At this noise intensity, $\sigma^2/\Delta U_{\text{eq}} \sim 1.2$. The deviation becomes more pronounced in Figure 7 for the Itô prescription at $\sigma^2 = 0.095$, as confirmed by the corresponding curve in Figure 8. This general behavior is expected because, as noted in Section 2, the barrier height decreases as $\alpha \to 0$ and for increasing values of noise. Indeed, there is a critical noise level at which the barrier completely disappears and the system dramatically changes its dynamics.

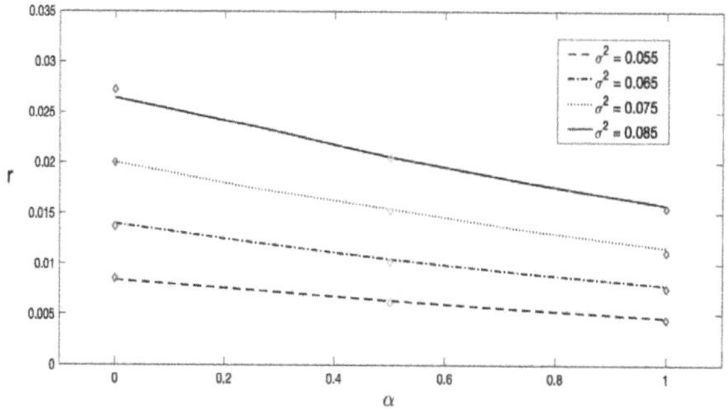

Figure 9. The decay rate r computed through Equation (35) for different values of σ^2, represented as a function of the stochastic prescription α. The solid line corresponds to $\sigma^2 = 0.085$ and the dashed one corresponds to $\sigma^2 = 0.075$, while the dash–dotted and dotted lines correspond to $\sigma^2 = 0.065$ and $\sigma^2 = 0.055$, respectively. The points (diamonds) were obtained via the least-squares fitting of $\ln[\langle x(t)\rangle - x_{eq}]$ based on numerical simulations for each case. The parameters $\lambda = 0.5$ and $p = 0.08$ were fixed for all the curves.

6. Discussion and Conclusions

In this paper, we explored the dynamics of bistable systems influenced by multiplicative noise, presenting both analytical and numerical approaches to understand escape rates within asymmetric potentials. Our work generalized Kramers's formula to systems with state-dependent diffusion, highlighting the impact of the stochastic prescription and asymmetry of the potential on equilibrium properties and transition rates. Among other applications, it serves as a starting point for studying stochastic resonance in systems governed by stochastic differential equations with multiplicative noise.

The main result of this paper is presented by Equation (35). The derived analytic expression for decay rates, validated through extensive numerical simulations, provides a robust framework for analyzing stochastic systems with multiplicative noise. Interestingly, the weak noise approximation was demonstrated to be effective, as evidenced by the excellent agreement between theoretical predictions and simulation data across various noise intensities and stochastic prescriptions.

In Figure 7, it can be clearly observed that the decay rate computed using the Itô prescription is greater than that computed with the Stratonovich interpretation, which in turn is greater than the rate computed with the Hänggi–Klimontovich prescription. Furthermore, in Figure 9, we demonstrate that the decay rate as a function of the prescription, $r(\alpha)$, is a monotonically decreasing function. This behavior is due to two main reasons. First, the decay rate is primarily related to the first moment of the probability distribution, as $rt \sim \ln(\langle x(t)\rangle)$ in the long-time limit where $rt \gg 1$. It has been analytically shown [18,19], at least for short times, that in systems with state-dependent diffusion the stochastic prescription significantly influences the mean value $\langle x(t)\rangle$ while leaving the mean square displacement (MSD) $\langle x^2(t)\rangle$ relatively unchanged. This agrees with numerical simulations, which show that the scaling properties of the MSD are independent of the stochastic prescription. Indeed, the prescription only affects the prefactor of the MSD, with a rather weak dependence [15]. Second, the most significant effect of the prescription lies in its modification of the equilibrium potential $U_{eq}(x)$, as described by Equation (8). In Figures 1a and 2a, we illustrate how heterogeneous diffusion alters the potential. It is evident that for the same noise level, the parameter α changes the height of the potential barrier. Specifically, in the Hänggi–Klimontovich prescription ($\alpha = 1$), the barrier is the highest. Thus, the particle remains more time in the well, resulting in the smallest decay rate. Conversely, in the Itô prescription, the barrier is the lowest, leading to the greater decay rate since the particle can escape from the well more easily. This effect can be so pronounced that it modifies the entire structure of the potential for sufficiently large noise levels, as shown in Figures 1b and 2b.

Equation (35) for the decay rate underscores the critical role of the equilibrium potential $U_{\rm eq}(x)$, which completely codifies the dependence of the dynamics on the stochastic discretization prescriptions. Our findings reveal that the choice of stochastic prescriptions directly influences decay rates, reflecting the underlying physics encoded in the critical points of the equilibrium potential and its local curvatures. Furthermore, the generalization to asymmetric potentials enriches the applicability of Kramers's framework, accommodating a broader class of real-world systems.

Future investigations could extend this work by exploring higher-order corrections to the weak noise approximation and considering multidimensional landscapes. Such endeavors would further deepen our understanding of noise-induced transitions in complex systems.

Author Contributions: Conceptualization, Z.G.A. and D.G.B.; Methodology, S.C.Q.V., R.d.C.L.B. and Z.G.A.; Software, S.C.Q.V., R.d.C.L.B. and Z.G.A.; Validation, S.C.Q.V., R.d.C.L.B. and Z.G.A.; Formal analysis, Z.G.A. and D.G.B.; Investigation, S.C.Q.V., R.d.C.L.B., Z.G.A. and D.G.B.; Writing—original draft, Z.G.A. and D.G.B.; Writing—review & editing, S.C.Q.V., R.d.C.L.B., Z.G.A. and D.G.B.; Supervision, Z.G.A. and D.G.B. All authors have read and agreed to the published version of the manuscript.

Funding: The Brazilian agencies Fundação Carlos Chagas Filho de Amparo à Pesquisa do Estado do Rio de Janeiro (FAPERJ), Conselho Nacional de Desenvolvimento Científico e Tecnológico (CNPq), and Coordenação de Aperfeiçoamento de Pessoal de Nível Superior—Brasil (CAPES)—Finance Code 001 are acknowledged for partial financial support.

Institutional Review Board Statement: Not applicable.

Data Availability Statement: Data supporting reported results are available from the corresponding authors, D.G.B. and Z.G.A., upon reasonable request.

Conflicts of Interest: The authors declare no conflicts of interest.

References

1. Fleming, G.R.; Hänggi, P. *Activated Barrier Crossing: Applications in Physics, Chemistry and Biology*; Advanced Database Research and Development Series; World Scientific Publishing Company: Singapore, 1993.
2. Kramers, H. Brownian motion in a field of force and the diffusion model of chemical reactions. *Physica* **1940**, *7*, 284–304. [CrossRef]
3. Moreno, M.V.; Barci, D.G.; Arenas, Z.G. State-dependent diffusion in a bistable potential: Conditional probabilities and escape rates. *Phys. Rev. E* **2020**, *101*, 062110. [CrossRef] [PubMed]
4. Øksendal, B.K. *Stochastic Differential Equations: An Introduction with Applications*; Springer: Berlin/Heidelberg, Germany; London, UK, 2003.
5. van Kampen, N.G. *Stochastic Processes in Physics and Chemistry*; Elsevier: London, UK, 2007.
6. Hänggi, P.; Thomas, H. Stochastic processes: Time evolution, symmetries and linear response. *Phys. Rep.* **1982**, *88*, 207. [CrossRef]
7. Gammaitoni, L.; Hänggi, P.; Jung, P.; Marchesoni, F. Stochastic resonance. *Rev. Mod. Phys.* **1998**, *70*, 223–287. [CrossRef]
8. Kærn, M.; Elston, T.C.; Blake, W.J.; Collins, J.J. Stochasticity in gene expression: From theories to phenotypes. *Nat. Rev. Genet.* **2005**, *6*, 451–464. [CrossRef]
9. Lançon, P.; Batrouni, G.; Lobry, L.; Ostrowsky, N. Drift without flux: Brownian walker with a space-dependent diffusion coefficient. *Europhys. Lett.* **2001**, *54*, 28. [CrossRef]
10. Lançon, P.; Batrouni, G.; Lobry, L.; Ostrowsky, N. Brownian walker in a confined geometry leading to a space-dependent diffusion coefficient. *Phys. A Stat. Mech. Its Appl.* **2002**, *304*, 65–76. [CrossRef]

11. Volpe, G.; Helden, L.; Brettschneider, T.; Wehr, J.; Bechinger, C. Influence of Noise on Force Measurements. *Phys. Rev. Lett.* **2010**, *104*, 170602. [CrossRef] [PubMed]
12. Brettschneider, T.; Volpe, G.; Helden, L.; Wehr, J.; Bechinger, C. Force measurement in the presence of Brownian noise: Equilibrium-distribution method versus drift method. *Phys. Rev. E* **2011**, *83*, 041113. [CrossRef] [PubMed]
13. Cherstvy, A.G.; Chechkin, A.V.; Metzler, R. Anomalous diffusion and ergodicity breaking in heterogeneous diffusion processes. *New J. Phys.* **2013**, *15*, 083039. [CrossRef]
14. Cherstvy, A.G.; Metzler, R. Population splitting, trapping, and non-ergodicity in heterogeneous diffusion processes. *Phys. Chem. Chem. Phys.* **2013**, *15*, 20220–20235. [CrossRef] [PubMed]
15. Cherstvy, A.G.; Chechkin, A.V.; Metzler, R. Particle invasion, survival, and non-ergodicity in 2D diffusion processes with space-dependent diffusivity. *Soft Matter* **2014**, *10*, 1591–1601. [CrossRef] [PubMed]
16. Sandev, T.; Kocarev, L.; Metzler, R.; Chechkin, A. Stochastic dynamics with multiplicative dichotomic noise: Heterogeneous telegrapher's equation, anomalous crossovers and resetting. *Chaos Solitons Fractals* **2022**, *165*, 112878. [CrossRef]
17. Van den Broeck, C.; Parrondo, J.M.R.; Toral, R.; Kawai, R. Nonequilibrium phase transitions induced by multiplicative noise. *Phys. Rev. E* **1997**, *55*, 4084–4094. [CrossRef]
18. Lau, A.W.C.; Lubensky, T.C. State-dependent diffusion: Thermodynamic consistency and its path integral formulation. *Phys. Rev. E* **2007**, *76*, 011123. [CrossRef] [PubMed]
19. Arenas, Z.G.; Barci, D.G. Hidden symmetries and equilibrium properties of multiplicative white-noise stochastic processes. *J. Stat. Mech.* **2012**, *2012*, P12005. [CrossRef]
20. Barci, D.G.; Arenas, Z.G.; Moreno, M.V. Path integral approach to nonequilibrium potentials in multiplicative Langevin dynamics. *EPL Europhys. Lett.* **2016**, *113*, 10009. [CrossRef]
21. Arnoulx de Pirey, T.; Cugliandolo, L.F.; Lecomte, V.; van Wijland, F. Path integrals and stochastic calculus. *Adv. Phys.* **2022**, *71*, 1–85. [CrossRef]
22. Aron, C.; Barci, D.G.; Cugliandolo, L.F.; Arenas, Z.G.; Lozano, G.S. Dynamical symmetries of Markov processes with multiplicative white noise. *J. Stat. Mech. Theory Exp.* **2016**, *2016*, 053207. [CrossRef]
23. Moreno, M.V.; Barci, D.G.; Arenas, Z.G. Conditional probabilities in multiplicative noise processes. *Phys. Rev. E* **2019**, *99*, 032125. [CrossRef]
24. Hänggi, P. Stochastic Processes I: Asymptotic Behaviour and Symmetries. *Helv. Phys. Acta* **1978**, *51*, 183. [CrossRef]
25. Janssen, H.K. *From Phase Transitions to Chaos: Topics in Modern Statistical Physics*; World Scientific: Singapore, 1992.
26. Klimontovich, Y.L. Nonlinear Brownian motion. *Phys. Uspekhi* **1994**, *37*, 737–766. [CrossRef]
27. Cugliandolo, L.F.; Lecomte, V.; van Wijland, F. Building a path-integral calculus: A covariant discretization approach. *J. Phys. A Math. Theor.* **2019**, *52*, 50LT01. [CrossRef]
28. Onsager, L.; Machlup, S. Fluctuations and Irreversible Processes. *Phys. Rev.* **1953**, *91*, 1505–1512. [CrossRef]
29. Arenas, Z.G.; Cugliandolo, L.F.; Barci, D.G. Covariant path integral for multiplicative noise Langevin equations. 2025, *In progress*.
30. Coleman, S. The Uses of Instantons. In *The Whys of Subnuclear Physics*; Zichichi, A., Ed.; Springer: Boston, MA, USA, 1979; pp. 805–941. [CrossRef]
31. Brézin, E.; Parisi, G.; Zinn-Justin, J. Perturbation theory at large orders for a potential with degenerate minima. *Phys. Rev. D* **1977**, *16*, 408–412. [CrossRef]
32. Bogomolny, E. Calculation of instanton-anti-instanton contributions in quantum mechanics. *Phys. Lett. B* **1980**, *91*, 431–435. [CrossRef]
33. Caroli, B.; Caroli, C.; Roulet, B. Diffusion in a bistable potential: The functional integral approach. *J. Stat. Phys.* **1981**, *26*, 83–111. [CrossRef]
34. Bray, A.J.; McKane, A.J. Instanton Calculation of the Escape Rate for Activation over a Potential Barrier Driven by Colored Noise. *Phys. Rev. Lett.* **1989**, *62*, 493–496. [CrossRef] [PubMed]
35. McKane, A.J.; Luckock, H.C.; Bray, A.J. Path integrals and non-Markov processes. I. General formalism. *Phys. Rev. A* **1990**, *41*, 644–656. [CrossRef]
36. Bray, A.J.; McKane, A.J.; Newman, T.J. Path integrals and non-Markov processes. II. Escape rates and stationary distributions in the weak-noise limit. *Phys. Rev. A* **1990**, *41*, 657–667. [CrossRef] [PubMed]
37. Luckock, H.C.; McKane, A.J. Path integrals and non-Markov processes. III. Calculation of the escape-rate prefactor in the weak-noise limit. *Phys. Rev. A* **1990**, *42*, 1982–1996. [CrossRef]

entropy

MDPI

Article

Sustainability of a Three-Species Predator–Prey Model in Tumor-Immune Dynamics with Periodic Treatment

Avan Al-Saffar [1,*] and Eun-jin Kim [2]

[1] Department of Statistics and Informatics, University of Duhok, Duhok 42001, Iraq
[2] Fluid and Complex Systems Research Centre, Coventry University, Coventry CV1 5FB, UK; ejk92122@gmail.com
* Correspondence: avan.al-saffar@uod.ac

Abstract: Using a tumor-immune growth model, we investigate how immunotherapy affects its dynamical characteristics. Specifically, we extend the prey–predator model of tumor cells and immune cells by including periodic immunotherapy, the nonlinear damping of cancer cells, and the dynamics of a healthy cell population, and investigate the effects of the model parameters. The ideal value of immunotherapy, which promotes the growth of immune (and healthy) cells while contributing to the elimination or control of the cancer cells, is determined by using Fisher information as a measure of variability throughout our study.

Keywords: dynamical systems; sustainability; fisher information; perturbations; probability density function (PDF)

1. Introduction

The predator–prey model provides a valuable framework for understanding the population dynamics of different species in nature [1–5]. The simplest form of this model is represented by the two-species Lotka–Volterra equations:

$$\begin{aligned} \frac{dx}{dt} &= ax - bxy, \\ \frac{dy}{dt} &= -cy + exy, \end{aligned} \tag{1}$$

which describe the interaction between the prey population x and the predator population y. Here, a, b, c, and e are positive parameters that represent the growth rate of prey (intrinsic rate of increase in the prey population), mortality of prey due to predation, the mortality rate of predators, and the growth of predators from the consumption of prey, respectively. This model was originally developed by Lotka (1925) [6] and Volterra (1926) [7] to describe the population dynamics of prey fish (x) and predator sharks (y). If competition between the two species is not excessively strong, the two populations can co-exist in a stable way and exhibit a limit cycle behavior [1,3,8–13].

This paper aims to extend the predator–prey model by incorporating an information-theoretic perspective. Specifically, we used Fisher information to analyze the sustainability of the model in Equation (1). Fisher information, a measure of system order and predictability, has been increasingly used in ecological and complex system analysis. Applying it to tumor-immune dynamics allows for a novel assessment of the sustainability and stability of the system. In our study, we reinterpret x and y as cancer and immune cells, respectively, and extend the model to include (i) the effect of immunotherapy modeled as an oscillatory forcing term, (ii) competition effects through logistic growth, and (iii) the dynamics of healthy cells. To better capture real-world tumor-immune interactions, we extend the model by incorporating an oscillatory immunotherapy term, competition effects, and the role of healthy cells, which play a critical role in tumor dynamics. A key focus of this study is immunotherapy, a promising therapeutic approach that improves the immune system's ability to eliminate cancer cells through external stimulation [14–19]. In our previous work, we conducted a detailed investigation of the two-species model, where x and y in Equation (1) were analyzed in depth [20]. Although predator–prey models provide a useful mathematical framework for studying interactions between cancer and immune cells, they must be modified to incorporate immune suppression, activation thresholds, and external interventions. The analogy is useful but requires adjustments to fully capture the biological complexity of the tumor-immune dynamics.

The remainder of this paper is organized as follows. Section 2 introduces the three-species model, while Section 3 presents the extended version. Finally, the conclusions are drawn in Section 4. The appendices provide additional details, including Fisher information calculations for the logistic equation and the extended three-species model.

2. Three-Species Model

We extend the two-species model [20] to the three-species model by including the dynamics of the population of healthy cells and look into how periodic disturbance affects the evolution of the system to figure out how the oscillatory behavior of immunotherapy treatment survives for a long time and remains unchanged at certain parameter values. By including immunotherapy ($F\sin^2(\omega t)$), we intend to show how the performance of the three-species model changes and to clarify how periodic therapy affects the behavior of the species from an information theory perspective.

Our model is given as follows:

$$\frac{dx}{dt} = ax\,(1-N_1x) - bxy - dxz, \tag{2a}$$

$$\frac{dy}{dt} = -cy + \frac{kyx}{g_2+x} - fyx + F\sin^2(\omega t), \tag{2b}$$

$$\frac{dz}{dt} = gz(1-N_2z) - hzx. \tag{2c}$$

Here, x, y, and z represents cancer cell population, the populations of healthy cells and immune cells, respectively. Taking into account two alternative carrying capacities, N_1 (>0) and N_2 (>0), respectively, we consider the growth of healthy populations and cancer to be logistic. Cancer cell growth is represented by a (>0), immune cell growth by k (>0), and healthy cell growth by g (>0). c (>0) is the natural mortality rate of immune cells. The rate at which immune cells prey on cancer cells is b (>0), while the rate at which cancer cells prey on immune cells is f (>0). The rate at which cancer cells prey on healthy cells is d (>0), while the rate at which cancer cells prey on healthy cells is h (>0). The periodic treatment amplitude is F, and its angular frequency is ω. We include the Michaelis–Menten form in the second term in Equation (2b) to model the saturated effects (with positive constants $g_2, \alpha > 0$) of the immune reaction (the finite interaction between immune cells and cancer cells) (see [9,21]).

We rescale the model in (2) to minimize the number of parameters in the following dimensionless form:

$$\begin{aligned} \frac{du}{d\mathcal{T}} &= u(1-u) - uv - a_{13}ur, \\ \frac{dv}{d\mathcal{T}} &= a_{21}[\frac{uv}{g_2+\alpha u} - v] - a_{22}uv + a_{23}\sin^2(\Omega\mathcal{T}), \\ \frac{dr}{d\mathcal{T}} &= a_{31}r(1-r) - a_{32}ru. \end{aligned} \tag{3}$$

Here, similarly, u, v and r represent the population of cancer cells, immune cells, and healthy cells, respectively. Note that a_{23} in Equation (3) corresponds to F in Equation (2). Equation (3) involves two competition modes. That is, in addition to the competition between cancer cells u and immune cells v in the predator–prey style, Equation (3) now has a new competition between healthy cells r and cancer cells u on the available resources. To understand the effect of periodic treatment $a_{23}\sin^2(\Omega\mathcal{T})$, we perform a set of numerical simulations of the Equation (3) considering a set of parameters such as $a_{13} = 1.2$, $a_{21} = 1.291$, $g_2 = 0.3$, $\alpha = 1$, $a_{22} = 1.1$, $a_{31} = 1.2$, $a_{32} = 4.8$, and $\Omega = 1$ together with the initial condition $u_0 = v_0 = r_0 = 1$ (which are similar to those used in [22]).

Local Stability Analysis and Equilibria

To gain a key insight into the dynamics of our model, we use the mean value of the periodic treatment and find the stationary points. $(0, \frac{a_{23}}{2a_{21}}, 0)$, $(0, \frac{a_{23}}{2a_{21}}, 1)$, $(u^*, v^*, r^*) = (1-v^*, v^*, 0)$, where v^* is the solution to the following equation:

$$a_{21}\left[\frac{v(1-v)}{g_2+\alpha(1-v)} - v\right] - a_{22}v(1-v) + \frac{a_{23}}{2} = 0.$$

The last effective equilibrium point is $(u^{**}, v^{**}, r^{**}) = (\frac{a_{31}v^{**}+a_{31}(a_{13}-1)}{a_{32}a_{13}-a_{31}}, v^{**}, \frac{a_{32}-a_{31}-a_{32}v^{**}}{a_{32}a_{13}-a_{31}})$, where v^{**} is the solution to the following equation:

$$a_{21}\left[\frac{v\left(\frac{a_{31}v+a_{31}(a_{13}-1)}{a_{32}a_{13}-a_{31}}\right)}{g_2+\alpha\left(\frac{a_{31}v+a_{31}(a_{13}-1)}{a_{32}a_{13}-a_{31}}\right)} - v\right] - a_{22}v\left(\frac{a_{31}v+a_{31}(a_{13}-1)}{a_{32}a_{13}-a_{31}}\right) + \frac{a_{23}}{2} = 0.$$

For the stability analysis of the above stationary points, it is useful to fix the values of the parameters and examine the effect of treatment by varying their a_{23}. The point $(0, 0.3873, 1)$ has three negative eigenvalues (smaller than zero), while the first point $(0, 0.3873, 0)$ is a saddle point with one negative eigenvalue and two positive eigenvalues when $a_{23} = 1$. This results in a stable solution where cancer cells are zero and only healthy and immune cells are present.

To understand the dynamics of the model in (3), we show time traces of u and v and phase portraits in Figure 1 by varying the amplitude of the periodic treatment a_{23}. Three rows in Figure 1 are for different values of a_{23}. When $a_{23} = 5$, we observe that the time trace and phase portrait in the bottom panels show a small value of the cancer cell in the limit of a long time. In comparison, for a smaller value of a_{23} shown in the top and middle panels, cancer cells take a finite value in the long time limit. These demonstrate the effect of periodic treatment on the elimination of the cancer population. Also, from Figure 1, we can observe that the higher the amplitude a_{23}, the better, as cancer cells begin to vanish and the immune system begins to increase. This suggests that it is advantageous to look at each species' average at various a_{23} values, as well as to determine the optimal value of a_{23} in which the cancer cell population is eliminated. The right panels are a phase portrait for each a_{23} after removing the initial transient.

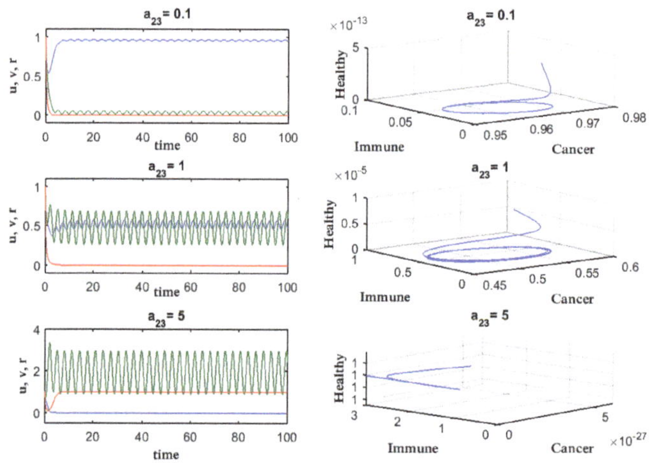

Figure 1. Phase portrait and time evolution of Equations (3) for various a_{23}, $(u_0, v_0, r_0) = (1, 1, 1)$. Green indicates immune cells, red indicates healthy cells, and a blue line indicates cancer.

To this end, in Figure 2, we present the mean values of u, v, r for different values of a_{23}. In Figure 2, the eradication of the cancer population occurs at an amplitude $a_{23} = 1.5$ that is smaller compared to the model of two species (without healthy cells) (amplitude $F = 2.62$). This means that healthy cells help to eliminate cancer cells and thus the optimal value of the amplitude of the treatment is smaller in the three-species model. It is also interesting to see that in Figure 3, the healthy population is almost in a steady state with the value 1 in the long time limit. On the other hand, the immune population increases as a_{23} increases. These results are the consequence of a stable effective equilibrium point $(0, \frac{a_{23}}{2a_{21}}, 1)$.

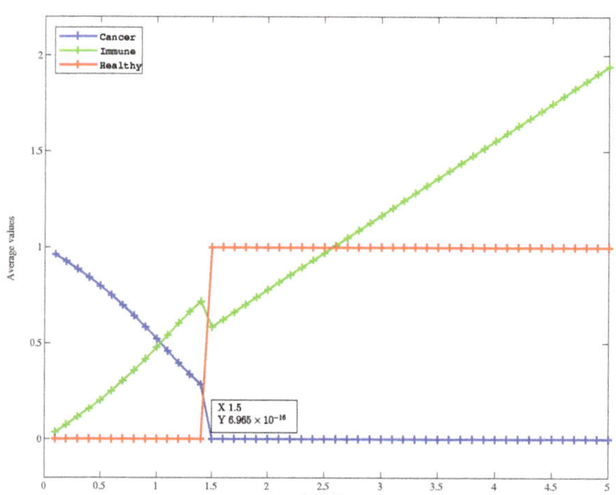

Figure 2. Average values as a function of periodic amplitude, with all parameter values fixed, $(u_0, v_0, r_0) = (1, 1, 1)$, and a_{23} changed.

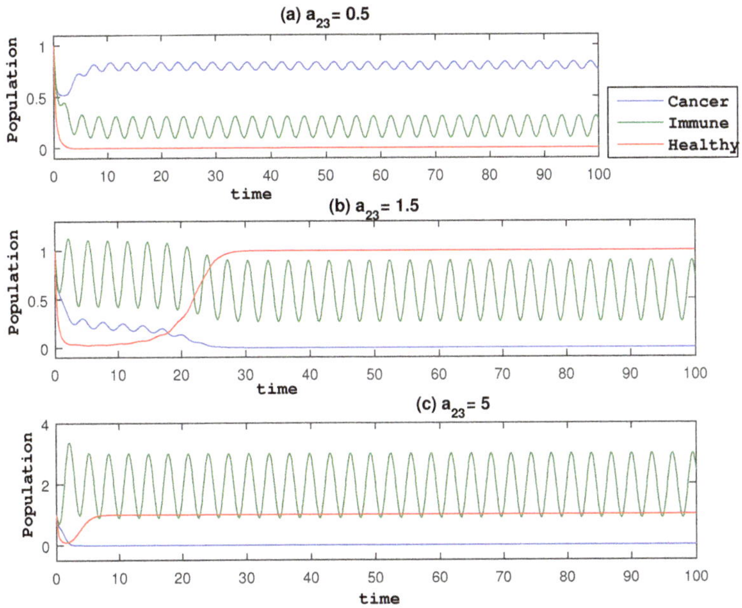

Figure 3. Time evolution of x using different a_{23} values, $(u_0, v_0, r_0) = (1, 1, 1)$.

For our main purpose of understanding the variability/sustainability of the tumor-immune system [23], we now compute the Fisher information F_T (see Appendices A.2 and A.4) at different values of amplitude a_{23}, and Figure 4 presents the results.

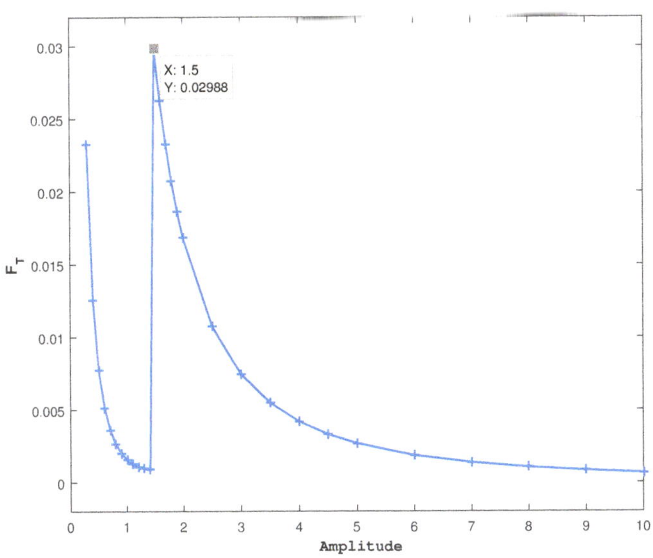

Figure 4. For $(u_0, v_0, r_0) = (1, 1, 1)$ and fixed other parameter values, F_T is shown against the amplitude. In Figures 2 and 3, a peak is observed at $a_{23} = 1.5$, which is also the value at which the cancer cells start to disappear.

In Figure 4, a prominent peak in $F_{\mathcal{T}}$ is observed in $a_{23} = 1.5$ which is the optimal value of the amplitude to eradicate tumor cells. This state is unsustainable and less variable because of the loss of one species. The large value of $F_{\mathcal{T}}$ means a low disorder and is related to a narrow PDF. Further increases in a_{23} cause a decrease in $F_{\mathcal{T}}$, similarly to what was observed in the two-species model.

3. Extended Three-Species Model

We further extend the model in Equation (3) including fluctuation (or variability) in the cancer growth rate modeled by $\epsilon sin(\Omega_1 \mathcal{T})$ as

$$\begin{aligned} \frac{du}{d\mathcal{T}} &= (1+\epsilon sin(\Omega_1 \mathcal{T}))\, u\, (1-u) - uv - a_{13} ur, \\ \frac{dv}{d\mathcal{T}} &= a_{21}[\frac{uv}{g_2+\alpha u} - v] - a_{22} uv \ + a_{23} \sin^2(\Omega \mathcal{T}), \\ \frac{dr}{d\mathcal{T}} &= a_{31} r\, (1-r) - a_{32} ur. \end{aligned} \quad (4)$$

Here, again, u, v and r represent the population of cancer cells, immune cells, and healthy cells, respectively. The only difference between the two models is in the equations. Equations (3) and (4) show the fluctuation in the growth of cancer cells.

To show the effect of fluctuation on the cancer growth rate, we fixed all parameter values to $a_{13} = 1.2$, $a_{21} = 1.291$, $g_2 = 0.3$, $\alpha = 1$, $a_{22} = 1.1$, $a_{31} = 1.2$, $a_{32} = 4.8$, in addition to $\Omega = \Omega_1 = 1$ and different values of a_{23}. In Figure 5, we present the results for different ϵ. Compared to Figure 2, we observe a similar behavior in Figure 5 without a significant effect of ϵ. In all panels, the cancer population starts to vanish roughly around $a_{23} = 1.5$. To examine its fluctuation at this value, we present $F_{\mathcal{T}}$ in Figure 6 using $\epsilon = 0.1$.

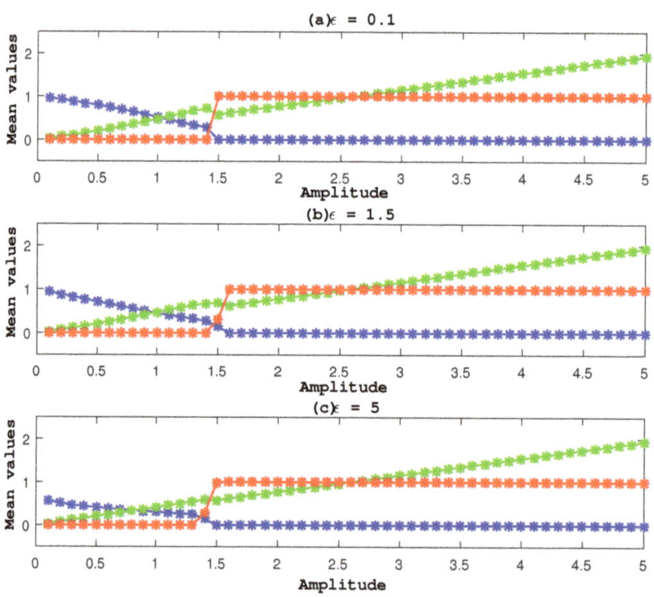

Figure 5. For $(u_0, v_0, r_0) = (1, 1, 1)$, populations' average values are shown against the amplitude a_{23}, with ϵ being different and the other parameter values fixed. A red line indicates healthy cells, a green line indicates immune cells, and a blue line indicates cancer cells.

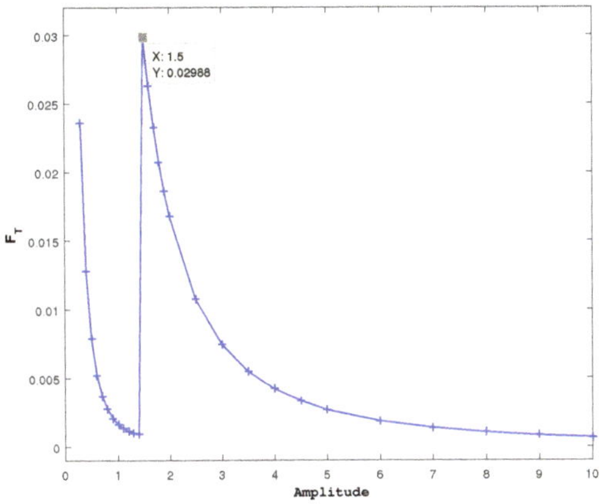

Figure 6. The amplitude a_{23} against F_T for $\Omega = \Omega_1 = 1$, $\epsilon = 0.1$, and $(u_0, v_0, r_0) = (1,1,1)$: At $a_{23} = 1.5$, a peak is shown, which is also the value at which the cancer cells begin to disappear in Figure 5.

In Figure 6, we again observe a distinct maximum of F_T at $a_{23} = 1.5$, quite similar to Figure 4. This leads us to the conclusion that there is no significant change induced by ϵ.

3.1. Effects of ϵ

To study the effect of ϵ systematically, it is useful to fix a_{23}, Ω, Ω_1 and vary ϵ.

Thus, we fix $a_{23} = 1.5$, $\Omega = \Omega_1 = 1$ and all other parameter values mentioned in the above text and vary ϵ, and show the results in Figure 7. In all cases, we see that the three species approach the same final states. We confirm that ϵ has no effect on long-term evolution by showing the plot of the average values of u, v, r in model (1) for different values of ϵ in Figure 8.

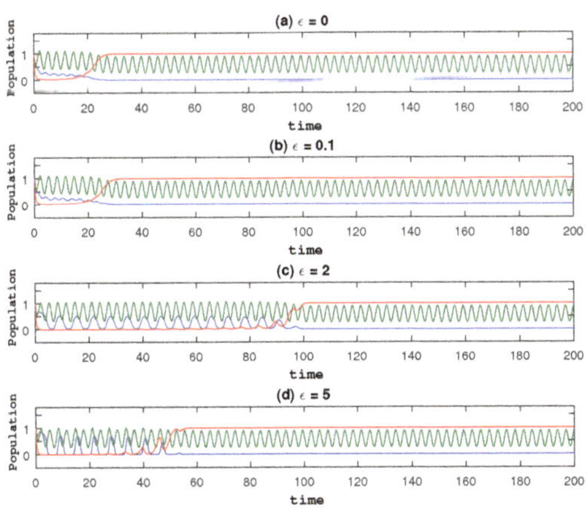

Figure 7. To show the dynamics of the model with fixed parameter values as $a_{23} = 1.5$, $\Omega = \Omega_1 = 1$, and $[u_0, v_0, r_0] = [1,1,1]$, we utilize different values of ϵ. Red indicates healthy cells, green indicates immune cells, and blue indicates cancer cells.

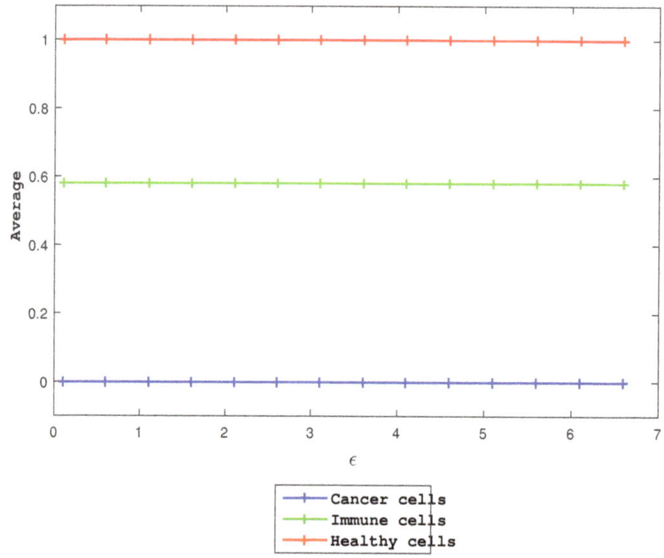

Figure 8. As a function of ϵ, we show the average of each species in various colors, while the other parameters remain unchanged. The average values show no changes, indicating that our model is independent of ϵ.

Using Fisher information, we compute F_T for different ϵ and fixed Ω_1 to be $\Omega_1 = 1$ and the amplitude to be $a_{23} = 1.5$ to determine the behavior of the model (4) in Figures 7 and 8. The system in a steady-state solution does not lose or acquire Fisher information, as we can see by the fact that the F_T values do not vary over time. Using $a_{23} = 4$, we obtain a similar result.

3.2. *Effects of* Ω_1

Finally, we examine the effect of Ω_1 in Equation (4). Specifically, we use the initial condition $u_0 = v_0 = r_0 = 1$ and the same other parameter values as mentioned above and vary Ω_1. Clearly, the average values for the three species do not change when Ω_1 increases.

4. Discussion

In this study, we investigated the impact of immunotherapy on the dynamical properties of a tumor-immune healthy-cell model by extending the classical predator–prey framework to a three-species system. The addition of periodic immunotherapy, non-linear damping of cancer cells, and the dynamics of healthy cells allowed us to explore a more comprehensive representation of tumor-immune interactions.

One of the key findings is that the introduction of immunotherapy leads to significant changes in the dynamics of the system. Using Fisher information as a measure of variability, we identified an optimal immunotherapy value, $a_{23} = 1.5$, which effectively reduces the tumor population while promoting the growth of healthy and immune cells. Compared to the two-species model, the three-species model demonstrates a more pronounced ability to suppress cancer growth, highlighting the importance of incorporating healthy cell dynamics into tumor-immune models.

Furthermore, our study revealed that varying the prey killing rate (b, cancer death rate) results in a monotonic decrease in the equilibrium point. At a specific value of b, Fisher information reaches a maximum, corresponding to a state where the probability density functions (PDF) of tumor and immune cells become narrow and do not overlap, indicating a minimal interaction between species. This suggests that at this parameter value, the system may be in an undesirable state due to reduced variability and interaction among species.

Furthermore, when the amplitude of immunotherapy (a_{23}) was varied, we observed that higher values led to greater tumor suppression but also resulted in reduced variability in the dynamics of the system. This suggests a trade-off between effective tumor suppression and maintaining system dynamism, which could have implications for treatment strategies.

In general, our results emphasize the importance of optimizing immunotherapy parameters to achieve the best therapeutic outcomes. The incorporation of healthy cell dynamics into the model significantly alters the behavior of the system and leads to a lower optimal immunotherapy value compared to the two-species model. These findings provide insight into how tumor-immune health interactions respond to external treatments and contribute to a broader understanding of cancer therapy dynamics.

5. Conclusions

This study examined the influence of immunotherapy on the dynamical properties of a tumor-immune healthy cell system using an extended predator–prey model. By incorporating periodic immunotherapy and the dynamics of healthy cells, we identified key parameter values that optimize treatment effectiveness.

The optimal value of immunotherapy ($a_{23} = 1.5$) promotes the growth of healthy and immune cells while reducing the tumor population, demonstrating the significant impact of including healthy cell dynamics in tumor modeling. Additionally, Fisher information analysis highlighted the importance of variability in system behavior, with specific parameter values leading to reduced interaction between species.

These findings provide valuable information on the complex interplay between tumor cells, immune responses, and treatment strategies. Our study could contribute to refining immunotherapy protocols and optimizing dosage strategies for cancer treatment, one of the world's leading causes of death.

While our study primarily focuses on analyzing the effects of immunotherapy using Fisher information, we acknowledge that averaging state variables may obscure complex dynamical behaviors. A more detailed bifurcation analysis could provide deeper insights into stability transitions, periodic oscillations, or chaotic dynamics in the system. Due to the scope of this work, a full bifurcation analysis was not included. However, future studies could incorporate bifurcation diagrams to complement our findings and further explore the nonlinear behavior of tumor-immune interactions under varying immunotherapy conditions.

Author Contributions: Conceptualization, A.A.-S. and E.-j.K.; Data curation, A.A.-S.; Formal analysis, A.A.-S. and E.-j.K.; Investigation, A.A.-S. and E.-j.K.; Methodology, E.-j.K.; Resources, A.A.-S.; Software, A.A.-S.; Supervision, E.-j.K.; Validation, A.A.-S. and E.-j.K.; Visualization, A.A.-S. and E.-j.K.; Writing—original draft, A.A.-S.; Writing—review and editing, E.-j.K. All authors have read and agreed to the published version of the manuscript.

Funding: This research received no external funding.

Institutional Review Board Statement: Not applicable.

Data Availability Statement: The datasets presented in this article are not readily available. Requests for access to data sets should be directed to the authors.

Conflicts of Interest: The authors declare no conflicts of interest.

Appendix A

Appendix A.1. Probability Density Function for the Logistic Equation [24,25]

Following the approach suggested by Cabezas et al. [2,3] and Rico-ramirez et al. [9], we determine the PDF of x by using the conservation of probability to relate the probability of observing the system at a specific value of x to the duration of time the system state spends at x:

$$p[x]\, dx = p[t]\, dt. \tag{A1}$$

Given that t has a uniform probability density and is a continuous variable,

$$p[t] = constant = A, \tag{A2}$$

From Equations (A1) and (A2), we derive the PDF of x as follows:

$$p[x] = p[t]\left|\frac{dt}{dx}\right| = A\left|\frac{dt}{dx}\right| = \frac{A}{u}, \tag{A3}$$

where

$$u = \frac{dx}{dt}. \tag{A4}$$

Since u is simply given by

$$\frac{dx}{dt} = [B + N_0 \sin(\omega t)]\, x\, (1 - \frac{x}{K}), \tag{A5}$$

in Equation (A3), we can write $p[x]$ as

$$p[x] = \frac{A}{(B + N_0 \sin(\omega t))\, x\, (1 - \frac{x}{K})}. \tag{A6}$$

Here, we need to consider the analytical solution for the logistic equation expressed in Equation (A5) as

$$x(t) = \frac{-Kx_0 \exp\left(Bt + \frac{N_0}{\omega}(1 - cos(\omega t))\right)}{(x_0 - K) - x_0 \exp\left(Bt + \frac{N_0}{\omega}(1 - cos(\omega t))\right)}, \tag{A7}$$

and substitute a function that just depends on x for it. since Equation (A6) includes the time-dependent function ($\sin \omega t$). To this end, we use Equation (A7) to express ($\cos \omega t$) as

$$\cos(\omega t) = 1 + \frac{\omega}{N_0} \ln\left[\frac{x_0\,(x - K)}{x\,(x_0 - K)}\right], \tag{A8}$$

as well as the identity $\sin(\omega t) = \sqrt{1 - \cos^2(\omega t)}$ and substitute the results in Equation (A6).

Appendix A.2. Obtaining Fisher Information for the Logistic Model [24,25]

We determine the Fisher information for the logistic model following the method proposed by a number of researchers. According to [24], the logistic model is as follows:

$$\frac{dx}{dt} = [B + N_0 \sin(\omega t)]\, x\, (1 - \frac{x}{K}). \tag{A9}$$

and the constant growth B is fixed at $B = 0$. Given that Fisher's information is

$$F_T = \frac{A}{T}\int_0^T \frac{\dot{u}^{\,2}}{u^{\,4}}\, dt, \tag{A10}$$

where $\dot{u} = \frac{du}{dt}$ and $u = \frac{dx}{dt}$, T is the total length of time, and A is a normalization constant, we express $\dot{u}$ as follows:

$$\dot{u} = \frac{\partial u}{\partial x}\frac{dx}{dt}. \tag{A11}$$

The following Fisher information is obtained by combining the outcomes of Equations (A9) and (A11) in Equation (A10):

$$F_T = \frac{A}{T}\int_0^T \frac{(N_0\, sin(\omega t)\ (\dot{x} - \frac{2x\dot{x}}{K})) + (N_0\, \omega\, cos(\omega t)\ (x - \frac{x^2}{K}))}{u^2}\, dt, \tag{A12}$$

Appendix A.3. Dimensionless Formulation for the Three-Species Model

We expand the predator–prey model by including another species which represents the healthy cells population as we can see from the following equations:

$$\frac{dx}{dt} = ax(1 - N_1 x) - bxy - dxz\,, \tag{A13a}$$

$$\frac{dy}{dt} = -cy + \frac{kyx}{g_2 + x} - fyx\ + F\sin^2(\omega t)\,, \tag{A13b}$$

$$\frac{dz}{dt} = gz(1 - N_2 z) - hzx. \tag{A13c}$$

By dividing the first Equation (A13a) by a and the second Equation (A13b) by c and the third Equation (A13c) by g, we will have

$$\begin{aligned}\frac{dx}{dat} &= x(1 - N_1 x) - \frac{b}{a}xy - \frac{d}{a}xz\,,\\ \frac{dy}{dct} &= -y + \frac{kyx}{c(g_2 + x)} - \frac{f}{c}yx\ + \frac{F}{c}\sin^2(\omega t)\,,\\ \frac{dz}{dgt} &= z(1 - N_2 z) - \frac{h}{g}zx.\end{aligned} \tag{A14}$$

In order to simplify the latter model in (A14), we introduce cancer, immune, and healthy populations in addition to the time in Equation (A14) in terms of dimensionless quantities as follows:

$$T = at, \quad u = \frac{k}{c}x, \qquad v = \frac{b}{a}y \quad and \quad r = \frac{h}{g}z.$$

As a result, we detect the following model after including all the above dimensionless quantities and more simplification:

$$\begin{aligned}\frac{c}{k}\frac{du}{dT} &= \frac{c}{k}u(1 - N_1\frac{c}{k}u) - \frac{c}{k}uv - \frac{cdg}{akh}ur\,,\\ \frac{a^2}{cb}\frac{dv}{dT} &= -\frac{a}{b}v + \frac{a}{b}\frac{vu}{g_2 + \frac{c}{k}u} - \frac{af}{bk}vu\ + \frac{F}{c}\sin^2(\frac{\omega}{a}T)\,,\\ \frac{a}{h}\frac{dr}{dT} &= \frac{g}{h}r(1 - N_2\frac{g}{h}r) - \frac{c}{k}ru\,.\end{aligned} \tag{A15}$$

Let us consider the following quantities in Equation (A15) in order to obtain a form with fewer parameters:

$$N_1 = \frac{k}{c}, \quad N_2 = \frac{h}{g}, \quad \Omega = \frac{\omega}{a} \quad \text{and} \quad \alpha = \frac{c}{k}.$$

The three-species model in its final formula can be displayed as follows after applying a dimensionless and simplification procedure:

$$\begin{aligned} \frac{du}{dT} &= u(1-u) - uv - a_{13}ur\,, \\ \frac{dv}{dT} &= a_{21}\left[\frac{vu}{g_2+\alpha u} - v\right] - a_{22}vu + a_{23}\sin^2(\Omega T)\,, \\ \frac{dr}{dT} &= a_{31}r(1-r) - a_{32}ru\,. \end{aligned} \tag{A16}$$

Appendix A.4. Fisher Information Construction for the Three-Species Model

The Fisher information may be calculated using an integral equation, which is subject to a collection of ordinary differential equations (ODE), according to the background theory described in several studies. Here, as an example, we demonstrate how to compute Fisher information for the three-species model. We use the following set of equations to illustrate the dimensionless formulation for the three-species model given in (4):

$$\begin{aligned} \frac{du}{dT} &= u(1-u) - uv - a_{13}ur\,, \\ \frac{dv}{dT} &= a_{21}\left[\frac{vu}{g_2+\alpha u} - v\right] - a_{22}vu + a_{23}\sin^2(\Omega T)\,, \\ \frac{dr}{dT} &= a_{31}r(1-r) - a_{32}ru\,. \end{aligned} \tag{A17}$$

Since the Fisher information quantity is as in the following equation,

$$F_T = \frac{A}{T}\int_0^T \frac{\dot{s}^{\,2}}{s^{\,4}}\,dt\,, \tag{A18}$$

where $s = \sqrt{\dot{u}^{\,2} + \dot{v}^{\,2} + \dot{r}^{\,2}}$, A is a normalization constant, and T is the total time duration. In the following, we need to find $\dot{s} = \frac{ds}{dt}$ as follows:

$$\frac{\partial s}{\partial t} = \frac{\partial s}{\partial u}\frac{\partial u}{\partial t} + \frac{\partial s}{\partial v}\frac{\partial v}{\partial t} + \frac{\partial s}{\partial r}\frac{\partial r}{\partial t}\,, \tag{A19}$$

or s's derivative in Equation (A19) could be expressed as follows:

$$\dot{s} = \frac{\partial s}{\partial u}\,\dot{u} + \frac{\partial s}{\partial v}\,\dot{v} + \frac{\partial s}{\partial r}\,\dot{r}\,,$$

With regard to the cancer population u, we can now use a basic differentiation's calculus for the function s as follows:

$$\frac{\partial s}{\partial u} = \frac{\dot{u}(1-2u-v-a_{13}r) + \dot{v}\left(\frac{a_{21}v(g_2+\alpha u) - a_{21}\alpha uv}{(g_2+\alpha u)^2} - a_{22}v\right) - a_{32}r\dot{r}}{s}\,, \tag{A20}$$

The following is the next derivative for s with regard to v:

$$\frac{\partial s}{\partial v} = \frac{\frac{a_{21}u\dot{v}}{g_2+\alpha u} - u\dot{u} - a_{21}\dot{v} - a_{22}u\dot{v}}{s}, \tag{A21}$$

The derivative of s with respect to r can evolve with the following equation:

$$\frac{\partial s}{\partial r} = \frac{a_{31}\dot{r} - a_{13}u\dot{u} - 2a_{31}r\dot{r} - a_{32}u\dot{r}}{s}, \tag{A22}$$

by substituting the results of (A20)–(A22) together in (A19) and finally including the outcome results in Equation (A18), the following Fisher information quantity is acquired:

$$F_T = \frac{A}{T}\int_0^T \frac{\left(\dot{u}^2 a + \dot{v}^2 b + \dot{r}^2 c + \dot{u}\dot{v}d - \dot{u}\dot{r}e\right)^2}{s^3}\, dt, \tag{A23}$$

where

$$a = 1 - 2u - v - a_{13}r,$$

$$b = \frac{a_{21}u}{g_2 + \alpha u} - a_{21} - a_{22}u,$$

$$c = a_{31} - 2a_{31}r - a_{32}u,$$

$$d = \frac{a_{21}v(g_2 + \alpha u) - a_{21}\alpha uv}{(g_2 + \alpha u)^2} - a_{22}v - u,$$

$$e = a_{32}r + a_{13}u.$$

This gives us the Fisher information index, which is then used to create the Fisher information in Figures 4 and 6. The same process may be used to calculate similar quantities for other dynamical systems.

References

1. Shastri, Y.; Diwekar, U. Sustainable ecosystem management using optimal control theory: Part 2 (stochastic systems). *J. Theor. Biol.* **2006**, *2006*, 522–532. [CrossRef]
2. Cabezas, H.; Fath, B.D. Towards a theory of sustainable systems. *Fluid Phase Equilibria* **2002**, *2002*, 3–14. [CrossRef]
3. Fath, B.D.; Cabezas, H.; Pawlowski, C.W. Regime changes in ecological systems: An information theory approach. *J. Theor. Biol.* **2003**, *2003*, 517–530. [CrossRef]
4. Abrams, P.A. Why don't predators have positive effects on prey populations? *Evol. Ecol.* **1992**, *1992*, 449–457. [CrossRef]
5. Kumar, S.M.; Sharma, A.; Sánchez-Ruiz, L.M. Dynamical Complexity of Modified Leslie—Gower Predator–Prey Model Incorporating Double Allee Effect and Fear Effect. *Symmetry* **2024**, *2024*, 1552. [CrossRef]
6. Lotka, A.J. *Elements of Physical Biology*; Hassell Street Press: Victoria, Australia, 1925.
7. Volterra, V. Fluctuations in the abundance of a species considered mathematically. *Nature* **1926**, *118*, 558–560. [CrossRef]
8. Mayer, A.L.; Pawlowski, C.W.; Cabezas, H. Fisher information and dynamic regime changes in ecological systems. *Ecol. Model.* **2006**, *2006*, 72–82. [CrossRef]
9. Rico-Ramirez, V.; Reyes-Mendoza, M.A.; Quintana-Hernez, P.A.; Ortiz-Cruz, J.A.; Hernez-Castro, S.; Diwekar, U.M. Fisher information on the performance of dynamic systems. *Ind. Eng. Chem. Res.* **2010**, *2010*, 1812–1821. [CrossRef]
10. Érika, Di.; Otero-Espinar, M.V. Predator–prey models: A review of some recent advances. *Mathematics* **2021**, *2021*, 1783. [CrossRef]
11. Murray, J.D. *Mathematical Biology: I. An Introduction*; Springer: New York, NY, USA, 2002.
12. Selin, N.E.; Giang, A.; Clark, W.C. Progress in modeling dynamic systems for sustainable development. *Proc. Natl. Acad. Sci. USA* **2023**, *120*, e2216656120. [CrossRef]

13. Banerjee, S.; Sarkar, R.R. Delay-induced model for tumor-immune interaction and control of malignant tumor growth. *Biosystems* **2008**, *2008*, 268–288. [CrossRef]
14. Joseph, M.K.; Ouifki, R.; Eladdadi, A.; de Pillis, L. Mathematical model of tumor–immune surveillance. *J. Theor. Biol.* **2016**, *2016*, 312–330.
15. d'Onofrio, A. A general framework for modeling tumor-immune system competition and immunotherapy: Mathematical analysis and biomedical inferences. *Phys. Nonlinear Phenom.* **2005**, *2005*, 220–235. [CrossRef]
16. Windarto, W.; Eridani, E. On modification and application of Lotka-Volterra competition model. *Aip Conf. Proc.* **2020**, *2268*, 050007.
17. Wu, Y.; Tur, A.; Wang, H. Sustainable Optimal Control for Switched Pollution-Control Problem with Random Duration. *Entropy* **2023**, *25*, 1426. [CrossRef]
18. Amany, N.; Albalawi, W.; Jazmati, M.S.; Elrashidi, A.; Ali, H.M.; Moaaz, O. Functional Differential Equations with an Advanced Neutral Term: New Monotonic Properties of Recursive Nature to Optimize Oscillation Criteria. *Axioms* **2024**, *2024*, 847. [CrossRef]
19. Kareva, I.; Luddy, K.A.; O'Farrelly, C.; Gatenby, R.A.; Brown, J.S. Predator-Prey in Tumor-Immune Interactions: A Wrong Model or Just an Incomplete One? *Front. Immunol.* **2021**, *12*, 668221. [CrossRef] [PubMed] [PubMed Central]
20. Al-Saffar, A.; Kim, E. Fisher Information Approach in Modified Predator—Prey Model. *Axioms* **2025**, *14*, 144. [CrossRef]
21. Kirschner, D.; Panetta, J.C. Modeling immunotherapy of the tumor-immune interaction. *J. Math. Biol.* **1998**, *1998*, 235–252. [CrossRef]
22. López, Á.G.; Sabuco, J.; Seoane, J.M.; Duarte, J.; Januário, C.; Sanjuán, M.A.F. Avoiding healthy cells extinction in a cancer model. *J. Theor. Biol.* **2014**, *2014*, 74–81. [CrossRef]
23. Sahbani, M.; Das, S.; Green, J.R. Classical Fisher information for differentiable dynamical systems. Chaos: An Interdisciplinary. *J. Nonlinear Sci.* **2023**, *33*, 103139.
24. Al-Saffar, A.; Kim, E. Sustainable theory of a logistic model-Fisher Information approach. *Math. Biosci.* **2017**, *2017*, 81–91. [CrossRef] [PubMed]
25. Al-Saffar, A.; Kim, E. Fisher information approach to understand the Gompertz model. *Math. Appl. Sci. Eng.* **2022**, *2022*, 269–279. [CrossRef]

entropy

MDPI

Article

Information Theoretical Analysis of Quantum Mixedness in a Finite Model of Interacting Fermions

Diana Monteoliva [1], Angelo Plastino [2,*] and Angel Ricardo Plastino [3]

[1] UNLP-Comisión de Investigaciones Científicas Provincia de Buenos Aires, La Plata 1900, Argentina; monteoli@fisica.unlp.edu.ar
[2] Instituto de Física La Plata, CCT-CONICET, Universidad Nacional de La Plata, La Plata 1900, Argentina
[3] CeBio-Departamento de Ciencias Básicas, Universidad Nacional del Noroeste Provincia de Buenos Aires (UNNOBA), CONICET, Junin 6000, Argentina; arplastino@unnoba.edu.ar
* Correspondence: plastino@fisica.unlp.edu.ar

Abstract: In this study, we utilize information theory tools to investigate notable features of the quantum degree of mixedness (C_f) in a finite model of N interacting fermions. This model serves as a simplified proxy for an atomic nucleus, capturing its essential features in a more manageable form compared to a realistic nuclear model, which would require the diagonalization of matrices with millions of elements, making the extraction of qualitative features a significant challenge. Specifically, we aim to correlate C_f with particle number fluctuations and temperature, using the paradigmatic Lipkin model. Our analysis reveals intriguing dependencies of C_f on the total fermion number, showcasing distinct behaviors at different temperatures. Notably, we find that the degree of quantum mixedness exhibits a strong dependence on the total fermion number, with varying trends across different temperature regimes. Remarkably, this dependence remains unaffected by the strength of the fermion–fermion interaction (as long as it is non-zero), underscoring the robustness of the observed phenomena. Through comprehensive numerical simulations, we provide illustrative graphs depicting these dependencies, offering valuable insights into the fundamental characteristics of quantum many-body fermion systems. Our findings illuminate the intricate dynamics of the degree of mixedness, a crucial quantum property, with potential implications for diverse fields ranging from condensed matter physics to quantum information science.

Keywords: Lipkin model; many fermion systems; mixedness-degree; finite temperature; SU2 symmetry

1. Introduction

One of the key aspects of quantum systems is the concept of quantum mixedness, which quantifies the degree to which a quantum state is a mixedness of different possible states, rather than a pure state. This property is crucial in various fields, from condensed matter physics to quantum information science, as it influences the behavior and characteristics of quantum systems.

The quantum notion of mixedness [1–5], in the context of many-body quantum systems [6–9], becomes particularly intriguing. Many-body systems, composed of interacting particles, exhibit complex behaviors that arise from the interplay of their individual components [6]. The degree of quantum mixedness (C_f) in such systems can reveal significant insights into their underlying physics. However, studying these systems poses substantial challenges, especially when dealing with realistic models that require the diagonalization of extremely large matrices [6–9].

To address these challenges, simplified models that capture the essential features of more complex systems are often employed. One such model is the Lipkin–Meshkov–Glick (LMG) model [6–9], which has been extensively used to study various aspects of interacting fermions. The LMG model provides a manageable yet insightful framework to explore the qualitative features of quantum many-body systems, making it an ideal choice for investigating the dependencies of quantum mixedness [6–9].

In this study, we utilize the LMG model to analyze the quantum degree of mixedness in a finite system of N interacting fermions. By correlating C_f with particle number fluctuations and temperature, we aim to uncover the fundamental dependencies and behaviors of quantum mixedness in many-body fermion systems. Our investigation reveals intriguing trends in C_f across different temperatures and total fermion numbers, highlighting the robustness of these phenomena against variations in the fermion–fermion interaction strength.

Through comprehensive numerical simulations, we provide a detailed theoretical framework and illustrative graphs that depict these dependencies. Our findings not only enhance the understanding of quantum mixedness in many-body systems but also have potential implications for various applications, from improving models of atomic nuclei to advancing quantum computing techniques.

The significance of this study lies in its ability to bridge the gap between simplified theoretical models and the complex reality of many-body quantum systems. By elucidating the behavior of quantum mixedness, we contribute to the broader comprehension of quantum mechanics and its applications, paving the way for future research in both theoretical and practical domains.

1.1. Our Present Protagonist: The Mixedness Notion

Mixedness plays an essential role in quantum tasks for various reasons. We list some of them.

- Understanding Quantum Correlations and Entanglement: The degree of mixedness provides insights into the quantum correlations and entanglement within a system. By studying the degree of mixedness, researchers can gain a deeper understanding of how entanglement is distributed and how it can be manipulated within a system.
- Characterizing Quantum States: The degree of mixedness helps characterize the purity of quantum states. Pure states, which have zero mixedness, are ideal for many quantum processes, while mixed states, which have non-zero mixedness, can result from decoherence and other environmental interactions. Understanding the mixedness can help in designing strategies to preserve quantum coherence and improve the performance of quantum devices.
- Analyzing Thermodynamic Properties: The degree of mixedness is linked to the thermodynamic properties of quantum systems. It can provide information about phase transitions, thermalization processes, and the overall behavior of a system at different temperatures. This is particularly important in understanding quantum statistical mechanics and the thermodynamics of small systems where quantum effects are significant.
- Exploring Quantum-to-Classical Transition: By studying how the degree of mixedness changes with various parameters, such as temperature and particle number, researchers can gain insights into the quantum-to-classical transition. This transition is crucial for understanding how classical behavior emerges from quantum systems, a fundamental question in quantum mechanics.

- Implications for Quantum Simulations: In quantum simulations of complex systems, knowing the degree of mixedness can help validate the accuracy of the simulations. It ensures that the simulated quantum states accurately reflect the intended physical systems, particularly when simulating open systems that interact with their environment.
- Overall, scrutinizing the features of the degree of mixedness in a quantum system is essential for advancing our understanding of quantum mechanics, improving quantum technologies and exploring new frontiers in quantum research. It provides a critical link between theoretical concepts and practical applications, enabling researchers to harness the full potential of quantum systems.

1.2. Our Purposes Here

The degree of quantum mixedness is influenced by several factors, including the total number of fermions and the temperature of the system [10–14]. Understanding how these factors **interact and affect the mixedness-degree** C_f is crucial for gaining deeper insights into the quantum dynamics of many-body systems. Through comprehensive numerical simulations, we investigate these dependencies and present our findings through illustrative graphs. Our results reveal that the degree of mixedness is strongly dependent on the total fermion number, with distinct trends emerging across different temperature regimes. Interestingly, this dependence seems to be unaffected by the strength of the fermion–fermion interaction, provided it is non-zero. Hopefully, our results will highlight the fundamental nature of the observed phenomena.

2. The SU2-Angular Momentum Lipkin Quasi-Spin Formalism [6–8]

We deal with N fermions and call $\Omega = N/2$; that is, the Lipkin model consists of $N = 2\Omega$ fermions that occupy two different N-fold degenerate single-particle (sp) energy levels. The two levels are separated by an energy gap ϵ. This entails 4Ω s.p. micro states. Two quantum numbers (called μ and p) are linked to a single micro state. The first one, μ, takes the values $\mu = -1$ (lower level) and $\mu = +1$ (upper level). The remaining quantum number is denoted the quasi spin p pertaining to the $2N$-fold degeneracy. The pair p, μ is viewed as a "site", which can be occupied (by a fermion) or be empty. Lipkin fixes

$$N = 2\mathcal{J}. \tag{1}$$

Here, $\mathcal{J}$ is a sort of angular momentum. Lipkin [7] uses special angular momentum operators called quasi-spin ones. These are

$$\mathcal{J}_z = \sum_{p,\mu} \mu \, C^{+}_{p,\mu} C_{p,\mu}, \tag{2}$$

$$\mathcal{J}_+ = \sum_{p} C^{+}_{p,+} C_{p,-}, \tag{3}$$

$$\mathcal{J}_- = \sum_{p} C^{+}_{p,-} C_{p,+}, \tag{4}$$

together with the Casimir operator

$$\mathcal{J}^2 = \mathcal{J}_z^2 + \frac{1}{2}(\mathcal{J}_+\mathcal{J}_- + \mathcal{J}_-\mathcal{J}_+). \tag{5}$$

The eigenvalues of $\mathcal{J}^2$ take form $\mathcal{J}(\mathcal{J}+1)$ and the Lipkin Hamiltonian reads (v is a coupling constant)

$$H = \epsilon \mathcal{J}_z + \frac{v}{4}(\mathcal{J}_+^2 + \mathcal{J}_-^2). \tag{6}$$

For the Lipkin Hamiltonian, we have the matrix [7]

$$\begin{aligned} \langle n'|H_L|n\rangle &= \left\{\frac{N}{2} - n + 1 - \left(Nn - \frac{N}{2} - n^2 + 2n - 1\right)\omega\right\}\delta_{n',n} - \\ &-\frac{v}{2}\sqrt{(N-n)(N-n+1)(n+1)n}\,\delta_{n',n+2} \\ &-\frac{v}{2}\sqrt{(N-n')(N-n'+1)(n'+1)n'}\,\delta_{n',n-2} \end{aligned} \tag{7}$$

with $n = 0, 1, \ldots, N$ for $\mathcal{J} = N/2$. Numerical diagonalization yields energy eigenvalues $E_n(v, \mathcal{J})$ for our Hamiltonian. These eigenvalues are needed to build the partition function Z in the canonical ensemble [13,14].

All thermal quantities of interest are deduced from the partition function Z [13,14]. We construct Z using probabilities assigned to the models' microscopic states. Their energies are E_i [13,14]. Some important macroscopic quantifiers are computed as in [13,14]. These indicators, together with Z, derive from the canonical probability distributions [13,14]. $P_n(v, \mathcal{J}, \beta)$. β is the inverse temperature. The pertinent expressions are given in [13,14]. If we call the mean energy U and the free energy F, we have

$$P_n(v, \mathcal{J}, \beta) = \frac{1}{Z(v, \mathcal{J}, \beta)} e^{-\beta E_n(v,\mathcal{J})} \tag{8}$$

$$Z(v, \mathcal{J}, \beta) = \sum_{n=0}^{N} e^{-\beta E_n(v,\mathcal{J})} \tag{9}$$

$$\begin{aligned} U(v, \mathcal{J}, \beta) &= \langle E\rangle = -\frac{\partial \ln Z(v, \mathcal{J}, \beta)}{\partial \beta} \\ &= \sum_{n=0}^{N} E_n(v, \mathcal{J}) P_n(v, \mathcal{J}, \beta) \\ &= \frac{1}{Z(v, \mathcal{J}, \beta)} \sum_{n=0}^{N} E_n(v, \mathcal{J}) e^{-\beta E_n(v,\mathcal{J})} \end{aligned} \tag{10}$$

$$S(v, \mathcal{J}, \beta) = -\sum_{n=0}^{N} P_n(v, \mathcal{J}, \beta) \ln[P_n(v, \mathcal{J}, \beta)] \tag{11}$$

$$F(v, \mathcal{J}, \beta) = U(v, \mathcal{J}, \beta) - T\, S(v, \mathcal{J}, \beta). \tag{12}$$

The thermal quantifiers above provide much more information than the one obtained via just the quantum resources of zero temperature T [13,14]. As stated above, taking a low enough T, our quantifiers above yield a good representation of the $T = 0$ scenario [13,14]. Below, we adopt the high enough $\beta = 20$ value.

The State's ρ Degree of Mixedness C_f

The concepts of purity and degree of mixedness are fundamental in quantum mechanics and play a crucial role in describing the behavior of quantum systems. They are particularly relevant in the study of quantum information, quantum computation, quantum entanglement, and quantum measurements. The distinction between pure states and mixed states allows for a comprehensive understanding of the coherence, superposition, and statistical behavior of quantum systems.

As is well known in quantum mechanics, the degree of mixedness C_f of a given state represented by ρ is given by

$$C_f = 1 - Tr\rho^2 = 1 - \sum_n P_n^2, \tag{13}$$

where, $Tr\rho^2$ is the so called "Purity" P_y. Note that we have $C_f = 0$ and $P_y = 1$ for pure states. C_f is a very important quantity for us here.

In probability terms, we have $P_y = \sum_{n=0}^{N}(P_n(v, \mathcal{J}, \beta))^2$ and $C_f = S_2 = 1 - P_y^2$.

3. First Results

3.1. C_f Versus Temperature T

The quantity C_f will be the protagonist of Figures 1–7. Strong changes in the system's dynamics emerge as N grows. We plot below the mixing C_f versus T and versus the inverse temperature β. There seems to be a critical temperature T_S at which C_f stabilizes itself and ceases growing with T. It is clear that when N reaches the value six, the system's behavior strongly changes as multiple artifacts arise, whereas for $N < 6$, the mixing grows smoothly with T. In somewhat more detail, we can make the following assertion.

The analysis of the degree of quantum mixedness (C_f) as a function of temperature (T) reveals significant insights into the system's thermodynamic behavior. As the number of particles (N) increases, pronounced changes in the system's dynamics become evident. In the provided plots of C_f versus T and versus the inverse temperature (β) [Figures 1–5], a critical temperature, T_S, emerges where C_f stabilizes and ceases to grow with increasing T. This indicates a phase transition-like behavior, where the system transitions from one state to another, exhibiting stability in the degree of mixedness beyond T_S. This critical temperature suggests a threshold above which the system reaches a state of equilibrium regarding its mixedness. For $N \geq 6$, the system's behavior changes markedly, introducing multiple artifacts and complexities. This indicates a possible shift in the underlying dynamics or interactions within the system as it scales. In contrast, for $N < 6$, C_f exhibits a smoother and more predictable growth with T, suggesting a more straightforward relationship between temperature and quantum mixedness in smaller systems. The observed artifacts for larger N could be attributed to increased interaction complexities and emergent phenomena that are not present in smaller systems. These artifacts might reflect underlying phase transitions, resonance effects, or other collective behaviors that only manifest at higher particle numbers. Overall, the behavior of C_f with temperature highlights the intricate dynamics of many-body quantum systems and underscores the importance of considering particle number effects when analyzing thermodynamic properties. These findings provide valuable insights into the quantum mechanical properties of the system and have potential implications for understanding more complex, realistic models in quantum physics.

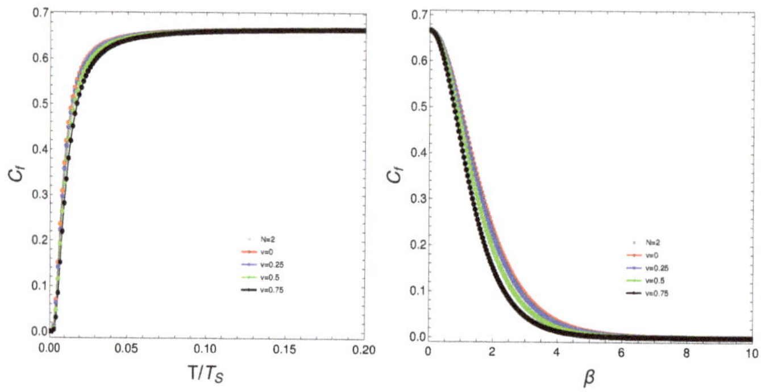

Figure 1. **(Left)** C_f versus T/T_S where T_S is that temperature for which the values of C_f stabilize. **(Right)** C_f versus β. For both graphs, N is fixed at $N = 2$, and each curve corresponds to a different v-value.

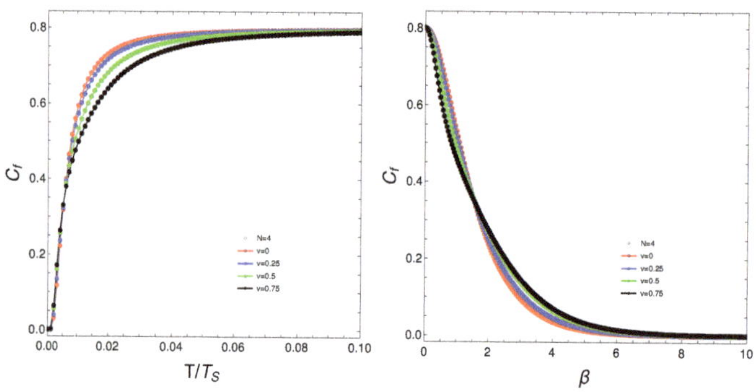

Figure 2. **(Left)** C_f versus T/T_S, where T_S is the temperature for which the values of C_f stabilize. **(Right)** C_f versus β. For both graphs, N is fixed at $N = 4$, and each curve corresponds to a different v-value.

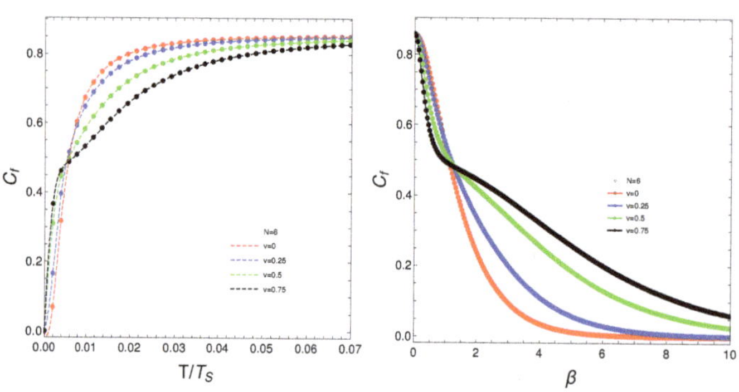

Figure 3. **(Left)** C_f versus T/T_S, where T_S is the temperature for which the values of C_f stabilize. **(Right)** C_f versus β. For both graphs, N is fixed at $N = 6$, and each curve corresponds to a different v-value.

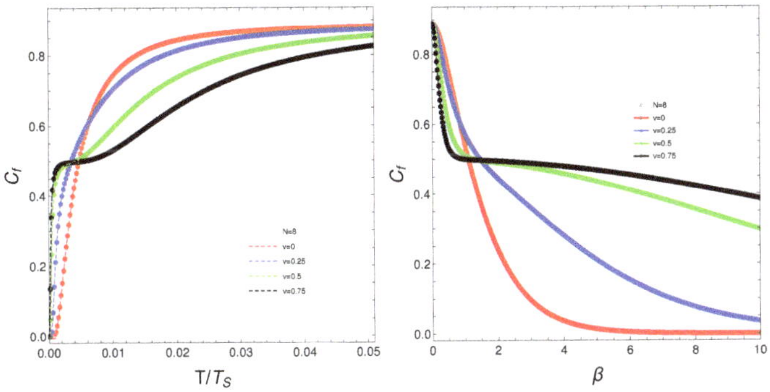

Figure 4. (**Left**) C_f versus T/T_S, where T_S is the temperature for which the values of C_f stabilize. (**Right**) C_f versus β. For both graphs, N is fixed at $N = 8$, and each curve corresponds to a different v-value.

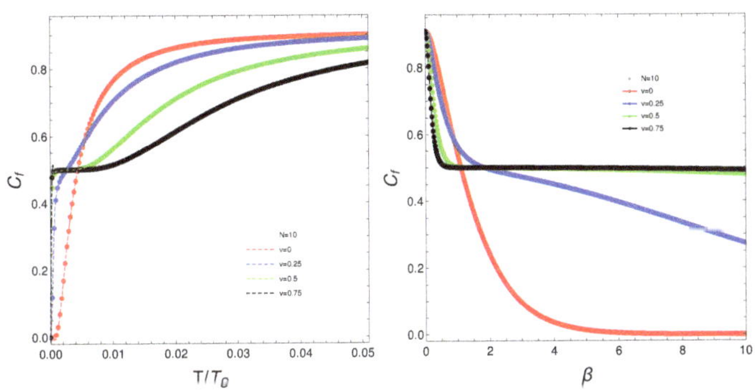

Figure 5. (**Left**) C_f versus T/T_S, where T_S is the temperature for which the values of C_f stabilize. (**Right**) C_f versus β. For both graphs, N is fixed at $N = 10$, and each curve corresponds to a different v-value.

What do we learn from these graphs? That the mixing strongly depends not only on the temperature but also, and in a very strong fashion, on the fermion number and the coupling constant. In summary, the degree of quantum mixing in the Lipkin model is a multifaceted phenomenon influenced by temperature, fermion number, and the coupling constant. Each of these factors contributes to the overall behavior of the system, determining how the quantum states are occupied and mixed. Understanding these dependencies is crucial for analyzing the dynamics and properties of many-fermion systems within the framework of the Lipkin model. Further analysis on these issues follows below. See Figures 6–9.

3.2. C_f Changes as Plotted Versus v or Versus N

The behavior of C_f in these circumstances is depicted in Figure 6. We again detect a critical v at which stability is reached. Same for N. Here, however, we see quite abrupt jumps in the mixedness-degree as the coupling constant grows for N large enough or for v finite.

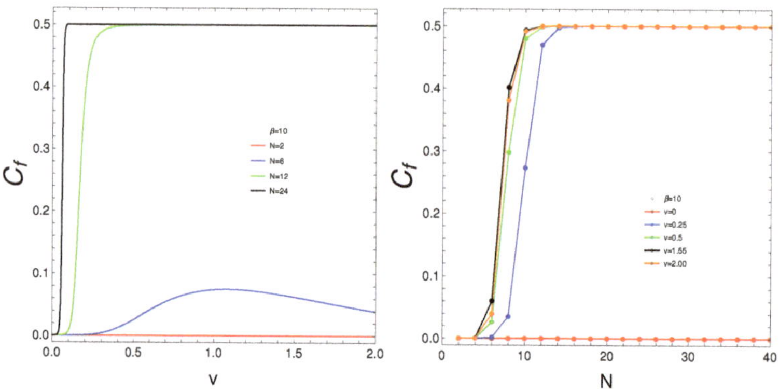

Figure 6. (**Left**) C_f versus v for different N-values. There seems to be a kind of phase transition for N large enough. (**Right**) C_f versus N for different v-values. Even for small N the mixing abruptly grows. For both graphs $\beta = 10$.

Let us reflect upon the two above graphs. The interaction between fermions acts as a perturbation that drives the system through a quantum phase transition. Before the interaction is introduced, the system may be in a non-mixed, ordered phase with a degree of mixing at zero.

- As soon as the interaction is turned on, it reaches a critical threshold where the system can no longer maintain this ordered phase, leading to an abrupt jump to a mixed phase with a degree of mixing at 0.5. This critical interaction threshold marks a point where the system's energy landscape changes significantly.
- At this critical point, the system's ground state undergoes a reconfiguration. The abrupt jump in the degree of mixing suggests that the system transitions from a non-mixed state to a highly mixed state in which the particles are now in a superposition of states.
- This reconfiguration minimizes the system's free energy under the new interaction regime, as we see below, leading to a more stable state with increased quantum coherence and entanglement among the fermions.

- The jump to a mixing of 0.5 indicates a sudden onset of quantum coherence. The system achieves a new equilibrium where the quantum states are coherently mixed, resulting in an optimal balance of energy.

- The high degree of mixing implies that the system has transitioned to a state where fermions are delocalized and strongly correlated, maximizing the entropy, as we confirm below, Thus, introducing fermion interactions induces strong correlation effects that are not present in the non-interacting system. These correlations enhance the mixing of quantum states, leading to a robust mixed phase.

- The interactions cause the particles to collectively behave in a way that drastically alters the macroscopic properties of the system, reflected in the sudden change in the degree of mixing. The abrupt change in the degree of mixing highlights the robustness of the quantum phase transition. It demonstrates that even a small interaction can lead to significant changes in the system's macroscopic properties when the number of particles is large enough.

- This robustness indicates that the system's behavior is dominated by collective effects rather than individual particle properties, a hallmark of many-body quantum systems. We can make here more elaborate reflections by analyzing and expanding on these points:

1. Interaction-Induced Phase Transition: The interaction between fermions acts as a perturbation that drives the system through a quantum phase transition. Initially, in the absence of interactions, the system is in a non-mixed, ordered phase with zero degree of mixing. When the interaction is introduced, it reaches a critical threshold where the system transitions abruptly to a mixed phase with a degree of mixing around 0.5. This critical interaction threshold signifies a significant change in the system's energy landscape.
2. Ground State Reconfiguration: At the critical point, the system's ground state undergoes a reconfiguration. The sudden jump in the degree of mixing from 0 to 0.5 suggests a rapid transition from a non-mixed to a highly mixed state, indicating that particles are now in a superposition of states. This reconfiguration minimizes the system's free energy under the new interaction regime, leading to a more stable state characterized by increased quantum coherence and entanglement among the fermions.
3. Quantum Coherence and Entanglement: The abrupt onset of a high degree of mixing (0.5) indicates the emergence of significant quantum coherence. The system achieves a new equilibrium where quantum states are coherently mixed, optimizing energy balance. This high degree of mixing implies that fermions are delocalized and strongly correlated, leading to maximized entropy. The introduction of fermion interactions induces strong correlation effects that enhance the mixing of quantum states, resulting in a robust mixed phase.
4. Collective Behavior and Macroscopic Property Changes: The interactions cause particles to behave collectively, dramatically altering the system's macroscopic properties. The sudden change in the degree of mixing reflects the robustness of the quantum phase transition. It demonstrates that even small interactions can lead to substantial changes in macroscopic properties when the particle number is sufficiently large. This robustness highlights that the system's behavior is governed by collective effects rather than individual particle properties, a hallmark of many-body quantum systems.
5. Implications for Many-Body Quantum Systems: The findings underscore the importance of collective behaviors in many-body quantum systems. The interaction-driven transition to a mixed phase with high quantum coherence and entanglement illustrates how many-body effects can dominate system dynamics. These results are crucial for understanding quantum phase transitions and the emergence of macroscopic quantum properties in interacting fermion systems.
6. In summary, the analysis presented in the remarks provides a comprehensive explanation of the effects of fermion interactions on the system's thermodynamic and quantum mechanical properties. The critical threshold, abrupt jumps in mixing, and the resulting coherent states offer valuable insights into the behavior of many-body quantum systems and their phase transitions. These findings have significant implications for theoretical models and practical applications in quantum physics.

4. Connection Between the Degree of Mixedness and the Differences Between Energy Levels

We discuss now results plotted in Figures 7–9 below. It is shown in reference [15], Figure 1, that the energy difference ΔE between those for the ground state energy and the first excited state diminishes as v grows, facilitating mixing. Diminishing the energy difference between the first excited state and the ground state facilitates mixing in the system due to increased thermal population and enhanced quantum mechanical coupling. In a system at thermal equilibrium, the population of states is governed by the Gibbs canonical distribution $P_i \propto \exp E_i/k_{BT}$, where E_i is the energy of state i, k_B is the Boltzmann constant, and T is the temperature.

When the energy difference ΔE between the ground state and the first excited state decreases, the exponential factor $\exp(\Delta E/k_{BT})$ increases. This means that at a given temperature, more particles are thermally excited to the first excited state. Furthermore, there is increased occupancy, as a higher thermal population in the first excited state leads to a greater number of particles available to transition between these states, facilitating interactions and mixing. Additionally, the probability of quantum transitions between states depends on the overlap of their wavefunctions and the energy separation. A smaller energy gap can enhance the coupling between the ground state and the first excited state, increasing the transition rate between these states.

Let us see what happens for $N = 2$, involving diagonalization of a 3×3 matrix. We look for an analytical expression for ΔE as a function of v and the plot C_f versus ΔE for several values of v. In diagonalizing the Hamiltonian H_L for $N = 2$, the energies are

$$E_0 = -\sqrt{1+v^2} \tag{14}$$
$$E_1 = 0 \tag{15}$$
$$E_2 = \sqrt{1+v^2} \tag{16}$$

Accordingly $\Delta E(v) = \sqrt{1+v^2}$ and $\Delta E(v=0) = 1$. The states' probabilities are

$$P_0 = \frac{1}{Z}e^{\beta\sqrt{1+v^2}} = \frac{1}{Z}e^{\beta\Delta E} \tag{17}$$
$$P_1 = \frac{1}{Z}e^{0} = \frac{1}{Z} \tag{18}$$
$$P_2 = \frac{1}{Z}e^{-\beta\sqrt{1+v^2}} = \frac{1}{Z}e^{-\beta\Delta E}, \tag{19}$$

with $Z = 1 + e^{\beta\sqrt{1+v^2}} + e^{-\beta\sqrt{1+v^2}} = 1 + 2\cosh(\beta\sqrt{1+v^2}) = 1 + 2\cosh(\beta\Delta E)$. As a consequence, we have for C_f

$$C_f = 1 - \sum_n P_n^2 = 1 - \frac{1 + 2\cosh(2\beta\Delta E)}{[1 + 2\cosh(\beta\Delta E)]^2}, \tag{20}$$

which is plotted in Figure 7 (C_f versus $\beta\Delta E$) for several numbers of fermions. $N = 2$ is a special instance. Other varieties of N exhibit collapse into a single curve. As expected, the degree of mixing increases as $\beta\Delta E$ diminishes.

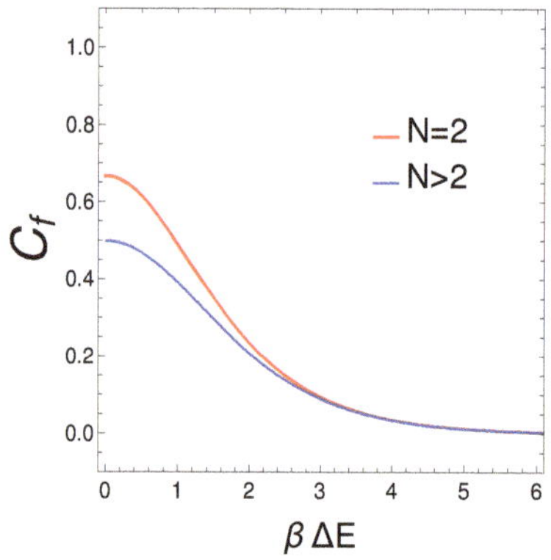

Figure 7. C_f versus $\beta\Delta E$ for $N = 2$ (red curve) and several other fermion numbers (4, 6, 8, 10) in the other curve, on which they coalesce.

We can easily think on reasons why, in an interacting many-fermion system, the degree of quantum mixing, or mixedness, tends to increase as the separation between energy levels diminishes. This phenomenon can be understood through several interrelated concepts in quantum mechanics and statistical mechanics.

1. Energy Level Density and Quantum States: As the separation between energy levels decreases, the density of available quantum states increases. When energy levels are closely spaced, fermions have a larger number of states to occupy within a small energy range. This increased state density enhances the probability of transitions between states, leading to greater mixing of quantum states.

2. Thermal Excitations: At low temperatures, fermions typically occupy the lowest available energy states due to the Pauli exclusion principle. However, as the energy level separation decreases, even small thermal excitations can cause fermions to transition between states. This results in a higher degree of occupation of excited states, contributing to quantum mixing.

3. Interaction-Induced Mixing: Interactions between fermions can lead to the hybridization of states, where the eigenstates of the system become superpositions of non-interacting states. When energy levels are closely spaced, interactions more readily cause mixing because the energy required to couple states is lower. This leads to an increased degree of quantum mixedness as interactions redistribute the fermions among the available states.

4. Quantum Fluctuations: In systems with closely spaced energy levels, quantum fluctuations become more significant. These fluctuations can induce transitions between states, further enhancing quantum mixing. The reduced energy gap means that even small perturbations (thermal or quantum) can cause changes in the occupation of states.

Now, let us take a look at the behavior of ΔE versus v for different N-values in Figure 8.

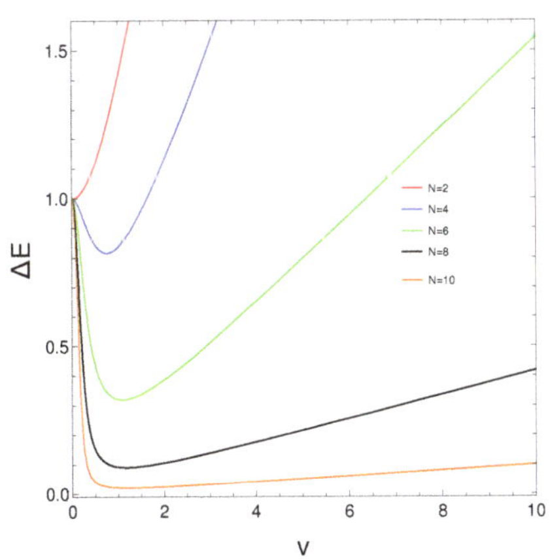

Figure 8. ΔE versus v for different Ns.

ΔE grows monotonously only for $N = 2$. We detect a local minimum for $N > 2$. The minima features for ΔE are clearly N-dependent.

5. Free Energy F Versus C_f

We pass to analyze the free energy behavior in relation to the mixedness degree. We plot F versus C_f considering either β, N, or v as the varying parameter.

Accordingly, we plot in Figure 9, for the Lipkin model (sub-index L), two things: (1) left: (β) versus $C_f(\beta)$ with $N = 12$ and $v = 1$; (2) center: $F(N)$ versus $C_f(N)$ with $\beta = 10$ and $v = 1$; (3) right: $F(v)$ versus $C_f(v)$ with $N = 12$ and $\beta = 10$.

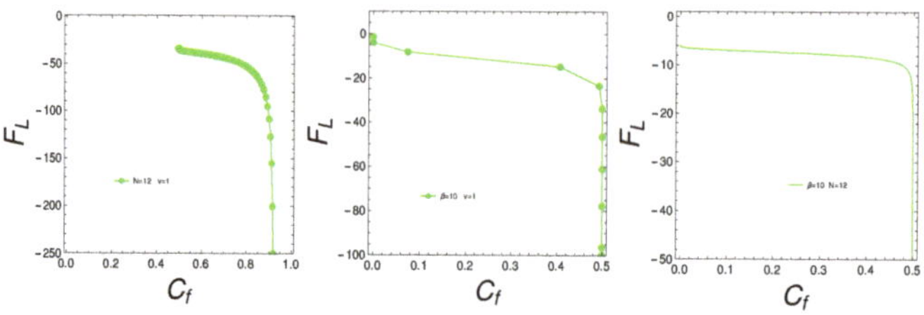

Figure 9. Lipkin free energy F_L versus C_f in various conditions described in the text.

The observation that the free energy becomes more negative with a larger degree of quantum mixedness in the Lipkin model can be attributed to increased entropy, lower effective energy, enhanced cooperative effects, and potential phase transitions. These factors collectively lead to a more energetically favorable state as the degree of quantum mixedness increases, thus lowering the free energy. See the plot of Figure 9. The increase in quantum mixedness can enhance correlations between particles. When these correlations reach a critical level, the system can reorganize into a more energetically favorable configuration. This reorganization would manifest itself as an abrupt decrease in the Lipkin mean energy U_L. This is what our $< U >_L$-plot in Figure 10 indeed shows. At a mixing of 0.5, the system seems to achieve a state of enhanced quantum coherence, where the superposition of states leads to a lower energy configuration. This coherence can stabilize the system, resulting in a sharp decrease in energy. The abrupt change could signify that the system is reaching an optimal balance between the different quantum states, minimizing energy through constructive interference and optimal state mixing. We next present a plot in Figure 10 for the man energy, which confirms some assertions made above.

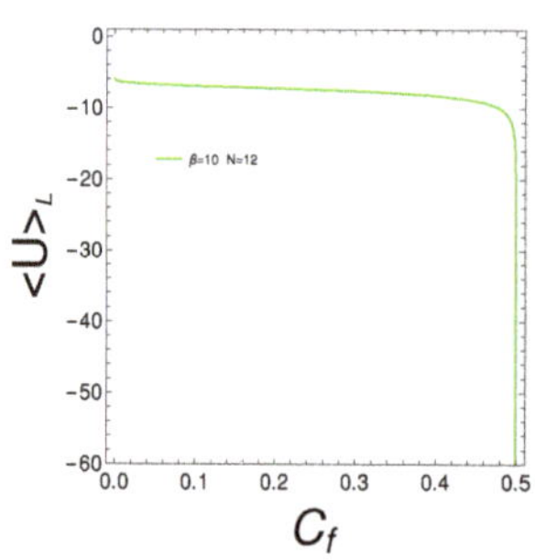

Figure 10. Lipkin's mean energy $< U >_L$ plotted against the mixedness indicator C_f for $\beta = 10$.

6. Conclusions

This study elucidates the intricate dynamics of quantum mixing, particle number fluctuations, and temperature variations within the Lipkin model, revealing profound insights into the behavior of many-fermion systems. Our findings underscore several key points:

- The degree of quantum mixedness displays a strong dependency on the total number of fermions, highlighting distinct behaviors across different temperature regimes. This emphasizes the importance of considering fermion number in analyzing quantum systems, as it directly influences the system's mixedness and overall state.

- Remarkably, the observed dependencies of quantum mixedness on fermion number and temperature are robust against variations in the fermion–fermion interaction strength, provided the interaction is non-zero. This robustness suggests that the fundamental properties of quantum mixedness are intrinsic to the system's structure rather than being heavily influenced by interaction specifics.
- The insights gained from this study have potential implications for various fields, including condensed matter physics and quantum information science. Understanding the dependencies and behaviors of quantum degree of mixedness in fermionic systems can inform the development of quantum technologies and enhance the theoretical models used to describe complex quantum systems.

Through a comprehensive numerical exploration of key features of the system, we have provided a detailed analysis of the Lipkin model, offering valuable perspectives on the fundamental characteristics of many-fermion systems. Our results contribute to a deeper understanding of quantum phenomena in fermionic systems, paving the way for future research and practical applications in related fields.

Author Contributions: Investigation, D.M., A.P. and A.R.P.; project administration, A.P.; writing—original draft, D.M., A.P. and A.R.P. All authors have read and agreed to the published version of the manuscript.

Funding: This research was partially supported by CONICET (Argentine Agency).

Data Availability Statement: The original contributions presented in this study are included in the article.

Conflicts of Interest: The authors declare no conflicts of interest.

References

1. Messiah, A. *Quantum Mechanics*; John Wiley: New York, NY, USA, 1966.
2. Cohen-Tannoudji, C.; Diu, B.; Laloë, F. *Quantum Mechanics*; John Wiley: New York, NY, USA, 2020.
3. Peres, A. *Quantum Theory: Concepts and Methods*; Kluwer Academic Publishers: New York, NY, USA, 1995.
4. Sakurai, J.J.; Napolitano, J. *Modern Quantum Mechanics*; Cambridge University Press: Cambridge, UK, 2020.
5. Dirac, P.A.M. *The Principles of Quantum Mechanics*; Oxford University Press: Oxford, UK, 1981.
6. Ring, P.; Schuck, P. *The Nuclear Many-Body Problem*; Springer: Berlin, Germany, 1980.
7. Lipkin; Meshkov, H.J.N.; Glick, A.J. Validity of many-body approximation methods for a solvable model: (I). Exact solutions and perturbation theory. *Nucl. Phys.* **1965**, *62*, 188–198. [CrossRef]
8. Lipkin; Meshkov, H.J.N.; Glick, A.J. Validity of many-body approximation methods for a solvable model: (II) Linearization procedure. *Nucl. Phys.* **1965**, *62*, 199–210. [CrossRef]
9. Bengtsson, I.; Zyczkowsi, K. *Geometry of Quantum States: An Introduction to Quantum Entanglement*; Cambridge University Press: Cambridge, UK, 2006.
10. Jaeger, G. *Quantum Information: An Overview*; Springer: Berlin, Germany, 2007.
11. Giulini, D.; Kiefer, C.; Kupsch, J. *Decoherence and the Appearance of a Classical World in Quantum Theory*; Springer: Berlin, Germany, 1996.
12. Facchi, P.; Mariano, A.; Pascazio, S. Decoherence versus entropy in neutron interferometry. *Phys. Rev. A* **2001**, *63*, 052108. [CrossRef]
13. Plastino, A.R.; Monteoliva, D.; Plastino, A. Information-theoretic features of many fermion systems: An exploration based on exactly solvable models. *Entropy* **2021**, *23*, 1488. [CrossRef] [PubMed]
14. Monteoliva, D.; Plastino, A.; Plastino, A.R. Statistical Quantifiers Resolve a Nuclear Theory Controversy. *Q. Rep.* **2022**, *4*, 127–134. [CrossRef]
15. Cambiaggio, M.C.; Klar, A.; Margetan, F.J.; Plastino, A.; Vary, J.P. Comparison of upper and lower bound methods using a soluble many-Fermion model. *Phys. Rev. C* **1983**, *27*, 2340. [CrossRef]

entropy

MDPI

Article

Irreversibility and Energy Transfer at Non-MHD Scales in a Magnetospheric Current Disruption Event

Giuseppe Consolini [1,*] and Paola De Michelis [2]

[1] INAF—Istituto di Astrofisica e Planetologia Spaziali, Via del Fosso del Cavaliere 100, 00133 Roma, Italy
[2] Istituto Nazionale di Geofisica e Vulcanologia, Via di Vigna Murata 605, 00143 Roma, Italy; paola.demichelis@ingv.it
* Correspondence: giuseppe.consolini@inaf.it

Abstract: Irreversibility and the processes occurring at ion and sub-ion scales are key challenges in understanding energy dissipation in non-collisional space plasmas. Recent advances have significantly improved the characterization of irreversibility and energy transfer across scales in turbulent fluid-like media, using high-order correlation functions and testing the validity of certain fluctuation relations. In this study, we explore irreversibility at non-MHD scales during a magnetospheric current disruption event. Our approach involves analyzing the asymmetric correlation function, assessing the validity of a fluctuation relation, and investigating delayed coupling between different scales to reveal evidence of a cascading mechanism. The results clearly demonstrate the irreversible nature of fluctuations at ion and sub-ion scales. Additionally, we provide potential evidence for an energy cascading mechanism occurring over short time delays.

Keywords: space plasma turbulence; irreversibility; energy transfer; magnetospheric current disruption

1. Introduction

Space plasmas are typically in an out-of-equilibrium state and are characterized by extremely low densities—on the order of just a few particles per cubic centimeter—making them effectively non-collisional. This non-collisional nature presents a fascinating challenge in understanding how such media relax toward equilibrium, i.e., how they dissipate energy. Another significant property of space plasmas is their turbulent nature [1,2]. For example, the solar wind, a turbulent plasma flow with a high Reynolds number, flows through the interplanetary medium with very low density. Despite its widespread observation, the mechanisms underlying its irreversible behavior and microscopic relaxation remain poorly understood. Unlike ordinary fluid turbulence, where viscosity dissipates energy below the inertial range and ensures irreversibility, space plasmas exhibit a different behavior. Below the inertial range, a new power-law domain emerges at ion and sub-ion scales, raising questions about whether dissipation occurs at these scales. Magnetic field fluctuations at ion/sub-ion scales follow a $\sim k^{-8/3}$ ($\sim f^{-8/3}$) spectrum, which suggests global scale invariance. This characteristic appears to contrast with the strong turbulent processes typically associated with dissipation [3]. Similar features have been observed during relaxation phenomena in space plasmas, such as magnetospheric current disruption events [4] and magnetic reconnection [5,6]. Furthermore, in the case of turbulent magnetic reconnection, the importance of filamentary current sheets in determining the observed spectral features has been widely investigated from both observational and theoretical perspectives [7–9].

Over the past decade, there has been renewed interest in exploring irreversibility in nonequilibrium systems. Many systems subjected to external constraints exist in nonequilibrium states, exhibiting currents that provide clear evidence of broken detailed balance [10,11]. Such systems are characterized by a positive entropy production rate, which is a hallmark of time-reversal symmetry breaking.

Evaluating the entropy production rate is, however, a highly complex task, especially for real-world systems where entropy-related quantities are not easily measurable (see Ref. [12] and the references therein for a detailed discussion). In this context, an alternative approach is to study specific time-asymmetric correlation functions, which can detect time-reversal symmetry breaking. This concept was first introduced in a seminal paper by Y. Pomeau in 1982 [13]. More recently, Josserand et al. [14] and Cocciaglia et al. [15] applied these methods to turbulent systems, demonstrating how asymmetric time-correlation functions can effectively probe the irreversible dynamics and energy cascades in such systems. Another prominent method for investigating the irreversibility of dynamical systems involves fluctuation relations (or theorems). In nonequilibrium stationary states (NESS), which are defined by nonzero entropy production rates, Gallavotti and Cohen [16,17] proposed a fluctuation theorem that describes a non-trivial symmetry property of the probability distribution of phase space contraction over long timescales [18,19]. This approach has been widely applied across various physical systems to test for the presence of NESS and irreversibility [20–23].

Recently, stochastic thermodynamics has provided additional tools for characterizing entropy production in nonequilibrium processes [24–26]. These methods have been successfully applied to space plasma dynamics, including solar, interplanetary, and magnetospheric plasma systems, to reveal their irreversible and nonequilibrium behaviors [23,27,28].

The Earth's magnetosphere is a complex, out-of-equilibrium system [27] that continuously interacts with the solar wind and ionosphere. This interaction gives rise to phenomena such as magnetospheric substorms and magnetic storms. These events involve various processes occurring across different regions of the magnetosphere, including plasma energization, reconfiguration, and intensification of currents.

One of the most significant phenomena occurring during the initial phase of a magnetospheric substorm is the *magnetotail current disruption* (CD) [29,30]. During this phase, a diversion or disruption of the near cross-tail current takes place, forming a current wedge that connects the central plasma sheet (CPS) of the Earth's magnetotail to the auroral ionosphere [31,32]. This current disruption leads to an increase in the magnetic field's B_Z-component, corresponding to a dipolarization of the magnetic field and a reduction in the cross-tail current.

Previous studies of current disruption events have demonstrated that CD is a multiscale phenomenon driven by an instability mechanism that produces broadband excitations spanning both lower [33,34] and higher frequencies. Specifically, CD involves a wide range of fluctuations, with magnetic field fluctuations displaying distinct spectral features depending on the frequency range.

At frequencies higher than the ion-cyclotron frequency f_{Ω}, the magnetic field fluctuations follow a power-law spectrum, $S(f) \sim f^{-\beta}$, with a spectral exponent $\beta \simeq 7/3 \div 8/3$ [4,35–37]. Conversely, at frequencies lower than Ω_c the spectral features resemble those of $1/f-$noise or MHD/HD turbulence. An inverse energy cascade may occur at scales and timescales larger than those associated with the drift-driven electromagnetic ion cyclotron (EMIC) instability, or ordinary mode instability, which is driven by cross-field ion drift and is thought to trigger the phenomenon [38]. Yoon et al. [39] further proposed that the observed fluctuation fields may be better explained by Alfvén ion cyclotron instability, characterized by propagation angles nearly perpendicular to the local magnetic field.

An important question arising from previous studies concerns the irreversible nature of the CD event and whether a cascading mechanism contributes to the spectral features observed at non-MHD ion and sub-ion scales. Specifically, do these spectral features result from an energy transfer from the instability scale to smaller scales via a cascading process, as seen in turbulence, or do they originate from a different mechanism?

To address these questions, methods based on the analysis of asymmetric correlation functions can be highly effective. In detail, we aim to characterize the nonequilibrium properties of CD fluctuations at sub-ion scales. In this study, we focus on how the system's variables evolve over time and whether their correlations indicate a broken temporal symmetry, which would signal the presence of irreversible processes. By analyzing these temporal correlations, we can gain valuable insights into the underlying dynamics of out-of-equilibrium systems. More in detail, we analyze the asymmetry of the correlation functions of time increments of the magnetic field across different temporal scales, focusing on frequencies higher than the ion-cyclotron scale. It is important to remark that—although studies based on actual space measurements generally assume Taylor's hypothesis to link temporal and spatial scales—in some situations, the validity of the correspondence between the two representations may be questionable. This is why we focus our analysis on temporal scales. Our findings reveal a clear breaking of time-reversal symmetry, which we discuss within the context of the irreversible nature of the observed phenomenon. Additionally, we investigate asymmetries in energy transfer across scales, seeking evidence of a cascading mechanism. This approach follows the methodology employed by Josserand et al. [14].

This paper is organized as follows: Section 2 presents the theoretical framework and the dataset employed in this study, detailing the methods used to analyze magnetic field fluctuations at ion and sub-ion scales during a CD event. Section 3 discusses the results, focusing on the asymmetric time-correlation functions, fluctuation theorem analyses, and potential evidence of cascading mechanisms. Finally, Section 4 provides a comprehensive discussion and concludes with insights into the irreversible dynamics and energy transfer processes, highlighting areas for future research.

2. Methods and Data

2.1. Methods: Theoretical Background

The hallmark of irreversibility in out-of-equilibrium systems is the violation of temporal symmetry, which reflects the system's inability to spontaneously return to its initial state once disturbed. This violation is mathematically represented by the absence of detailed balance, leading to a positive entropy production rate σ_X, i.e.,

$$\sigma_X = \sum_k J_k X_k \geq 0 \tag{1}$$

where J_k and X_k are the thermodynamic fluxes and forces, respectively [10]. This concept is central to our study, as we aim to identify and quantify the irreversible dynamics of complex systems by examining temporal asymmetries. The violation of temporal symmetry, leading to a positive entropy production rate, is a fundamental characteristic of both stationary and non-stationary nonequilibrium systems. As such, a nonzero entropy production rate serves as evidence of irreversibility. However, direct measurements of the entropy production rate σ_X are typically not feasible in most real-world situations. As a result, an alternative approach to testing for the occurrence of irreversibility is to investigate whether temporal symmetry is broken. This can provide valuable insights into the system's irreversible dynamics without the need for direct entropy measurements.

In the past, Pomeau [13] introduced a method based on asymmetric correlation functions to test for temporal symmetry breaking (see also Ref. [15]).

The basic idea is rooted in the violation of the *detailed balance relation* [40,41]. Specifically, for a system in equilibrium, given two different observable state functions, f and g, the following relationship should hold due to reversibility:

$$\langle f(t)g(0)\rangle = \langle f(0)g(t)\rangle \tag{2}$$

Thus, if this equation is not true, i.e., if,

$$\langle f(t)g(0)\rangle \neq \langle f(0)g(t)\rangle \tag{3}$$

then we are in an out-of-equilibrium situation and temporal symmetry is expected to be broken. Moving from Equation (3), we could use the following difference

$$\Delta(t) = \langle f(t)g(0)\rangle - \langle f(0)g(t)\rangle \tag{4}$$

as a measure of the distance from equilibrium and an evidence for the temporal symmetry breaking.

Following this very simple reasoning and the formalism introduced by Pomeau [13] (see also Jossrand et al. [14]), Cocciaglia et al. [15] considered the following asymmetric time-correlation function to detect the temporal symmetry breaking,

$$\Psi_x(\tau) = \langle x^2(t)x(t+\tau)\rangle - \langle x(t)x^2(t+\tau)\rangle \tag{5}$$

where $x(t)$ is a signal associated with a system observable. They applied this method to test irreversibility in fluid shell models by assuming as observable $x(t) = e_n(t) \equiv | u_n(t) |^2 /2$. While in the case of inviscid shell models the *detailed balance* is locally satisfied in the Fourier space,

$$\Psi_{e_n}(\tau) = 0, \tag{6}$$

for the forced and viscid case there is a clear violation of the *detailed balance* [15] being,

$$\Psi_{e_n}(\tau) \neq 0, \tag{7}$$

The dynamics observed for the viscous case align with the concept that energy decreases over short timescales, then increases over longer times. As highlighted by Cocciaglia et al. [15], this behavior is consistent with the occurrence of *flight-crash events* [22], where *fluid elements decelerate more quickly than they accelerate*, a phenomenon familiar to anyone who has driven in heavy traffic. The trend observed in the asymmetric time-correlation function, $\Psi_{e_n}(\tau)$, further supports the irreversible nature of energy transfer within the cascade mechanism.

To investigate the irreversible nature of a system's dynamics, another important quantity to examine is the statistics of energy increments, defined as $W(\tau) = E(t+\tau) - E(t)$. For stationary, homogeneous flows, the first moment of this quantity is expected to vanish, i.e., $\langle W(\tau)\rangle = 0$, while the third moment is the first nonzero odd moment, meaning that $-\langle W^3(\tau)\rangle \neq 0$. However, an alternative and potentially more effective method for characterizing the irreversibility of the observed phenomenon is through the probability distribution function (PDF) of $W(\tau)$, which encapsulates all the relevant information about the energy fluctuations. To gain a deeper understanding of irreversibility, one could apply an appropriate fluctuation theorem [16,17,19] to analyze the skewness of the PDF $p[W(\tau)]$. In systems with a nonzero entropy production rate, which are out of equilibrium, fluctuation theorems assert that the probabilities of energy gain and energy loss are related by the following equation:

$$\ln\left[\frac{p(-W)}{p(W)}\right] \propto W \tag{8}$$

which, at a first glance, is related to the shape of the tails of PDFs.

Another important aspect to explore in the context of turbulent fluctuations is the presence of a cascading process [14]. This can be examined by analyzing the following test function:

$$H(k_1, k_2; \tau) = \langle E(k_1, t_1)E(k_2, t_2)\rangle - \langle E(k_1, t_2)E(k_2, t_1)\rangle \tag{9}$$

where $E(k, t)$ is the energy at a wavenumber k at time t, $\tau = t2 - t1$, and $k_2 \gg k_1$.

This quantity is expected to vanish both at $\tau = 0$ and as $\tau \to \infty$, when the signals become completely decorrelated. Additionally, if $H > 0$, this implies that k_2 is correlated with k_1 with a time delay of $\tau > 0$, meaning that energy transfers in the direction from k_1 to k_2. Conversely, for $H < 0$, the correlation occurs in the opposite direction, with energy transferring from k_2 to k_1. By using the previously defined asymmetric correlation function H and considering three or more well-separated wavenumbers such that $k_3 \gg k_2 \gg k_1$, it becomes possible to investigate whether energy takes more time to cascade from k_1 to k_2 than from k_1 to k_2 [14]. This analysis provides valuable insights into the dynamics of energy transfer across different scales, offering potential evidence for the occurrence of a cascading mechanism. Specifically, it helps identify whether energy transfer between scales follows a hierarchical structure, which is characteristic of a turbulence-driven cascade.

2.2. Data

The theoretical framework outlined above was applied to analyze the statistics of magnetic field increments at ion/sub-ion scales during a specific CD event that occurred on 14 August 2013. This event was observed by the P5 (A) satellite of the THEMIS mission, which was positioned at a distance of approximately nine Earth radii (R_E) in the magnetotail, within the neutral sheet, and in the pre-midnight sector. The analysis of this event builds upon previous work by Lui et al. [42], who studied the same event in the context of the dynamics of the magnetospheric current disruption. The plasma velocity is generally very small before and after CD, i.e., the plasma can be considered to be practically stagnant. However, during CD, the plasma is accelerated in a bursty manner, with a typical ion velocity on the order of $\sim 200 \div 300$ km/s [43], which is close to the Alfvén velocity $c_A \simeq 400$ km/s, assuming a plasma density of the order of 1 cm^{-3}. Furthermore, during the CD event, the breakdown of the frozen-in condition has been observed [43].

Figure 1 displays the three components of the magnetic field (Bx, By, Bz) measured by the THEMIS P5 satellite during the CD event on 14 August 2013. During the event, the Bz component exhibits a marked increase, indicating the onset of a dipolarization process associated with the disruption of the cross-tail current, a hallmark of CD events. Before the event, the magnetic field was nearly zero, reflecting the satellite's position within the neutral sheet, where the field is inherently weak. After the CD, the magnetic field predominantly aligns along the Z-axis, indicating a transition to a more dipolar geomagnetic configuration in the magnetotail, signifying substantial reorganization of the field structure.

To analyze short-timescale fluctuations, we use burst-mode measurements from the fluxgate magnetometer (FGM), offering a temporal resolution of approximately 7.8 ms. These high-cadence data, central to this study, correspond to a specific sub-interval of the CD event, as shown in Figure 2. This dataset enables a detailed investigation of rapid magnetic field fluctuations during the event.

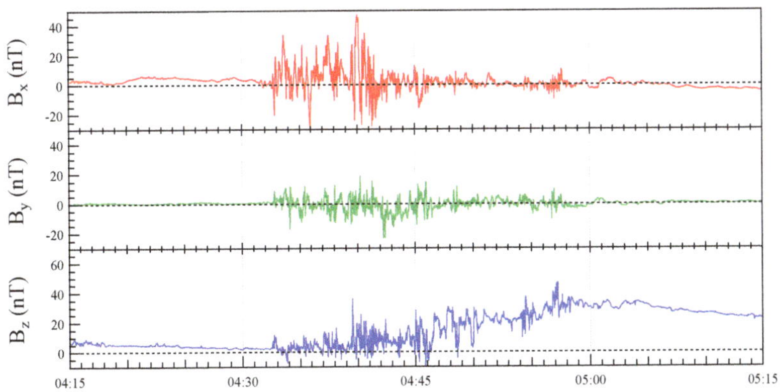

Figure 1. Time series of the magnetic field components (B_x, B_y, B_z) measured by the THEMIS P5 satellite during the CD event on 14 August 2013.

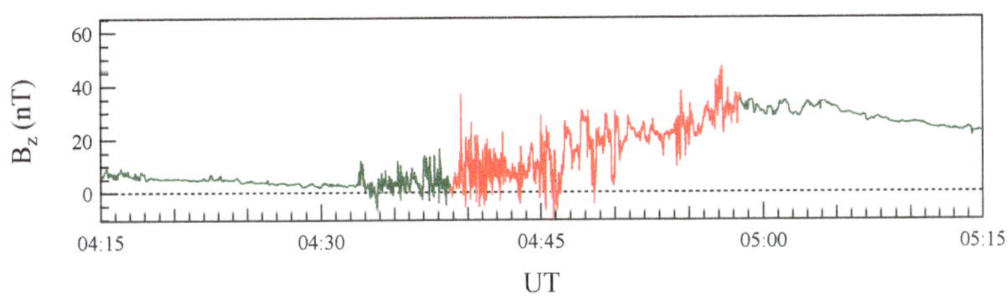

Figure 2. High-cadence measurements of the B_z component of the magnetic field during the CD event, recorded by the THEMIS P5 satellite. The red trace marks the sub-interval selected for detailed analysis.

To focus on the properties of magnetic field fluctuations, long-term trends were removed from the magnetic field components using the empirical mode decomposition (EMD) technique [44]. This method ensures that each component of the magnetic field has a zero-mean value, effectively isolating the short-term fluctuations from any underlying trends. This preprocessing step is crucial for accurately analyzing the fluctuations at the ion/sub-ion scales, as it eliminates any low-frequency components that might obscure the more rapid, short timescale dynamics. Henceforward, all analyses will be performed using the detrended magnetic field time series.

In Figure 3, we present a comparison of the trace of the power spectral density (PSD), defined as $S(f) = \sum_i PSD_i(f)$, where $PSD_i(f)$ represents the spectral density of the i-th component of the magnetic field, between the slow and burst mode measurements. This comparison highlights the differences in the spectral characteristics captured by the two measurement modes, providing insight into the behavior of the magnetic field fluctuations across different timescales.

The trace of the power spectral density (PSD), $S(f)$, of the high-cadence data reveals a distinct power-law region, $S(f) \sim f^{-\alpha}$, below the ion-cyclotron frequency ($f_\Omega \sim 0.1$ Hz). This region is characterized by a spectral exponent $\alpha \simeq -8/3$ and extends up to approximately 8 Hz. The observed behavior is strikingly similar to the power spectrum observed in the solar wind [3]. The possible origin of the observed similarity with kinetic domain spectrum in solar wind could be an effect of large-scale random sweeping as described in Servidio et al. [45]. However, it is important to emphasize that, in the present case, we are dealing with real-time fluctuations. This distinction is crucial because, unlike in cases where Taylor's hypothesis applies, we cannot assume the usual assumptions of spatial homogeneity or isotropy to relate the temporal and spatial characteristics of the fluctuations. This issue is critical because, when analyzing temporal fluctuations directly, we can investigate the occurrence of irreversibility, a property that has been explored in the context of shell-model turbulence [15]. This opens up the possibility of studying the irreversible nature of the system from a different perspective compared to the traditional spatial turbulence models.

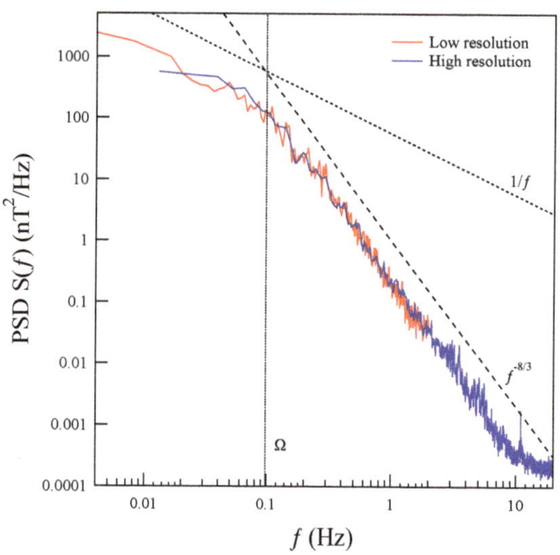

Figure 3. Comparison of the trace of the power spectral density (PSD), $S(f)$, obtained from low-resolution and high-resolution (burst-mode) measurements during the CD event. The dotted and dashed lines represent power-law behaviors characterized by a $1/f$ and $f^{-8/3}$ behavior, respectively.

3. Results

Following the approach outlined by Cocciaglia et al. [15] for investigating temporal symmetry breaking, we concentrate on computing the energy content at a fixed timescale, denoted as δt, in our analysis. To efficiently extract the energy corresponding to a specific timescale, we employ the wavelet transform, which is a highly effective tool for this purpose. The wavelet transform allows for the analysis of localized fluctuations in both time and frequency, making it particularly suited for capturing energy at different scales. In our analysis, we apply the continuous wavelet transform (CWT) to each detrended magnetic field component, b_i, to compute the energy content across different timescales. Specifically, the CWT of a $b_i(t)$ is given by the following:

$$W_{\delta t}(t_0) = \frac{1}{\sqrt{\delta t}} \int_{-\infty}^{\infty} \psi\left(\frac{t - t_0}{\delta t}\right) b_i(t) dt, \tag{10}$$

where $\psi\left(\frac{t-t_0}{\delta t}\right)$ is a chosen wavelet function (the complex Morlet wavelet in our case [46,47]), δt is the timescale, and t_0 is the time translation. By applying this transform to the detrended magnetic field components, we can then evaluate the energy associated with each timescale, δt, and analyze how it varies over time. This process allows us to examine how energy is distributed across different timescales, which is key for studying temporal symmetry breaking and identifying potential irreversible processes in the system. In addition, by assessing the energy content at various timescales, we can explore how energy transfer occurs across different scales, which is a crucial step in understanding the dynamics of the system, particularly in the context of cascade processes in turbulence. We remark that differently from Cocciaglia et al. [15] we do not work in the Fourier k-space and that an association between temporal and spatial scales could be possible only knowing the corresponding dispersion relation for the excited modes (e.g., whistler waves [48]) and/or assuming that Taylor's hypothesis may be applicable to some extent due to the high ion velocity observed during CD [43]. This issue could make more difficult the comparison of our results with previous works [15]. However, we believe that the above method remains valid regardless in terms of temporal evolution of spectral features.

In particular, we defined the total energy at the timescale δt as

$$E_i(t \mid \delta t) = \mid W_{\delta t}(t) \mid^2 . \tag{11}$$

Using this quantity, we define the observable quantity, $x(t)$ as the total energy of the fluctuation field at the timescale δt,

$$x(t \mid \delta t) = \sum_{i=x,y,z} E_i(t \mid \delta t), \tag{12}$$

which represents the main focus of our study.

It is important to note that, in the case of the complex Morlet wavelet, a direct correspondence exists between the timescale δt and the frequency f. Specifically, by choosing as mother wavelet,

$$\psi(t \mid \delta t) = \frac{1}{\sqrt{\delta t}} \exp\left[i\frac{2\pi t}{\delta t} - \frac{1}{2}\left(\frac{t}{\delta t}\right)^2 \right], \tag{13}$$

this wavelet exhibits a peak at $f = 1/\delta t$ with a full width at half maximum (FWHM) given by $\Delta f = f/4$.

The quantity $x(t)$ can be assumed to be equivalent to the energy of the fluctuations at the timescale δt as performed by Cocciaglia et al. [15] when analyzing the behavior of fluctuations at a fixed scale in turbulent shell models.

Successively, we selected a series of timescales, δt, corresponding to the frequencies in the range $f \in [0.5, 4]$ Hz (refer to Table 1 and Figure 4), i.e., in the non-MHD domain.

Table 1. The selected timescales (δt) and the corresponding frequencies f, computed as $f = 1/\delta t$.

Label	δt (s)	$f = 1/\delta t$ (Hz)
(1)	0.25	4
(2)	0.375	8/3
(3)	0.5	2
(4)	0.75	4/3
(5)	1	1
(6)	2	0.5

For each selected timescale, δt, we first compute the corresponding $x(t \mid \delta t)$, and then calculate the following third-order correlation function:

$$\Psi_{\delta t}(\tau) = \frac{\langle x^2(t)x(t+\tau)\rangle - \langle x(t)x^2(t+\tau)\rangle}{\langle x^3(t)\rangle}. \tag{14}$$

Figure 5 illustrates the behavior of the third-order correlation function, $\Psi_{\delta t}(\tau)$, for the selected timescales listed in Table 1. The correlation function exhibits a clear time-dependent trend, which reinforces the irreversible nature of the fluctuations observed at non-MHD scales. Notably, for all the timescales under investigation, $\Psi_{\delta t}(\tau)$ shows an initial increase, followed by a rapid decrease and a subsequent rise. This pattern suggests an energy transfer originating from the characteristic timescale of the instability, which likely contributes to the broadband spectrum observed at non-MHD scales. The observed trend supports the hypothesis that the fluctuations undergo complex energy exchanges, with the energy first being pumped into the system and then released or redistributed across different timescales. This kind of behavior is characteristic of nonequilibrium processes, further confirming the irreversibility of the system dynamics. We remark that the observed trends exhibit some differences compared to the results of Cocciaglia et al. [15]. In particular, the 3rd asymmetric correlation function shows a different trend at certain temporal scales when compared with results reported in Figure 2 of Cocciaglia et al. [15]. The origin of these differences could be due to three possible different origins: (i) the phenomenon under investigation is not stationary, (ii) the spectral decomposition we use is temporal, and the applicability of Taylor's hypothesis to associate a spatial scale may be questionable at these scales, making a direct comparison with the spatial-scale results of Cocciaglia et al. [15] not straightforward, and (iii) the statistical sample in our case is clearly more limited.

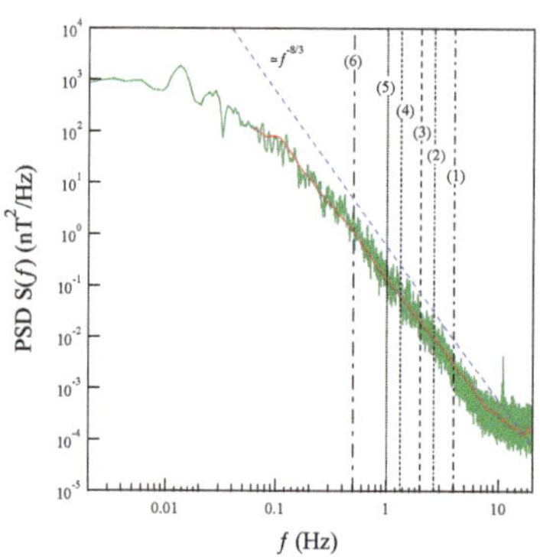

Figure 4. The trace of the power spectral density matrix, $S(f)$, of the magnetic field. The vertical lines indicate the frequencies corresponding to the selected timescales τ. The red trace marks the spectra as obtained by averaging the wavelet transform in time.

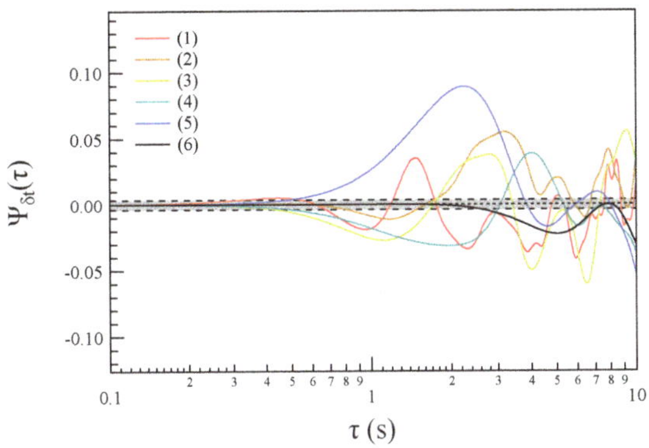

Figure 5. The behavior of the third-order asymmetric correlation function $\Psi_{\delta t}(\tau)$ for the selected timescales. The gray region refers to the interval for non significance.

To verify the significance of the temporal correlations observed in $\Psi_{\delta t}(\tau)$, we compute the corresponding third-order asymmetric time-correlation functions after shuffling the $x(t \mid \delta t)$ values. The results obtained from this procedure are presented in Figure 6.

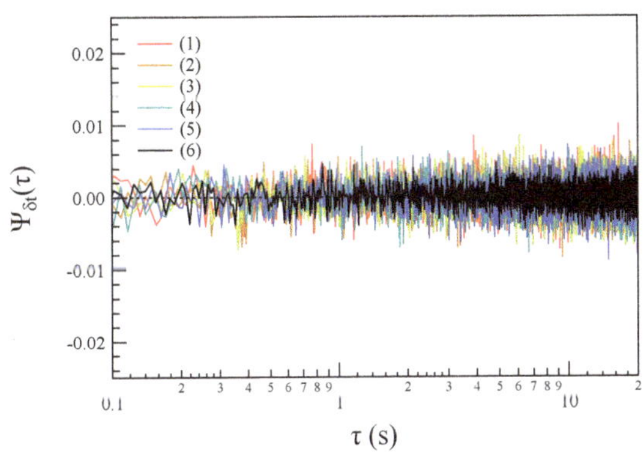

Figure 6. The behavior of the third-order asymmetric correlation function $\Psi_{\delta t}(\tau)$ for shuffled data at the different considered timescales.

A comparison of the results in Figures 5 and 6 reveals clear evidence of temporal symmetry breaking. This observation strongly suggests the irreversible nature of the fluctuations occurring at non-MHD timescales. Additionally, the presence of temporal asymmetry in the third-order correlation function, which vanishes after data shuffling, further supports the idea that the underlying processes driving these non-MHD magnetic field fluctuations are inherently irreversible.

As a next step, we investigate the potential scaling behavior of the first minima in the asymmetric correlation function, $\Psi_{\delta t}(\tau)$. Figure 7 presents the trend of the delay time, τ_{min}, which corresponds to the first minimum of the asymmetric correlation functions at different timescales, plotted as a function of the frequency, f. The trend of $\tau_{min}(f)$ is not constant; instead, it exhibits a transition-like behavior as the frequency changes. The transition occurs around a frequency of $f^* \sim 1.2$ Hz. This suggests that a dynamical transition may take place in the intermediate range between the ion and electron scales.

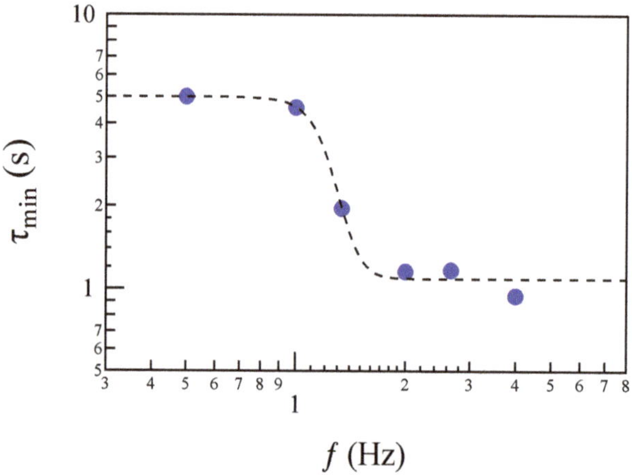

Figure 7. The behavior of the delay τ_{min} as a function of the corresponding frequency.

Following Xu et al. [22], we now examine whether the energy fluctuations at non-MHD timescales adhere to the validity of the fluctuation theorem for the magnetic field power density at timescale τ (see Equation (8)), defined as follows:

$$W_\tau = E_b(t+\tau) - E_b(t), \quad (15)$$

where magnetic field power density is defined as $E_b(t) = \sum_i b_i^2/\sqrt{4\pi\rho}$ with ρ plasma density, which is assumed to be constant (so that, $E_b \simeq b^2$). The timescale τ is assumed to range from 0.125 to 1 s. Figure 8 illustrates the evolution of the probability density functions (PDFs) of the magnetic field power density W_τ at different timescales. These PDFs are leptokurtic and exhibit skewness. The positive skewness indicates that large energy increases are more probable than energy decreases. This behavior may be attributed to the explosive nature of the instability associated with the current disruption event.

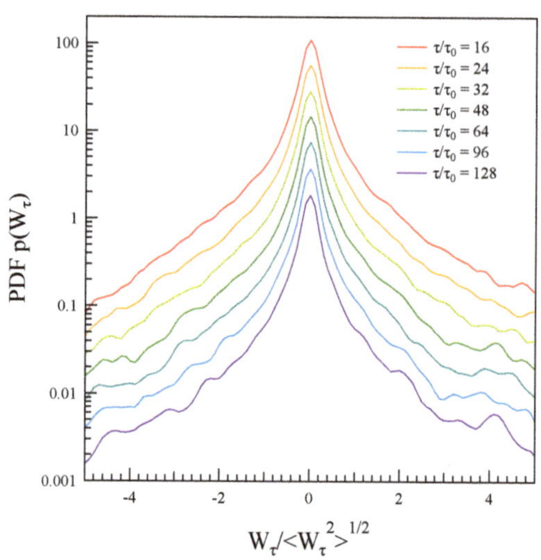

Figure 8. Evolution of the probability density function of the power density $p(W_\tau)$ at different timescales τ, with the PDFs shifted for display purposes.

By analyzing the evolution of the probability distribution function $p(W_\tau)$ of the power density at different timescales (see Figure 8), we can attempt to estimate the prediction of the fluctuation theorem, i.e.,

$$\ln\left[\frac{p(-W_\tau)}{p(W_\tau)}\right] \simeq c_+(\tau)\frac{W_\tau}{\langle W_\tau^2\rangle^{1/2}}, \tag{16}$$

where $c_+(\tau) = c_+\tau$ and $W_\tau \le 0$.

In Figure 9, we report the behavior of $\ln\left[\frac{p(-W_\tau)}{p(W_\tau)}\right]$ for the case of $\tau/\tau_0 = 24$ where $\tau_0 \simeq 7.8$ ms is the resolution scale of magnetic field measurements. As predicted by the fluctuation theorem, a linear trend is observed on average.

For small systems in contact with thermostats, Derrida [19] suggests that the quantity c_+ should remain constant. Figure 10 shows the behavior of $c_+ = c_+(\tau)/\tau$ as a function of τ. However, in contrast to the fluctuation theorem's prediction, c_+ is not constant across the entire range of investigated scales. Instead, it decreases for $\tau > 0.5$ s. This indicates that the asymmetry in the power density distribution function diminishes as the timescale increases, approaching either the ion-cyclotron frequency f_Ω or the timescale of the CD instability. Consequently, the conditions outlined by the fluctuation theorem appear to hold true at scales well below the typical ion scales.

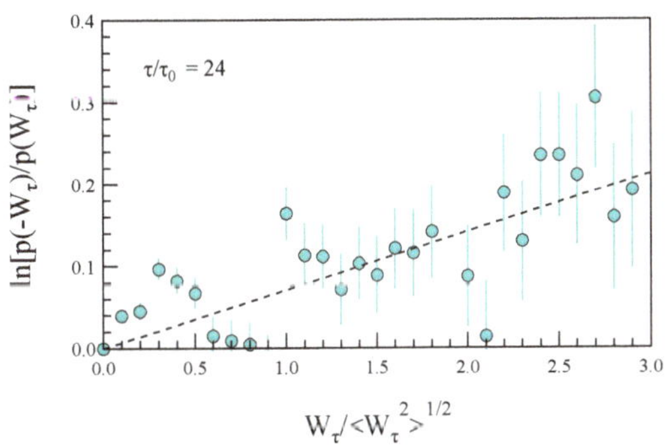

Figure 9. The behavior of $\ln\left[\frac{p(-W_\tau)}{p(W_\tau)}\right]$ for the case of $\tau/\tau_0 = 24$. The dashed line is a linear best fit.

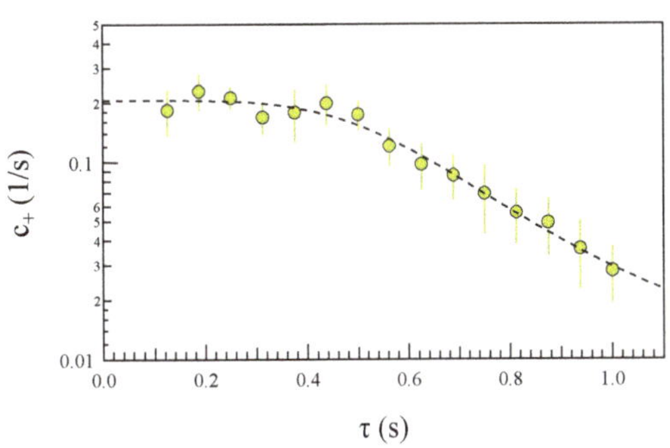

Figure 10. The behavior of c_+ as a function of τ. The dashed line is a guide for the eye.

We observe that the change in the asymmetry of the distribution occurs at a timescale that aligns with the transition frequency identified in the dependence of $\tau_{min}(f)$ for the asymmetric correlation function. Specifically, assuming $f \sim 1/2\tau$, we find that the scenario is consistent.

The next aspect we examine is the potential occurrence of a cascading mechanism in the energy transfer at non-MHD scales. To investigate this, we use the test function outlined in Equation (17). In our case, instead of using the wavenumber k, we work with the frequency f, and thus we compute

$$H(f_1, f_2; \tau) = \langle E(f_1, t_1)E(f_2, t_2)\rangle - \langle E(f_1, t_2)E(f_2, t_1)\rangle \tag{17}$$

where $E(f_i, t) = x(t \mid \delta t_i)$ with $f_i = 1/\delta t_i$, and $t_2 - t_1 = \tau$. Clearly, our analysis is strictly temporal so that we cannot exclude the possibility that space-time correlation might be affected by a simply temporal-based approach. However, if we take into account that during CD plasma acceleration events are observed [43] and the plasma cannot be considered stagnant, then we are confident that the investigation of $H(f_1, f_2; \tau)$ could provide some information on the occurrence of cascading (energy transfer) mechanisms also in k-space. A multipoint analysis would be very useful in this framework.

In our analysis, we set $f_1 = 1$ Hz and consider $f_2 = 2, 4$, and 8 Hz. Figure 11 illustrates the test function $H(f_1, f_2; \tau)$ for the three cases examined. We limit our analysis to the first maxima of the test function $H(f_1, f_2; \tau)$. The position of the maxima, τ_{max}, as a function of the frequency difference $\delta f = f_2 - f_1$, provides evidence for the potential occurrence of a cascading energy transfer mechanism at non-MHD scales, at least on short timescales.

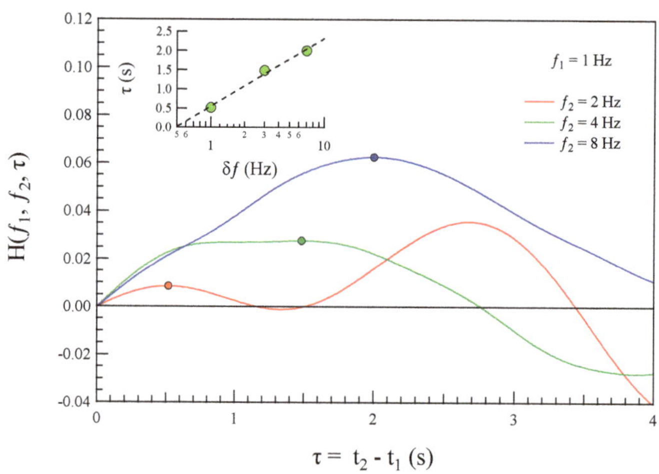

Figure 11. The behavior of the test function $H(f_1, f_2; \tau)$ for three cases studied. The dots indicate the position of the maxima. The inset shows $\tau(\delta f)$ where $\delta f = f_2 - f_1$. Dashed line is a logarithmic fit.

Indeed, we found that the first maxima occur in a sequence of increasing frequencies, $f_2 < f_3 < f_4$. This suggests that the initial energy transfer from f_1 to f_2 is followed by a second transfer from f_1 to $f_3 > f_2$, and a third transfer from f_1 to $f_4 > f_3$. Additionally, as shown in the inset of Figure 11, the timings corresponding to the first maxima at different frequencies follow a logarithmic trend as a function of the frequency difference $\Delta f = f_i - f_1$.

4. Discussion and Conclusions

The analysis of the asymmetric time-correlation function, $\Psi_{\delta t}(\tau)$, across various timescales (δt) within the ion and sub-ion scales (non-MHD domain, i.e., $f > \Omega_c$), reveals a clear violation of the detailed balance equation at these scales. This violation is a strong indicator of the irreversible nature of the fluctuations in this frequency range, suggesting that the system is far from equilibrium and dominated by nonequilibrium dynamics.

This irreversible behavior is consistent with the expected characteristics of energy transfer or dissipation processes occurring at ion/sub-ion scales. Moreover, these findings are corroborated by results obtained from shuffled data, which serve as a control test to confirm that the observed asymmetry is not due to random noise or statistical artifacts. The combined evidence points to the presence of fundamental mechanisms breaking time-reversal symmetry, further underscoring the unique and complex dynamical processes occurring in this non-MHD regime.

However, in contrast to the behavior observed in the inertial range by Cocciaglia et al. [15], the asymmetric time-correlation function $\Psi_{\delta t}(\tau)$ at certain timescales δt initially shows an increase, followed by a decrease (see Figure 5). This behavior could be a consequence of the explosive nature of the CD instability, which initially pumps energy into the system. Following this initial energy injection, the energy is then released through a damping or transfer mechanism. This is particularly evident in Figure 12, where for the smallest timescale studied ($\delta t = 0.25$ s), the time-correlation function $\Psi_{\delta t}(\tau))$ is initially positive, suggesting an increase in energy at this timescale. Subsequently, it becomes negative, indicating that the energy is released. This transition from positive to negative could reflect a typical energy redistribution process in nonequilibrium systems, where energy is initially accumulated and then dissipated or transferred across scales. Such behavior highlights the dynamic and transient nature of the processes occurring during the current disruption event.

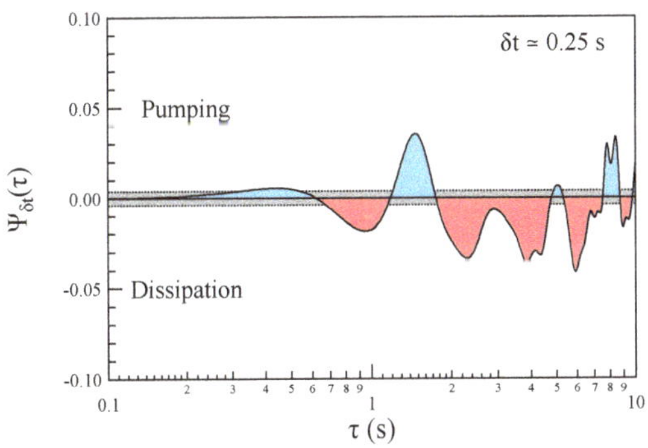

Figure 12. The behavior of the asymmetric correlation function for $\delta t = 0.125$ s. The dashed lines delimit the non-significance range, which is marked in gray.

In a previous study, Consolini and Lui [49] investigated the occurrence of three-wave coupling in magnetic field fluctuations and found evidence of short-lived nonlinear interactions as current disruption progresses. This intermittent coupling aligns well with the sporadic nature of energy pumping observed through the asymmetric correlation function. Specifically, the energy transfer from the instability timescale to the non-MHD scales appears to occur intermittently, rather than in a continuous manner. This intermittent behavior suggests that the process of energy transfer is not steady but fluctuates over time, possibly driven by the dynamics and evolving nature of the current disruption event. The periodicity and short-lived nature of the coupling imply that the system undergoes transient states, where energy is rapidly transferred across different scales, before being damped or redistributed. These findings reinforce the idea that the energy dynamics during the CD event are complex and governed by nonlinear processes that are not only spatially multiscale but also temporally intermittent.

An investigation into the validity of the fluctuation theorem (FT) for the probability distribution function $p(W_\tau)$ of the power density revealed that the FT requirements for NESS are only partially met. Specifically, while the scaling of the relative probabilities of positive and negative W_τ is satisfied on average, the coefficient $c_+(\tau)$ does not exhibit a linear trend with respect to τ. Notably, linearity seems to hold only for timescales $\tau < 0.5$ s. This finding suggests the possible occurrence of a dynamical transition, as indicated by the behavior of the first minima in the asymmetric correlation functions.

Assuming that the frequency $f^* \simeq 1.2$ Hz at which we observe the dynamical transition corresponds to the ion inertial length via the Alfvén velocity c_A, i.e., $c_A = d_i f^*$, we can estimate a magnetic field intensity $B \simeq 10$ nT. This value is of the same order as the magnetic field intensity observed during the initial phase of the current disruption (CD). Based on this, we can hypothesize that the dynamical transition occurs at a scale close to the ion inertial length d_i. This implies that the dissipation and irreversibility associated with the CD event are primarily occurring at scales smaller than d_i, i.e., at the sub-ion scales. At these smaller scales, the physical processes responsible for energy dissipation and the breakdown of detailed balance become more pronounced. Consequently, the transition to irreversible dynamics could be linked to the scale where the ion inertial length becomes relevant, marking a shift in the energy redistribution mechanisms and the onset of turbulence-like behavior. This observation further supports the idea that dissipation is predominantly occurring in the non-MHD regime, where interactions at sub-ion scales play a key role.

Furthermore, the fact that the symmetries predicted by the fluctuation theorem (FT) are only partially satisfied, a property already observed in turbulence [22], could be related to the nature of the forces acting on the plasma during the CD event. A key point to emphasize is the potential role of nonstationarity in causing the deviation from the FT predictions. Indeed, the CD event may be viewed as a transition between two nonequilibrium stationary states, where the system is evolving and not in a steady state, potentially leading to the observed discrepancies.

Another important consideration is whether a cascading mechanism is responsible for transferring energy at sub-ion scales. Although based on temporal measurements our preliminary analysis suggests that such a mechanism might be occurring. Specifically, by using the test function $H(f_1, f_i, \tau)$, we found evidence of time-ordering in the coupling between frequencies, with f_1 coupling to $f_2 < f_3 < f_4$. However, it is crucial to emphasize that a more thorough assessment of the potential cascading mechanism requires further analysis [14], which could also be capable of resolving spatial scales. We are confident that multipoint measurements from future space missions, such as the NASA HelioSwarm mission or the proposed ESA Plasma Observatory, can help overcome some of the limitations imposed here by purely temporal measurements in unveiling the occurrence of a clear cascading process.

For instance, when examining the test function over a longer time interval, such as τ up to 30 s, we observed moments where the coupling between different timescales appeared nearly instantaneous. This suggests a rapid energy transfer from f_1 to the other frequencies, which would contradict the existence of a cascading process, as the time-ordering would be lost. One possible explanation for this behavior is the role of instability, and potentially secondary instabilities, in the current disruption event (CD). These instabilities could be driving the observed rapid coupling between timescales, further complicating the identification of a cascading mechanism. However, given our analysis on the time domain, we cannot exclude that non-trivial space time correlations might affect the observed behavior. Thus, a confirmation of this point requires more accurate studies.

In conclusion, this study explores the potential of the asymmetric time-correlation function $\Psi_{\delta t}(\tau)$ as a tool for detecting irreversibility in space plasma processes at the ion and sub-ion scales. Our findings confirm the emergence of irreversibility in the fluctuations occurring during CD. However, we observed some key differences when compared to well-studied cases of systems in contact with thermostats, particularly in the application of fluctuation theorems (FTs). These differences suggest that nonequilibrium processes, such as turbulence which involves multiple scales, may exhibit distinct features and symmetries in their irreversible dynamics. Additionally, our results point to the possible occurrence of a cascading mechanism that transfers energy across sub-ion scales over short time intervals. While this observation provides promising evidence of such a mechanism, further investigation is required to fully validate this hypothesis. Future studies will need to explore this phenomenon in more detail, taking into account the complexities of the underlying dynamics and the potential influence of instability mechanisms, which may play a crucial role in shaping the energy transfer processes. Overall, this work highlights the value of using time-asymmetric correlation functions to probe the irreversible nature of space plasma phenomena and opens avenues for further research into the multiscale processes driving energy dissipation and transfer in space plasma systems. In particular, we believe that approaches like the ones presented in this work may be relevant in several different plasma context, such as the study of magnetic reconnection in decaying turbulence (see, e.g., [50,51]), as well as in the investigation of time-symmetry breaking phenomena in the analysis of nonlinear kinetic Vlasov–Maxwell simulations, which should formally preserve entropy [52,53].

Author Contributions: Conceptualization, G.C.; methodology, G.C. and P.D.M.; formal analysis, G.C.; investigation, G.C. and P.D.M.; writing—original draft preparation, G.C. and P.D.M.; writing—review and editing, G.C. and P.D.M. All authors have read and agreed to the published version of the manuscript.

Funding: This research was funded by the *Space It Up!* project funded by the Italian Space Agency, ASI, and the Ministry of University and Research, MUR, under contract n. 2024-5-E.0—CUP n. I53D24000060005.

Institutional Review Board Statement: Not applicable.

Informed Consent Statement: Not applicable.

Data Availability Statement: THEMIS data used in this study are available at the NASA Space Physics Data Facility (SPDF) (https://cdaweb.gsfc.nasa.gov/index.html) accessed on 12 September 2024.

Acknowledgments: The authors acknowledge the contributions of the FGM team to THEMIS mission for making available the data used in this work. We thank the two anonymous reviewers for their useful comments and suggestions, which helped us to improve our work. We extend a special thought for Tom T.S. Chang, who passed away last January, for his continued invitation to take untrodden paths and for the exciting discussions over the years between us.

Conflicts of Interest: The authors declare no conflicts of interest.

Abbreviations

The following abbreviations are used in this manuscript:

CD	Current disruption
CPS	Central plasma sheet

FT	Fluctuation Theorem
FWHM	Full width half maximum
HD	Hydro-Dynamic
MHD	Magneto-Hydro-Dynamic
NESS	Nonequilibrium stationary state
PDF	Probability density function
PSD	Power spectral density

References

1. Bruno, R.; Carbone, V. The solar wind as a turbulence laboratory. *Living Rev. Sol. Phys.* **2013**, *10*, 1–208. [CrossRef]
2. Zhou, Y.; Matthaeus, W.H.; Dmitruk, P. Colloquium: Magnetohydrodynamic turbulence and time scales in astrophysical and space plasmas. *Rev. Mod. Phys.* **2004**, *76*, 1015–1035. [CrossRef]
3. Kiyani, K.H.; Chapman, S.C.; Khotyaintsev, Y.V.; Dunlop, M.W.; Sahraoui, F. Global Scale-Invariant Dissipation in Collisionless Plasma Turbulence. *Phys. Rev. Lett.* **2009**, *103*, 075006. [CrossRef]
4. Consolini, G.; Kretzschmar, M.; Lui, A.; Zimbardo, G.; Macek, W. On the magnetic field fluctuations during magnetospheric tail current disruption: A statistical approach. *J. Geophys. Res. Space Phys.* **2005**, *110*. [CrossRef]
5. Eastwood, J.; Phan, T.; Bale, S.; Tjulin, A. Observations of turbulence generated by magnetic reconnection. *Phys. Rev. Lett.* **2009**, *102*, 035001. [CrossRef] [PubMed]
6. Consolini, G.; Grandioso, S.; Yordanova, E.; Marcucci, M.; Pallocchia, G. Statistical and scaling features of fluctuations in the dissipation range during a reconnection event. *Astrophys. J.* **2015**, *804*, 19. [CrossRef]
7. Servidio, S.; Carbone, V.; Primavera, L.; Veltri, P.; Stasiewicz, K. Compressible turbulence in Hall magnetohydrodynamics. *Planet. Space Sci.* **2007**, *55*, 2239–2243. [CrossRef]
8. Greco, A.; Perri, S.; Servidio, S.; Yordanova, E.; Veltri, P. The complex structure of magnetic field discontinuities in the turbulent solar wind. *Astrophys. J. Lett.* **2016**, *823*, L39. [CrossRef]
9. Consolini, G.; Alberti, T.; Benella, S.; Papini, E.; Pezzi, O. On the fractal pattern of the current structure at ion scales in turbulent space plasmas. *Chaos Solitons Fractals* **2023**, *177*, 114253. [CrossRef]
10. De Groot, S.; Mazur, P. *Non-Equilibrium Thermodynamics*; Dover Publications Inc.: New York, NY, USA, 1984.
11. Livi, R.; Politi, P. *Nonequilibrium Statistical Physics: A Modern Perspective*; Cambridge University Press: Cambridge, UK, 2017.
12. Lucente, D.; Baldovin, M.; Cecconi, F.; Cencini, M.; Cocciaglia, N.; Puglisi, A.; Viale, M.; Vulpiani, A. Conceptual and practical approaches for investigating irreversible processes. *arXiv* **2024**, arXiv:2410.15925.
13. Pomeau, Y. Symétrie des fluctuations dans le renversement du temps. *J. Phys.* **1982**, *43*, 859–867. [CrossRef]
14. Josserand, C.; Le Berre, M.; Lehner, T.; Pomeau, Y. Turbulence: Does energy cascade exist? *J. Stat. Phys.* **2017**, *167*, 596–625. [CrossRef]
15. Cocciaglia, N.; Cencini, M.; Vulpiani, A. Nonequilibrium statistical mechanics of the turbulent energy cascade: Irreversibility and response functions. *Phys. Rev. E* **2024**, *109*, 014113. [CrossRef]
16. Gallavotti, G.; Cohen, E.G.D. Dynamical ensembles in nonequilibrium statistical mechanics. *Phys. Rev. Lett.* **1995**, *74*, 2694. [CrossRef] [PubMed]
17. Gallavotti, G.; Cohen, E.G.D. Dynamical ensembles in stationary states. *J. Stat. Phys.* **1995**, *80*, 931–970. [CrossRef]
18. Lebowitz, J.L.; Spohn, H. A Gallavotti–Cohen-type symmetry in the large deviation functional for stochastic dynamics. *J. Stat. Phys.* **1999**, *95*, 333–365. [CrossRef]
19. Derrida, B. Non-equilibrium steady states: Fluctuations and large deviations of the density and of the current. *J. Stat. Mech. Theory Exp.* **2007**, *2007*, P07023. [CrossRef]
20. Ciliberto, S.; Garnier, N.; Hernandez, S.; Lacpatia, C.; Pinton, J.F.; Chavarria, G.R. Experimental test of the Gallavotti–Cohen fluctuation theorem in turbulent flows. *Phys. A Stat. Mech. Its Appl.* **2004**, *340*, 240–250. [CrossRef]
21. Shang, X.D.; Tong, P.; Xia, K.Q. Test of steady-state fluctuation theorem in turbulent Rayleigh-Bénard convection. *Phys. Rev. E* **2005**, *72*, 015301. [CrossRef]
22. Xu, H.; Pumir, A.; Falkovich, G.; Bodenschatz, E.; Shats, M.; Xia, H.; Francois, N.; Boffetta, G. Flight–crash events in turbulence. *Proc. Natl. Acad. Sci. USA* **2014**, *111*, 7558–7563. [CrossRef]
23. Viavattene, G.; Consolini, G.; Giovannelli, L.; Berrilli, F.; Del Moro, D.; Giannattasio, F.; Penza, V.; Calchetti, D. Testing the Steady-State Fluctuation Relation in the Solar Photospheric Convection. *Entropy* **2020**, *22*, 716. [CrossRef]
24. Sekimoto, K. *Stochastic Energetics*; Springer: Berlin, Germany, 2010.
25. Seifert, U. Stochastic thermodynamics, fluctuation theorems and molecular machines. *Rep. Prog. Phys.* **2012**, *75*, 126001. [CrossRef] [PubMed]
26. Peliti, L.; Pigolotti, S. *Stochastic Thermodynamics: An Introduction*; Princeton University Press: Princeton, NJ, USA, 2021.

27. Consolini, G.; De Michelis, P.; Tozzi, R. On the Earth's magnetospheric dynamics: Nonequilibrium evolution and the fluctuation theorem. *J. Geophys. Res. Space Phys.* **2008**, *113*. [CrossRef]
28. Stumpo, M.; Benella, S.; Alberti, T.; Pezzi, O.; Papini, E.; Consolini, G. Relating Intermittency and Inverse Cascade to Stochastic Entropy in Solar Wind Turbulence. *Astrophys. J. Lett.* **2023**, *959*, L20. [CrossRef]
29. Takahashi, K.; Zanetti, L.; Lopez, R.; McEntire, R.; Potemra, T.; Yumoto, K. Disruption of the magnetotail current sheet observed by AMPTE/CCE. *Geophys. Res. Lett.* **1987**, *14*, 1019–1022. [CrossRef]
30. Lui, A.; Lopez, R.; Krimigis, S.; McEntire, R.; Zanetti, L.; Potemra, T. A case study of magnetotail current sheet disruption and diversion. *Geophys. Res. Lett.* **1988**, *15*, 721–724. [CrossRef]
31. Ahn, B.H.; Kamide, Y.; Kroehl, H.; Candidi, M.; Murphree, J.S. Substorm changes of the electrodynamic quantities in the polar ionosphere: CDAW 9. *J. Geophys. Res. Space Phys.* **1995**, *100*, 23845–23856. [CrossRef]
32. Lui, A.; Volwerk, M.; Dunlop, M.; Alexeev, I.; Fazakerley, A.; Walsh, A.; Lester, M.; Grocott, A.; Mouikis, C.; Henderson, M.; et al. Near-Earth substorm features from multiple satellite observations. *J. Geophys. Res. Space Phys.* **2008**, *113*. [CrossRef]
33. Ohtani, S.; Higuchi, T.; Lui, A.; Takahashi, K. Magnetic fluctuations associated with tail current disruption: Fractal analysis. *J. Geophys. Res. Space Phys.* **1995**, *100*, 19135–19145. [CrossRef]
34. Lui, A.; Najmi, A.H. Time-frequency decomposition of signals in a current disruption event. *Geophys. Res. Lett.* **1997**, *24*, 3157–3160. [CrossRef]
35. Kozak, L.; Petrenko, B.; Lui, A.; Kronberg, E.; Grigorenko, E.; Prokhorenkov, A. Turbulent processes in the Earth's magnetotail: Spectral and statistical research. *Ann. Geophys.* **2018**, *36*, 1303–1318. [CrossRef]
36. Kozak, L.; Petrenko, B.; Kronberg, E.; Grigorenko, E.; Lui, E.; Cheremnykh, S. Spectra of Turbulence during the Dipolarization of the Magnetic Field. *Kinemat. Phys. Celest. Bodies* **2018**, *34*, 258–269. [CrossRef]
37. Kozak, L.; Petrenko, B.; Lui, A.; Kronberg, E.; Daly, P. Processes in the current disruption region: From turbulence to dispersion relation. *J. Geophys. Res. Space Phys.* **2021**, *126*, e2020JA028404. [CrossRef]
38. Lui, A.; Yoon, P.; Mok, C.; Ryu, C.M. Inverse cascade feature in current disruption. *J. Geophys. Res. Space Phys.* **2008**, *113*. [CrossRef]
39. Yoon, P.; Lui, A.; Bonnell, J. Identification of plasma instability from wavelet spectra in a current disruption event. *J. Geophys. Res. Space Phys.* **2009**, *114*. [CrossRef]
40. Onsager, L. Reciprocal relations in irreversible processes. I. *Phys. Rev.* **1931**, *37*, 405. [CrossRef]
41. Onsager, L. Reciprocal relations in irreversible processes. II. *Phys. Rev.* **1931**, *38*, 2265. [CrossRef]
42. Lui, A.; Mitchell, D.; Lanzerotti, L. Comparison of energetic electron intensities outside and inside the radiation belts. *J. Geophys. Res. Space Phys.* **2014**, *119*, 6213–6230. [CrossRef]
43. Lui, A. Reduction of the cross-tail current during near-Earth dipolarization with multisatellite observations. *J. Geophys. Res. Space Phys.* **2011**, *116*. [CrossRef]
44. Huang, N.E.; Wu, Z. A review on Hilbert-Huang transform: Method and its applications to geophysical studies. *Rev. Geophys.* **2008**, *46*, RG2006. [CrossRef]
45. Servidio, S.; Carbone, V.; Dmitruk, P.; Matthaeus, W. Time decorrelation in isotropic magnetohydrodynamic turbulence. *Europhys. Lett.* **2011**, *96*, 55003. [CrossRef]
46. Torrence, C.; Compo, G.P. A practical guide to wavelet analysis. *Bull. Am. Meteorol. Soc.* **1998**, *79*, 61–78. [CrossRef]
47. Mallat, S. *A Wavelet Tour of Signal Processing*; Academic Press: San Diego, CA, USA, 1999.
48. Zhang, W.; Fu, H.; Cao, J.; Wang, Z.; Fu, W.; Liu, Y.; Yu, Y. First determination of whistler wave dispersion relation in superhot (Te > 5 keV) plasmas. *Phys. Rev. Res.* **2024**, *6*, L012047. [CrossRef]
49. Consolini, G.; Lui, A. Symmetry breaking and nonlinear wave-wave interaction in current disruption: Possible evidence for a phase transition. *Geophys. Monogr.-Am. Geophys. Union* **2000**, *118*, 395–401.
50. Franci, L.; Cerri, S.S.; Califano, F.; Landi, S.; Papini, E.; Verdini, A.; Matteini, L.; Jenko, F.; Hellinger, P. Magnetic reconnection as a driver for a sub-ion-scale cascade in plasma turbulence. *Astrophys. J. Lett.* **2017**, *850*, L16. [CrossRef]
51. Cerri, S.S.; Califano, F. Reconnection and small-scale fields in 2D-3V hybrid-kinetic driven turbulence simulations. *New J. Phys.* **2017**, *19*, 025007. [CrossRef]
52. Ghizzo, A.; Del Sarto, D. Momentum transfer driven by fluctuations in relativistic counter-propagating electron beams. *Plasma Phys. Control. Fusion* **2021**, *63*, 055007. [CrossRef]
53. Ghizzo, A.; Del Sarto, D.; Betar, H. Collisionless heating driven by Vlasov filamentation in a counterstreaming beams configuration. *Phys. Rev. Lett.* **2023**, *131*, 035101. [CrossRef]

entropy

MDPI

Article

Turbulent Flow in Street Canyons: A Complexity Approach

Csanád Árpád Hubay *, Bálint Papp and Tamás Kalmár-Nagy

Department of Fluid Mechanics, Faculty of Mechanical Engineering, Budapest University of Technology and Economics, 1111 Budapest, Hungary; papp.balint@gpk.bme.hu (B.P.); kalmar.nagy.tamas@gpk.bme.hu (T.K.-N.)
* Correspondence: hubay.csanad@gpk.bme.hu

Abstract: Velocity measurements and simulations in an idealized urban environment were studied, focusing on turbulent flow over street canyons. Time series of fluctuating velocities were considered as marked point processes, and the distribution of mean residence times was characterized using a lognormal fit. The quadrant method was applied to transform time series into symbolic sequences, enabling the investigation of their information content. By analyzing word frequency and normalized entropy levels, we compared measured and simulated sequences with periodic symbol sequences with and without noise. Our results indicate that noisy periodic sequences exhibit entropy distributions qualitatively similar to those of the measured and simulated data. Surrogate sequences generated using first-, and higher-order Markov statistics also displayed similarity. Higher-order Markov chains provide a more accurate representation of the information content of velocity fluctuation series. These findings contribute to the comparison of experimental and simulation techniques in the investigation of turbulence.

Keywords: street canyon; turbulence; quadrant method; word statistics; entropy; Markov statistics

1. Introduction

Air pollution is closely linked to public health and climate issues. Understanding pollutant dispersion in urban environments is key for development of turbulence models capturing this phenomenon. Atmospheric boundary layer turbulence is a complex process of interacting coherent structures (spatial regions where flow properties are strongly correlated); thus, we seek to identify tools to detect and characterize coherent structures.

Robinson [1] presented a review of major findings and challenges concerning coherent structures in turbulent boundary layers, specifically those developing over flat surfaces with zero pressure gradient and at relatively low Reynolds numbers (Re < 5000). A study by Kline et al. [2] revealed the presence of near-wall low-speed "streaks" originating in the laminar sublayer. The dynamic sequence involving the generation, upward movement, and disintegration of these structures is widely referred to as the "bursting" process. Offen and Kline [3] characterized this behavior as a "shear layer interface sandwiched between an upstream, high-speed sweep and a downstream, low-speed ejection".

Bursts are essential contributors to momentum transfer over short durations due to their dominant role in producing turbulent kinetic energy within shear-driven flows. The study of burst-related phenomena generally focuses on three main areas: (i) techniques for detecting burst intervals, (ii) the formulation of appropriate scaling laws for burst duration, and (iii) exploring how bursts relate to coherent structures. An overview of research addressing these themes was provided by Metzger et al. [4]. In [5], Mahrt investigated intermittent turbulence bursts in the atmospheric boundary layer, emphasizing that higher-order statistics and event-based statistical analysis is needed to characterize non-Gaussian behavior of burst-dominated flows. Later, Mahrt focused on turbulence under stable and stratified conditions [6] and noted that turbulence is patchy and intermittent, especially in strongly stratified conditions.

Renewal processes are important to our understanding of intermittent and burst-like transport phenomena. Higbie [7] and Danckwerts [8] showed that interfacial transfer cannot be understood purely from a static perspective; rather, stochastic and dynamic surface processes should be used (eddies rise to or fall from the interface, carrying new fluid volumes). Perlmutter [9] introduced realistic eddy age/lifetime distributions into surface renewal models.

Katul et al. [10] applied renewal theory to model passive scalar transport in turbulent flows. Paradisi and collaborators [11–13] further incorporated renewal concepts to describe scalar dissipation, turbulent structure, and memory effects in stratified boundary layers. Cesari et al. [14] and Luo et al. [15] extended renewal-based models to environmental turbulence and wind energy applications, respectively, highlighting the model's versatility across complex boundary-layer phenomena.

Using direct numerical simulations, Itano and Toh [16] examined burst dynamics in turbulent channel flows and identified a traveling wave solution corresponding to a saddle point in the low-dimensional near-wall system, shedding light on the self-sustaining nature of wall turbulence. Similarly, Jiménez et al. [17] employed direct numerical simulations to investigate near-wall turbulence in Couette and Poiseuille flows, revealing that intermittent bursting is primarily associated with vortex-dominated quiescent structures.

Kawahara and Kida [18] discovered an exact time-periodic solution for Couette flow, offering a simplified model for the bursting process. The nonlinear growth and transition mechanisms in Couette flow were further explored by Cherubini and De Palma [19], focusing on the evolution of optimal disturbances. Bomminayuni and Stoesser [20] analyzed dominant turbulent events, such as sweeps and ejections, and their contributions to Reynolds stress, also quantifying turbulence anisotropy in rough-bed open-channel flows.

Quantitative insights into fluctuating velocity components were historically obtained through the development of conditional sampling methods, designed to enhance flow visualization by providing statistical context. Foundational contributions to this approach include the early work of Wallace [21] and Willmarth [22].

The *quadrant method* provides a useful framework for analyzing turbulent flows by transforming velocity time series into symbolic sequences, enabling the investigation of their information content. The streamwise and vertical velocity components are decomposed into their mean and fluctuating parts. The signs of the fluctuating streamwise and vertical components distinguish four distinct quadrants, each representing different turbulent events—*outward interaction*, corresponding to the upward motion of faster-than-average fluid; *ejection*, representing the upward motion of slower-than-average fluid; *inward interaction*, indicating the downward motion of slower-than-average fluid; and *sweep*, describing the downward motion of faster-than-average fluid.

From the standpoint of turbulent momentum exchange, ejection and sweep events are essential mechanisms that facilitate the mixing of slow-moving near-wall fluid with faster fluid from above. An extensive account of quadrant analysis and its applications is presented by Wallace [23]. Further discussion on conditional sampling approaches, including variable-interval time averaging, can be found in Antonia et al. [24].

Quadrant-based methods have been employed for various turbulent flows, including river hydraulics, the atmospheric boundary layer, and wind tunnel studies. For instance,

Kirkbride et al. [25] performed simultaneous measurements of vertical and streamwise velocities at multiple points above a gravel riverbed. By applying a simplified quadrant analysis, they categorized turbulence states and analyzed their transitions using a Markov chain model. Their results revealed the presence of large-scale structures.

Using a similar experimental setup, Buffin-Bélanger et al. [26] corroborated the role of large-scale coherent structures in shaping river turbulence. They demonstrated that quadrant events frequently occur in cyclic or oscillatory groupings, likely driven by the passage of alternating high- and low-momentum fluid regions.

To further understand the spatiotemporal properties of such coherent structures, Roy et al. [27] utilized an array of electromagnetic flow meters to characterize their scale, lifespan, and dynamics. Their statistical analysis of burst durations revealed that lab-scale observations align closely with the scaling behavior of natural river flows.

Field measurements have also identified coherent structures in atmospheric flows. Finnigan [28] demonstrated that large-scale coherent structures dominate flow dynamics above plant canopies. Quadrant analysis showed that the sweep quadrant contributes most to momentum flux, as fast downward gusts penetrate lower boundary layer levels. In urban settings, Christen et al. [29] used anemometers in a vertical grid across an urban canyon to study momentum and scalar transport. They found that sweep events dominate transport in the upper canyon, while ejection events follow sweeps in a cyclical pattern. Nelson et al. [30] demonstrated that all four quadrants contribute to momentum transfer near the roof level within street canyons. They introduced the concept of a disorganized canopy layer, where the aerodynamic influence of individual buildings outweighs that of broader atmospheric flow structures.

Since direct turbulence measurements in urban environments are costly and complex, wind tunnel models are widely used [31]. Kukačka et al. [32] investigated flow and tracer gas concentration above an X-shaped street intersection, analyzing the dominant role of sweep events in impulse transfer. Nosek et al. [33] applied quadrant analysis to study turbulence above a 3D street canyon, showing that:

- Sweeps correlate with entrainment of clean air;
- Ejections are associated with ventilation of polluted air;
- Coherent structures passing over the canyon drive this process.

Kellnerová et al. [34] examined the velocity field within a two-dimensional idealized street canyon using Particle Image Velocimetry. Their findings indicate that the street canyon vortex is frequently disturbed, a process which can significantly influence the efficiency of urban ventilation. Additionally, Di Bernardino et al. [35] reported that the shear layer above the canyon roof exhibits a periodic flapping motion.

Conditional sampling techniques are commonly used for event detection in turbulent flows. This inspired our research into symbolic turbulence data analysis. A comprehensive review of symbolic analysis and its applications is provided by Daw et al. [36]. The key step in symbolic analysis involves converting the original dataset into a sequence of symbols. For continuous datasets, this is typically achieved by partitioning the range of observed values into a finite number of discrete regions. Each data point is then replaced by the symbol corresponding to its respective region. This process, known as coarse graining or quantization, reduces high-resolution data to a lower-resolution representation.

Lehrman and Rechester [37] developed a symbolic cycle extraction method for turbulent fluctuations in pipe flows. If simultaneous velocity component measurements are available, quadrant-based symbolization is a viable approach.

Daw et al. [38] studied pressure fluctuations from air bubbles released through a nozzle into a water tank, aiming to detect periodic behavior in a two-phase fluid system under varying Reynolds numbers. They converted the time series into binary symbol sequences using the dataset median as a threshold and introduced a modified Shannon entropy, referred to as *normalized entropy*, to evaluate the informational complexity of the symbolized data. To characterize the bubble formation process, they used word statistic histograms, where "words" represented smaller partitions of the symbolic sequence.

Using a comparable symbolic framework, Finney et al. [39] analyzed heat release patterns across combustion engine cycles. Their results indicated that normalized entropy reaches a minimum at an optimal word length, beyond which the symbolic sequences either fail to encapsulate meaningful temporal structure (if too short) or become dominated by noise and limited sample size (if too long). This highlights the delicate balance between resolution and statistical reliability in symbolic analysis.

Keshavarzi et al. [40] investigated the turbulent bursting process and its link to sediment transport using velocity measurements from an open water channel. By applying the quadrant method, they encoded instantaneous vertical and streamwise velocity components into discrete symbolic states. The sequence of transitions between these states was modeled using a first-order Markov chain.

Gheisi et al. [41] expanded upon this approach by exploring the three-dimensional bursting process near the bed of a settling chamber. Using synchronized measurements of all three velocity components, they developed the *octant method*, which symbolized the data according to the sign of velocity fluctuations along all three axes. Their analysis confirmed that a first-order Markov model was sufficient to capture the dominant dynamics of the bursting process. A similar Markov-based approach was employed by Jin et al. [42], who simulated particle deposition near the wall by incorporating flow eddies, effectively modeling the impact of coherent structures on heavy particle trajectories.

In a wind tunnel study aimed at understanding the temporal organization of turbulence in urban-like environments, Kalmár-Nagy and Varga [43] analyzed two-component velocity signals. Their methodology, closely followed in the present work, treated velocity fluctuations as a marked point process and used the quadrant method to construct symbolic sequences. Entropy analysis revealed that these sequences exhibited traits similar to noisy periodic signals, necessitating higher-order Markov chains to accurately capture the information content. Furthermore, performing quadrant analysis with rotated coordinate systems minimized entropy in the principal axes system of the velocity fluctuation cloud, suggesting a fundamental structure in turbulence dynamics. In this paper, we extend the analysis of the streamwise and vertical velocity components by including both measurements and simulations of street canyon flows, i.e., an idealized urban environment. The novelty of our work lies in the direct comparison between experimental and simulated symbolic sequences.

Further contributions by Chowdhuri et al. [44–46] provided detailed persistence-based analyses of velocity and temperature fluctuations in convective turbulence. Their work explored intermittent heat transport, Reynolds stress anisotropy, and temperature variability during gust front events, offering insights into the structure of atmospheric turbulence.

The paper is organized as follows. Section 2 details the experimental and numerical setups, along with the signal acquisition and processing methodologies. In Section 3, we explore statistical characteristics of velocity fluctuations, comparing the experimental observations with simulation results. Section 4 introduces a quadrant-based symbolic encoding of both datasets and the marked point process used in the study. Section 5 delves into word statistics derived from the symbolic sequences and quantifies their information content through normalized entropy. In Section 6, we examine the Markovian properties of both the symbolic series and the quadrant-based marked point process. We also compare these datasets with surrogate sequences that mimic noisy periodic behavior and demonstrate how higher-order Markov models can replicate the symbolic data's informational structure. Section 7 focuses on the two-dimensional distribution of velocity fluctuations, and Section 8 summarizes the key findings.

2. Data

Street canyons are widely used as a minimalistic model of the urban environment. The literature of street canyon flows is reviewed extensively by Vardoulakis et al. [47], Ahmad et al. [48], Li et al. [49], Yazid et al. [50], Zhang et al. [51], and Voordeckers et al. [52]. In the present study, the flow field within and above a series of parallel street canyons with uniform building height and perpendicular wind conditions were analyzed via wind tunnel experiments and Computational Fluid Dynamics (CFD) simulations.

2.1. Wind Tunnel Measurements

2.1.1. Experimental Setup

The wind tunnel experiments were performed using the closed-circuit horizontal (Göttingen-type) wind tunnel of the Theodore von Kármán Wind Tunnel Laboratory at the Department of Fluid Mechanics of the Budapest University of Technology and Economics. The wind tunnel has a circular cross-section of 2.6 m in diameter at the open test section of 3.8 m in length, and it is equipped with a 2.5 m wide horizontal table (Figure 1).

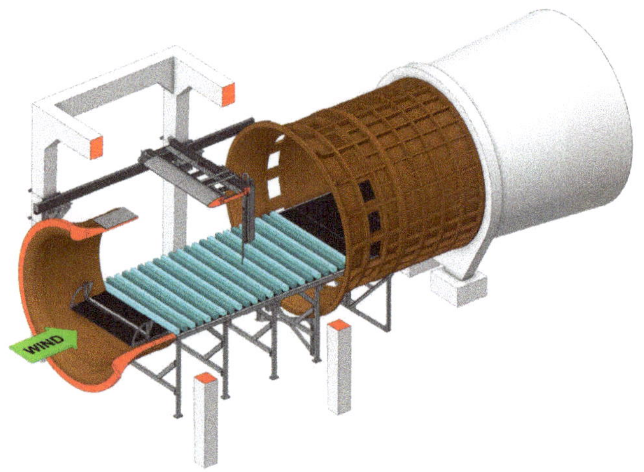

Figure 1. The wind tunnel setup used for the acquisition of the experimental time series.

In total, 23 rows of prismatic building blocks were constructed from styrofoam and were mounted to the floor of the test section, forming 22 consecutive street canyons, with their axis oriented perpendicular to the incoming flow. The height-to-width aspect ratio of the canyons was $H/W = 1.0$, with the roof height $H = 0.1$ m being equal to both the street width W and the building breadth B. The dimensions of the geometry are shown in detail in Figure 2. The incoming flow of the wind tunnel was homogeneous with a free-stream velocity of $u_\infty = 9$ m/s, and below 1% turbulence intensity. No additional roughness elements were placed upstream of the building models; this way, the boundary layer could adapt to the investigated building configuration. It was shown by Papp et al. [53,54] that the flow field is fully developed after the first 11 canyons.

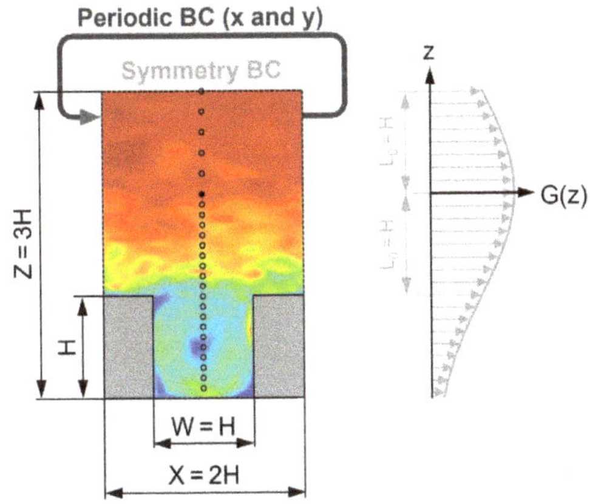

Figure 2. Layout and dimensions of the computational domain, containing a street canyon with $H = 0.1$ m building height. The size of the computational domain in the y direction is $Y = 1.04H$. $G(z)$ denotes the vertical distribution of the driving force of the TWF model with $L_0/H = 1.0$ radius. The reference location, i.e., the center of the TWF propulsion at $z_0/H = 2.0$, is denoted by a filled marker. The sampling points used for recording the velocity time series are denoted by hollow markers. The background is colored by the instantaneous velocity magnitude.

The Reynolds number based on the building height and the free-stream velocity yields $Re = u_\infty H/\nu = 5 \cdot 10^4$, in which ν is the kinematic viscosity of air. This value is high enough for the velocity field to be considered independent of the Reynolds number according to the results of Chew et al. [55] for $H/W = 1.0$ street canyons. Papp et al. [53] have pointed out that the velocity field showed indeed very little variation as the function of the Reynolds number in the $Re = 2.5 \ldots 7.5 \cdot 10^4$ range ($u_\infty = 4.5 \ldots 13.5$ m/s); therefore, the results obtained in model scale can be considered relevant in full scale as well.

2.1.2. Measurement Techniques and Signal Processing

The time histories of the streamwise and vertical velocity components denoted by $u(t)$ and $w(t)$ hereinafter, were measured simultaneously, using a two-component Laser Doppler Anemometer by TSI. For details of the measurement instrumentation, the reader is referred to Papp et al. [53,54].

The velocity sampling points were located in a vertical array in the middle of the 11th street canyon. The height of the ith measurement point can be calculated as $z_i = iH/10$, with $i = 1, 2, \ldots, 19, 20, 22, 24, 26, 28, 30$.

The velocity measurement at each point took 150 s with a sampling frequency of 100...1000+ Hz, depending on how effectively the seeding particles released upstream could reach each point. The free-stream velocity ($u_\infty = 9$ m/s) of the wind tunnel was monitored using a Pitot-static probe, and its minor changes were compensated for during the normalization process. The absolute measurement uncertainty of the LDA system was 0.1 m/s at 5 m/s flow velocity.

Generally, the data sampling frequency in LDA measurements is not constant. Moreover, the so-called coincidence mode of the LDA system was not enabled in order to achieve as high sampling rate as possible; therefore, the LDA bursts corresponding to the $u(t)$ and $w(t)$ velocity components were not recorded at the same time instances. To overcome the temporal irregularity of the measured data series, the time series between 4 s and 34 s were re-sampled to the temporal resolution of the CFD time series using linear interpolation for further data processing (see Table 1 and Section 2.2.4).

Table 1. Summary of the numerical simulations carried out at three different spatial resolutions.

Parameters of Numerical Meshes	Coarse	Medium	Fine
Mesh resolution ($\Delta x = \Delta x = \Delta z$)	$H/22$	$H/32$	$H/48$
Total cell count (N)	111k	343k	1312k
Time step size (Δt)	0.45 ms	0.3 ms	0.2 ms
Total number of data points ($T_s/\Delta\text{t}$)	67k	100k	150k

2.2. *Computational Fluid Dynamics Simulations*

2.2.1. Boundary Conditions and Numerical Mesh

The computational domain used in the numerical simulations consists of a three-dimensional section of a single street canyon (see Figure 2).

The size of the computational domain is $X \times Y \times Z = 2H \times 2.0833H \times 3H$. On the lateral boundaries, periodic boundary conditions (BCs) are assumed (in both x and y directions), and at the top of the domain, slip symmetry (zero gradient BC) is assumed. At the solid surfaces, i.e., the ground and the building walls, smooth, no-slip walls are applied.

The computational domain was discretized using equidistant Cartesian meshes of three different spatial resolutions $H/\Delta x = 22$, 32, and 48. Note that according to Xi and Castro [56], a mesh uniformly resolving the characteristic length by 16 cells can be considered sufficiently dense for capturing the large-scale eddies governing the dispersion processes in urban-like obstacle arrays using large eddy simulation (LES)—the same turbulence model as used for the present investigation. The mesh dependence of the results was assessed based on the descriptive statistics (mean, standard deviation, skewness, kurtosis) as well as the residence times (Equation (6)). It was found that although the results are not independent of the mesh resolution, they show monotonous convergence. This is expected behavior, as LES run on a finer mesh can resolve more turbulent kinetic energy. In this paper, the results obtained on the finest mesh are shown only.

2.2.2. Transient Wind Forcing: Modeling Large-Scale Turbulence in a Small Periodic Domain

Importantly, the main driving force of atmospheric flows is the momentum exchange at the top of the atmospheric boundary layer (ABL), resulting in the formation of large-scale eddies, which cause sudden changes in the wind direction and magnitude near the surface. The review article by [51] highlighted that accurately modeling the resultant time-varying inflow conditions can have a significant effect on the governing flow regimes observed in street canyons; therefore, it is important to consider the effects of large-scale turbulence in microscale meteorological models.

The Transient Wind Forcing (TWF) method was introduced by Kristóf et al. [57] to model the effects of the changing wind direction and magnitude in a small periodic computational domain using velocity time series recorded in field experiments or mesoscale simulations. Moreover, their model was capable of tracking particle trajectories in a horizontally limitless space based on the periodic velocity field. Papp et al. [53] added the capability of calculating a spatially continuous concentration field derived from discrete particle tracks to the TWF model, and the latest version published by Koren and Kristóf [58] allows for even modeling thermal stratification and the Coriolis force.

According to the TWF approach, turbulence can be divided into the following categories based on the size of the flow structures:

1. Macroscopic turbulence, i.e., eddies exceeding the size of the computational domain;
2. Mesoscopic turbulence, i.e., eddies that are resolved in the LES calculations;
3. Microscopic turbulence, i.e., vortices that are smaller than the mesh resolution; therefore, their effects are modeled by the subgrid-scale stress (SGS) model.

In the TWF model, the effect of macroscopic turbulence is taken into account as a time-dependent driving force, implemented as sources terms in the x and y components of the momentum equation. The source terms are formulated as

$$S_u(z,t) = \rho \cdot a_u(t) \cdot G(z) = \rho \cdot \frac{u_0(t) - u(t)}{\tau(t)} \cdot e^{-\frac{1}{2}\left(\frac{z-z_0}{L_0}\right)^2}, \tag{1}$$

$$S_v(z,t) = \rho \cdot a_v(t) \cdot G(z) = \rho \cdot \frac{v_0(t) - v(t)}{\tau(t)} \cdot e^{-\frac{1}{2}\left(\frac{z-z_0}{L_0}\right)^2}. \tag{2}$$

In the above equations $S_u(z,t)$ and $S_v(z,t)$ denote the volume source intensities, ρ is the constant air density, $u_0(t)$ and $v_0(t)$ are the measured velocity time series (reference time series), $u(t)$ and $v(t)$ are the velocity components in the CFD model, and $\tau(t)$ is the relaxation time of the velocity control. Furthermore, t is time, z is the vertical coordinate, and $z_0/H = 2.0$ is the reference height. As can be seen in the above formulas, the vertical distribution of the driving force is Gaussian, with its center being at z_0, and with a characteristic radius of $L_0/H = 1.0$, shown in Figure 2. The direction of the resultant body force can take any horizontal direction, determined by the values of the so-called control terms $a_u(t)$ and $a_v(t)$, which are responsible for imposing the measured velocity time series at the reference point.

The relaxation time in the denominator of the control terms is increased exponentially within the startup transient expressed by the formula

$$\tau(t) = \Delta t_0 + (\Delta t - \Delta t_0) \cdot e^{-\frac{t-t_{\text{start}}}{\Delta t_0}}, \tag{3}$$

where $\Delta t_0 = 0.0195$ s is the propulsion time scale of the TWF model (half of the flow-through time calculated as X/u_0, in which $u_0 = 5.117$ m/s is the mean streamwise velocity at z_0), t_{start} is the starting time, and Δt is the time step size of the simulation. The latter was fixed, with its value chosen to satisfy the Courant–Friedrichs–Lewy (CFL) criterion, i.e., the maximum of the Courant number ($C = u\Delta x/\Delta t$) in the entire computational domain should be around 1.

An appropriately chosen relaxation time allows for the applied large eddy simulation (LES) turbulence model, to generate realistic mesoscopic turbulence around the reference point, with the large-scale trends following the velocities prescribed via $u_0(t)$ and $v_0(t)$ accurately, but it prevents the TWF driving force from acting as a direct velocity constraint, resulting in an unrealistic flow field. Imposing the reference time series recorded in the wind tunnel experiments is effectively an immersed boundary condition. Therefore, reproducing the velocity field above the reference height is out of the scope of the TWF approach; however, to allow free eddy motion, the simulation domain was kept taller.

The effects of microscopic turbulence was taken into account using the Smagorinsky–Lilly subgrid-scale stress (SGS) model [59,60], using $C_s = 0.1$, following the recommendations of Shah [61] for a flow past a blunt obstacle.

2.2.3. Solver Setup

For the numerical solution of the flow equations, Ansys Fluent [62], a general-purpose CFD solver, was utilized, applying the Bounded Central Differencing Scheme flux formulation and the SIMPLE (Semi-Implicit Method for Pressure Linked Equations) algorithm, along with the bounded second-order implicit scheme for temporal discretization. Iterative solution was carried out at each time step until the residuals of the flow equations were reduced by at least three orders of magnitude compared to the initial state, but in a maximum of 20 iterations per time step.

2.2.4. Time Series Acquisition

The time series used for the analysis of the flow field were recorded in each time step for all three velocity components, at 25 gauging points placed in a vertical array identical to that of the wind tunnel experiments described in Section 2.1.2 and shown in Figure 3. The length of the sampling interval was 34 s in each case, but the initial 4 s of the simulations were considered part of the startup transient based on the visual assessment of the time series; therefore, the $t < 4$ s intervals were discarded. Further information about the time series is included in Table 1.

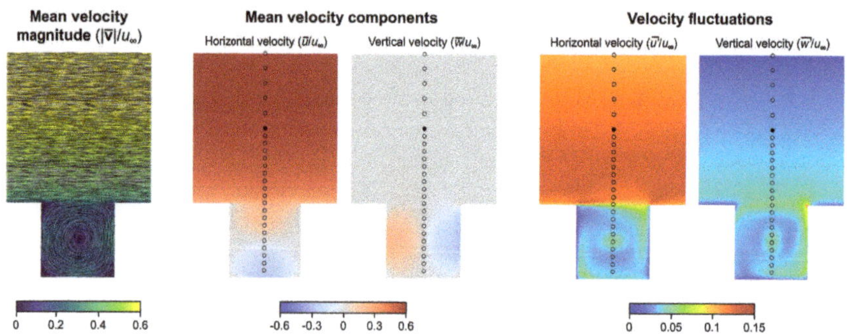

Figure 3. CFD results: time averaged streamlines plotted over the mean normalized velocity magnitude; mean velocity components (streamwise $u(t)$ and vertical $w(t)$); and the time-averages of the velocity fluctuations. The velocity statistics are averaged over time and the spanwise direction, i.e., the length of the street canyon, and are normalized by the free-stream velocity u_∞. (The sampling points used for recording the velocity time series are denoted by hollow markers. The filled marker denotes the center of the TWF propulsion at $z_0/H = 2.0$.)

2.2.5. General Description of the Flow Field

Figure 3 shows the velocity statistics. The streamlines plotted over the normalized mean velocity magnitude reveal that below roof height, a closed vortex structure (with an axis of rotation parallel to the y axis, rotating in the clockwise direction) is formed, called the canyon vortex. Above roof height, the velocity magnitude gradually increases in the boundary layer as we move away vertically from the buildings. As shown by the spatial distribution of the time-averaged velocity components, the gradual increment of the velocity magnitude comes from the increasing streamwise velocity ($\partial u/\partial z > 0$). It is not pictured in this figure, but due to the prismatic geometry and the perpendicular wind direction, the time-average of the spanwise velocity component $v(t)$ is zero in the entire computational domain. Importantly, the maximum normalized streamwise velocity is around 0.6, as the computational domain does not include the full depth of the atmospheric boundary layer, but uses an immersed boundary condition in the form of the TWF propulsion at $z_0/H = 2.0$. It can also be observed in Figure 3 that the mean of the vertical velocity components $w(t)$ are uniformly close to zero above roof height. Within the canyon, the formation of the clockwise-rotating canyon vortex can also be identified in the contour plots of the mean $\overline{u}$ and $\overline{w}$ velocity components.

It can also be observed in Figure 3 that above roof height, the streamwise velocity fluctuations are 3–4 times higher than the vertical ones, which is more than the 2:1 ratio expected in atmospheric boundary layer flows, and might be attributed to the fact that in the TWF method, the volume source terms act in the horizontal directions, but not in the vertical one; hence, the vertical velocity fluctuations recorded in the experiments are not directly imposed in the CFD model. The vertical fluctuations are further attenuated as we approach the top of the domain, where the applied symmetry boundary condition damps any fluctuations perpendicular to it. Below roof height, the magnitude of the streamwise and vertical velocity fluctuations—similarly to the time-averaged values—are close to one another. Distinct local maxima can be found at the vortex core near $z/H = 0.5$, and at the shear layer at $z/H = 1.0$, which is responsible for driving the cavity flow within the canyon.

3. Descriptive Statistics of the Experimental and Simulation Data

The basic statistics of the measured and simulated velocity components $u(t)$ and $w(t)$ at different locations are given in Table 2.

Table 2. Statistical properties of $u(t)$ and $w(t)$ from experimental (Exp.) and simulation (Sim.) data.

	$u(t)$											
	Min [m/s]		Mean [m/s]		Max [m/s]		σ[m/s]		Skewness [−]		Kurtosis [−]	
Location	Exp.	Sim.	Exp.	Sim.	Exp.	Sim.	Exp.	Sim.	Exp.	Sim.	Exp.	Sim.
$z/H = 0.1$	−3.588	−3.293	−1.574	−2.119	0.011	−1.124	0.531	0.261	−0.387	−0.109	0.320	0.332
$z/H = 0.5$	−1.862	−1.092	0.013	0.021	1.584	1.275	0.418	0.348	0.037	0.151	0.157	−0.102
$z/H = 1.0$	−0.131	1.088	2.494	2.138	6.744	5.850	0.992	0.425	0.716	1.350	0.183	3.930
$z/H = 2.0$	1.373	2.360	5.015	5.078	9.815	8.260	1.296	0.933	0.131	0.062	−0.577	−0.334
$z/H = 3.0$	2.877	3.587	7.008	5.676	10.66	8.086	1.222	0.666	−0.378	0.054	−0.205	−0.287
	$w(t)$											
	Min [m/s]		Mean [m/s]		Max [m/s]		σ[m/s]		Skewness [−]		Kurtosis [−]	
Location	Exp.	Sim.	Exp.	Sim.	Exp.	Sim.	Exp.	Sim.	Exp.	Sim.	Exp.	Sim.
$z/H = 0.1$	−2.176	−1.589	−0.027	0.014	1.707	1.370	0.474	0.318	−0.138	0.254	0.878	0.840
$z/H = 0.5$	−1.972	−1.210	−0.072	−0.041	2.029	0.943	0.664	0.311	0.067	−0.073	−0.442	0.052
$z/H = 1.0$	−3.329	−1.364	−0.097	0.014	3.538	1.125	0.721	0.322	−0.036	−0.364	0.762	−0.067
$z/H = 2.0$	−3.541	−0.933	−0.043	−0.014	5.237	1.438	1.074	0.267	0.055	0.408	−0.175	0.466
$z/H = 3.0$	−3.315	−0.185	0.124	0.001	3.956	0.189	0.960	0.030	−0.044	0.490	0.251	2.212

The mean value of the vertical velocity component $w(t)$ is close to $\overline{w} = 0$ for all heights, meaning that the data represents fluctuations rather than persistent up/down motions. Table 2 shows that the probability density histograms of $u(t)$ and $w(t)$ for both cases have mainly normal distribution (note that Table 2 contains the excess kurtosis values). The skewness and kurtosis values are similar to those given in Czernuszenko et al. [63]. Comparing the experiment and simulation the most significant difference can be seen at $z/H = 1.0$, where the distribution of the simulated $u(t)$ is more skewed towards smaller values and "skinnier" than for the experimental data. The mean of the simulated $u(t)$ at $z/H = 3.0$ is lower than that of the experimental data. Furthermore, the range of the measured $w(t)$ is bigger and the corresponding distribution is less skewed than for the simulated vertical velocity component.

The fluctuating components of the streamwise and vertical velocity, u' and w', respectively, are

$$u' = u(t) - \overline{u}, \quad w' = w(t) - \overline{w}. \tag{4}$$

where $\overline{u}$ and $\overline{w}$, i.e., the time-averaged velocity components over an observation period T_p are given by

$$\overline{u} = \frac{1}{T_p}\int_t^{t+T_p} u(\tau)\mathrm{d}\tau, \quad \overline{w} = \frac{1}{T_p}\int_t^{t+T_p} w(\tau)\mathrm{d}\tau. \tag{5}$$

The velocity fluctuations u' and w' can be analyzed by identifying the location of their sign changes (note that the $u' < 0 \leftrightarrow u' > 0$ and $w' < 0 \leftrightarrow w' > 0$ switches are treated separately). These sign changes are called *events*. The set of times $\{t_i\}$ at which these events occur provides a so-called random point process [64–66]. Other examples of random point processes are neuronal spike patterns (spike trains) and customer arrival times.
Let us define the residence times

$$T_i = t_i - t_{i-1}, \quad i > 1, \tag{6}$$

to express the duration of an event, i.e., the duration the system spends in state $u' > 0$, $u' < 0$, $w' > 0$ or $w' < 0$. The mean residence times T_{mean} as a function of height are shown in Figure 4.

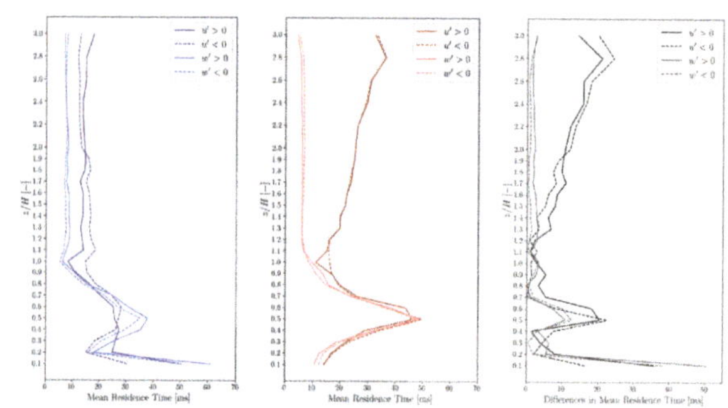

Figure 4. Residence time statistics of states $u' > 0$, $u' < 0$, $w' > 0$, $w' < 0$ in the experimental (**left**, blue) and simulation (**middle**, red) data. The differences (**right**, black) between the experimental and simulation data are also shown.

The mean residence times are between 5 and 60 ms for both the experimental and simulation data. However, the simulation data show high discrepancies in the residence times of the horizontal fluctuations u' compared to the experimental results.

In general, the values of T_{mean} for $w' < 0$ and $w' > 0$ are smaller than those for $u' < 0$ and $u' > 0$ for locations above the height of the building, showing that at these locations the fluctuation of the vertical velocity component $w(t)$ is more rapid than that of the horizontal velocity component $u(t)$. However, close to the ground (only for the experimental data) and in the vortex core the values of T_{mean} for the measured $w' < 0$ and $w' > 0$ are larger than those for $u' < 0$ and $u' > 0$.

The mean residence times of the vertical fluctuations w' show acceptable correspondence above roof height. Furthermore, the simulations yield mean residence times corresponding to the streamwise velocity substantially larger than the experimental observations, which can be attributed to the fact that the TWF control filters the streamwise velocity fluctuations above the frequency corresponding to the propulsion time scale ($\Delta t_0 = 0.0195$ s).

4. Statistics of "Quadrantified" Experimental and Simulation Data

The *quadrant method* distinguish four different states based on the signs of u' and w', such as

- Outward interaction ($Q1$): $u' > 0, w' \geq 0$;
- Ejection ($Q2$): $u' \leq 0, w' > 0$;
- Inward interaction ($Q3$): $u' < 0, w' \leq 0$;
- Sweep ($Q4$): $u' \geq 0, w' < 0$.

Note that as the mean of the vertical velocity components are close to zero ($\overline{w} = 0$) in all 25 gauging points (see Figure 3), it can be stated that $Q1$ and $Q2$ are associated with upward motion and $Q3$ and $Q4$ are associated with downward motion at those specific locations. The visualization of the four quadrants is shown in Figure 5.

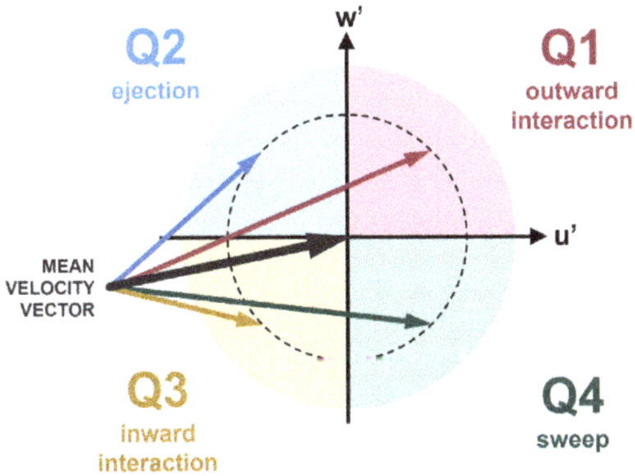

Figure 5. Visualization of the four quadrants.

We now apply the quadrant method to the experimental and CFD data. To the joint velocity fluctuation pair $\mathbf{v} = (u', w')$, we assign a quadrant index according to

$$Q(\mathbf{v}) = \begin{cases} 1, & u' > 0, \quad w' \geq 0 \\ 2, & u' \leq 0, \quad w' > 0 \\ 3, & u' < 0, \quad w' \leq 0 \\ 4, & u' \geq 0, \quad w' < 0 \end{cases}. \tag{7}$$

These 4 numbers can be thought as symbols/letters of the alphabet $A = \{1, 2, 3, 4\}$ with 4 letters. The joint velocity fluctuation time series is transformed to a sequence of time-symbol pairs $\{(t_i, s_i) \mid s_i \in A,\ s_i \neq s_{i+1} \text{ and } i = 0, 1, 2, \ldots\}$, denoting the occurrence of the distinct quadrant events at time t_i. This is a *marked point process*, an extension of the simple point process described in Section 3. In our quadrant-event model, outward interaction ($u' > 0$, $w' \geq 0$), ejection ($u' \leq 0$, $w' > 0$), inward interaction ($u' < 0$, $w' \leq 0$), and sweep ($u' \geq 0$, $w' < 0$) are the discrete events. The residence times T_i's express the average duration that the system spends in a given quadrant.

The comparison of the mean residence times for all 4 events are shown in Figure 6. The mean residence time are 2 ms $< T_{mean} <$ 30 ms for both cases. In case of the simulation, the mean residence time has a higher peak at $z/H = 0.5$ than for the experimental data, which can be explained by the fact that in the simulation the vortex core is substantially more enclosed than what was observed in the experiment. Corresponding evidence can be found in the concentration plots shown in Papp et al. [53].

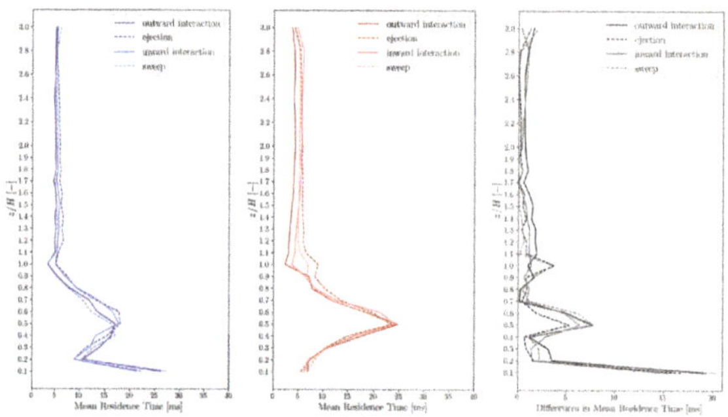

Figure 6. Mean residence times of the "quadrantified" experimental (**left**, blue) and simulation (**middle**, red) data. The differences (**right**, black) between the experimental and simulation data are also shown. Outward interaction: $u' > 0, w' \geq 0$; ejection: $u' \leq 0, w' > 0$; inward interaction: $u' < 0, w' \leq 0$; sweep: $u' \geq 0, w' < 0$.

Figure 7 shows the distribution (relative occurrence) of ejection ($u' \leq 0, w' > 0$) residence times for both the experimental and simulation data at heights $z/H = 0.1, 0.5, 1.0$ and 2.0. The relative occurrences of residence time intervals up to 20 ms are plotted. Bogard et al. [67] reported that the time intervals between ejection events follow a distribution that is approximately exponential. However, several studies have shown that the durations between mean-level crossings in turbulent signals are better described by a lognormal distribution [68–71]. Based on this observation, we fit the following lognormal probability density function (PDF) to the distribution of relative occurrences of residence times T_{res}

$$P(T_{res}) = \frac{1}{z_2 T_{res} \sqrt{2\pi}} \exp\left(-\frac{1}{2} \left(\frac{\ln T_{res} - z_1}{z_2} \right)^2 \right). \tag{8}$$

where $P(T_{\text{res}})$ denotes the relative frequency associated with residence times T_{res} across the four quadrant event types (outward interaction, ejection, inward interaction, sweep), and z_1 and z_2 are the parameters of the lognormal distribution. The fitted values of these parameters are presented in Table 3.

Table 3. Lognormal fit parameters for residence time distributions belonging to ejection ($u' \leq 0$, $w' > 0$) at different heights.

	$u' \leq 0, w' > 0$			
	z_1		z_2	
Location	Exp.	Sim.	Exp.	Sim.
$z/H = 0.1$	1.869	1.640	0.668	0.769
$z/H = 0.5$	1.908	2.311	0.878	0.539
$z/H = 1.0$	1.560	1.757	1.085	0.798
$z/H = 2.0$	1.560	1.575	1.085	0.814

The fitted lognormal curves are also shown in Figure 7 with their coefficient of determination [72]

$$R^2 = 1 - \frac{\sum_i (y_i - P_i)^2}{\sum_i (y_i - \overline{y})^2}. \tag{9}$$

where y_i values are the calculated relative occurrence values of residence time T_{res} and $\overline{y}$ is the mean of the relative occurrence values y_i's. The predicted relative occurrence values P_i are obtained from the fitted lognormal model described in Equation (8). It is important to note that the coefficient of determination R^2 can occasionally be negative. This occurs when the mean value $\overline{y}$ provides a better fit to the data than the nonlinear model predictions P_i, as indicated by Equation (9).

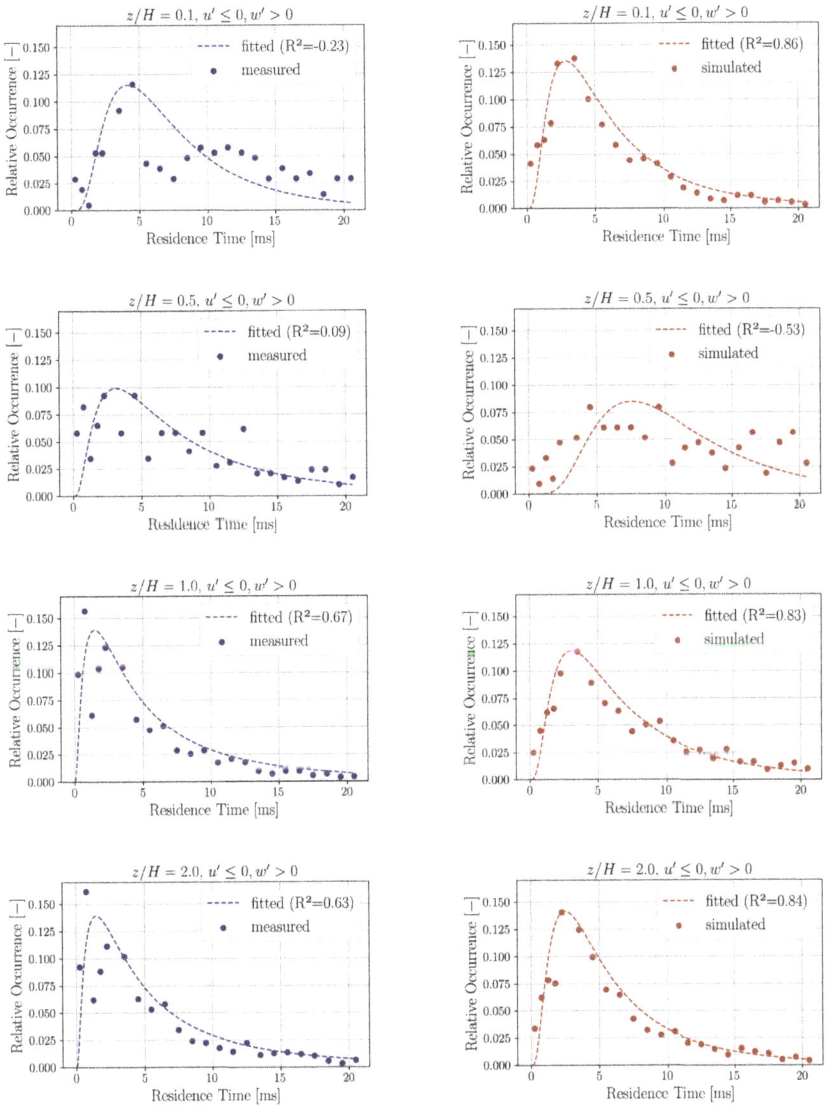

Figure 7. Relative occurrence of ejection ($u' \leq 0, w' > 0$) in the experimental (**left**, blue) and the simulation (**right**, red) data measured at different z/H heights.

The distributions for the other quadrants are qualitatively similar to the ones shown in Figure 7 for ejection. The relative occurrence results show fairly good qualitative agreement between the experimental and simulation data in all analyzed measurement points.

The relative occurrences of residence times for a quadrant transitioning to another specific quadrant is shown in Figure 8. The residence time distributions corresponding to one quadrant transitioning to another specific quadrant obtained here are comparable with the ones obtained in [43].

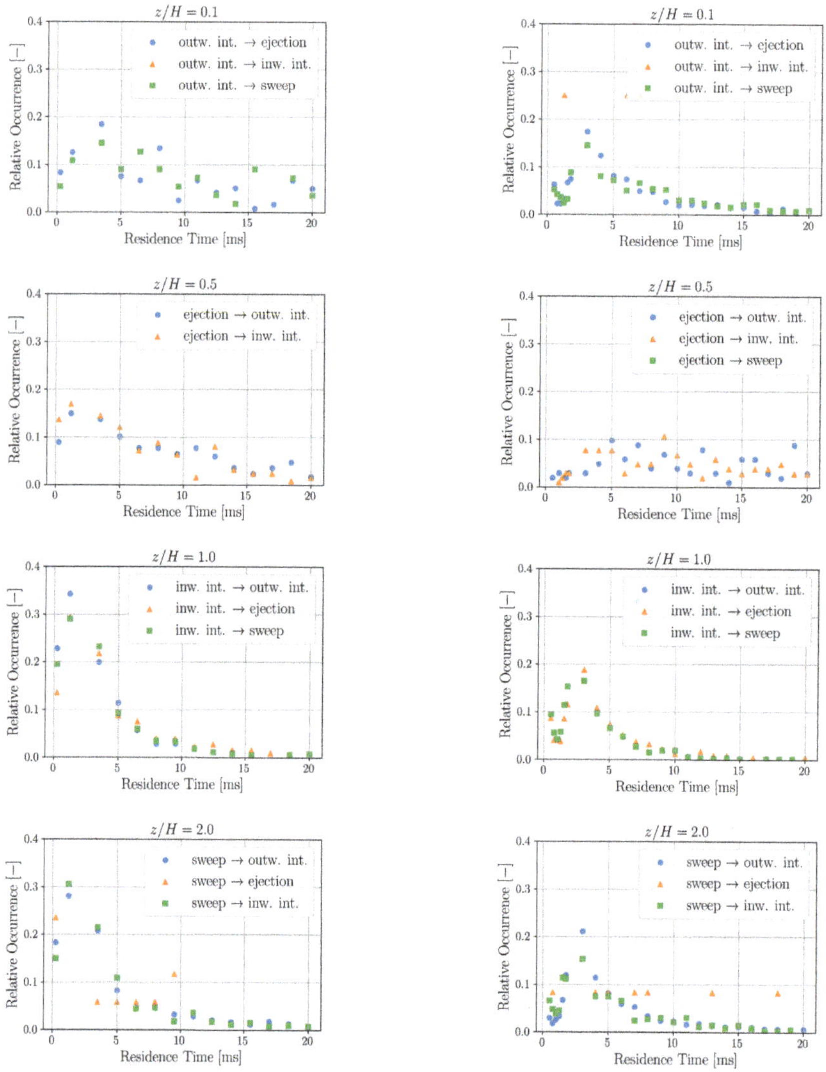

Figure 8. Relative occurrence of one quadrant transitioning into another specific quadrant in the experimental (**left**) and simulation (**right**) data measured at different z/H heights.

Finally, let us utilize the fine spatial resolution of the CFD simulations to gain further insight into the flow field. Figure 9 shows the quadrant statistics based on the CFD simulations, averaged in time and in the spanwise direction, i.e., over the length of the street canyon. The presence of the shear layer at roof height is clearly visible in the quadrant statistics: at $z/H = 1.0$, the normalized total residence time (probability) corresponding to $Q2$ (ejection) peaks at a value of above 0.4. Correspondingly, the probabilities of the other quadrants must decrease locally. Interestingly, $Q1$ (outward interaction) shows a probability below 0.1 in the shear layer. This, combined with the fact that the mean vertical velocity is around zero in the shear layer ($\overline{w} = 0$, see Figure 3), means that while downward motion ($w' < 0$) is equally distributed over $Q3$ (inward interaction) and $Q4$ (sweep), upward motion ($w' > 0$) dominantly happens in $Q2$ (ejection). In other words, air parcels (or transported pollutants) are likely to escape the canyon when the streamwise velocity component in the shear layer is lower than its time-average.

Furthermore, let us observe the path of a fluid parcel through the clockwise-rotating canyon vortex. Importantly, due to the fact that the mean velocity vector can significantly deviate from the positive x direction below roof height, the meaning of the quadrants must be assessed with special attention. Around the 3 o'clock position in the downward-moving fluid ($\overline{u} \approx 0$, $\overline{w} < 0$), $Q1$ is dominant (to the detriment of $Q2$), meaning that the most probable state of the current velocity vector is of a smaller magnitude than the mean velocity vector, and it is pointing outwards. Further ahead, around the 6 o'clock position, i.e., in the street-level backflow ($\overline{u} < 0$, $\overline{w} \approx 0$), $Q3$ is dominant, and in the upward-moving fluid ($\overline{u} \approx 0$, $\overline{w} > 0$) around the 9 o'clock position, the most probable quadrants are $Q3$ and $Q2$, respectively. These findings indicate that outward drift is the most likely state in these regions too.

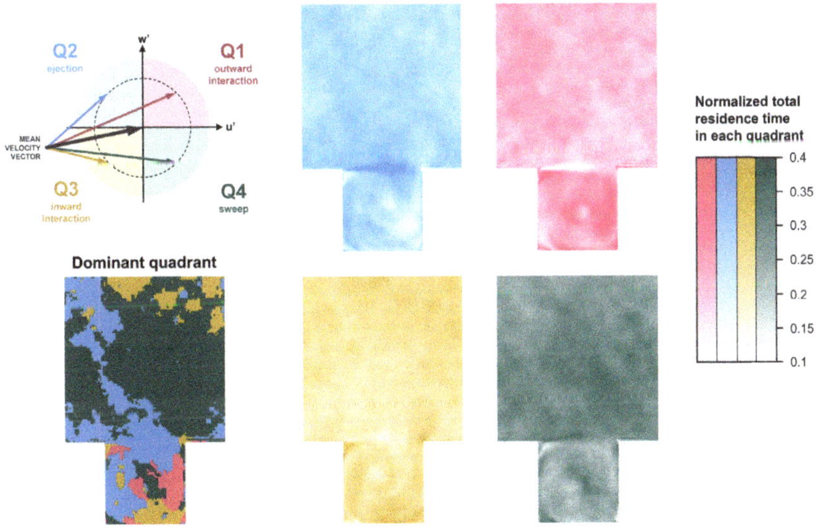

Figure 9. Quadrant statistics based on the CFD results averaged over time and the spanwise direction. The total residence time in each quadrant is normalized by the total residence time in all quadrants combined; therefore, the normalized value must be between 0 and 1. The dominant quadrant is the one with the highest total residence time at each location.

5. Word Statistics and Entropy

Following the approach described by Daw et al. [36], symbolic sequences S, composed of individual symbols s_i, are transformed into sequences of words of fixed length L. This transformation is accomplished by sliding a window of length L across the original sequence.

In formal terms, the sequence S is transformed into a multiset of words W^L as follows:

$$W^L = \{w_i\}^{|S|-L+1} = \left\{w_1, w_2, \ldots, w_{|S|-L+1}\right\} = \left\{s_1 s_2 \ldots s_L, s_2 s_3 \ldots s_{L+1}, \ldots, s_{|S|-L+1} s_{|S|-L+2} \ldots s_{|S|}\right\}. \quad (10)$$

The word list W^L has $n_L = \left|W^L\right| = |S| - L + 1$ elements and is a multiset, i.e., it can contain identical words multiple times. We denote with $\widehat{W}^L$ the set of unique words in W^L (obtained by removing duplicate words W^L), with cardinality denoted by $\hat{n}_L = \left|\widehat{W}^L\right|$.

In Figure 10, the normalized quantities (dimensionless word number) $\hat{n}_L/n_L$ are shown for different heights. For word lengths approximately $L = 12$, the number of unique words $\hat{n}_L$ in the measured series is approximately the half of the length n_L of the measured word lists. This is different for the simulation data, especially for heights $z/H = 2.0$ and 3.0 in the simulation data. We attribute this to the driving force being applied at height $z/H = 2.0$ and the slip symmetry boundary condition at $z/H = 3.0$.

For comparison, the dimensionless word number for both infinite (theoretically possible) and finite random sequences is also shown in Figure 10. One can see that the dimensionless word number curve of the measurement at $z/H = 0.1$, i.e., close to the ground in the canyon, is in-between the curve of the random sequence and the sequences corresponding to the other measurement heights. The random sequence represents a fully disordered, uncorrelated flow; hence, the data above $z/H = 0.1$ may show the presence of coherent structures.

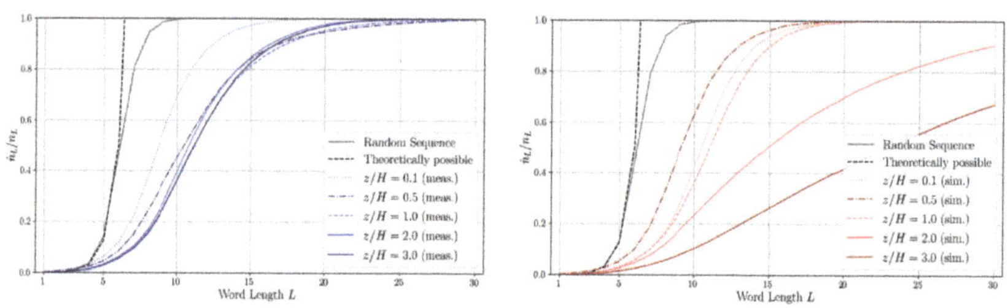

Figure 10. Dimensionless word number $\hat{n}_L/n_L$ calculated from the number of unique words $\hat{n}_L$ found in the word list W_L of measured (**left**, blue), simulated (**right**, red), and random generated symbolic sequences for different word length L at different measurement heights. The number of words in word list W_L is n_L.

A comparison of the symbolic sequences up to word length $L = 3$ reveals that the number of unique words increases at a nearly identical rate for both measured and randomly generated sequences. However, beyond $L = 3$, the number of unique words in the random sequences grows significantly faster. Due to the finite length of the sequences (an infinite sequence is also shown for comparison), the growth of unique words is naturally bounded by the total number of possible words n_L. This saturation occurs at L = 9–10 for the random sequence, between L = 16–30 for the measured sequence and around L = 20–50 for the simulated sequences.

Figure 11 shows at which word length value L is the inflection point of the experimental and dimensionless word number curves. We can state that the inflection points for the experimental and simulation sequences show a good agreement setting aside a few "jumps" in the inflection points of the simulation sequences. We believe the shift of inflection point outside the canyon means that the symbolic sequence—exhibiting a more ordered flow—is diverging from the reference random sequence, i.e., the number of unique symbolic words increases more slowly with the word length.

The lower number of unique words in the measured sequence, relative to the random counterpart, is attributable to the recurrence of specific words, indicating that some word patterns appear with higher frequency. For instance, the relative frequency distribution of words of length $L = 4$ is illustrated in Figures 12–15.

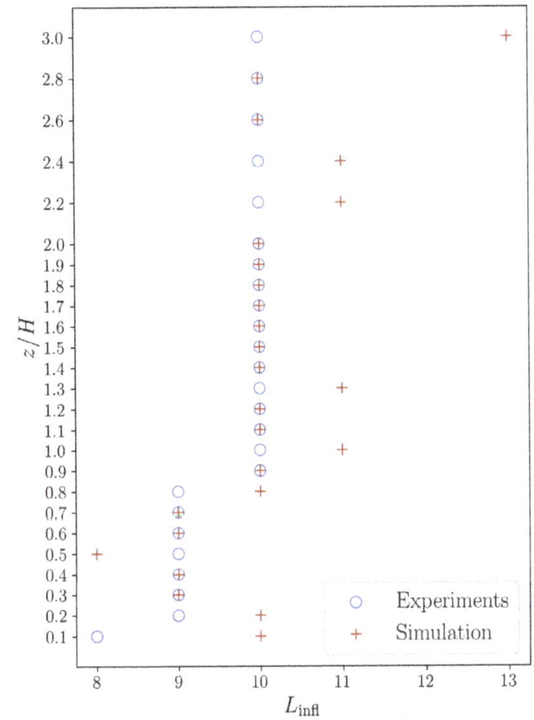

Figure 11. Word length positions of inflection points (blue—experiment, red—simulation) of the dimensional word number curves (see Figure 10).

In the canyon at $z/H = 0.5$, the four most common words are 1212/2121 and 3434/4343 in the experimental sequence and 1432/3214 and 4321/2143 in the simulation sequence. Outside the canyon, for both experimental and simulation data, the four most common words are 1414/4141 and 2323/3232.

At longer word lengths, the most frequently observed words are in which symbol pairs 12/21 and 34/43 (within the canyon) and 14/41 and 23/32 (outside the canyon) follow one another. These alternating patterns indicate that pairs of outward interaction–ejection events ($1 \leftrightarrow 2$) and inward interaction–sweep events ($3 \leftrightarrow 4$) commonly follow one another inside the canyon, possibly due to the influence of the no-slip boundary condition at the canyon bottom.

In contrast, outside the canyon, the repetition of symbol pairs 1, 4 and 2, 3 suggests a tendency for outward interaction–sweep and ejection–inward interaction pairs to alternate. This behavior implies that the reduced quadrantified signal often becomes trapped in cycles involving rapid switching between two dominant turbulent event states.

Note that moving away from the bottom of the canyon, the most common unique words represent almost $40 - 50\%$ of the word list. The word frequency distributions show a somewhat central symmetry that suggests a self-similar (multi-fractal) nature of the data [73].

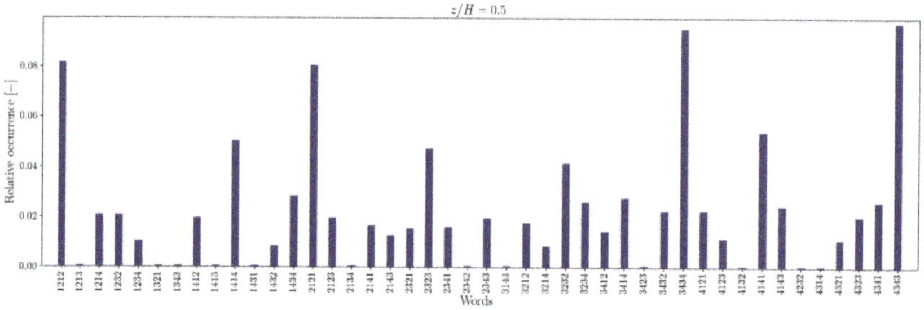

Figure 12. Relative occurrence of unique words of length $L = 4$ in the experimental data at measurement height $z/H = 0.5$.

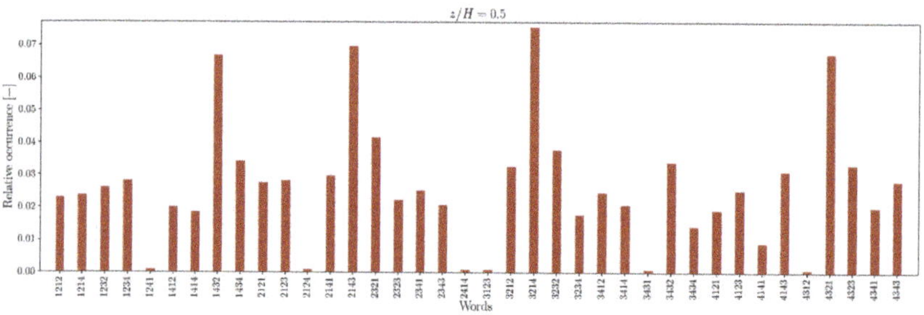

Figure 13. Relative occurrence of unique words of length $L = 4$ in the simulation data at height $z/H = 0.5$.

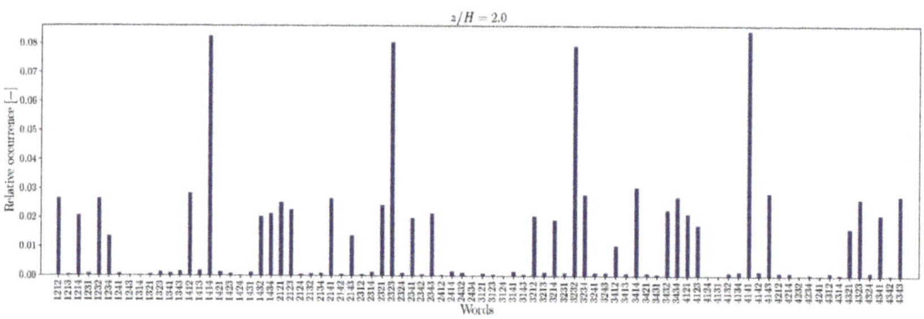

Figure 14. Relative occurrence of unique words of length $L = 4$ in the experimental data at measurement height $z/H = 2.0$.

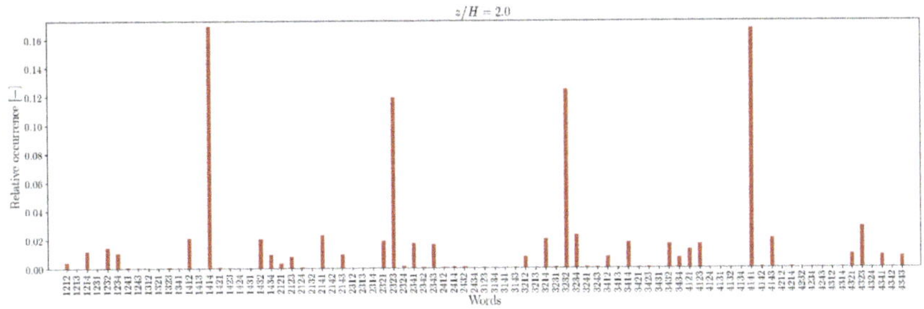

Figure 15. Relative occurrence of unique words of length $L = 4$ in the simulation data at height $z/H = 2.0$.

We characterize the degree of regularity and information content within the measured reduced symbolic sequences using the normalized entropy H^L_{norm}, a modified form of Shannon entropy as introduced by Finney et al. [39]. For a given word length L, it is defined as:

$$H^L_{\text{norm}} = -\frac{1}{\ln(\hat{n}_L)} \sum_{1}^{n_L} P(w_i) \ln P(w_i). \tag{11}$$

where $P(w_i)$ represents the relative frequency of the word w_i in the word list W^L. Notably, this formulation parallels the concept used in Kolmogorov–Sinai entropy for evaluating the information rate in dynamical systems. A review on entropy in various fields of science is given in [74].

The normalized entropy values for various word lengths and heights are plotted in Figure 16.

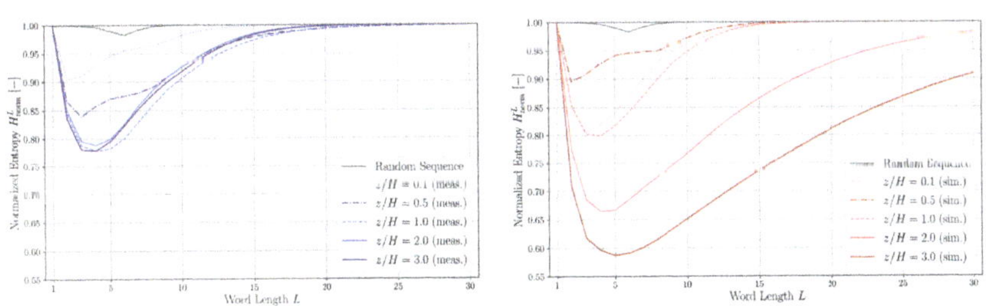

Figure 16. Normalized entropy H^L_{norm} as a function of word length L for experimental (**left**, blue) data and simulation (**right**, red) data at different measurement heights and for a random sequence.

The entropy curves exhibit trends consistent with those reported by Finney et al. [39]: a significant decrease in entropy occurs for the shortest word lengths, reaching a minimum around L = 2–5, followed by a gradual approach toward $H^L_{\text{norm}} = 1$ at larger L.

Higher entropy values indicate greater disorder and randomness, whereas lower entropy suggests more correlated, deterministic behavior, the dip of entropy suggest the presence of structured patterns or intermittency. For the random sequence, H^L_{norm} remains close to 1 across all word lengths, reflecting the absence of meaningful structure. The small dip near $L = 6$ is attributed to the finite length of the random dataset.

The smallest entropy value is obtained at $z/H = 1.0, 3.0$ for the measurement data and at $z/H = 2.0, 3.0$ for the simulation data. The entropies of the experimental and simulation sequences at $z/H = 1.0$ show good agreement; at $z/H = 0.5$, the flow is more disordered in the simulation data, and at $z/H = 2.0$ and 3.0, one can again see the effect of driving force and slip symmetry boundary condition the makes the flow more ordered than under real-world conditions.

6. Surrogate Series

In Section 5, we compare the normalized entropy H_{norm}^{L} derived from symbolic analysis of measured and simulated velocity fluctuation data, as well as from a random sequence. To assess the variation of entropy as a function of word length across different types of symbolic sequences, we generated the following artificial sequences, all having the same length as the measured dataset:

- **Noisy periodic sequence:** A periodic symbolic pattern constructed by repeating the block $\{1,3,1,3,2,3,1,4\}$, with a subset of symbols randomly replaced at a controlled rate. Replacements are constrained to differ from both the preceding and following symbols.
- **Markov chain-generated sequences:** Symbolic sequences generated using Markov models of orders up to five, constructed from transition probabilities estimated from both measured and simulated data.

To construct these Markov-based surrogate sequences, we define transition probabilities between quadrant states. Transitions between the four quadrants are categorized as follows: *stable* (remaining in the same quadrant), *cross* (transitions such as 1↔3, 2↔4), and *marginal* (transitions such as 1↔2, 3↔4). Kalmár-Nagy and Varga [43] observed that stable transitions have the highest probability. This finding is in agreement with earlier observations by Keshavarzi et al. [75,76], who identified similar patterns of transition preference near different types of boundaries, including plain beds, ripple beds, and obstacles such as bridge piers.

When the time stamps of events are discarded from the marked point process, the result is a sequence of symbolic words of length L, each composed of four-symbol combinations $s_i s_{i+1} \ldots s_{i+L-1}$. A key constraint on these sequences is that no two adjacent symbols are identical, i.e., $s_i \neq s_{i+1}$.

If the symbolic sequence arises from a Markovian process, transitions between symbols can be characterized by transition probabilities $p_{i\to j}$ (denoted simply as p_{ij}). Under this assumption, the symbolic quadrant sequence S forms a first-order Markov chain with the transition probability matrix:

$$M^1 = \begin{pmatrix} p_{11} & p_{12} & p_{13} & p_{14} \\ p_{21} & p_{22} & p_{23} & p_{24} \\ p_{31} & p_{32} & p_{33} & p_{34} \\ p_{41} & p_{42} & p_{43} & p_{44} \end{pmatrix}. \tag{12}$$

where p_{ij} denotes the probability of transitioning from quadrant i to quadrant j. For our quadrant-based symbolic process, self-transitions are not permitted, so the diagonal elements $p_{11}, p_{22}, p_{33}, p_{44}$ are all zero. These stable movements instead correspond to the residence times spent within each state.

To generalize the size of the transition matrix for a Markov model of order m, we consider an alphabet A of λ symbols. The number of valid symbolic words of length L, where adjacent symbols differ, is given by:

$$|W_{\lambda}^{\mathrm{pos}}| = \lambda(\lambda-1)^{L-1}. \tag{13}$$

In our case with $\lambda = 4$, this evaluates to $|W_{4}^{\mathrm{pos}}| = 4 \cdot 3^{L-1}$.

The elements of the transition probability matrix M^m for a Markov chain of order $m = L - 1$ are defined as follows:

$$M^m_{s_{j-L+1} \cdots s_{j-2} s_{j-1} \to s_j} = p\big(s_j = i | s_{j-L+1} \cdots s_{j-2} s_{j-1}\big), \quad j = L-1, \ldots, |W^{\text{pos}}_{\lambda}|, \quad i = 1, \ldots, \lambda. \tag{14}$$

This expression specifies the conditional probability that a symbol $s_j = i$ follows the prefix sequence $s_{j-L+1} \cdots s_{j-1}$ within a valid word $w_j \in W^{\text{pos}} = \{w_i^{\text{pos}}\}$.

Figure 17 presents a comparison of normalized entropy values H^L_{norm} as a function of word length L for both the measured and artificially generated noisy periodic sequences. For the purely periodic sequence, H^L_{norm} remains close to 1 across nearly all word lengths. This behavior is attributed to the nearly uniform distribution of words in the partitioned word list W^L, which mimics the properties of a random sequence where all unique words have approximately equal probability of occurrence.

Entropy curves for several noisy periodic sequences, derived by corrupting the base periodic pattern with noise levels of 35%, 50%, and 90% (relative to the length of the measured sequence) and 5%, 15%, and 35% (relative to the length of the simulated sequence), are also shown in Figure 17. These entropy profiles exhibit characteristics similar to those observed in measured and simulated data: they are unimodal and gradually converge toward $H^L_{\text{norm}} = 1$. Additionally, the location of the entropy minimum shifts to higher word lengths with increasing noise. However, the entropy minima for the experimental and simulated datasets tend to occur at shorter word lengths compared to those of the noisy periodic sequences.

The transition probability matrices M^m were computed up to order $m = 5$ (see Equation (14)) using experimental and simulation data at heights $z/H = 0.5$ and $z/H = 2.0$. These matrices were then used to generate surrogate symbolic sequences. The corresponding normalized entropy values H^L_{norm} for both the original and the surrogate sequences are compared in Figure 18.

In general, surrogate sequences generated using fifth-order Markov models provide the closest match to the entropy curves of the measured and simulated data. An exception arises at $z/H = 0.5$ in the experimental dataset, where the characteristic dip in the entropy curve is not replicated even by the fifth-order Markov chain. This discrepancy suggests that the symbolic quadrant sequence possesses a level of temporal correlation or structural complexity that exceeds what can be captured by a fifth-order Markov process.

Such behavior points to the influence of temporally coherent structures within the flow field, and supports the interpretation that the quadrant series retains memory effects over longer word lengths, indicative of organized turbulence phenomena.

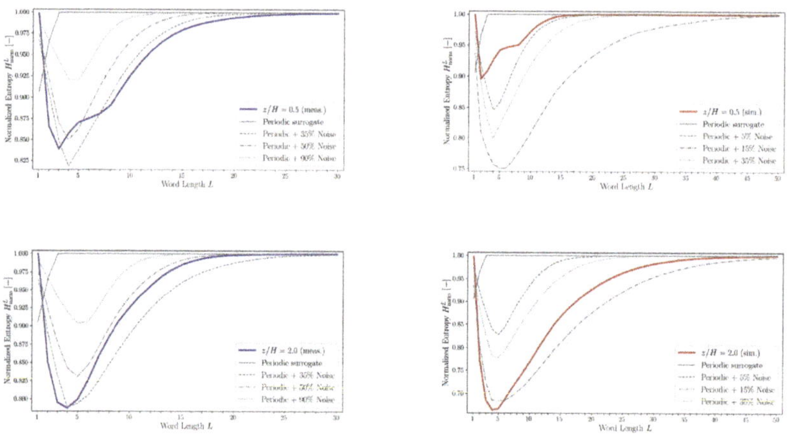

Figure 17. Comparisons of normalized entropy H^L_{norm} belonging to the experimental (blue)/simulation (red) data and the artificial noisy periodic sequences as the function of word length L.

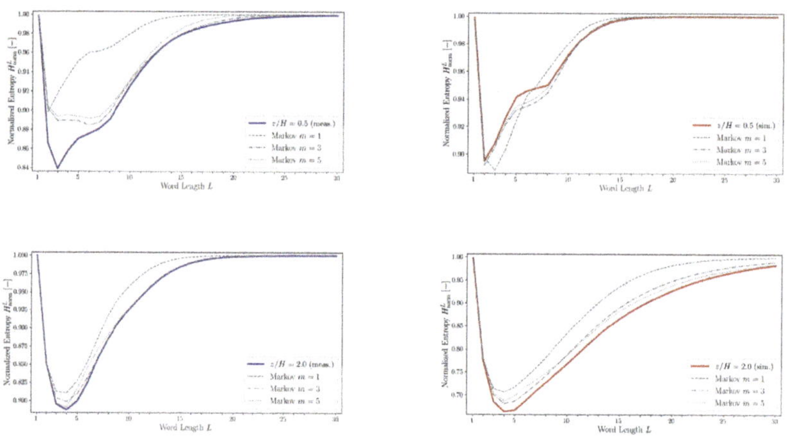

Figure 18. Comparisons of normalized entropy H^L_{norm} belonging to the experimental (blue)/simulation (red) data and the artificial sequences generated using Markov chains (transition probabilities obtained from the experimental/simulation data).

7. 2D Distributions of Velocity Fluctuations

Figure 19 shows the velocity fluctuation pairs $(u'_{\text{norm}}, w'_{\text{norm}})$ normalized with the free stream velocity $u_\infty = 9$ m/s colored by the probability density distribution of the fluctuation pairs at heights $z/H = 0.5$ and 2.0. The probability distributions were calculated by the Gaussian kernel-density estimation utilizing a 100×100 grid over the point cloud range. The probability distributions of the point clouds are colored by point density.

Comparing the probability distribution in Figure 19 to the relative occurrence of words in Figures 12–15, one can see that for z/H, the "horizontal" transitions $1 \leftrightarrow 2$ and $3 \leftrightarrow 4$ are really more frequent at height $z/H = 5$ in the experimental data. However, for larger heights like $z/H = 2.0$, the transitions $1 \leftrightarrow 4$ and $2 \leftrightarrow 3$ happen more. This means that while the fluctuations may display isotropy at some point, at the investigated heights, there is a preferential orientation of the velocity fluctuations. A comparable pattern of anisotropy was reported in open-channel flow over a rough bed by [20].

Note that at $z/H = 2.0$ the distribution of the simulation data shows a remarkably flat ellipse compared to the experimental result. This also highlights that there is a noticeable deficit in the vertical fluctuations above roof height, a finding consistent with the results shown in Figure 3, and is a shortcoming of the applied model (also shown in Figure 20).

For all velocity fluctuation pairs $(u'_{\mathrm{norm}}, w'_{\mathrm{norm}})$ the moment of inertia matrices,

$$I = \begin{pmatrix} \sum_1^N w'^2_{\mathrm{norm}} & \sum_1^N u'_{\mathrm{norm}} w'_{\mathrm{norm}} \\ \sum_1^N u'_{\mathrm{norm}} w'_{\mathrm{norm}} & \sum_1^N u'^2_{\mathrm{norm}} \end{pmatrix} \tag{15}$$

determines the principal axis (dashed straight lines in Figure 20) of the point clouds. The angles between the u' axis and the principal axis relative to u' are $\alpha_{0.5} = 0.24°$ and $\alpha_{2.0} = -13.85°$ for the experimental data and $\alpha_{0.5} = -40.20°$ and $\alpha_{2.0} = -1.43°$ for the simulation data. Obviously, the calculated angles for both cases are different, showing the difference between a vortex core at $z/H = 0.5$ in the CFD model and the measurement, as well as the effect of driving force at $z/H = 2.0$.

From the inertia matrices, the radii of gyration R_g were determined, too, for each measurement and simulation height (represented by the dashed circles in Figure 20). The radii of gyration values show better agreement for the experimental and simulation data.

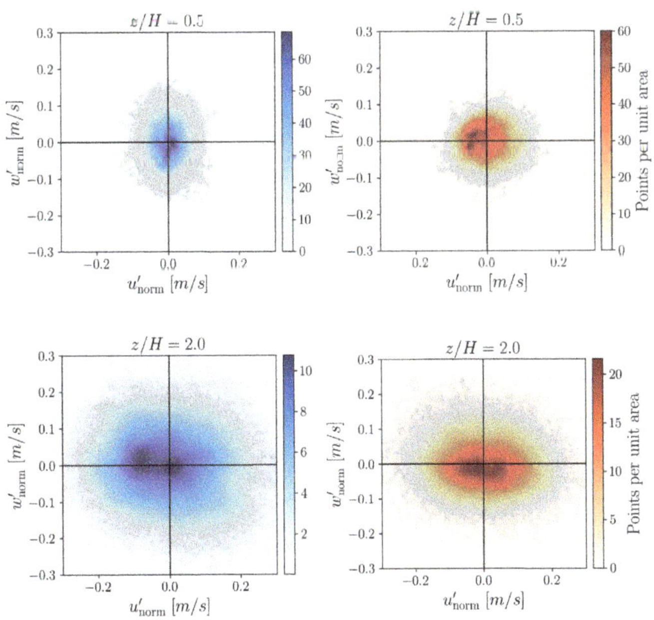

Figure 19. Velocity fluctuation pairs $(u'_{\mathrm{norm}}, w'_{\mathrm{norm}})$ and their probability density distribution of the fluctuation pairs at heights $z/H = 0.1$ (**first row**) and $z/H = 2.0$ (**second row**). The experimental data on the left are denoted by blue and the simulation data on the right with red.

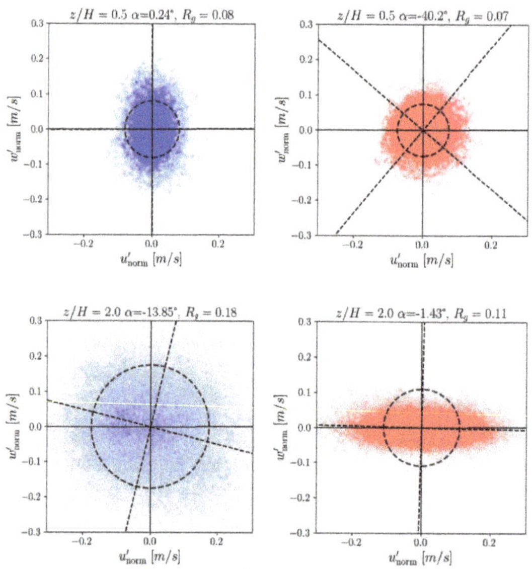

Figure 20. Velocity fluctuation pairs $(u'_{\text{norm}}, w'_{\text{norm}})$ and their probability density distribution of the fluctuation pairs at heights $z/H = 0.1$ (**first row**) and $z/H = 2.0$ (**second row**). The experimental data on the left are denoted by blue and the simulation data on the right with red. The angle between the vertical axis and the principal axis of the point cloud is denoted by α. The values of the radii of gyration R_g are also given.

Since the proper partitioning for the use of symbolic dynamics is still an open question [77], we also investigated the influence of coordinate rotation on the entropy. We considered the coordinate system (u'^*, w'^*) generated from the original $(u'_{\text{norm}}, w'_{\text{norm}})$ system by a counterclockwise rotation by angle θ as

$$\begin{pmatrix} u'^* \\ w'^* \end{pmatrix} = \mathbf{R}(\theta) \begin{pmatrix} u' \\ w' \end{pmatrix}, \tag{16}$$

$$\mathbf{R}(\theta) = \begin{pmatrix} \cos\theta & \sin\theta \\ -\sin\theta & \cos\theta \end{pmatrix}. \tag{17}$$

For a given rotation angle θ, the rotation (16) is applied to each pair of normalized velocity fluctuations $(u'_{\text{norm}}, w'_{\text{norm}})$. The resulting rotated fluctuation components (u'^*, w'^*) are then classified into symbolic sequences through quadrant analysis, yielding the series $\{Q(u'^*, w'^*)\}$.

Using this procedure, the normalized entropy H^4_{norm} (see Equation (11)) was computed as a function of the rotation angle θ. Figure 21 displays the resulting entropy profiles for measurement heights $z/H = 0.1$ and $z/H = 2.0$, with angular resolution $\Delta\theta = 1.5°$.

As expected, the function $H^4_{\text{norm}}(\theta)$ exhibits $\pi/2$-periodicity because a $\pm 90°$ rotation simply permutes the quadrant labels. Kalmár-Nagy and Varga [43] observed that the locations of entropy extrema with respect to θ are invariant under changes in word length. They further reported that the rotation angle $\theta_{\min}$, corresponding to the minimum of the smoothed entropy curve (indicated by vertical dashed lines), tends to align closely with the principal axis directions (marked by vertical solid lines). However, in our case, this holds only outside the canyon region.

The hypothesis of Kalmár-Nagy and Varga was that the observed entropy minimum results from the most frequent transitions. In the original coordinate system, these transitions primarily cross the u'_{norm} axis. However, transitions more frequently align with the principal axis, so when the system is rotated accordingly, the number of transitions increases, leading to lower entropy values. This interpretation, however, does not hold at height $z/H = 0.1$, as evident from Figure 21.

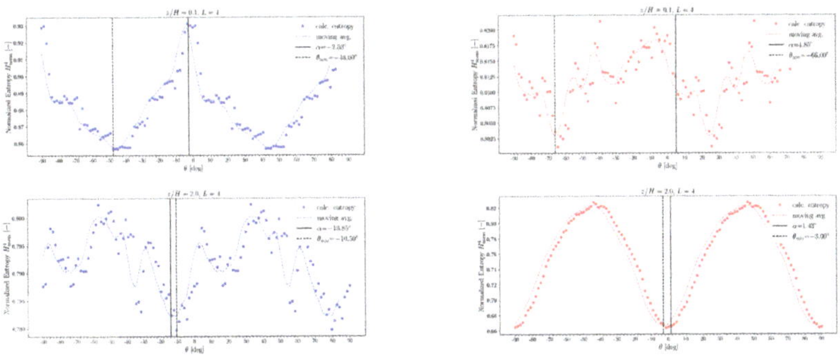

Figure 21. H^4_{norm} normalized entropy values for word length $L = 4$ at heights $z/H = 0.1$ and 2.0. On the left, blue figures correspond to the experimental data, while on the right, red figures correspond to the simulation data.

8. Conclusions

We analyzed and compared time series data from wind tunnel measurement and CFD simulation of flows over street canyons through the use of the quadrant method, calculating entropies, Markov matrices and generating surrogate series.

We studied the idea that coherent structures in the atmospheric boundary layer manifest as "almost periodic temporal patterns", characterized by the imperfect repetition of specific turbulent events or states. The time series of fluctuating velocities were converted into symbolic sequences using the quadrant method and treated as marked point processes. Analysis of the residence times—defined as the duration spent in each quadrant state—revealed distributions consistent with a lognormal form. To evaluate the information content of the symbolic sequences, normalized entropy was calculated across a range of word lengths. The resulting entropy profiles and the number of unique symbolic words were then compared to those derived from random sequences. Interestingly, artificially generated noisy periodic sequences produced entropy trends that qualitatively matched those of the experimental data.

Furthermore, surrogate sequences were created using Markov processes (orders 1 through 5). These surrogates reproduced entropy distributions close to those of the measured and simulated series, supporting the suitability of higher-order Markov models for characterizing the observed symbolic dynamics. Analysis of symbolic word histograms indicated that much of the information content in both measured and simulated sequences arises from the elevated frequency of words composed of alternating pairs of symbols.

Our investigation showed that simulation results are in qualitative agreement with measurements in several aspects, including the general trends observed in the residence time distributions of quadrants, the inflection points of dimensionless word number curves, as well as word frequencies. The vortex structures within the canyon, as the 2D distributions of the velocity fluctuations show, are well captured, demonstrating the capability of the model to reproduce the dominant flow patterns. It should be noted that some discrepancies exist, such as the differences in the mean residence time of the states above roof level ($z/H > 1.0$) and in the entropy minima. Another discrepancy is the underestimation of the vertical velocity fluctuation (w') magnitude, but it is a known imperfection of the simulation model used in this work. Note that an important advantage of the CFD model used in the present work is that it provides a valuable tool for analyzing the entire flow field across the computational domain, offering insight beyond the discrete measurement points.

Author Contributions: Conceptualization, T.K.-N.; methodology, T.K.-N.; software, C.Á.H. and B.P.; validation, C.Á.H. and B.P.; formal analysis, C.Á.H.; investigation, B.P.; resources, B.P. and T.K.-N.; data curation, B.P.; writing—original draft preparation, C.Á.H.; writing—review and editing, B.P. and T.K.-N.; visualization, C.Á.H. and B.P.; supervision, T.K.-N.; project administration, T.K.-N.; funding acquisition, C.Á.H., T.K.-N. and B.P. All authors have read and agreed to the published version of the manuscript.

Funding: The research reported in this paper is part of project no. TKP-6-6/PALY-2021. Project no. TKP-6-6/PALY-2021 has been implemented with the support provided by the Ministry of Culture and Innovation of Hungary from the National Research, Development and Innovation Fund, financed under the TKP2021-NVA funding scheme. This work has been supported by the Hungarian National Research, Development and Innovation Fund under contract NKFI K137726. The authors acknowledge the Digital Government Development and Project Management Ltd. for awarding access to the Komondor HPC facility based in Hungary. The work of Csanád Árpád Hubay supported by the Doctoral Excellence Fellowship Programme (DKÖP-25-1-BME-54) is funded by the National Research Development and Innovation Fund of the Ministry of Culture and Innovation and the Budapest University of Technology and Economics. The work of Bálint Papp was supported by the EKÖP-24-4-I-BME-357 University Research Scholarship Program of the Ministry for Culture and Innovation from the source of the National Research, Development, and Innovation Fund.

Data Availability Statement: The data presented in this study are available on request from the corresponding author.

Acknowledgments: The authors thank the reviewers for their insightful and constructive comments.

Conflicts of Interest: The authors declare no conflicts of interest.

References

1. Robinson, S.K. Coherent motions in the turbulent boundary layer. *Annu. Rev. Fluid Mech.* **1991**, *23*, 601–639. [CrossRef]
2. Kline, S.J.; Reynolds, W.C.; Schraub, F.A.; Runstadler, P.W. The structure of turbulent boundary layers. *J. Fluid Mech.* **1967**, *30*, 741–773. [CrossRef]
3. Offen, G.R.; Kline, S.J. A proposed model of the bursting process in turbulent boundary layers. *J. Fluid Mech.* **1975**, *70*, 209–228. [CrossRef]
4. Metzger, M.; McKeon, B.; Arce-Larreta, E. Scaling the characteristic time of the bursting process in the turbulent boundary layer. *Phys. D Nonlinear Phenom.* **2010**, *239*, 1296–1304. [CrossRef]
5. Mahrt, L. Intermittency of atmospheric turbulence. *J. Atmos. Sci.* **1989**, *46*, 79–95. [CrossRef]
6. Mahrt, L. Stratified atmospheric boundary layers. *Bound.-Layer Meteorol.* **1999**, *90*, 375–396. [CrossRef]
7. Higbie, R. The rate of absorption of pure gas into a still liquid during short periods of exposure. *Trans. Am. Inst. Chem. Eng.* **1935**, *31*, 365–389.
8. Danckwerts, P.V. Significance of liquid-film coefficients in gas absorption. *Ind. Eng. Chem.* **1951**, *43*, 1460–1467. [CrossRef]
9. Perlmutter, D.D. Surface-renewal models in mass transfer. *Chem. Eng. Sci.* **1961**, *16*, 287–296. [CrossRef]
10. Katul, G.; Porporato, A.; Cava, D.; Siqueira, M. An analysis of intermittency, scaling, and surface renewal in atmospheric surface layer turbulence. *Phys. D Nonlinear Phenom.* **2006**, *215*, 117–126. [CrossRef]
11. Paradisi, P.; Cesari, R.; Contini, D.; Donateo, A.; Palatella, L.N.A. Characterizing memory in atmospheric time series: An alternative approach based on renewal theory. *Eur. Phys. J. Spec. Top.* **2009**, *174*, 207–218. [CrossRef]
12. Paradisi, P.; Cesari, R.; Donateo, A.; Contini, D.; Allegrini, P. Diffusion scaling in event-driven random walks: An application to turbulence. *Rep. Math. Phys.* **2012**, *70*, 205–220. [CrossRef]

13. Paradisi, P.; Cesari, R.; Donateo, A.; Contini, D.; Allegrini, P. Scaling laws of diffusion and time intermittency generated by coherent structures in atmospheric turbulence. *Nonlinear Process. Geophys.* **2012**, *19*, 113–126. [CrossRef]
14. Cesari, R.; Paradisi, P.; Allegrini, P. Source identification by a statistical analysis of backward trajectories based on peak pollution events. *Int. J. Environ. Pollut.* **2014**, *55*, 94–103. [CrossRef]
15. Luo, X.; Wang, M.; Lee, J.; Hendry, J. Dynamic modelling based on surface renewal theory, model validation and process analysis of rotating packed bed absorber for carbon capture. *Appl. Energy* **2021**, *301*, 117462. [CrossRef]
16. Itano, T.; Toh, S. The dynamics of bursting process in wall turbulence. *J. Phys. Soc. Jpn.* **2001**, *70*, 703–716. [CrossRef]
17. Jiménez, J.; Kawahara, G.; Simens, M.P.; Nagata, M.; Shiba, M. Characterization of near-wall turbulence in terms of equilibrium and "bursting" solutions. *Phys. Fluids* **2005**, *17*, 015105. [CrossRef]
18. Kawahara, G.; Kida, S. Periodic motion embedded in plane Couette turbulence: Regeneration cycle and burst. *J. Fluid Mech.* **2001**, *449*, 291–300. [CrossRef]
19. Cherubini, S.; De Palma, P. Nonlinear optimal perturbations in a Couette flow: Bursting and transition. *J. Fluid Mech.* **2013**, *716*, 251–279. [CrossRef]
20. Bomminayuni, S.; Stoesser, T. Turbulence statistics in an open-channel flow over a rough bed. *J. Hydraul. Eng.* **2011**, *137*, 1347–1358. [CrossRef]
21. Wallace, J.M.; Eckelmann, H.; Brodkey, R.S. The wall region in turbulent shear flow. *J. Fluid Mech.* **1972**, *54*, 39–48. [CrossRef]
22. Willmarth, W.W. Structure of turbulence in boundary layers. *Adv. Appl. Mech.* **1975**, *15*, 159–254.
23. Wallace, J.M. Quadrant analysis in turbulence research: History and evolution. *Annu. Rev. Fluid Mech.* **2016**, *48*, 131–158. [CrossRef]
24. Antonia, R.A. Conditional sampling in turbulence measurement. *Annu. Rev. Fluid Mech.* **1981**, *13*, 131–156. [CrossRef]
25. Kirkbride, A.D.; Ferguson, R. Turbulent flow structure in a gravel-bed river: Markov chain analysis of the fluctuating velocity profile. *Earth Surf. Process. Landforms* **1995**, *20*, 721–733. [CrossRef]
26. Buffin-Bélanger, T.; Roy, A.G.; Kirkbride, A.D. On large-scale flow structures in a gravel-bed river. *Geomorphology* **2000**, *32*, 417–435. [CrossRef]
27. Roy, A.G.; Buffin-Belanger, T.; Lamarre, H.; Kirkbride, A.D. Size, shape and dynamics of large-scale turbulent flow structures in a gravel-bed river. *J. Fluid Mech.* **2004**, *500*, 1–27. [CrossRef]
28. Finnigan, J. Turbulence in plant canopies. *Annu. Rev. Fluid Mech.* **2000**, *32*, 519–571. [CrossRef]
29. Christen, A.; van Gorsel, E.; Vogt, R. Coherent structures in urban roughness sublayer turbulence. *Int. J. Climatol.* **2007**, *27*, 1955–1968. [CrossRef]
30. Nelson, M.A.; Pardyjak, E.R.; Brown, M.J.; Klewicki, J.C. Properties of the wind field within the Oklahoma City Park Avenue street canyon. Part II: Spectra, cospectra, and quadrant analyses. *J. Appl. Meteorol. Climatol.* **2007**, *46*, 2055–2073. [CrossRef]
31. Kastner-Klein, P.; Berkowicz, R.; Britter, R. The influence of street architecture on flow and dispersion in street canyons. *Meteorol. Atmos. Phys.* **2004**, *87*, 121–131. [CrossRef]
32. Kukačka, L.; Nosek, Š.; Kellnerová, R.; Jurčáková, K.; Jaňour, Z. Wind tunnel measurement of turbulent and advective scalar fluxes: A case study on intersection ventilation. *Sci. World J.* **2012**, *2012*, 381357. [CrossRef] [PubMed]
33. Nosek, Š.; Kukačka, L.; Kellnerová, R.; Jurčáková, K.; Jaňour, Z. Ventilation processes in a three-dimensional street canyon. *Bound.-Layer Meteorol.* **2016**, *159*, 259–284. [CrossRef]
34. Kellnerová, R.; Kukačka, L.; Jurčáková, K.; Uruba, V.; Jaňour, Z. PIV measurement of turbulent flow within a street canyon: Detection of coherent motion. *J. Wind Eng. Ind. Aerodyn.* **2012**, *104*, 302–313. [CrossRef]
35. Di Bernardino, A.; Monti, P.; Leuzzi, G.; Querzoli, G. Water-channel study of flow and turbulence past a two-dimensional array of obstacles. *Bound.-Layer Meteorol.* **2015**, *155*, 73–85. [CrossRef]
36. Daw, C.S.; Finney, C.E.A.; Tracy, E.R. A review of symbolic analysis of experimental data. *Rev. Sci. Instruments* **2003**, *74*, 915–930. [CrossRef]
37. Lehrman, M.; Rechester, A.B. Extracting symbolic cycles from turbulent fluctuation data. *Phys. Rev. Lett.* **2001**, *87*, 164501. [CrossRef]
38. Daw, C.S.; Finney, C.E.A.; Nguyen, K.; Halow, J.S. Symbol statistics: A new tool for understanding multiphase flow phenomena. *ASME Int. Mech. Eng. Congr. Expo. Am. Soc. Mech. Eng.* **1998**, *26744*, 221–229.
39. Finney, C.E.A.; Green, J.B., Jr.; Daw, C.S. Symbolic time-series analysis of engine combustion measurements. *SAE Trans.* **1998**, pp. 888–897.
40. Keshavarzi, A.R.; Ziaei, A.N.; Homayoun, E.; Shirvani, A. Fractal-Markovian scaling of turbulent bursting process in open channel flow. *Chaos Solitons Fractals* **2005**, *25*, 307–318. [CrossRef]
41. Gheisi, A.R.; Alavimoghaddam, M.R.; Dadrasmoghaddam, A. Markovian–Octant analysis based stable turbulent shear stresses in near-bed bursting phenomena of vortex settling chamber. *Environ. Fluid Mech.* **2006**, *6*, 549–572. [CrossRef]

42. Jin, C.; Potts, I.; Reeks, M.W. A simple stochastic quadrant model for the transport and deposition of particles in turbulent boundary layers. *Phys. Fluids* **2015**, *27*, 053305. [CrossRef]
43. Kalmár-Nagy, T.; Varga, Á. Complexity analysis of turbulent flow around a street canyon. *Chaos Solitons Fractals* **2019**, *119*, 102–117. [CrossRef]
44. Chowdhuri, S.; Kalmár-Nagy, T.; Banerjee, T. Persistence analysis of velocity and temperature fluctuations in convective surface layer turbulence. *Phys. Fluids* **2020**, *32*, 076601. [CrossRef]
45. Chowdhuri, S.; Kumar, S.; Banerjee, T. Revisiting the role of intermittent heat transport towards Reynolds stress anisotropy in convective turbulence. *J. Fluid Mech.* **2020**, *899*, A26. [CrossRef]
46. Chowdhuri, S.; Todekar, K.; Murugavel, P.; Karipot, A.; Prabha, T.V. Unravelling the turbulent structures of temperature variations during a gust front event: A case study. *Environ. Fluid Mech.* **2021**, *21*, 263–281. [CrossRef]
47. Vardoulakis, S.; Fisher, B.E.A.; Pericleous, K.; Gonzalez-Flesca, N. Modelling air quality in street canyons: A review. *Atmos. Environ.* **2003**, *37*, 155–182. [CrossRef]
48. Ahmad, K.; Khare, M.; Chaudhry, K.K. Wind tunnel simulation studies on dispersion at urban street canyons and intersections—a review. *J. Wind Eng. Ind. Aerodyn.* **2005**, *93*, 697–717. [CrossRef]
49. Li, X.X.; Liu, C.H.; Leung, D.Y.; Lam, K.M. Recent progress in CFD modelling of wind field and pollutant transport in street canyons. *Atmos. Environ.* **2006**, *40*, 5640–5658. [CrossRef]
50. Yazid, A.W.M.; Sidik, N.A.C.; Salim, S.M.; Saqr, K.M. A review on the flow structure and pollutant dispersion in urban street canyons for urban planning strategies. *Simulation* **2014**, *90*, 892–916. [CrossRef]
51. Zhang, Y.; Gu, Z.; Yu, C.W. Impact factors on airflow and pollutant dispersion in urban street canyons and comprehensive simulations: A review. *Curr. Pollut. Rep.* **2020**, *6*, 425–439. [CrossRef]
52. Voordeckers, D.; Lauriks, T.; Denys, S.; Billen, P.; Tytgat, T.; Van Acker, M. Guidelines for passive control of traffic-related air pollution in street canyons: An overview for urban planning. *Landsc. Urban Plan.* **2021**, *207*, 103980. [CrossRef]
53. Papp, B.; Kristóf, G.; Istók, B.; Koren, M.; Balczó, M.; Balogh, M. Measurement-driven Large Eddy Simulation of dispersion in street canyons of variable building height. *J. Wind Eng. Ind. Aerodyn.* **2021**, *211*, 104495. [CrossRef]
54. Papp, B.; Istók, B.; Koren, M.; Balczó, M.; Kristóf, G. Statistical assessment of the concentration fluctuations in street canyons via time-resolved wind tunnel experiments. *J. Wind Eng. Ind. Aerodyn* **2024**, *246*, 105665. [CrossRef]
55. Chew, L.W.; Aliabadi, A.A.; Norford, L.K. Flows across high aspect ratio street canyons: Reynolds number independence revisited. *Environ. Fluid Mech.* **2018**, *18*, 1275–1291. [CrossRef]
56. Xie, Z.; Castro, I.P. LES and RANS for turbulent flow over arrays of wall-mounted obstacles. *Flow Turbul. Combust.* **2006**, *76*, 291–312. [CrossRef]
57. Kristóf, G.; Papp, B.; Wang, H.; Hang, J. Investigation of the flow and dispersion characteristics of repeated orographic structures by assuming transient wind forcing. *J. Wind Eng. Ind. Aerodyn.* **2020**, *197*, 104087. [CrossRef]
58. Koren, M.; Kristóf, G. Investigation of the accuracy of the transient wind forcing model applied for a shear-driven LES. *J. Wind Eng. Ind. Aerodyn.* **2023**, *236*, 105363. [CrossRef]
59. Smagorinsky, J. General circulation experiments with the primitive equations: I. The basic experiment. *Mon. Weather Rev.* **1963**, *91*, 99–164. [CrossRef]
60. Lilly, D.K. A proposed modification of the germano sugrid-scale closure method. *Phys Fluids A* **1992**, *4*, 633–635. [CrossRef]
61. Shah, K.B.; Ferziger, J.H. A fluid mechanicians view of wind engineering: Large eddy simulation of flow past a cubic obstacle. *J. Wind Eng. Ind. Aerodyn.* **1997**, *67*, 211–224. [CrossRef]
62. ANSYS Inc. *ANSYS Mechanical User's Guide, Release 2019.* Canonsburg, PA, 2019.
63. Czernuszenko, W.; Rowiński, P.M. Shear stress statistics in a compound channel flow. *Arch. Hydro-Eng. Environ. Mech.* **2008**, *55*, 3–27.
64. Snyder, D.L.; Miller, M.I. *Random Point Processes in Time and Space*; Springer Science & Business Media: Berlin/Heidelberg, Germany, 2012.
65. Van Lieshout, M.C. *Markov Point Processes and Their Applications*; World Scientific: Singapore, 2000.
66. Segundo, J.P. Nonlinear dynamics of point process systems and data. *Int. J. Bifurc. Chaos* **2003**, *13*, 2035–2116. [CrossRef]
67. Bogard, D.G.; Tiederman, W.G. Burst detection with single-point velocity measurements. *J. Fluid Mech.* **1986**, *162*, 389–413. [CrossRef]
68. Rao, K.N.; Narasimha, R.; Narayanan, M.A.B. The 'bursting' phenomenon in a turbulent boundary layer. *J. Fluid Mech.* **1971**, *48*, 339–352. [CrossRef]
69. Sreenivasan, K.R.; Prabhu, A.; Narasimha, R. Zero-crossings in turbulent signals. *J. Fluid Mech.* **1983**, *137*, 251–272. [CrossRef]
70. Kailasnath, P.; Sreenivasan, K.R. Zero crossings of velocity fluctuations in turbulent boundary layers. *Phys. Fluids A Fluid Dyn.* **1993**, *5*, 2879–2885. [CrossRef]

71. Meng, J. Wall layer microturbulence phenomenological model and a semi-Markov probability predictive model for active control of turbulent boundary layers. In Proceedings of the 29th AIAA, Fluid Dynamics Conference, New Orleans, LA, USA, 17–20 June 1996; p. 2995.
72. Colin Cameron, A.; Windmeijer, F.A.G. An R-squared measure of goodness of fit for some common nonlinear regression models. *J. Econom.* **1997**, *77*, 329–342. [CrossRef]
73. Meneveau, C.; Sreenivasan, K.R. The multifractal nature of turbulent energy dissipation. *J. Fluid Mech.* **1991**, *224*, 429–484. [CrossRef]
74. Amigó, J.M.; Keller, K.; Unakafova, V.A. On entropy, entropy-like quantities, and applications. *Discret. Contin. Dyn. Syst.-Ser. B* **2015**, *20*, 3301–3343. [CrossRef]
75. Keshavarzi, A.; Ball, J.E.; Nabavi, H. Frequency pattern of turbulent flow and sediment entrainment over ripples using image processing. *Hydrol. Earth Syst. Sci.* **2012**, *16*, 147–156. [CrossRef]
76. Keshavarzi, A.; Melville, B.; Ball, J.E. Three-dimensional analysis of coherent turbulent flow structure around a single circular bridge pier. *Environ. Fluid Mech.* **2014**, *14*, 821–847. [CrossRef]
77. Bollt, E.M.; Stanford, T.; Lai, Y.C.; Życzkowski, K. What symbolic dynamics do we get with a misplaced partition? On the validity of threshold crossings analysis of chaotic time-series. *Phys. D Nonlinear Phenom.* **2001**, *154*, 259–286. [CrossRef]

entropy

MDPI

Article

A Chaos Synchronization Diagnostic: Difference Time Series Peak Complexity (DTSPC)

Zhe Lin [1] and Arjendu K. Pattanayak [2,*]

[1] United World College Changshu China, Suzhou 215500, China; zlin22@uwcchina.org
[2] Department of Physics and Astronomy, Carleton College, Northfield, MN 55057, USA
* Correspondence: arjendu@carleton.edu

Abstract: Chaotic systems can exhibit completely different behaviors given only slightly different initial conditions, yet it is possible to synchronize them through appropriate coupling. A wide variety of behaviors—complete chaos, complete synchronization, phase synchronization, etc.—across a variety of systems have been identified but rely on systems' phase space trajectories, which suppress important distinctions between very different behaviors and require access to the differential equations. In this paper, we introduce the Difference Time Series Peak Complexity (DTSPC) algorithm, a technique using entropy as a tool to quantitatively measure synchronization. Specifically, this uses peak pattern complexity created from sampled time series, focusing on the behavior of ringing patterns in the difference time series to distinguish a variety of synchronization behaviors based on the entropic complexity of the populations of various patterns. We present results from the paradigmatic case of coupled Lorenz systems, both identical and non-identical, and across a range of parameters and show that this technique captures the diversity of possible synchronization, including non-monotonicity as a function of parameter as well as complicated boundaries between different regimes. Thus, this peak pattern entropic analysis algorithm reveals and quantifies the complexity of chaos synchronization dynamics, and in particular captures transitional behaviors between different regimes.

Keywords: chaos; synchronization; entropy

1. Introduction

Certain dynamical systems have been known since Poincaré to be chaotic [1,2], i.e., they are highly sensitive to initial conditions and parameters. Classical models of deterministic chaos include the Lorenz system modeling weather [3] and the logistic map modeling population growth [4,5]. Dynamical systems can also display synchronization, when two or more different systems exhibit similar dynamics, either behaving identically as a limiting example, or more broadly possessing similarities in some key aspects, such as sharing the same frequencies or a constant phase lag [6]. Remarkably, though chaotic systems exhibit completely different behaviors with only slightly different initial conditions, adding a small amount of coupling can cause them to be synchronized. More broadly, chaos synchronization occurs when two (or many) chaotic systems (either identical, with the same equations, parameters and initial conditions, or dissimilar, with different equations, parameters or initial conditions) adjust a given property of their motion to a common behavior due to a coupling or a forcing (periodical or noisy) [7]. Chaos synchronization has found applications in a wide variety of fields such as electrical circuits [8–11], power systems [12–14], medicine [15,16], and chemical reactors [17,18].

Historically, chaos synchronization has been classified into different categories including complete synchronization, practical synchronization, and phase synchronization [19,20]. Such categorization, however, only provides qualitative classification of chaos synchronization, so research in quantitative complexity measurement of chaos synchronization started to develop over the last two decades. Classic methods of complexity measurement include phase portraits, bifurcation diagrams [21], and Lyapunov exponents [22], which require access to the dynamical equations. A growing body of work uses broader and more abstract techniques relying on entropic analysis such as using Kolmogorov entropy to quantify the level of unpredictability in a chaotic system over time [23], spectral entropy to diagnose multi-stable fractional-order chaotic systems [24], entanglement and relative entropy to investigate the chaotic behaviors in quantum systems [25], topological entropy to characterize synchronization in piecewise linear maps [26,27], and wavelet entropy to study brain electrical signals [28]. One advantage of these entropy-based analyses is that they can often be applied to systems where only the dynamical data are available, in the absence of the actual equations, such as in experimental situations, and the results are typically robust to the presence of noise in the data or the dynamics.

In this paper, we present numerical results from applying an entropic analysis to discrete peak patterns visible in the difference time series between systems to quantitatively measure chaos synchronization broadly, where such peak patterns should be visible across all basic models of chaos, including the Lorenz system, the Duffing oscillator, the Chua circuit, and the Rössler attractor [29]. Specifically, we use the paradigmatic model of the basic coupled Lorenz systems, applying entropic analysis to patterns extracted from peak dynamics. We show that this entropy-based analysis in quantitatively measuring chaos synchronization not only captures overall similarities between two synchronized systems but also reveals a diversity of possible synchronization, non-monotonic changes in behavior with respect to linear change in parameters during transition from one regime to another, and complicated boundaries between different synchronization regimes.

Below, in Section 2, coupled identical Lorenz systems with dissimilar initial conditions are introduced, as well as coupled dissimilar Lorenz systems with identical initial conditions, the synchronization effects of which vary with the change in coupling parameters. Section 3 describes the specific methodology of the new synchronization diagnostic, focusing on extracting from the raw time series information about patterns in the peak dynamics before computing the Shannon entropy of the pattern populations thus obtained. Results and findings about the coupled Lorenz systems after applying the new diagnostic are presented in Section 4 and are compared with the conventional methods, showing values of our new synchronization diagnostic. Finally, discussions and conclusions are presented in Section 5, revealing findings on other interesting behaviors and pointing out directions of further research.

2. Comparing Synchronization in Coupled Identical and Dissimilar Lorenz Systems

In this section, we discuss various details about different kinds of synchronization behaviors with the examples of coupled identical and dissimilar Lorenz systems, initially studied in weather forecasting. Under certain initial conditions and parameter values, the Lorenz system exhibits chaotic behavior. Before we introduce the more complex dynamics in the coupled dissimilar Lorenz systems, we first examine the coupled identical Lorenz systems.

2.1. Identical Lorenz Systems with Different Initial Conditions

Consider two Lorenz systems defined by

$$\begin{aligned} \dot{x_1} &= \sigma_1(x_2 - x_1) \\ \dot{x_2} &= -x_1x_3 + \rho_1x_1 - x_2 \\ \dot{x_3} &= x_1x_2 - \beta_1x_3 \end{aligned} \tag{1}$$

and

$$\begin{aligned} \dot{y_1} &= \sigma_2(y_2 - y_1) \\ \dot{y_2} &= -y_1y_3 + \rho_2y_1 - y_2 \\ \dot{y_3} &= y_1y_2 - \beta_2y_3 \end{aligned} \tag{2}$$

where $\sigma_1 = \sigma_2 = 10, \rho_1 = \rho_2 = 50, \beta_1 = \beta_2 = 8/3$. The initial conditions for the first Lorenz are $x_1 = 0, x_2 = 1, x_3 = 0$; the initial conditions for the second Lorenz are $y_1 = 1, y_2 = 1, y_3 = 1$.

Throughout this paper, we use the ODEINT solver from SciPy, with time interval $[0, 100]$ and time step 0.001, in Python on a personal computing device to simulate solutions to the differential equations. Figure 1a,c show the 3D phase space diagrams of the two identical systems with different initial conditions. Although the two Lorenz systems have the same attractor, Figure 1e showing x_1 against y_1 makes it clear that the specific trajectories are quite different, demonstrated in time series data as well in Figure 1b,d and the difference time series of $x_1 - y_1$ (Figure 1f), which behaves rather randomly. This sets our baseline to observe how coupling alters these differences.

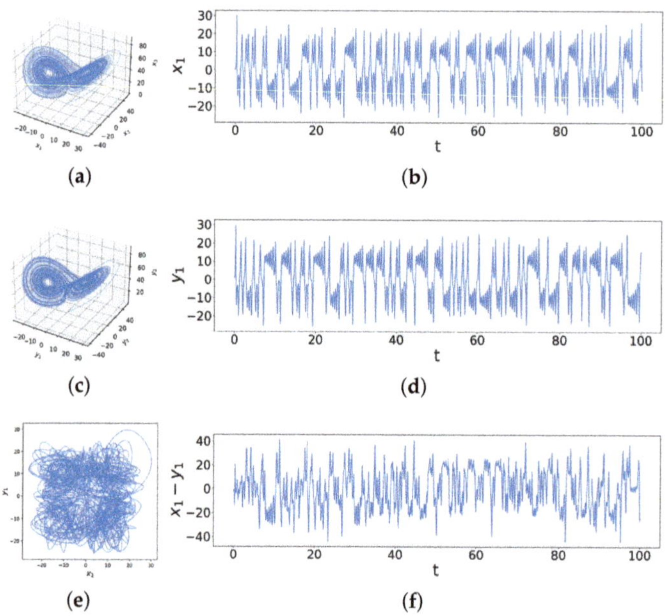

Figure 1. These figures show the behavior of two uncoupled identical Lorenz systems with different initial conditions. In (**a**,**b**) we see the phase space trajectory and time series for initial conditions $x_1 = 0, x_2 = 1, x_3 = 0$; (**c**,**d**) show the same for initial conditions $y_1 = 1, y_2 = 1, y_3 = 1$, while (**e**) shows x_1 vs y_1 phase space behaviors and (**f**) is the behavior of the difference time series $x_1 - y_1$ for these two systems.

We use the linear coupling model [7]

$$
\begin{aligned}
\dot{x_1} &= \sigma_1(x_2 - x_1) + d_1(y_1 - y_2) \\
\dot{x_2} &= -x_1x_3 + \rho_1x_1 - x_2 + d_2(y_2 - x_2) \\
\dot{x_3} &= x_1x_2 - \beta_1x_3 + d_3(y_3 - x_3) \\
\dot{y_1} &= \sigma_2(y_2 - y_1) + d_1(x_1 - y_1) \\
\dot{y_2} &= -y_1y_3 + \rho_2y_1 - y_2 + d_2(x_2 - y_2) \\
\dot{y_3} &= y_1y_2 - \beta_2y_3 + d_3(x_3 - y_3)
\end{aligned}
\tag{3}
$$

where d_1, d_2, d_3 are coupling strengths, which indeed yields that the trajectories of the two chaotic Lorenz systems do overlap. The parameters and initial conditions are the same as above: $\sigma_1 = \sigma_2 = 10$, $\rho_1 = \rho_2 = 50$, $\beta_1 = \beta_2 = 8/3$, $x_1 = 0, x_2 = 1, x_3 = 0$ and $y_1 = 1, y_2 = 1, y_3 = 1$.

While Figure 2a,c look quite similar to the uncoupled systems (Figure 1a,c), the relationship between the two systems is now quite different, as visualized in the x_1 vs. y_1 phase space trajectories. In Figure 2e, the values of x_1 and y_1 are nearly identical, suggesting that the trajectories of the two systems almost completely overlap. Moreover, as shown in Figure 2f, the difference between x_1 and y_1 gradually reduce to zero over time. In fact, $x_2 - y_2$ and $x_3 - y_3$ are also decreasing down to zero. This proves that the trajectories of the two coupled Lorenz systems indeed gradually overlap, so complete synchronization between two identical chaotic systems with different initial conditions is achieved through linear coupling. Note that though the systems synchronize, each of them separately remain chaotic, as can be seen from the attractors in Figure 2a,c. The situation, however, becomes far more complicated in the case of coupled dissimilar Lorenz systems which possess identical initial conditions yet dissimilar parameters.

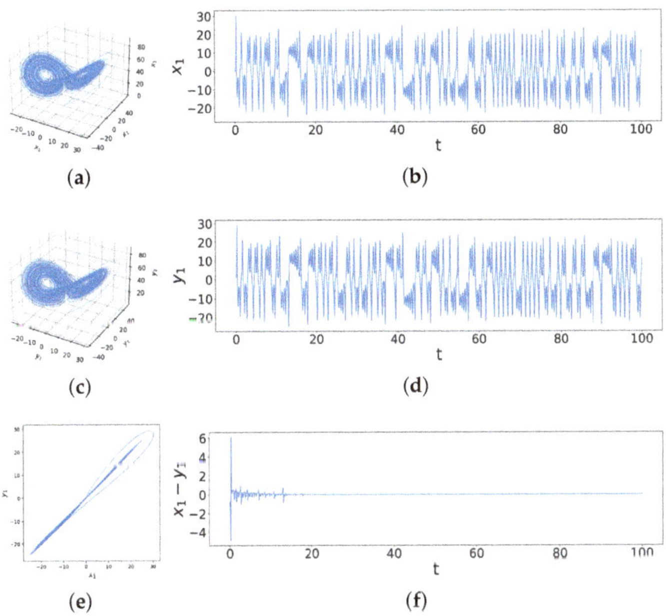

Figure 2. These figures show the behavior of two coupled identical Lorenz systems with different initial conditions under coupling strengths $d_1 = 3.0, d_2 = 1.2, d_3 = 0.0$. In (**a**,**b**) we see the phase space trajectory and time series for initial conditions $x_1 = 0, x_2 = 1, x_3 = 0$ under coupling; (**c**,**d**) show the same for initial conditions $y_1 = 1, y_2 = 1, y_3 = 1$ under coupling, while (**e**) shows x_1 vs y_1 phase space behaviors and (**f**) is the behavior of the difference time series $x_1 - y_1$ for these two systems under coupling.

2.2. Dissimilar Lorenz Systems with Identical Initial Conditions

Above, we saw examples of non-synchronization before and complete synchronization after coupling. However, such clear distinctions do not completely cover the range of behaviors, as we show below; even a simple division into practical synchronization or phase synchronization does not suffice to describe the complex behavior seen between non-synchronization and complete synchronization. We can see this when we consider two slightly different Lorenz systems, one with parameters $\sigma_1 = 10, \rho_1 = 50, \beta_1 = 8/3$, the other with parameters $\sigma_2 = 10, \rho_2 = 40, \beta_2 = 8/3$. The initial conditions are now identical, i.e., $x_1 = y_1 = 0, x_2 = y_2 = 1, x_3 = y_3 = 0$. Each of the two systems are chaotic, and they are unsynchronized. On introducing linear coupling, we see a wide range of different synchronization behaviors with more variety, as shown in Figure 3, compared to the coupled identical Lorenz systems.

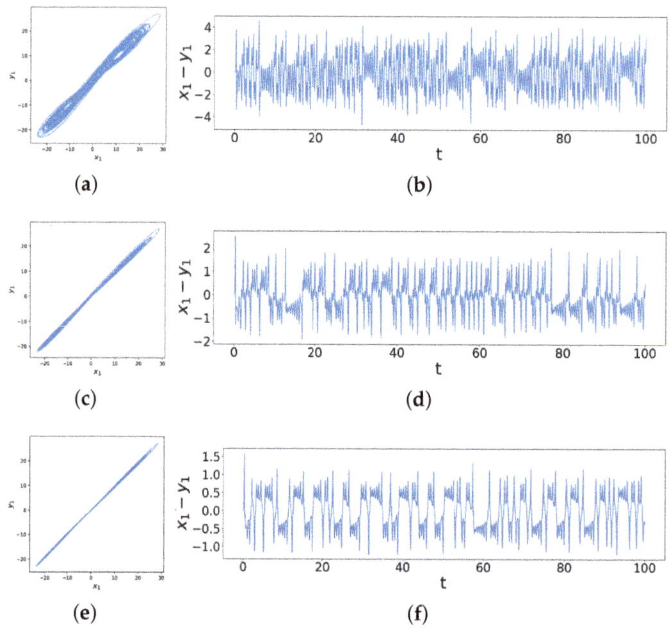

Figure 3. Phase space trajectories and $x_1 - y_1$ difference time series of coupled dissimilar Lorenz systems with coupling strengths (**a**,**b**) $d_1 = 8.0, d_2 = 0.6$, (**c**,**d**) $d_1 = 8.0, d_2 = 4.0$, (**e**,**f**) $d_1 = 9.0$, $d_2 = 9.0$, showing weak synchronization under weak coupling and practical synchronization after strong coupling.

In particular, we see in Figure 3 that as the coupling increases, the range of $x_1 - y_1$ decreases from roughly ± 4 to roughly ± 2. Notice, in comparing Figure 3a–d with Figure 3e,f (at relatively different coupling strengths), we see that $x_1 - y_1$ behaves similarly in phase space but displays more variety in the time series. Specifically, notice that there are peaks and dips above and below the horizontal line of $x_1 - y_1 = 0$ in all three cases, but the last two cases also exhibit ringing patterns (corresponding to entrainment at certain differences) in practical synchronization, where the systems only display small variations in phase and amplitude, which means their trajectories' difference converges to a neighborhood around the origin [20]. Moreover, notice that as the coupling strengths increase from the second case to the third case, these ringing patterns change character—specifically, oscillating for longer durations around locations further from the horizontal axis (zero difference). This change in ringing patterns as a function of coupling strengths is what distinguishes behaviors where phase-portraits could be a misleading indicator. We now quantitatively measure the behavior of the difference time series to more rigorously identify different synchronization patterns across changing coupling strengths, using an analysis of peak patterns via its computed Shannon entropy, a technique which we term the Difference Time Series Peak Complexity (DTSPC). This quantification allow us to map carefully, via a heat map of the DTSPC vs. coupling strengths, different kinds of synchronization and transitions as well as the complex shape of the transition regimes. We now turn to describing the method.

3. Methods: Quantitatively Measuring Synchronization

Although the time series patterns are quite different across the various types of synchronization shown in Figure 3, they all have the common element that the difference between x_1 and y_1 is centered around 0, i.e., the time series is centered around the horizontal axis $x_1 - y_1 = 0$. We focus on behaviors relative to this horizontal axis, specifically considering the behavior of the ringing patterns previously discussed in the context of the difference time series for $x_1 - y_1$. We count the number of peaks between zero-crossings in the difference time series, noting either the number of maxima if $x_1 - y_1 > 0$ or the number of minima if $x_1 - y_1 < 0$. This compresses the time series data into a list of the observed populations of number of consecutive peaks between zero-crossings in the original difference time series.

We now examine the $x_1 - y_1$ difference time series of time interval $[0, 10]$ of a coupled dissimilar Lorenz system when $d_1 = 1.2, d_2 = 9.4$, shown in Figure 4a, and apply our method to create a peak series out of it. We first note that $x_1 - y_1$ starts at 0 at $t = 0$, and the first zero-crossing occurs after one maximum, marked in green, above the central red line. Hence, our first element of the peak series is 1. Then, between the second and third zero-crossing, there are three minima, marked in orange, below $x_1 - y_1 = 0$, so our second element of the peak series is -4. Similarly, we can convert the number of consecutive peaks between each zero-crossing into a numerical element in the peak series. The peak series, then, of this difference time series from time interval $[0, 10]$ is

$$1, -4, 2, -2, 2, -1, 1, -1, 5, -1, 1, -3, 2.$$

Note that the last two minima are not counted because there does not exist another zero-crossing until beyond $t = 10$, which is not included in this truncated difference time series. To visualize the distribution of number of consecutive peaks in the list, we create a bar chart, as shown in Figure 4b below, where the x-axis denotes the elements in the peak series and the y-axis denotes the relative frequency.

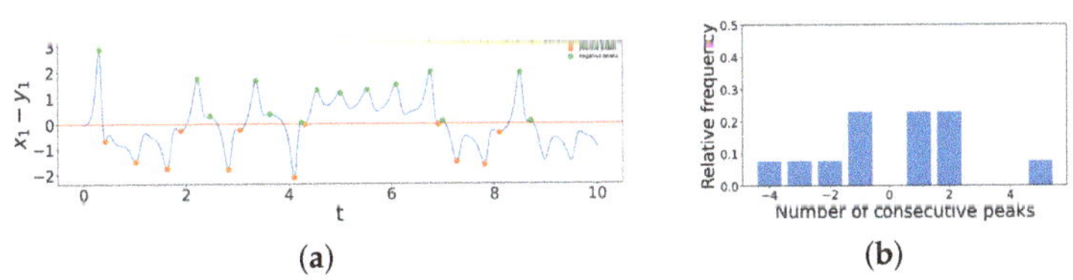

Figure 4. (**a**) Illustration of constructing peak series from difference time series for time interval $[0, 10]$ of a coupled dissimilar Lorenz system when $d_1 = 1.2, d_2 = 9.4$. (**b**) Bar chart of distribution of number of consecutive peaks in the difference time series.

Now, we extend the method illustrated using the truncated difference time series above to a longer time interval and a wider variety of cases, i.e., different coupling strengths d_1 and d_2, as shown in Figure 5 below. The figure shows the phase space trajectories (left column), the difference time series (middle column), and the distribution of consecutive peaks (right column) of coupled dissimilar Lorenz systems under $d_1 = 1.2$ and $d_2 = 0, 1.0, 6.4, 8.0, 9.4$, respectively.

As we look down the rows, the coupling strength d_2 increases while d_1 stays unchanged, and the phase portraits become more linear. Correspondingly, the bar charts in the right column show that the distribution of consecutive peaks in the difference time series also changes with the increase in d_2. When the system is not synchronized, there are no ringing patterns but only single spikes, so there are usually very few instances of more than one consecutive peak, as shown in Figure 5c. As the transition begins, the range of possible values for numbers of consecutive peaks increases, whence the bar charts show a corresponding wide range of values on the x-axis. At the early stage of such a transition, single peaks are still the majority, as shown in Figure 5f. We find here that the ringing pattern that exists often crosses the horizontal axis such that only the last peak in a ringing pattern counts as a single 'peak' by our algorithm. Then, as the system becomes more synchronized with the increase in coupling strengths, the ringing patterns diverge from the horizontal axis, so the consecutive peaks of all lengths are distributed more evenly, as shown in Figure 5o. All of these observations motivate us to quantify the complexity by computing the Shannon entropy of the distribution of peak populations [30], thus yielding the Difference Time Series Peak Complexity (DTSPC) metric,

$$\zeta(d_1, d_2) = -\sum_{i=1}^{n} p_i \log_2 p_i \tag{4}$$

to quantitatively measure the level of synchronization.

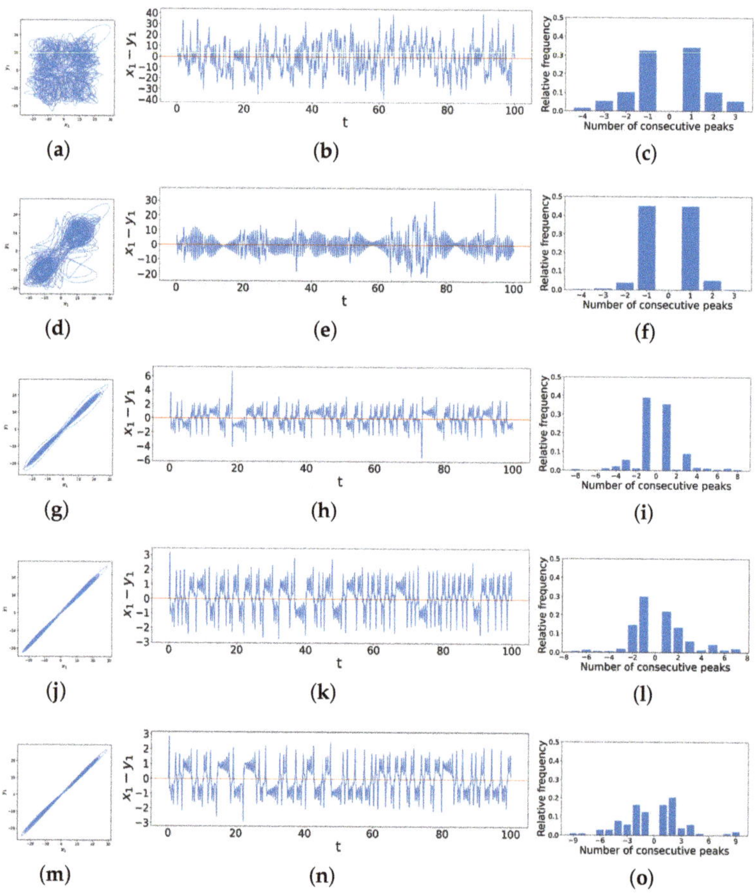

Figure 5. Phase space trajectories, $x_1 - y_1$ difference time series and peak distribution bar charts for coupled dissimilar Lorenz systems with coupling strengths $d_1 = 1.2$ and **(a–c)** $d_2 = 0$, **(d–f)** $d_2 = 1.0$, **(g–i)** $d_2 = 6.4$, **(j–l)** $d_2 = 8.0$, and **(m–o)** $d_2 = 9.4$, showing correlations between phase space trajectories, difference time series and distribution of consecutive peaks.

Although normalized entropy is often used for statistical measures, we prefer unnormalized DTSPC to normalized DTSPC. The nature of our DTSPC technique, i.e., the fact that we utilized the natural peak pattern of difference time series to generate our peak series, results in varying possibilities of consecutive peaks under each pair of coupling strengths. Hence, normalization greatly affects the trend of color change in the heat map, as shown in Appendix A. After examining the phase space trajectories, difference time series and peak distribution bar charts corresponding to different regions in the heat maps, which we will further discuss in the next section, we believe that unnormalized DTSPC more accurately captures the trend of change in level of synchronization with the increase in coupling strengths.

4. Results

We now show how DSTPC ($\zeta(d_1, d_2)$) captures behavioral changes as d_2 increases for different constant d_1 values under the initial conditions $x_1 = y_1 = 0, x_2 = y_2 = 1, x_3 = y_3 = 0$. The line graph of ζ versus d_2 when $d_1 = 1.2$ is plotted in Figure 6a. As we see in the graph, there is a significant drop in ζ between $d_2 = 0.0$ and $d_2 = 1.0$, a sudden increase between $d_1 = 1.0$ and $d_1 = 2.0$, a steady increase between $d_2 = 2.0$ and $d_2 = 6.0$, a sudden increase between $d_2 = 6.0$ and $d_2 = 7.0$, and a steady increase after $d_2 = 7.0$. Other colored lines in the figure also show ζ as a function of d_2 under other conditions—$d_1 = 0.8, 4.0, 8.0$, respectively. These four line graphs share similarities. The two respective periods of sudden and steady increases are present in all graphs. Moreover, when d_1 is larger, the initial sudden drop in ζ at $d_2 \in [0, 1]$ disappears, and there starts to emerge a final drop in ζ at large d_2 values. The gradual changes in the plots with the increase in d_1, as well as the changes in ζ with the increase in d_2 (Figure 6b), motivate us to create a heat map of $\zeta(d_1, d_2)$, i.e., displaying the DTSPC as a function of both coupling strength parameters.

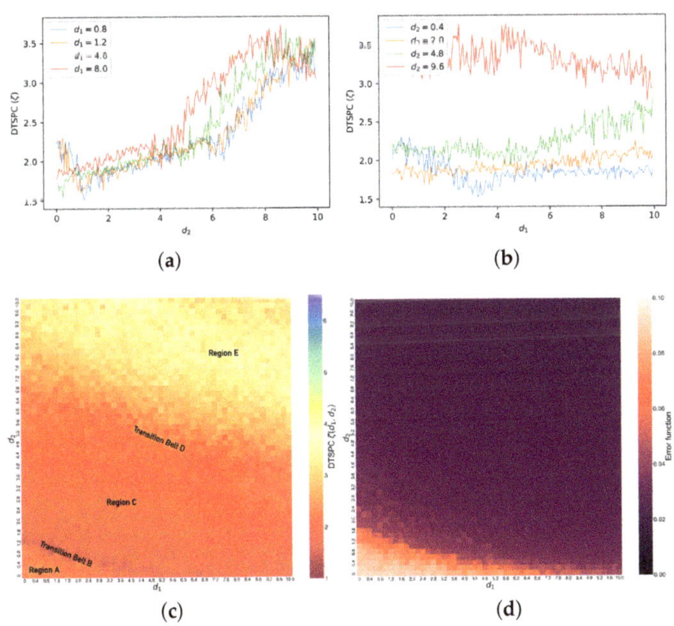

Figure 6. Line graphs of (**a**) ζ versus d_2 when $d_1 = 0.8, 1.2, 4.0, 8.0$ and (**b**) ζ versus d_1 when $d_2 = 0.4, 2.0, 4.8, 9.6$. (**c**) Heat map of $\zeta(d_1, d_2)$ (DTSPC) with changing coupling strengths in coupled dissimilar Lorenz systems. Region *A*: unsynchronized chaos. Transition Belt *B* and Region *C*: abrupt and then gradual decrease in $x_1 - y_1$, enhanced synchronization. Transition Belt *D* and Region *E*: emergence and enhancement of separated ringing patterns, practical synchronization. (**d**) Heat map of error $e(d_1, d_2)$ with changing coupling strengths in coupled dissimilar Lorenz systems.

In Figure 6c, we show the DTSPC heat map for the coupled dissimilar Lorenz systems under initial conditions $x_1 = y_1 = 0, x_2 = y_2 = 1, x_3 = y_3 = 0$. The generation of such a heat map is within a reasonable time frame, i.e., between 10^3 and 10^4 seconds on a personal computing device using Python. We note that we have not conducted exhaustive convergence analysis for our results, since our goal is to explore if our technique can distinguish between different kinds of observed synchronization from time series rather than comprehensively studying the Lorenz system itself. To show the advantages and accuracy of our technique, we compare our DTSPC heat map against the conventional error function (the time-averaged distance between the oscillator state vectors defined by $e = \sqrt{(x_1 - y_1)^2 + (x_2 - y_2)^2 + (x_3 - y_3)^2}$ [19]) heat map shown in Figure 6d. We see that our DTSPC heat map captures not only the main transitions in the bottom-left corner of the error function heat map, but it also reveals details in different levels of synchronization across the rest of the heat map. We find distinct regimes of different levels and types of synchronization via our DTSPC heat map. The color coding for ζ ranges from warmer dark red to light yellow, the warmer color representing a smaller ζ and hence a less complex set of data from the synchronization dynamics. Though the range of ζ is roughly between 1 and 4, we choose the range $[1, 6]$ for the color bar for convenience in comparison with the coupled identical case presented later. The heat map starts from orange in the bottom-left corner, indicated as Region A, and transitions through the dark red Belt B, beyond which is Region C, where $\zeta(d_1, d_2)$ changes gradually. Beyond this Region C is the orange-yellow Transition Belt D, separating it from Region E above, which is bright yellow.

While it is important to emphasize that we can construct such a heat map given just the peak data without access to either the differential equations or the time series themselves, to understand what these various regimes represent, in Figure 7, we explore details of these different synchronization behaviors, showing examples from each region of the heat map. Region A represents chaotic dynamics for any measure of difference between the two systems, and we can see this, for example, in Figure 7a, in which the phase space trajectories of y_1 vs. x_1 are chaotic and patternless. Figure 7b also displays the $x_1 - y_1$ difference, which wanders between ± 30. Because of this lack of pattern in the difference time series, it frequently crosses zero, so observations of the number of consecutive peaks between zero-crossings are therefore mainly centered at -1 and 1, as shown in Figure 7c, resulting in a low ζ. Transition Belt B signals a rapid onset of synchronization as a function of coupling strengths. In the phase-portrait (Figure 7d), we see that the in-phase component of the trajectories—that is, trajectory lines being traced from the bottom-left to top-right diagonal—increases, while the out-of-phase component—that is, trajectory lines being traced from the bottom-right to top-left—decreases. This also means that x_1 and y_1 are more directly correlated, and in general, the phase-space trajectories become more linear. We see this in the difference time series (Figure 7e), which abruptly decreases to lie between ± 20 most of the time and is more symmetric around zero. This leads to the data in the bar chart (Figure 7f) becoming even more centered at -1 and 1 consecutive peaks, resulting in a decrease in ζ.

In Region C, the differences continue to decrease, with the in-phase component of the phase space trajectories (Figure 7g) continuing to become sharper and more linear. In the difference time series (Figure 7h), there start to appear some ringing patterns above and below the horizontal axis, though not yet very obvious. The difference between x_1 and y_1 further decreases to lie roughly between ± 10. Transition Belt D is where practical synchronization becomes rather apparent. As shown clearly in the phase space trajectories (Figure 7j), x_1 and y_1 follow a linear relationship. Accordingly, the difference between these two observables (Figure 7k) reduces to between ± 3, and clear ringing patterns start to appear that lie strictly above and below the horizontal axis of $x_1 - y_1 = 0$. However, before this clear separation above and below the horizontal axis, we also see some small-scale dynamics in this difference variable, which presents as 'signal fuzziness' and a 'stickiness' when this difference crosses zero, resulting from the slight entanglement of the two oscillators' trajectories when they are exchanging positions in phase space during synchronization, and often resulting in a small single peak for our criterion. This results in many −1 and 1 consecutive peaks (Figure 7l), so the ζ remains low. Finally, Region E is where the two systems are more synchronized, as the ringing patterns appear further apart from each other, shown in Figure 7n, with fewer crossings at zero. This indicates that there is less stagnation during a phase change, such that the number of −1 and 1 consecutive peaks reduces. Thus, the data are more spread out in the bar chart, as shown in Figure 7o, and ζ consequently experiences a significant increase. Note that there is a region of slightly darker yellow, or light orange, in the top-right corner of Region E. This slight drop in ζ, also reflected in Figure 6a when $d_1 = 4.0, 8.0$, is a result of a reduced range in the numbers of consecutive peaks as the coupling strengths reach large values around 10, leading to smaller entropy values. Thus, to summarize, ζ starts off low in Region A, then decreases and increases with the increase in coupling strengths, as before for identical systems. However, the transition to synchronization now carries more regions of distinct dynamical regimes: Transition Belt B, Region C, and Transition Belt D. Within them, ζ first decreases a little, then increases, and slightly decreases again, each part corresponding to different dynamics of the system, as we have just seen. This shows again the value of using ζ to quantitatively identify and classify synchronization behaviors.

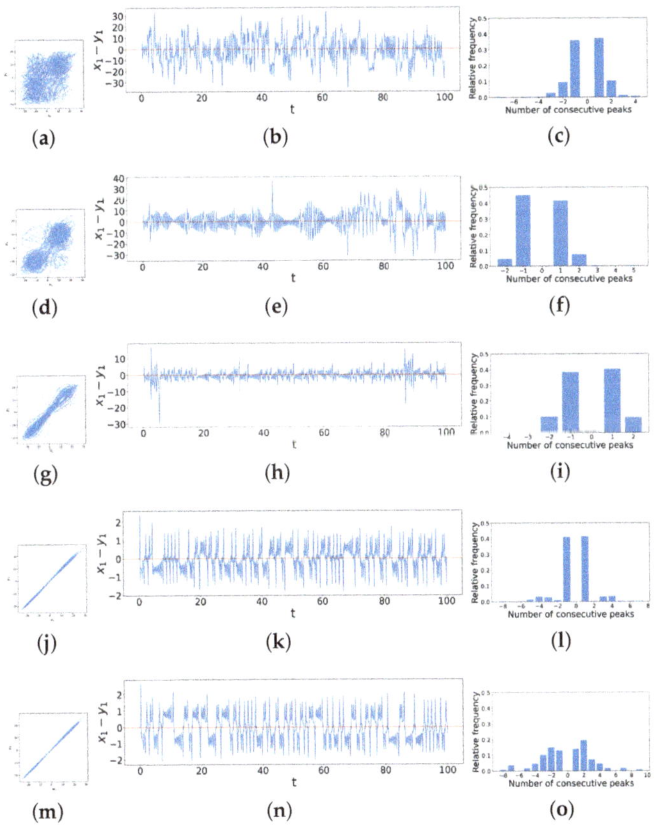

Figure 7. Phase space trajectories, $x_1 - y_1$ difference time series and peak distribution bar charts for different regimes of coupled dissimilar Lorenz systems. (**a–c**) correspond to a case in Region A in Figure 6c when $d_1 = 1.0, d_2 = 0.4$; (**d–f**) correspond to Transition Belt B in Figure 6c when $d_1 = 0.6, d_2 = 1.2$; (**g–i**) correspond to Region C in Figure 6c when $d_1 = 1.2, d_2 = 2.6$; (**j–l**) correspond to Transition Belt D in Figure 6c when $d_1 = 8.8, d_2 = 4.0$; (**m–o**) correspond to Region E in Figure 6c when $d_1 = 2.2, d_2 = 9.0$.

In Figure 8c, we show the values of $\zeta(d_1, d_2)$ with changes in d_1 and d_2 in the coupled identical Lorenz systems with initial conditions $x_1 = 0, x_2 = 1, x_3 = 0, y_1 = y_2 = y_3 = 1$. Comparing against the error function heat map to its right, our DTSPC heat map reveals more clearly a gradual transition in ζ values across regimes of synchronization, indicating diverse synchronization behaviors with changes in coupling parameters. In particular, we can identify different regimes across the map based on the change in color patterns in the heat map. Specifically, moving diagonally starting from the bottom left corner, we label the dominantly orange-red region as Region A'; this is followed by a distinct thin ridge of dark red, which we label as Transition Belt B'. Beyond this Transition Belt B' is Region C', where the color steadily changes from yellow to bluish-green. We note immediately that we can visually identify a speckled pattern, suggesting that the transition need not be smooth as a function of coupling. We also see visually that these transitions are less sensitive to coupling (are slower) in Regions A' and C', while they are more rapid in Transition Belt B'.

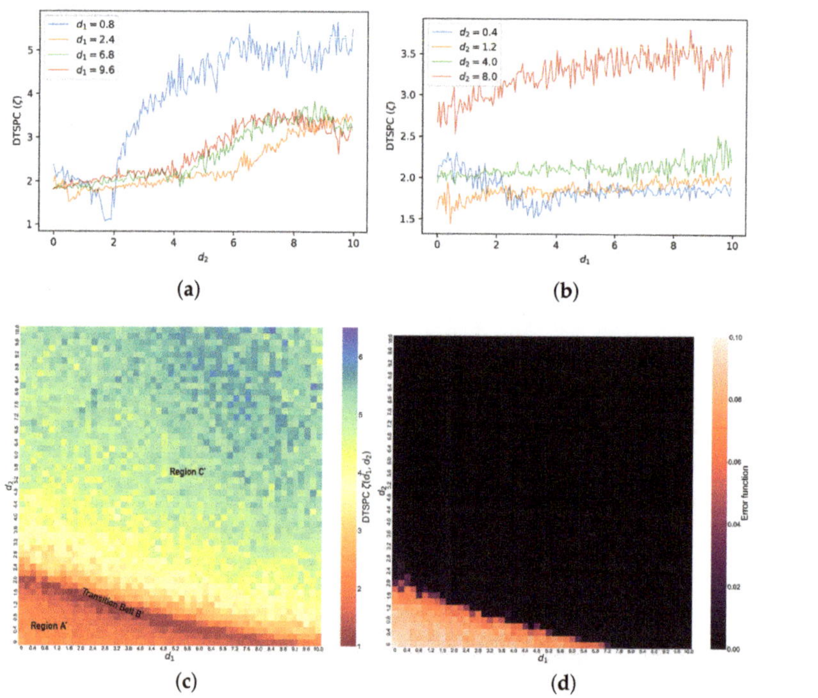

Figure 8. Line graphs of (**a**) ζ versus d_2 when $d_1 = 0.8, 2.4, 6.8, 9.6$ and (**b**) ζ versus d_1 when $d_2 = 0.4, 1.2, 4.0, 8.0$. (**c**) Heat map of ζ (DTSPC) with changing coupling strengths in coupled identical Lorenz systems, showing unsynchronized chaos in Region A', transition to synchronization in Transition Belt B', and complete synchronization in Region C'. (**d**) Heat map of error function $e(d_1, d_2)$ with changing coupling strengths in coupled identical Lorenz systems.

As before, we specify the corresponding dynamical behaviors of the system under different regimes by examining examples from each, as shown in Figure 9. The first example is when $d_1 = 1.4, d_2 = 0.8$ in Region A', shown in Figure 9a–c. Region A' represents unsynchronized chaos, where x_1 and y_1 display the random relationship shown in the phase space trajectories (Figure 9a). Similarly, in the time series (Figure 9b), the difference between x_1 and y_1 remains significant and irregular. This results in its constant crossing of zeros, with few segments of the time series showing larger numbers of consecutive peaks. Thus, there are mainly 0 or 1 consecutive peaks according to our criterion, indeed, as shown in Figure 9c, which leads to the low ζ value. In Transition Belt B', dynamics change quickly with parameters, as shown in Figure 9d–f when $d_1 = 3.0, d_2 = 1.2$. The phase space trajectories (Figure 9d) show that the relationship between x_1 and y_1 becomes mainly linear, and in time series (Figure 9e), accordingly, the difference between x_1 and y_1 gradually decreases to zero, where there remains some bursts at the beginning in the first 20 time steps. This time series frequently crosses the horizontal axes such that the count for consecutive peaks remains even more concentrated at 0 or 1 than in Region A', so ζ for systems in the Transition Belt B is even lower. The third example when $d_1 = 5.0, d_2 = 5.0$ is chosen from Region C' and represents complete synchronization. In the phase space trajectories (Figure 9g), the relationship between x_1 and y_1 is even more strictly linear, and the difference time series (Figure 9h) reduces to zero immediately after the two Lorenz systems are coupled. However, there could be nuanced and extremely frequent ringing patterns above or below the zero line (red horizontal line), as the number of consecutive peaks can range from −600 to 600, as shown in Figure 9i. Hence, this analysis reveals overall that as a function of coupling parameters, the dynamical synchronization as quantified by ζ starts low, and then with the increase in coupling strengths, it experiences first a decrease then an increase, which matches the general trend of the heat map (Figure 8c). In summary, we see that coupling in general increases synchronization, though the transition can be abrupt and non-monotonic in this case, and that the DTSPC captures this appropriately.

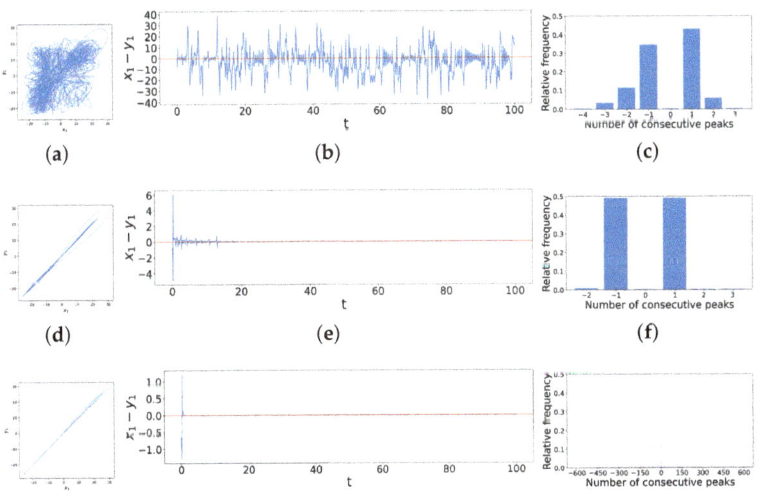

Figure 9. Phase space trajectories, $x_1 - y_1$ difference time series and peak distribution bar charts for different regimes of coupled identical Lorenz systems. (**a–c**) correspond to a case in Region A' in Figure 8c when $d_1 = 1.4, d_2 = 0.8$; (**d–f**) correspond to Transition Belt B' in Figure 8c when $d_1 = 3.0, d_2 = 1.2$; (**g–i**) correspond to Region C' in Figure 8c when $d_1 = 5.0, d_2 = 5.0$.

It is worth noting that our results show that the relationship between the level of synchronization and the value of $\zeta(d_1, d_2)$ is rather unconventional, namely, when the entropy ζ increases, the level of synchronization increases as well. This is in fact due to the reason that practical synchronization (Figure 7n), the ultimate state for coupled dissimilar systems, has ringing patterns around 0, but complete synchronization (Figure 9h), the ultimate state for coupled identical systems, might not, resulting in a lack of zero-crossings and therefore huge values in peak series, leading to a larger ζ value. Hence, $\zeta(d_1, d_2)$ is not a direct quantification of synchronization that draws a linear relationship between the two, but is instead a measure of the complexity of the differences in trajectories between two synchronized chaotic systems.

We also note that both heat maps (as shown in Figures 6c and 8c) show that the synchronization is not locally smooth as a function of coupling strength, given the speckled pattern or occasional grids visibly brighter or darker than their surroundings. However, if we increase the time interval for which we generate the peak series and to which we apply our DTSPC technique, the speckled patterns become significantly less obvious. Below, we show the DTSPC heat maps with a time interval 10 times longer ($t \in [0, 1000]$) than what we took before ($t \in [0, 100]$). Notice that we use the same color bar scale for the two heat maps for coupled dissimilar Lorenz (Figure 10a,b). Figure 10a is in fact an enhanced version of Figure 10b with more refined transition boundaries and less obvious speckled patterns. The case is similar for coupled identical Lorenz (Figure 10c,d). You may notice that in Figure 10c, we allow for a bigger scale of ζ values on the color bar because some values reach beyond 6.5, which was our previous color bar maximum. This, we believe, is due to the fact that complete synchronization is achieved after coupling identical Lorenz systems, so a longer time interval taken will result in consecutive peaks even more dispersed than the case in Figure 9i. However, a darker color in the region for complete synchronization does not affect how accurate our DTSPC heat map, using any sufficiently long time interval, can capture the overall regimes of different synchronization behaviors. Heat maps generated after taking a 10 times longer time interval for analysis prove to be more precise, yet we would like to point out that the computational demand is 10 times higher than our previous heat maps generated using a shorter time interval, while both are accurate enough for reflecting and distinguishing the diverse dynamics at different levels of synchronization.

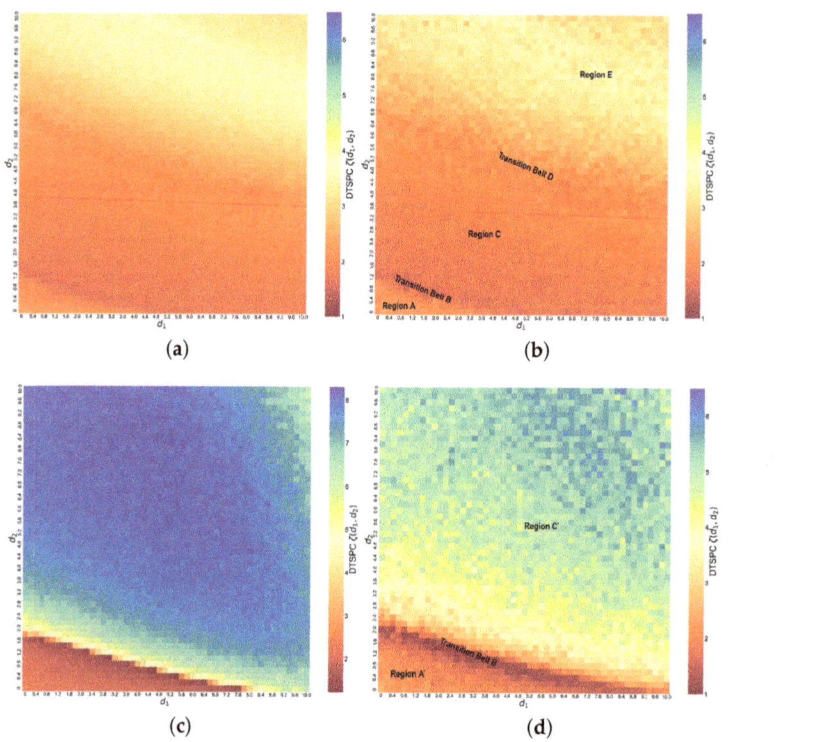

Figure 10. (**a**) Heat map for coupled dissimilar Lorenz systems with $\zeta(d_1, d_2)$ computed using time interval $[0, 1000]$ with original time step 0.001. (**b**) Original heat map for coupled dissimilar Lorenz systems (same as Figure 6c, with time interval $[0, 100]$ and time step 0.001). (**c**) Heat map for coupled identical Lorenz systems with $\zeta(d_1, d_2)$ computed using time interval $[0, 1000]$ with original time step 0.001. (**d**) Original heat map for coupled identical Lorenz systems (same as Figure 8c, with time interval $[0, 100]$ and time step 0.001).

With our DTSPC technique, we can also investigate the synchronization of coupled Lorenz systems with other different initial conditions or different parameters. The DTSPC metric, while detecting slight fluctuations in the synchronization dynamics for systems with other initial conditions, reveals that initial conditions do not affect synchronization regimes much, while the systems' parameter values do have a nontrivial impact. Further detailed discussion of these can be found in Appendix B. However, we would like to emphasize that these discoveries are beyond the scope of this paper, which aims to provide a new metric to quantitatively measure chaos synchronization and verify its accuracy.

5. Discussion and Conclusions

In this paper, we computed the Shannon entropy of peak population distributions in the difference time series of chaotic coupled Lorenz systems, introducing a Difference Time Series Peak Complexity (DTSPC) diagnostic for chaos synchronization. This digitizes a characteristic ringing pattern that changes as a function of parameter in the dynamics of the difference variable. We find that the DTSPC entropic analysis of extracted peak data compresses useful information and in particular is able to distinguish and quantitatively measure via a single metric the differences in system behavior during the transition from non-synchronization to synchronization. We are thus able to more systematically study variations in synchronization across parameter values and initial conditions, and present both the range of behaviors and its parametric dependence via a single heat map. In particular, in the test cases of our coupled identical and dissimilar Lorenz systems, we can distinguish a variety of synchronization behaviors, clarifying that though system behaviors can be macroscopically classified into non-synchronization or complete synchronization, such rough classification omits detailed behaviors, including non-monotonic parameter dependence. We again emphasize that these behaviors are not visible when examining phase space trajectories and can be constructed using substantially less access to the complete time series; these results are also robust to minor noisy perturbations.

In conclusion, the success of our Difference Time Series Peak Complexity (DTSPC) diagnostic in creating digitized representation of diverse synchronization behaviors from phase space trajectories and difference time series demonstrates the effectiveness of using Shannon entropy to quantitatively measure chaos in systems with peak dynamics, specifically utilizing the statistical information behind peak distribution. For further development of this research, one can test and evaluate the applicability of entropy and its variations on chaotic systems without significant peak dynamics or of higher dimensions. Additionally, the speckled patterns in the heat maps (Figures 6c and 8c) suggest complex boundaries between different behaviors as a function of coupling parameters when using a finite time interval; however, from Figure 10a,c, we conclude that such complex boundaries become smoother when using a longer time interval, though doing so is computationally expensive. Finally, while we note that it would be possible to fit the boundaries between regimes with approximate curves that would allow us to extract parametric boundaries, it is not our goal to specify these details for any specific system. We hope that this inspires those studying or using synchronization in Lorenz—and other—systems to further develop the connection of our technique to system particulars and analytical details.

Author Contributions: Conceptualization, Z.L. and A.K.P.; methodology, Z.L.; formal analysis: Z.L.; software, Z.L.; validation, Z.L. and A.K.P.; formal analysis, Z.L.; investigation, Z.L.; writing—original draft preparation, Z.L.; writing—review and editing, A.K.P.; visualization, Z.L.; supervision, A.K.P.; project administration, A.K.P. All authors have read and agreed to the published version of the manuscript.

Funding: This research received no external funding.

Institutional Review Board Statement: Not applicable.

Informed Consent Statement: Not applicable.

Data Availability Statement: The computer code and data used in this manuscript are available from the authors on reasonable request.

Conflicts of Interest: The authors declare no conflicts of interest.

Abbreviations

The following abbreviations are used in this manuscript:

DTSPC Difference Time Series Peak Complexity

Appendix A. Heat Maps of Normalized DTSPC

Based on the definition of normalized entropy, the normalized DTSPC is defined by

$$\begin{aligned}\overline{\zeta(d_1,d_2)} &= \frac{-\sum_{i=1}^{n} p_i \log_2 p_i}{-\sum_{i=1}^{n} \frac{1}{n} \log_2 \frac{1}{n}} \\ &= -\frac{\sum_{i=1}^{n} p_i \log_2 p_i}{\log_2 n} \\ &= \frac{\zeta(d_1,d_2)}{\log_2 n}\end{aligned}$$

Below, we show the heat map figures we obtain by plotting $\overline{\zeta(d_1,d_2)}$ over d_1 and d_2. Comparing Figure 1a with Figure 6c and Figure 1b with Figure 8c, we notice that the different regimes in the heat maps of unnormalized DTSPC are equally distinguished in the normalized DTSPC, but the coloring in the normalized DTSPC heat maps is inconsistent with the dynamics of the Lorenz systems. For example, both the thin transition belt at the bottom of Figure 1a and the top-right region are colored in blue, meaning that they have similar $\overline{\zeta(d_1,d_2)}$ values, yet in Section 4, we could see that the two regions represent very different behaviors of the coupled dissimilar Lorenz systems, corresponding to Figure 7d–f and Figure 7m–o, respectively. Similar issues arise when we notice that both regions below and above the light orange transition belt in Figure 1b are colored in the same green, yet they also represent different behaviors of the systems. The colors in the heat maps of normalized DTSPC do not change uni-directionally with the change in level of synchronization, leading to ambiguity in interpreting the behaviors of the systems using the heat maps. Hence, we prefer unnormalized DTSPC.

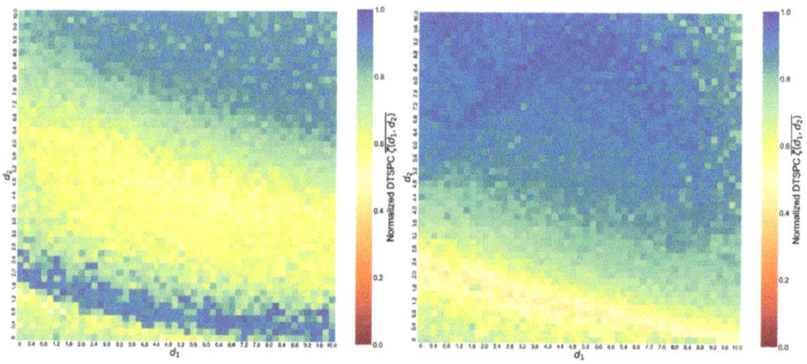

(a) **(b)**

Figure A1. Heat maps of normalized DTSPC ($\overline{\zeta(d_1,d_2)}$ for (**a**) coupled dissimilar Lorenz systems with initial conditions $x_1 = y_1 = 0, x_2 = y_2 = 1, x_3 = y_3 = 0$ and (**b**) coupled identical Lorenz systems with initial conditions $x_1 = 0, x_2 = 1, x_3 = 0, y_1 = y_2 = y_3 = 1$. Initial conditions are consistent with those used throughout the paper.

Appendix B. Heat Maps Generated Under Other Initial Conditions and Parameters

Below, we present our findings after generating heat maps of $\zeta(d_1, d_2)$ for coupled dissimilar Lorenz systems with other initial conditions (Figure 2a), coupled identical Lorenz systems with other initial conditions (Figure 2b) and coupled dissimilar Lorenz systems with different parameter values than what we have used throughout this paper (Figure 2c). Interestingly, Figure 2a,b are generally similar to the heat maps generated using our original initial conditions, where although locally there are different speckled patterns, the overall trends of coloring of the heat maps are preserved. This reveals that the differences in initial conditions barely affect the dynamics of synchronization of coupled Lorenz systems as long as the coupling method and strengths are kept identical. Figure 2c is the heat map for coupled dissimilar systems with parameters $\rho_1 = 28, \rho_2 = 29$, while we have used $\rho_1 = 50, \rho_2 = 40$ for coupled dissimilar Lorenz systems throughout the paper. This heat map presents a slightly different overall trend in the coloring than that in Figure 6c, the former with a larger orange region in the top-right corner. After verifying using the phase space trajectories, the difference time series and the peak distribution bar charts, we confirm that the orange region in the top-right corner in Figure 2c does correspond to when there start to occur more single peaks with large coupling strengths.

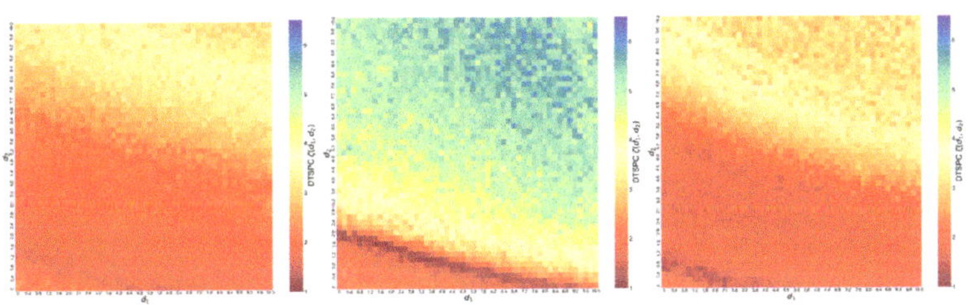

(a) (b) (c)

Figure A2. Heat maps of (**a**) coupled dissimilar Lorenz systems with initial conditions $x_1 = 0, x_2 = 1, x_3 = 0, y_1 = 5, y_2 = 6, y_3 = 5$, (**b**) coupled identical Lorenz systems with initial conditions $x_1 = 3, x_2 = 1, x_3 = 2, y_1 = 8, y_2 = 6, y_3 = 5$ and (**c**) coupled dissimilar Lorenz systems with parameters $\rho_1 = 28, \rho_2 = 29$ and same initial conditions $x_1 = y_1 = 0, x_2 = y_2 = 1, x_3 = y_3 = 0$.

References

1. Poincaré, H. Introduction. *Acta Math.* **1890**, *13*, 5–7. [CrossRef]
2. Poincaré, H. *The Three-Body Problem and the Equations of Dynamics*; Springer: Berlin/Heidelberg, Germany, 2017.
3. Lorenz, E.N. Deterministic nonperiodic flow. *J. Atmos. Sci.* **1963**, *20*, 130–141. [CrossRef]
4. Lorenz, E.N. The problem of deducing the climate from the governing equations. *Tellus* **1964**, *16*, 1–11. [CrossRef]
5. May, R.M. Simple mathematical models with very complicated dynamics. *Nature* **1976**, *261*, 459–467. [CrossRef] [PubMed]
6. Eroğlu, D.; Lamb, J.S.W.; Pereira, T. Synchronisation of chaos and its applications. *Contemp. Phys.* **2017**, *58*, 207–243. [CrossRef]
7. Lü, J.; Zhou, T.; Zhang, S. Chaos synchronization between linearly coupled chaotic systems. *Chaos Solitons Fractals* **2002**, *14*, 529–541. [CrossRef]
8. Liu, Y. Circuit implementation and finite-time synchronization of the 4D Rabinovich hyperchaotic system. *Nonlinear Dyn.* **2011**, *67*, 89–96. [CrossRef]
9. Carroll, T.; Pecora, L. Synchronizing chaotic circuits. *IEEE Trans. Circuits Syst.* **1991**, *38*, 453–456. [CrossRef]
10. Elsonbaty, A.; Elsaid, A.; Nour, H. Circuit realization, chaos synchronization and estimation of parameters of a hyperchaotic system with unknown parameters. *J. Egypt. Math. Soc.* **2014**, *22*, 550–557. [CrossRef]
11. Petrzela, J. Chaos in analog Electronic Circuits: Comprehensive review, solved problems, open topics and small example. *Mathematics* **2022**, *10*, 4108. [CrossRef]
12. Harb, A.; Jabbar, N.A. Controlling Hopf bifurcation and chaos in a small power system. *Chaos Solitons Fractals* **2003**, *18*, 1055–1063. [CrossRef]
13. Shahverdiev, E.; Hashimova, L.; Hashimova, N. Chaos synchronization in some power systems. *Chaos Solitons Fractals* **2008**, *37*, 827–834. [CrossRef]
14. Ivanov, I.L. Synchronization of Chaotic Power System with Delay Under Impulsive Perturbations. *Int. Appl. Mech.* **2023**, *59*, 480–492. [CrossRef]
15. Ditto, W.L. Applications of chaos in biology and medicine. *AIP Conf. Proc.* **1996**, *376* , 175–201. [CrossRef]

16. Javan, A.A.K.; Jafari, M.; Shoeibi, A.; Zare, A.; Khodatars, M.; Ghassemi, N.; Alizadehsani, R.; Gorriz, J.M. Medical images encryption based on Adaptive-Robust Multi-Mode synchronization of CHEN Hyper-Chaotic Systems. *Sensors* **2021**, *21*, 3925. [CrossRef]
17. Lamba, P.; Hudson, J.L. Experiments on bifurcations to chaos in a forced chemical reactor. *Chem. Eng. Sci.* **1987**, *42*, 1–8. [CrossRef]
18. Khan, A.; Khan, T.; Chaudhary, H. Chaos controllability in chemical reactor system via active controlled hybrid projective synchronization method. *AIP Conf. Proc.* **2022**, *2435*, 020054. [CrossRef]
19. Boccaletti, S.; Kurths, J.; Osipov, G.V.; Valladares, D.; Zhou, C. The synchronization of chaotic systems. *Phys. Rep.* **2002**, *366*, 1–101. [CrossRef]
20. Femat, R.; Solís-Perales, G. On the chaos synchronization phenomena. *Phys. Lett. A* **1999**, *262*, 50–60. [CrossRef]
21. Glendinning, P. *Stability, Instability and Chaos*; Cambridge University Press: Cambridge, UK, 1994. [CrossRef]
22. Lyapunov, A.M. The general problem of the stability of motion. *Int. J. Control* **1992**, *55*, 531–534. [CrossRef]
23. Shiryayev, A.N. *New Metric Invariant of Transitive Dynamical Systems and Automorphisms of Lebesgue Spaces*; Springer Nature: Berlin/Heidelberg, Germany, 1993; pp. 57–61. [CrossRef]
24. Xiong, P.Y.; Jahanshahi, H.; Alcaraz, R.; Chu, Y.M.; Gómez-Aguilar, J.; Alsaadi, F.E. Spectral Entropy Analysis and Synchronization of a Multi-Stable Fractional-Order Chaotic System using a Novel Neural Network-Based Chattering-Free Sliding Mode Technique. *Chaos Solitons Fractals* **2021**, *144*, 110576. [CrossRef]
25. Nakagawa, Y.O.; Sárosi, G.; Ugajin, T. Chaos and relative entropy. *J. High Energy Phys.* **2018**, *2018*, 2. [CrossRef]
26. Rocha, J.L.; Carvalho, S. Information theory, synchronization and topological order in complete dynamical networks of discontinuous maps. *Math. Comput. Simul.* **2020**, *182*, 340–352. [CrossRef]
27. Caneco, A.; Rocha, J.L.; Grácio, C. Topological entropy in the synchronization of piecewise linear and monotone maps: coupled duffing oscillators. *Int. J. Bifurc. Chaos* **2009**, *19*, 3855–3868. [CrossRef]
28. Rosso, O.A.; Blanco, S.; Yordanova, J.; Kolev, V.; Figliola, A.; Schürmann, M.; Başar, E. Wavelet entropy: A new tool for analysis of short duration brain electrical signals. *J. Neurosci. Methods* **2001**, *105*, 65–75. [CrossRef]
29. Piccardi, C.; Rinaldi, S. Optimal control of chaotic systems via peak-to-peak maps. *Phys. D Nonlinear Phenom.* **2000**, *144*, 298–308. [CrossRef]
30. Shannon, C.E. A mathematical theory of communication. *Bell Syst. Tech. J.* **1948**, *27*, 379–423. [CrossRef]

entropy

MDPI

Article

Local Predictors of Explosive Synchronization with Ordinal Methods

I. Leyva [1,2], Juan A. Almendral [1,2], Christophe Letellier [3] and Irene Sendiña-Nadal [1,2,*]

1 Complex Systems Group & GISC, Universidad Rey Juan Carlos, 28933 Móstoles, Spain; inmaculada.leyva@urjc.es (I.L.); juan.almendral@urjc.es (J.A.A.)
2 Center for Biomedical Technology, Universidad Politécnica de Madrid, 28223 Pozuelo de Alarcón, Spain
3 Campus Universitaire du Madrillet, Rouen Normandie Université-CORIA, F-76800 Saint-Etienne du Rouvray, France; christophe.letellier@coria.fr
* Correspondence: irene.sendina@urjc.es

Abstract: We propose using the ordinal pattern transition (OPT) entropy measured at sentinel central nodes as a potential predictor of explosive transitions to synchronization in networks of various dynamical systems with increasing complexity. Our results demonstrate that the OPT entropic measure surpasses traditional early warning signal (EWS) measures and could be valuable to the tools available for predicting critical transitions. In particular, we investigate networks of diffusively coupled phase oscillators and chaotic Rössler systems. As maps, we consider a neural network of Chialvo maps coupled in star and scale-free configurations. Furthermore, we apply this measure to time series data obtained from a network of electronic circuits operating in the chaotic regime.

Keywords: ordinal patterns; ordinal permutation entropy; ordinal transition network; chaotic synchronization; complex networks; early warning signals

1. Introduction

Critical transitions, or tipping points, refer to sudden and often irreversible changes in the behavior of systems occurring in various natural and engineered contexts. These phenomena are especially concerning in fields such as ecology [1,2], neuroscience [3–5], climate science [6–8], and even financial markets, where predicting such transitions is crucial for mitigating adverse outcomes [9,10]. Researchers have extensively studied early warning signals (EWSs) to identify impending transitions [11,12]. These signals include critical slowing down [13–16], intermittency [17,18], and pattern formation [15,19], which appear in many systems.

Complex networks, which describe interconnected systems such as ecosystems, brain networks, and social interactions, are particularly challenging for transition prediction. In these systems, critical transitions often manifest as changes in the collective synchronization state. The shift to a coherent state can enhance the functionality of networked systems, as seen in brain networks [20] and power grids [21]. However, this transition to synchronization can also have catastrophic consequences, such as epileptic seizures [3,5], fibromyalgia [4], or species extinction [22,23]. Understanding how network topology influences system dynamics is essential for predicting synchronization transitions, and several studies have shown that features of the structural [24,25] or functional [26] network can serve as indicators for the transition. Identification of sentinel nodes, displaying EWSs more prominently than others, has significantly improved prediction accuracy [22,27]. Sentinel nodes have been utilized in ecological networks to detect potential regime shifts and in brain networks to pinpoint critical regions for neuromodulation [4,24].

However, traditional EWS measures such as autocorrelation, fluctuation variability, or kurtosis are not universally reliable. In highly nonlinear or stochastic systems, their effectiveness is significantly diminished. For instance, as systems approach critical transitions, higher-order nonlinear effects, often overlooked in linear approximations, become increasingly important [11,12,28–30]. Additionally, traditional methods may struggle to distinguish meaningful signals from noise in systems with high-dimensional dynamics, emphasizing the need for more robust techniques [25,27,31].

Recently, machine learning (ML) methods have demonstrated the potential to address challenges by identifying complex patterns in data that conventional EWSs may overlook. Studies have demonstrated the utility of ML in predicting tipping points in oscillator ensembles [10,32–34], ecology systems [23,35] or brain networks [32,36]. Additionally, ordinal methods, which analyze the relative order of data, have emerged as a promising tool for predicting transitions [19,37,38]. By focusing on the sequence of events rather than absolute values, ordinal analysis proves to be robust against noise and non-stationarity, making it well suited for detecting subtle precursors to changes in evolving systems [39–42].

A further challenge in prediction is the occurrence of explosive transitions. This term encompasses various phenomena, such as explosive synchronization (ES) [43], sudden structural changes in adaptive systems [44], and abrupt shifts in social dynamics [45,46]. Explosive transitions have profound implications for real-world systems, as they have been linked to conditions such as epilepsy [47], hypersensitivity in brain networks [4], and massive extinctions [35]. Predicting these sudden transitions is particularly challenging because collective observables tend to remain stable until the moment of transition. Traditional EWSs often struggle to predict explosive synchronization due to its rapid onset [48]. In addition, EWSs are studied and designed for forecasting transitions mainly in stochastic systems [11]. For systems embedded in a weakly coupled network, the inputs from their neighborhood act as an incoherent source in each node's dynamics, playing a perturbation role, but cannot be considered a stochastic source. Although machine learning approaches have proven helpful in detecting explosive synchronization in small ensembles [23,35], there is still a need for alternative methods that are computationally less demanding.

Recently, ordinal analysis applied to time series from single nodes of dynamical networks [40,49] has revealed a strong correlation between ordinal pattern transition (OPT) entropy and structural and dynamical centrality in a wide range of weakly coupled heterogeneous ensembles. This suggests that ordinal local measures, particularly OPT, can effectively rank the sensitivity of nodes based on their proximity to transitions [25,27]. This work explores the ordinal transition entropy measured at sentinel central nodes as a potential predictor of explosive transitions to synchronization. Our findings indicate that the OPT entropy adds more information to traditional EWS measures and could be a valuable addition to the tools available to predict critical transitions.

2. Methods

2.1. *Ordinal Patterns and Permutation Entropy Measures*

Ordinal pattern-based methods are a family of simple and robust techniques that characterize time series simply by comparing neighbouring relative values [39,50–52]. Given any time series $x = \{x_t; t = 1, \ldots, T\}$, data are projected into a sequence of symbols of ordinal patterns of a given length $D > 1$ obtained from the comparison of consecutive ($\tau = 1$) or non-consecutive ($\tau > 1$) points in the following manner. First, the time series is divided into blocks $v_t = (x_t, x_{t+\tau}, \ldots, x_{t+(D-1)\tau})$ of size D, and then each element of the block is replaced by a number in $[1, \ldots, D]$ corresponding to its relative position when arranged in ascending order. In this way, each block v_t is assigned to one of the $D!$ possible orderings (permutations) π_ℓ in which the D elements can be ordered. Finally, the probability at which the ordinal pattern π_ℓ is found in the time series is computed as $p(\pi_\ell) = \#\{S_t \text{ is mapped into symbol } \pi_\ell\}/L$, with $L = \lfloor T/(\tau D) \rfloor$ being the total number of blocks S_t in which the time series is divided ($\lfloor\rfloor$ is the floor function). This procedure enables a reliable symbolic dynamics mapping a time series to a symbolic sequence of ordinal patterns (where each sequence of D integers receives a specific symbol). The time series has to be sufficiently long, $L \gg D!$, to obtain reliable statistics and a valuable ordinal pattern probability distribution $P = \{p(\pi_\ell), \ell = 1, \ldots, D!\}$ [50]. Throughout this work, we will use $D = 3$ (Bandt and Pompe suggest using $3 \leq D \leq 7$ for practical purposes [50]) and $T = 2000$. Although the time lag τ at which the data are sampled is not critical, this parameter could be relevant when the system under study has intrinsic time scales [53,54]. Typically, to overcome the dependency of the ordinal pattern probability distribution on the sampling time τ, its selection is usually based on time delay embedding criteria [51,55]. In addition to using a fix sampling time applied to the original time series, we will also consider a Poincaré section approach by retaining the relative maxima of the signal [56,57].

Our goal is to determine whether the probability distribution of ordinal patterns associated with the time series of a dynamical network is able to detect hidden clues in its collective state, pointing to the occurrence of an eventual abrupt phase transition. To achieve this, we will compare two entropic quantities estimated from the sequence of the permutation patterns defined above: the permutation entropy and the transition permutation entropy.

2.1.1. Ordinal Permutation Entropy

Given the D-order permutation probability distribution $P = (p(\pi_1), \ldots, p(\pi_{D!}))$, the Bandt–Pompe permutation entropy is the corresponding Shannon entropy evaluated as

$$\mathcal{S}[P] = -\sum_{\ell=1}^{D!} p_\ell \ln p_\ell, \tag{1}$$

with the criterion $0^0 = 1$. We define a normalized permutation entropy as

$$\mathcal{H}[P] = \frac{\mathcal{S}}{\mathcal{S}_{\max}} \tag{2}$$

where $\mathcal{S}_{\max}$ is the Shannon entropy of the uniform permutation probability distribution—that is, $\mathcal{S}_{\max} = \mathcal{S}[P_e]$ with $P_e = (\frac{1}{D!}, \ldots, \frac{1}{D!})$.

2.1.2. Ordinal Permutation Transition Entropy

While the previous entropy measure is solely based on the appearance of permutation patterns, transitions between patterns may reveal information about the finer temporal organization of a dynamical system [58]. We define the ordinal transition probability matrix $O_T := (p_{\ell m})$ as

$$p_{\ell m} = \frac{\#(\pi_\ell, \pi_m)}{\#(\pi_\ell)} \tag{3}$$

with $p_{\ell m}$ being the probability that pattern m follows pattern ℓ. In case $\#(\pi_\ell) = 0$ for some π_ℓ, we assume $p_{\ell m} = 0$. We move from having $D!$ patterns to $D!^2$ transitions. Thus, the time series must be longer to make the transition matrix O_T statistically significant.

Since $\sum_m p_{\ell m} = 1$, O_T is a column-stochastic matrix (weighted and directed) whose coefficients are the transition probabilities among ordinal patterns, including self-transitions.

Thus, we can characterize the dynamics of the transition patterns at the local and global level of a time series. At the local level, we can define the entropy of each pattern π_ℓ through the distribution probability of the patterns that follow it as

$$\mathcal{H}_{\pi_\ell} = -\frac{1}{lnD!}\sum_{m=1}^{D!} p_{\ell m} \ln p_{\ell m}. \tag{4}$$

which quantifies the predictability of the local transitions from the ordinal pattern π_ℓ to any other pattern [59,60].

At the global level, we measure the transitional complexity of the whole O_T as the average of the local (pattern) entropies $\mathcal{H}_{\pi_\ell}$:

$$\mathcal{H}_{\mathrm{T}} = \frac{1}{D!}\sum_{\ell=1}^{D!} \mathcal{H}_{\pi_\ell} \tag{5}$$

2.2. Dynamical Networks and Local Early Warning Indicators

We consider networks whose nodes are either continuous (a flow) or discrete (a map) dynamical systems on the phase space. In particular, as an example of flows, we will investigate networks of N diffusively coupled phase oscillators and chaotic Rössler systems, while for maps, we consider a neural network of Chialvo maps. The general form of the equations governing the dynamics in each case are, respectively,

$$\dot{\mathbf{x}}_i = \mathbf{F}_i(\mathbf{x}_i) + d\sum_{j=1}^{N} a_{ij}\mathbf{H}(\mathbf{x}_j - \mathbf{x}_i), \tag{6}$$

where $\mathbf{x}_i(t)$ is the vector state of node i at time t and $\mathbf{F}_i$ is the local vector field, and for the map,

$$\mathbf{x}_i(t+1) = \mathbf{M}_i(\mathbf{x}_i(t)) + d\sum_{j=1}^{N} a_{ij}\mathbf{H}(\mathbf{x}_j(t) - \mathbf{x}_i(t)), \tag{7}$$

where $\mathbf{x}_i(t)$ is a m-dimensional vector describing the state of each node i at the t-th iteration of the map $\mathbf{M}_i$. In both cases, $\mathbf{H}$ is the output function describing the units' interaction. The coupling architecture among them is defined by the adjacency matrix A of the graph whose coefficients $A := (a_{ij})$ are $a_{ij} = 1$, if i and j are connected, and $a_{ij} = 0$ otherwise, such that, the degree of each node $k_i = \sum_j a_{ij}$. The parameter $d = \sigma/k_{\max}$ acts as a normalized coupling strength by the maximum degree of the network $k_{\max} = \max(k_i)$ to compare different realizations of the network.

Depending on the local functions $\mathbf{F}$ and $\mathbf{M}$, the output function $\mathbf{H}$, and on the adjacency matrix A, the dynamical networks described by Equations (6) and (7) may display a transition from a non-synchronous state to a synchronized state when the strength of the coupling d is above a critical value [61]. To monitor the collective state of the network for increasing values of the control order parameter d, we compute the time-averaged stationary value of the phase order parameter,

$$R = \frac{1}{N}\left\langle \left| \sum_{j=1}^{N} e^{i\theta_j(t)} \right| \right\rangle_t \tag{8}$$

where θ_j is the associated phase of the oscillator whose state is $\mathbf{x}_j(t)$. The phase-order parameter R accounts for the level of phase synchronization ($0 \leq R \leq 1$), and $\langle\rangle_t$ stands for the time average along a sufficiently large time series once a stationary state is reached.

In general, we compute two synchronization curves for the phase order parameter R, the forward and backward continuations. The forward curve is performed by gradually increasing the value of d, starting from random initial conditions, and computing the stationary value of R, and then taking the final stationary state as the initial state for the next value of d. For the backward curve, d is gradually decreased starting from initial conditions close to a stable synchronous state and the network eventually returns to a non synchronous state when $d = 0$.

Transitions to synchronization can be abrupt, also called explosive transitions, when the order parameter remains $R \simeq 0$ as the coupling d increases until suddenly it jumps to $R \simeq 1$ and the network stays in a synchronized state. They often exhibit hysteresis with the backward transition occurring at lower values of the coupling strength. Explosive phase synchronization appears when the interaction between the network structure and the local node dynamics hinder the formation of microscopic seeds of synchronization as the coupling increases, a mechanism underlying the emergence of a collective coherent state in smooth, second-order transitions. As a result, the system remains incoherent despite very high coupling strengths, only to abruptly transition to a fully synchronized state [43].

Most practical ways to create this effect are based on inducing *frequency dissasortativity*, ensuring that each node is dynamically isolated [62]. When the network is structurally heterogeneous, this dynamical separation is easily achieved by imposing a degree–frequency correlation [63,64]. Therefore, in this work, we study highly heterogeneous topologies (star and scale-free (SF) networks), where the nodes are distributed such that their inner temporal scale is proportional to their degree. In the case of the star configuration, this rule is slightly modified by adding some randomness to the leaves, that is, nodes with $k_i = 1$. This quenched disorder is drawn from a uniform distribution $\mathcal{U}(0, \xi)$ and avoids a persistent synchronous state when reducing the coupling strength in the backward route to desynchronization. This parametric uncertainty is very small, producing slight variations of the critical coupling strength for the ES when performing different simulation instances, and preserves the frequency gap between the hub and the leaves; otherwise, the transition to synchronization is no longer abrupt. Other sources of stochasticity come from the unavoidable tolerances on the electronic component parameters and the noise from uncontrolled sources in the experimental setup that will be described in Section 3.4. Details of the dynamical ranges used in each system are provided in their respective sections.

As advanced in Section 2.1, we propose using a local entropic measure as an EWS of an abrupt transition in a dynamic network. For each node i, we consider one of the accessible scalar variables of the vector state $x_i(t)$, or, when convenient, a function of that variable, and compute the corresponding permutation transition entropy $\mathcal{H}_{\mathrm{T}}^i$. We expect that nodes with the same degree k will have similar dynamical patterns within the network [49], allowing us to define the averaged $k-$class permutation transition entropy

$$\langle \mathcal{H}_{\mathrm{T}} \rangle_k = \frac{1}{N_k} \sum_{i|k_i=k} \mathcal{H}_{\mathrm{T}}^i, \tag{9}$$

where N_k is the number of nodes having degree k, and $\langle \rangle_k$ is an ensemble average. Similarly, we define a k-class average for the permutation entropies of those nodes with the same degree $\langle \mathcal{H} \rangle_k$.

Finally, to compare the performance of the proposed local entropic measure given by Equation (9) with other common early warning indicators for critical transitions used in the literature [1], in some cases, we will monitor the amplitude of the fluctuations of each node time series $x_i(t)$, $\sigma_{f,i} = \sqrt{\langle x_i(t)^2\rangle - \langle x_i(t)\rangle^2}$, and the normalized autocorrelation

$$AC_i(l) = \frac{\sum_{t=1}^{T-l} x_i(t) \cdot x_i(t+l)}{\sqrt{\sum_{t=1}^{T-l} x_i^2(t) \cdot \sum_{t=1}^{T-l} x_i^2(t+l)}} \tag{10}$$

with lag $l = 1$.

3. Results

In the following sections, we present results demonstrating that the transition entropy measured at the hubs of a network is a highly effective metric anticipating the onset of abrupt synchronization transitions in networks of various dynamical systems with increasing complexity. These systems include phase oscillators, a map emulating neuronal firing dynamics, and chaotic amplitude oscillators. Additionally, we apply this measure to experimental time series data obtained from a network of electronic circuits.

3.1. Abrupt Synchronization Transition in Kuramoto Phase Oscillators

The first studied system is the Kuramoto network,

$$\dot{\theta}_i = \omega_{o,i} + d \sum_{j=1}^{N} a_{ij} \sin(\theta_j - \theta_i) \tag{11}$$

where the coupling d is the coupling strength and $\omega_{o,i}$ is the natural frequency of node i. In the example in Figure 1, we use a $N = 31$ star-like structure, and to achieve explosive synchronization, we impose a frequency–degree correlation in the nodes [43,63], with $\omega_{o,h} = 1.3$ for the hub and $\omega_{o,l} = 1 + 0.005\epsilon$ for the leaves with ϵ a $[0, 1]$.

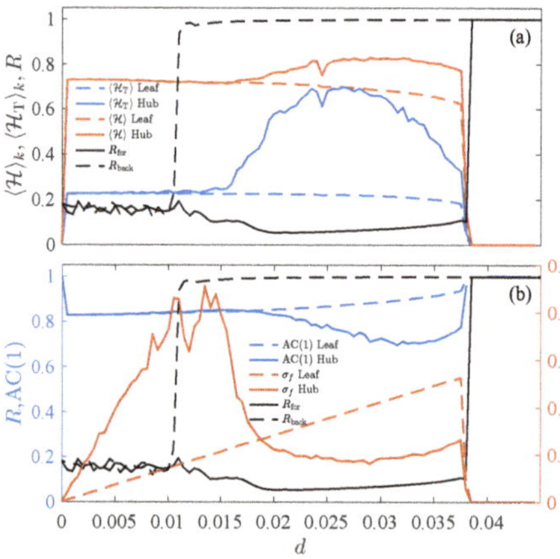

Figure 1. Abrupt synchronization transition in Kuramoto phase oscillators. (**a**) $\langle\mathcal{H}\rangle_k$ and $\langle\mathcal{H}_T\rangle_k$ as a function of the normalized coupling $d = \sigma/(N-1)$ in a Kuramoto star network of size $N = 31$ with $\omega_{o,h} = 1.3$ for the hub and $\omega_{o,l} = 1 + 0.005\epsilon$ for the leaves with ϵ a $[0, 1]$ random number. Black lines correspond to the Kuramoto order parameter R for the forward (continuous lines) and the backward (dashed lines) transitions. (**b**) Normalized 1-Lag autocorrelation $AC(1)$ (blue lines, left y-axis for scale) and standard deviation σ_f (orange lines, right y-axis for scale) of the signal fluctuations. Input data are a $\tau = 200$ periodic sampling of the instantaneous frequency $\omega_i(t)$ series. Results averaged over 10 random instances.

When searching for signals that indicate an imminent transition, one important decision is determining which observable to monitor, as not all the system outputs are equally sensitive to changes in the state [13,24,65,66]. In networked systems, not all nodes are equally effective as predictors of transitions, and their sensitivities may vary due to internal dynamics or their position within the network structure [10,27]. In a Kuramoto network, using the global instantaneous synchronization state $R(t)$ as an EWS could provide moderate utility [48], but this requires access to a global variable that may not be readily available. In our case, we are more interested in local measures. Therefore, in Figure 1, we use the instantaneous frequency of each oscillator $\omega_i(t)$, which is recorded with a sampling time of $\tau = 200$ integration steps.

As seen in ES transitions, the average order parameter R does not provide information about the transition and remains close to its initial value until it suddenly jumps to its maximum. We then evaluate the ordinal permutation entropy $\langle \mathcal{H} \rangle_k$ and the ordinal permutation transition entropy $\langle \mathcal{H}_\mathrm{T} \rangle_k$ at both $k_\mathrm{hub} = N-1$ for the hub and $k_\mathrm{leaf} = 1$ for the leaves [Figure 1a]. For the leaves, the entropy measurements show minor sensitivity to the proximity of the transition. However, at the hub, both entropies increase noticeably when the coupling is still one-third of its critical synchronization value. Notably, the relative increase in $\mathcal{H}_\mathrm{T}$ is much more significant than that of $\mathcal{H}$, providing a clear advantage for generating an early detection alarm. Given this advantage, the remaining sections will focus on exploring $\mathcal{H}_\mathrm{T}$ as a more promising measure. Specifically, we propose as EWS measure the increase in the relative difference between the OPT entropies of the most connected and less connected nodes in the network.

We now compare the result of the local entropy measures with some of those traditionally used as early warning predictors (Figure 1), the standard deviation of the fluctuations $\sigma_{f,i}$ and the 1-lag autocorrelation AC$_i$(1) of the same data analyzed in panel (a). As for the entropies, for the leaf, none of these observables present significant variations that can be considered EWSs. However, the dispersion of fluctuations exhibits a linear growth proportional to the coupling that results from the increase in the amplitude of the frequency beat, before sharply dropping simultaneously to the transition, recovering the value for the uncoupled system $d = 0$ as expected. The AC(1) measure for the hub produces a result comparable to $\mathcal{H}$.

On the other hand, the fluctuation dispersion σ_f of the hub shows a relatively significant maximum (note that the scale of this measure refers to the right Y-axis). However, this maximum does not occur near the synchronous forward transition we are studying, but rather in the region associated with the critical coupling for *desynchronization* in the backward transition, a much smaller coupling due to the wide hysteresis. This interesting effect would, however, lead to a false alarm regarding the ES transition.

3.2. Abrupt Synchronization Transitions in Coupled Chialvo Maps

To illustrate the performance of the ordinal transition entropy to foresee an upcoming network synchronization event in networks of coupled maps, we consider a neural network of Chialvo maps. The Chialvo map is a two-dimensional neural model that produces periodic or chaotic burst-spike dynamics [67], and it has the following form:

$$\begin{aligned} x_i(t+1) &= x_i^2(t)\exp\left(y_i(t) - x_i(t)\right) + I_i + d\sum_{j=1}^{N} a_{ij}(x_j(t) - x_i(t)), \\ y_i(t+1) &= a y_i(t) - b x_i(t) + c \end{aligned} \tag{12}$$

where $x_i(t)$ acts as a membrane potential and $y_i(t)$ as a restoring variable. The subscripts t represent a discretized time evolution ($t = 1, \ldots, T$), while subscript i refers to a node (neuron) of the network ($i = 1, \ldots, N$). Parameters $a = 0.89$, $b = 0.6$, and $c = 0.28$ are chosen such that the neural dynamics are periodic or quasi-periodic, whose frequency and amplitude is modulated by the parameter I, which plays the role of an ion current injected into the neuron. As I increases, the frequency increases monotonically [67] (an extensive study of the parameter space of this map is reported in Ref. [68]). Here, we impose a linear dependence of the constant bias parameter I with the neuron connectivity as $I_i = 0.049 + \alpha k_i$ to induce an explosive synchronization when increasing the control order parameter d.

Figure 2a,b shows an example of the time evolution of the activation variable x_t and phase portrait (x_t, y_t) of an isolated neuron for $I_i = 0.05$. The red segments connect the spike maxima with $x_t > 0.5$, used to define a phase $\theta_i(t)$ and be able to monitor the network synchronization as described in Section 2.2. The phase increases by 2π each time there is a spike, and it is linearly interpolated between spikes [69] as

$$\theta_i(t) = 2\pi n_i + 2\pi \frac{t - t_{i,n}}{t_{i,n+1} - t_{i,n}}, \quad t_{i,n} \leqslant t < t_{i,n+1} \tag{13}$$

where $t_{i,n}$ is the time at which the i-th neuron fires its n_i-th spike.

The sequence of maxima is used to construct the ordinal patterns and compute the ordinal transition entropy of Chialvo maps coupled in star and scale-free configurations. Figure 2 shows the evolution of the k-class transition entropy $\langle \mathcal{H}_T \rangle_k$ for nodes of increasing degree (colored curves) belonging to a star of size $N = 31$ (panel c) and to a scale-free network of size $N = 100$ (panel d) as the coupling d increases. In a star configuration, there are only two classes of sensors: the hub and the leaves. In contrast, a scale-free network exhibits a more diverse degree distribution, where many nodes have lower degrees while only a few have significantly larger degrees. The nodes with higher degrees are particularly good candidates for signaling an abrupt phase transition. As in the Kuramoto case, we observe how, in the two configurations, the largest degree class nodes exhibit a pronounced growth of their transition entropy long before the sudden jump in the order phase parameter R (black continuous line), while the lowest degree ones do not provide information about the abrupt change occurring in the network phase synchronization.

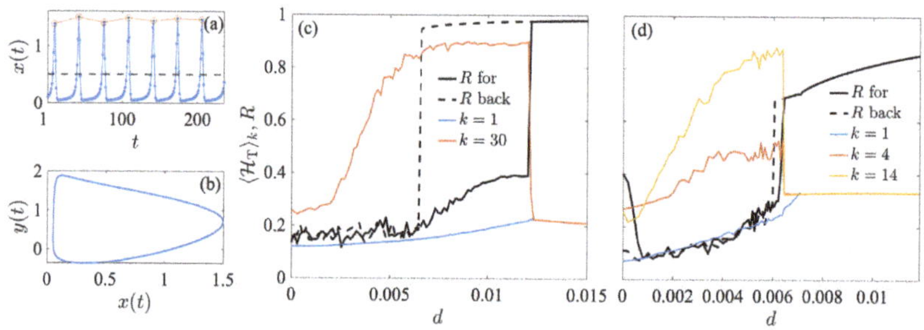

Figure 2. Anticipating abrupt synchronization in networks of coupled Chialvo maps. (**a**) Time series for x_t and (**b**) phase portrait $x_t - y_t$ of an isolated neural model (12) for $a = 0.89$, $b = 0.6$, $c = 0.28$, and $I = 0.05$. Red segments in (**a**) connect spike maxima with $x_t > 0.5$. (**c**,**d**) $\langle \mathcal{H}_T \rangle_k$ for several node degrees (see legend) and Kuramoto order parameter R as a function of the coupling strength d for (**c**) a star graph of size $N = 31$ with $I = 0.050$ for the hub and $I = 0.049 + 10^{-4}\epsilon$ for the leaves, with ϵ, a random number uniformly distributed in $[0, 1]$, and (**d**) a scale-free network of $N = 100$ nodes of mean degree 4, with $I(k) = 0.049 + \alpha k$ and $\alpha = 3 \times 10^{-5}$. In each case, there is an abrupt transition to phase synchronization (black solid line) and an abrupt transition back to desynchronization (black dashed line) with a large hysteresis behavior for the star configuration.

3.3. Predicting Explosive Transitions in Networks of Rössler Oscillators

We now consider a dynamical network composed of chaotic oscillators, adding an extra level of complexity to the local dynamics. We chose Rössler systems [70] coupled through the y variable, whose governing equations are the following:

$$\begin{aligned} \dot{x}_i &= -w_i y_i - z_i, \\ \dot{y}_i &= w_i x_i + a y_i + d \sum_{j=1}^{N} a_{ij}(y_j - y_i), \\ \dot{z}_i &= b + z_i(x_i - c), \end{aligned} \tag{14}$$

where $a = 0.165$, $b = 0.4$, and $c = 8.5$ are set to obtain a phase-coherent chaotic attractor, and w_i is a control parameter that directly tunes the main frequency of the chaotic oscillator. We set the linear frequency–degree correlation as $\omega_i(k) = 1.06 + \alpha k_i$ with $\alpha = 2.73 \times 10^{-4}$ to induce an explosive synchronization transition. To monitor the network state of phase synchronization, we define the phase of each Rössler oscillator as $\theta_i = \arctan(y_i/x_i)$ and compute the phase order parameter as given by Equation (8).

To investigate this dynamical network, we chose to map each node's vector state $\mathbf{x}_i = (x_i, y_i, z_i)$ to the one-dimensional time series $y_i(t_m), m = 1, \ldots, T$ generated by the Poincaré section $\mathcal{P} \equiv \{[x_i(t_m), z_i(t_m)] \in \mathbb{R}^2 \mid \dot{y}_i(t_m) = 0, \ddot{y}_i(t_m) > 0\}$ [37]—that is, we take the minima of the y time series to build the $D = 3$ permutation patterns.

Figure 3 condenses the main results, evidencing again how the hubs' transition entropy $\mathcal{H}_\mathrm{T}$ detects an explosive synchronization event in advance while the phase order parameter R does not sense any collective effect. In detail, Figure 3a shows a markedly distinct behavior of the OPT entropies for the leaf and hub of a star network. While the $\langle \mathcal{H}_\mathrm{T} \rangle$ of the leaf remains almost flat along the synchronization path, the hub's entropy rises at coupling values even before the hysteresis window starts, defined by the continuous and dashed R curves for increasing (R_for) and decreasing (R_back) strength of the coupling. In more complex situations with a heterogeneous degree distribution, as shown in Figure 3b, we observe a similar hierarchical trend where nodes in higher-degree classes begin to gain entropy at smaller coupling values. Figure 3c shows the sensitivity of the OPT entropy as a function of the node's degree depending on the distance from the abrupt transition. While for very low coupling values ($d = 0.08 \times 10^{-3}$), the $\langle \mathcal{H}_\mathrm{T} \rangle_k$ monotonously decreases with k, for values closer to the hysteresis window ($d = 0.6 \times 10^{-3}$), there is a clear cutoff around $k_c \sim 50$ indicating that nodes with $k > k_c$ are potential EWS sensors.

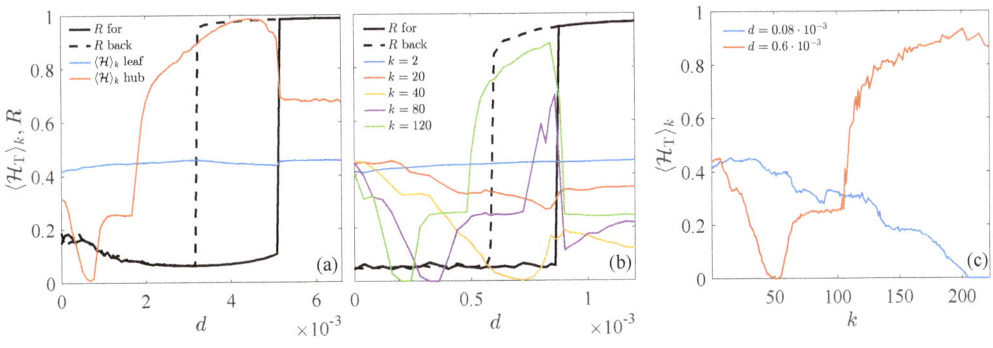

Figure 3. Predicting explosive transitions in networks of Rössler oscillators. (**a**,**b**) $\langle \mathcal{H}_\mathrm{T} \rangle_k$ for different k values (colored lines) and order parameter R for the forward (black continuous line) and backward

(black dashed line) routes to explosive synchronization/desynchronization as the coupling strength d increases/decreases for (**a**) a star of size $N = 31$ and (**b**) a $N = 500$ scale-free network ($\langle k \rangle = 4$, $\gamma = 2.25$). (**c**) $\langle \mathcal{H}_{\mathrm{T}} \rangle_k$ is a function of k for two values of the coupling strength in (**b**): a relatively small coupling value, $d = 0.08 \times 10^{-3}$ (blue line), and a value still far from the transition, $d = 0.6 \times 10^{-3} \ll d_t \sim 0.9 \times 10^{-3}$ (red line). In (**b**,**c**), each point averages the result of 64 independent network realizations.

3.4. Predicting Explosive Transitions in Networks of Electronic Circuits

Finally, we test the performance of the OPT entropy $\mathcal{H}_{\mathrm{T}}$ as an early warning measure for an experimental case of explosive synchronization. The experiment consists of an $N = 6$ star network of piecewise Rössler electronic circuits operating in the chaotic regime, as the numerical counterpart in the previous section. The circuits are configured such that the central node oscillates with a mean frequency of 3333 Hz, and the leaf nodes are set with frequencies in the range of 2240 ± 200 Hz. The transition to synchrony of this ensemble was first analyzed in Ref. [64], where the rest of the experimental details can be found and the dataset can be downloaded from [71].

In the numerical analysis presented in Section 3.3, the node coupling and the variable being analyzed correspond to y_i, while in the experiment, this role is assigned to x. To align our analysis with the results above, we extract the Poincarè section from the experimental data to generate the $D = 3$ permutation patterns. An experimental parameter controls the node's dynamics so that transition can be tuned from second-order (Figure 4a) to explosive (Figure 4b) without affecting the node's frequencies or the chaotic state. The phase reconstruction and synchronization monitoring are performed as in Section 3.3.

For a better comparison, in Figure 4, we normalize the OPT entropies of each node $\langle \mathcal{H}_{\mathrm{T}} \rangle_k$ to their respective value when the system is uncoupled, $\langle \mathcal{H}(0)_{\mathrm{T}} \rangle_k$. In the continuous transition (a), neither the hub nor the leaves show changes in their initial entropy value along the synchronization process. In contrast, during the ES transition plotted in Figure 4b, the hub OPT entropy increases by up to a factor of five when the coupling value is still far from the transition. Meanwhile, the leaf experiences a slight reduction before the transition. This result is similar to the one obtained in Figure 3a for the numerical case, which confirms the robustness of this ordinal method as an EWS for the explosive transition even in this case subject to the node's experimental heterogeneity and noisy environment.

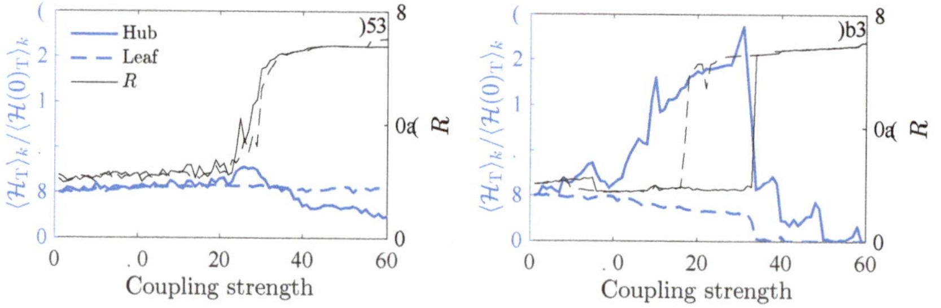

Figure 4. Normalized ordinal pattern transition entropies $\langle \mathcal{H}_{\mathrm{T}} \rangle_k / \langle \mathcal{H}(0)_{\mathrm{T}} \rangle_k$ (left y-axis) and phase order parameter R (right y-axis) as a function of the coupling strength in a star network of $N = 6$ piecewise Rössler electronic circuits for the case of (**a**) continuous and (**b**) explosive transitions to synchronization. Input data are Poincaré sections series of a voltage local maxima.

4. Conclusions

Predicting state transitions in complex systems is a difficult challenge due to the enormous variety of systems, forms of transition, and diversity of data. In the case of abrupt transitions, it is even more difficult, since many of the system's observables do not show changes during the stages before the transition. In this work, we explore the efficacy of ordinal pattern transition (OPT) entropy as a predictive tool for explosive synchronization transitions in various dynamical networks. Through simulations and experimental validations, OPT entropy can identify subtle signals of the proximity of the transitions across diverse network configurations and dynamical regimes, even outperforming traditional EWSs.

Our approach differs from previous studies because the observable used as input—such as the instantaneous frequencies of individual nodes in the case of the Kuramoto model, or the local Poincaré sections for the Rössler and Chialvo maps—is not the same as the one that undergoes ES, which is the global phase synchronization value R. In extended networked systems, it is more common to have access to local measurements rather than global synchronization data.

Our results indicate that the sensitivity of OPT entropy is particularly strong in central nodes, which can function as sentinel nodes based on this observable measure. We successfully applied this method to chaotic networks of electronic circuits, demonstrating its versatility and practical utility in real-world applications. One promising application is the prediction of epileptic seizures, yet the performance of permutation entropy from electroencephalogram recordings still remains unclear [72–75]. To our knowledge, OPT entropy has not yet been explored to analyze the transition process from normal to seizure state in spatially distributed EEG signals. The combination of a complexity measure that captures subtle changes in the temporal organization of a time series while considering the underlying network structure of the EEG signals deserves attention and further investigation. These findings highlight the potential to incorporate ordinal methods into the toolkit for studying critical transitions in complex systems, especially when traditional EWS measures struggle under high dimensionality or non-linearity conditions.

Future research could focus on refining ordinal measures to enhance their sensitivity and applicability to larger and more diverse networks or to analyze the influence in the prediction of different types of observational and dynamical noise [11,12]. Additionally, integrating machine learning with ordinal methods may prove beneficial in extending their predictive capabilities in various fields such as neuroscience, ecology, and engineering. This study paves the way for broader applications of ordinal methodologies in forecasting and mitigating the effects of critical transitions in complex dynamical systems.

Author Contributions: Conceptualization, I.L. and I.S.-N.; Methodology, I.L., J.A.A., C.L. and I.S.-N.; Investigation, I.L., J.A.A., C.L. and I.S.-N.; Writing original draft, I.L. and I.S.-N.; Writing—review & editing, I.L., J.A.A., C.L. and I.S.-N.; Visualization, I.L., J.A.A., C.L. and I.S.-N. All authors have read and agreed to the published version of the manuscript.

Funding: This research was supported by the Spanish Ministerio de Ciencia e Innovación under Projects PID2020-113737GB-I00 and PID2023-147827NB-I00.

Institutional Review Board Statement: Not applicable.

Data Availability Statement: The experimental dataset used in this study is available at https://doi.org/10.5281/zenodo.14146078 (see Ref. [71]).

Conflicts of Interest: The authors declare no conflicts of interest.

References

1. Scheffer, M.; Bascompte, J.; Brock, W.A.; Brovkin, V.; Carpenter, S.R.; Dakos, V.; Held, H.; Nes, E.H.V.; Rietkerk, M.; Sugihara, G. Early-warning signals for critical transitions. *Nature* **2009**, *461*, 53–59. [CrossRef]
2. Scheffer, M.; Carpenter, S.R.; Lenton, T.M.; Bascompte, J.; Brock, W.; Dakos, V.; van de Koppel, J.; van de Leemput, I.A.; Levin, S.A.; van Nes, E.H.; et al. Anticipating Critical Transitions. *Science* **2012**, *338*, 344–348. [CrossRef]
3. Kuhlmann, L.; Lehnertz, K.; Richardson, M.P.; Schelter, B.; Zaveri, H.P. Seizure prediction — ready for a new era. *Nat. Rev. Neurol.* **2018**, *14*, 618–630. [CrossRef]
4. Kim, M.K.; Harris, R.E.; DaSilva, A.F.; Lee, U.C. Explosive Synchronization-Based Brain Modulation Reduces Hypersensitivity in the Brain Network: A Computational Model Study. *Front. Comput. Neurosci.* **2022**, *16*, 815099. [CrossRef]
5. Costa, G.; Teixeira, C.; Pinto, M.F. Comparison between epileptic seizure prediction and forecasting based on machine learning. *Sci. Rep.* **2024**, *14*, 5653. [CrossRef]
6. Livina, V.N.; Lenton, T.M. A modified method for detecting incipient bifurcations in a dynamical system. *Geophys. Res. Lett.* **2007**, *34*, L03712. [CrossRef]
7. Lenton, T.M.; Held, H.; Kriegler, E.; Hall, J.W.; Lucht, W.; Rahmstorf, S.; Schellnhuber, H.J. Tipping elements in the Earth's climate system. *Proc. Natl. Acad. Sci. USA* **2008**, *105*, 1786–1793. [CrossRef]
8. Lenton, T. Early warning of climate tipping points. *Nat. Clim. Chang.* **2011**, *1*, 201–209. [CrossRef]
9. Moore, J.C. Predicting tipping points in complex environmental systems. *Proc. Natl. Acad. Sci. USA* **2018**, *115*, 635–636. [CrossRef]
10. George, S.V.; Kachhara, S.; Ambika, G. Early warning signals for critical transitions in complex systems. *Phys. Scr.* **2023**, *98*, 072002. [CrossRef]
11. Proverbio, D.; Skupin, A.; Gonçalves, J. Systematic analysis and optimization of early warning signals for critical transitions using distribution data. *iScience* **2023**, *26*, 107156. [CrossRef]
12. Kuehn, C.; Lux, K.; Neamţu, A. Warning signs for non-Markovian bifurcations: Colour blindness and scaling laws. *Proc. R. Soc. A Math. Phys. Eng. Sci.* **2022**, *478*, 20210740. [CrossRef]
13. Patterson, A.C.; Strang, A.G.; Abbott, K.C. When and where we can expect to see early warning signals in multispecies systems approaching tipping points: Insights from theory. *Am. Nat.* **2021**, *198*, E12–E26. [CrossRef]
14. Tirabassi, G.; Masoller, C. Correlation lags give early warning signals of approaching bifurcations. *Chaos Solitons Fractals* **2022**, *155*, 111720. [CrossRef]
15. Sardanyés, J.; Ivančić, F.; Vidiella, B. Identifying regime shifts, transients and late warning signals for proactive ecosystem management. *Biol. Conserv.* **2024**, *290*, 110433. [CrossRef]
16. Evers, K.; Borsboom, D.; Fried, E.I.; Hasselman, F.; Waldorp, L. Early warning signals of complex critical transitions in deterministic dynamics. *Nonlinear Dyn.* **2024**, *112*, 19071–19094. [CrossRef]
17. Kalitzin, S.; Petkov, G.; Suffczynski, P.; Grigorovsky, V.; Bardakjian, B.L.; Silva, F.L.D.; Carlen, P.L. Epilepsy as a manifestation of a multistate network of oscillatory systems. *Neurobiol. Dis.* **2019**, *130*, 104488. [CrossRef]
18. Vera-Ávila, V.P.; Sevilla-Escoboza, J.R.; Leyva, I. Complex networks exhibit intermittent synchronization. *Chaos* **2020**, *30*, 103119. [CrossRef]
19. Tirabassi, G. Linear theory of the spatial signatures of critical slowing down. *Phys. Rev. Res.* **2024**, *6*, 023228. [CrossRef]
20. Papo, D.; Buldú, J. Does the brain behave like a (complex) network? I. Dynamics. *Phys. Life Rev.* **2024**, *48*, 47–98. [CrossRef]
21. Motter, A.E.; Myers, S.A.; Anghel, M.; Nishikawa, T. Spontaneous synchrony in power-grid networks. *Nat. Phys.* **2013**, *9*, 191–197. [CrossRef]
22. Aparicio, A.; Velasco-Hernández, J.X.; Moog, C.H.; Liu, Y.Y.; Angulo, M.T. Structure-based identification of sensor species for anticipating critical transitions. *Proc. Natl. Acad. Sci. USA* **2021**, *118*, e2104732118. [CrossRef]
23. Fan, H.; Kong, L.W.; Lai, Y.C.; Wang, X. Anticipating synchronization with machine learning. *Phys. Rev. Res.* **2021**, *3*, 023237. [CrossRef]
24. Masuda, N.; Aihara, K.; MacLaren, N.G. Anticipating regime shifts by mixing early warning signals from different nodes. *Nat. Commun.* **2024**, *15*, 1086. [CrossRef]
25. MacLaren, N.G.; Kundu, P.; Masuda, N. Early warnings for multi-stage transitions in dynamics on networks. *J. R. Soc. Interface* **2023**, *20*, 20220743. [CrossRef]
26. Ehstand, N.; Donner, R.V.; López, C.; Hernández-García, E. Network percolation provides early warnings of abrupt changes in coupled oscillatory systems: An explanatory analysis. *Phys. Rev. E* **2023**, *108*, 054207. [CrossRef]
27. MacLaren, N.G.; Aihara, K.; Masuda, N. Applicability of spatial early warning signals to complex network dynamics. *arXiv* **2024**, arXiv:2410.04303. [CrossRef]
28. Bury, T.M.; Sujith, R.I.; Pavithran, I.; Scheffer, M.; Lenton, T.M.; Anand, M.; Bauch, C.T. Deep learning for early warning signals of tipping points. *Proc. Natl. Acad. Sci. USA* **2021**, *118*, e2106140118. [CrossRef]

29. Kong, L.W.; Fan, H.W.; Grebogi, C.; Lai, Y.C. Machine learning prediction of critical transition and system collapse. *Phys. Rev. Res.* **2021**, *3*, 013090. [CrossRef]
30. Vishnoi, N.; Gupta, V.; Saurabh, A.; Kabiraj, L. Reliability of early warning indicators of critical transition in stochastic Van der Pol oscillators with additive correlated noise. *Nonlinear Dyn.* **2024**, *112*, 15193–15217. [CrossRef]
31. Tarigo, J.P.; Stari, C.; Masoller, C.; Martí, A.C. Basin entropy as an indicator of a bifurcation in a time-delayed system. *Chaos* **2024**, *34*, 053113. [CrossRef] [PubMed]
32. Bassi, H.; Yim, R.P.; Vendrow, J.; Koduluka, R.; Zhu, C.; Lyu, H. Learning to predict synchronization of coupled oscillators on randomly generated graphs. *Sci. Rep.* **2022**, *12*, 15056. [CrossRef]
33. Ma, R.; Dai, Q.; Li, H.; Yang, J. Dynamics reconstruction in the presence of bistability by using reservoir computer. *Chaos Solitons Fractals* **2023**, *172*, 113523. [CrossRef]
34. Zhang, H.; Fan, H.; Du, Y.; Wang, L.; Wang, X. Anticipating measure synchronization in coupled Hamiltonian systems with machine learning. *Chaos* **2022**, *32*, 083136. [CrossRef]
35. Liu, Z.; Zhang, X.; Ru, X.; Gao, T.T.; Moore, J.M.; Yan, G. Early Predictor for the Onset of Critical Transitions in Networked Dynamical Systems. *Phys. Rev. X* **2024**, *14*, 031009. [CrossRef]
36. Roy, M.; Senapati, A.; Poria, S.; Mishra, A.; Hens, C. Role of assortativity in predicting burst synchronization using echo state network. *Phys. Rev. E* **2022**, *105*, 064205. [CrossRef]
37. Shahriari, Z.; Algar, S.D.; Walker, D.M.; Small, M. Ordinal Poincaré sections: Reconstructing the first return map from an ordinal segmentation of time series. *Chaos* **2023**, *33*, 053109. [CrossRef]
38. Leyva, I.; Masoller, C. Inferring the connectivity of coupled oscillators and anticipating their transition to synchrony through lag-time analysis. *Chaos Solitons Fractals* **2020**, *133*, 109604. [CrossRef]
39. Leyva, I.; Martínez, J.H.; Masoller, C.; Rosso, O.A.; Zanin, M. 20 years of ordinal patterns: Perspectives and challenges. *Europhys. Lett.* **2022**, *138*, 31001. [CrossRef]
40. Almendral, J.A.; Leyva, I.; Sendiña-Nadal, I. Unveiling the connectivity of complex networks using ordinal transition methods. *Entropy* **2023**, *25*, 1079. [CrossRef]
41. Tirabassi, G.; Masoller, C. Entropy-based early detection of critical transitions in spatial vegetation fields. *Proc. Natl. Acad. Sci. USA* **2023**, *120*, e2215667120. [CrossRef]
42. Lehnertz, K. Ordinal methods for a characterization of evolving functional brain networks. *Chaos* **2023**, *33*, 022101. [CrossRef]
43. Boccaletti, S.; Almendral, J.; Guan, S.; Leyva, I.; Liu, Z.; Sendiña-Nadal, I.; Wang, Z.; Zou, Y. Explosive transitions in complex networks' structure and dynamics: Percolation and synchronization. *Phys. Rep.* **2016**, *660*, 1–94. [CrossRef]
44. Avalos-Gaytán, V.; Almendral, J.A.; Leyva, I.; Battiston, F.; Nicosia, V.; Latora, V.; Boccaletti, S. Emergent explosive synchronization in adaptive complex networks. *Phys. Rev. E* **2018**, *97*, 042301. [CrossRef]
45. Soriano-Paños, D.; Guo, Q.; Latora, V.; Gómez-Gardeñes, J. Explosive transitions induced by interdependent contagion-consensus dynamics in multiplex networks. *Phys. Rev. E* **2019**, *99*, 062311. [CrossRef]
46. Khalil, N.; Leyva, I.; Almendral, J.A.; Sendiña Nadal, I. Deterministic and stochastic cooperation transitions in evolutionary games on networks. *Phys. Rev. E* **2023**, *107*, 054302. [CrossRef]
47. Ranjan, A.; Gandhi, S.R. Propagation of transient explosive synchronization in a mesoscale mouse brain network model of epilepsy. *Netw. Neurosci.* **2024**, *8*, 883–901. [CrossRef]
48. Rahjerdi, B.K.; Ramamoorthy, R.; Nazarimehr, F.; Rajagopal, K.; Jafari, S. Indicating the synchronization bifurcation points using the early warning signals in two case studies: Continuous and explosive synchronization. *Chaos Solitons Fractals* **2022**, *164*, 112656. [CrossRef]
49. Tlaie, A.; Leyva, I.; Sevilla-Escoboza, R.; Vera-Avila, V.P.; Sendiña Nadal, I. Dynamical complexity as a proxy for the network degree distribution. *Phys. Rev. E* **2019**, *99*, 012310. [CrossRef]
50. Bandt, C.; Pompe, B. Permutation Entropy: A Natural Complexity Measure for Time Series. *Phys. Rev. Lett.* **2002**, *88*, 174102. [CrossRef]
51. McCullough, M.; Small, M.; Stemler, T.; Ho, H.; Iu, C.; Iu, H.H.C. Time lagged ordinal partition networks for capturing dynamics of continuous dynamical systems. *Chaos* **2015**, *25*, 53101. [CrossRef]
52. Zanin, M.; Olivares, F. Ordinal patterns-based methodologies for distinguishing chaos from noise in discrete time series. *Commun. Phys.* **2021**, *4*, 190. [CrossRef]
53. Zunino, L.; Soriano, M.C.; Fischer, I.; Rosso, O.A.; Mirasso, C.R. Permutation-information-theory approach to unveil delay dynamics from time-series analysis. *Phys. Rev. E* **2010**, *82*, 046212. [CrossRef]
54. Soriano, M.C.; Zunino, L.; Rosso, O.A.; Fischer, I.; Mirasso, C.R. Time Scales of a Chaotic Semiconductor Laser With Optical Feedback Under the Lens of a Permutation Information Analysis. *IEEE J. Quantum Electron.* **2011**, *47*, 252–261. [CrossRef]

55. De Micco, L.; Fernández, J.G.; Larrondo, H.A.; Plastino, A.; Rosso, O.A. Sampling period, statistical complexity, and chaotic attractors. *Phys. A Stat. Mech. Its Appl.* **2012**, *391*, 2564–2575. [CrossRef]
56. Chrisment, A.M.; Firpo, M.C. Entropy–complexity analysis in some globally-coupled systems. *Phys. A Stat. Mech. Its Appl.* **2016**, *460*, 162–173. [CrossRef]
57. Letellier, C. Estimating the Shannon entropy: Recurrence plots versus symbolic dynamics. *Phys. Rev. Lett.* **2006**, *96*, 254102. [CrossRef]
58. Small, M. Complex networks from time series: Capturing dynamics. In Proceedings of the 2013 IEEE International Symposium on Circuits and Systems (ISCAS), Beijing, China, 19–23 May 2013; pp. 2509–2512. [CrossRef]
59. Masoller, C.; Hong, Y.; Ayad, S.; Gustave, F.; Barland, S.; Pons, A.J.; Gómez, S.; Arenas, A. Quantifying sudden changes in dynamical systems using symbolic networks. *New J. Phys.* **2015**, *17*, 023068. [CrossRef]
60. McCullough, M.; Small, M.; Iu, H.H.C.; Stemler, T. Multiscale ordinal network analysis of human cardiac dynamics. *Philos. Trans. R. Soc. A Math. Phys. Eng. Sci.* **2017**, *375*, 20160292. [CrossRef]
61. Boccaletti, S.; Latora, V.; Moreno, Y.; Chavez, M.; Hwang, D.U. Complex networks: Structure and dynamics. *Phys. Rep.* **2006**, *424*, 175–308. [CrossRef]
62. Leyva, I.; Navas, A.; Sendina-Nadal, I.; Almendral, J.; Buldú, J.; Zanin, M.; Papo, D.; Boccaletti, S. Explosive transitions to synchronization in networks of phase oscillators. *Sci. Rep.* **2013**, *3*, 1281. [CrossRef]
63. Gómez-Gardenes, J.; Gómez, S.; Arenas, A.; Moreno, Y. Explosive synchronization transitions in scale-free networks. *Phys. Rev. Lett.* **2011**, *106*, 128701. [CrossRef] [PubMed]
64. Leyva, I.; Sevilla-Escoboza, R.; Buldú, J.M.; Sendiña Nadal, I.; Gómez-Gardeñes, J.; Arenas, A.; Moreno, Y.; Gómez, S.; Jaimes-Reátegui, R.; Boccaletti, S. Explosive first-order transition to synchrony in networked chaotic oscillators. *Phys. Rev. Lett.* **2012**, *108*, 168702. [CrossRef]
65. Letellier, C.; Aguirre, L.A. Investigating nonlinear dynamics from time series: The influence of symmetries and the choice of observables. *Chaos* **2002**, *12*, 549–558. [CrossRef]
66. Sendiña-Nadal, I.; Letellier, C. Observability analysis and state reconstruction for networks of nonlinear systems. *Chaos* **2022**, *32*, 083109. [CrossRef]
67. Chialvo, D.R. Generic excitable dynamics on a two-dimensional map. *Chaos Solitons Fractals* **1995**, *5*, 461–479. [CrossRef]
68. Stankevich, N.V.; Gonchenko, A.S.; Popova, E.S.; Gonchenko, S.V. Complex dynamics of the simplest neuron model: Singular chaotic Shilnikov attractor as specific oscillatory neuron activity. *Chaos Solitons Fractals* **2023**, *172*, 113565. [CrossRef]
69. Boccaletti, S.; Kurths, J.; Osipov, G.; Valladares, D.; Zhou, C. The synchronization of chaotic systems. *Phys. Rep.* **2002**, *366*, 1–101. [CrossRef]
70. Rössler, O.E. An equation for continuous chaos. *Phys. Lett. A* **1976**, *57*, 397–398. [CrossRef]
71. Sevilla Escoboza, J.R.; Leyva Calleja, I.; Sendiña Nadal, I.; Martin Buldú, J. *Dataset on: "Explosive First-Order Transition to Synchrony in Networked Chaotic Oscillators"*; Zenodo: Geneva, Switzerland, 2024. [CrossRef]
72. Li, X.; Ouyang, G.; Richards, D.A. Predictability analysis of absence seizures with permutation entropy. *Epilepsy Res.* **2007**, *77*, 70–74. [CrossRef]
73. Mammone, N.; Duun-Henriksen, J.; Kjaer, T.W.; Morabito, F.C. Differentiating Interictal and Ictal States in Childhood Absence Epilepsy through Permutation Rényi Entropy. *Entropy* **2015**, *17*, 4627–4643. [CrossRef]
74. Yang, Y.; Zhou, M.; Niu, Y.; Li, C.; Cao, R.; Wang, B.; Yan, P.; Ma, Y.; Xiang, J. Epileptic Seizure Prediction Based on Permutation Entropy. *Front. Comput. Neurosci.* **2018**, *12*, 55. [CrossRef]
75. Bratu, I.F.; Makhalova, J.; Garnier, E.; Villalon, S.M.; Jegou, A.; Bonini, F.; Lagarde, S.; Pizzo, F.; Trébuchon, A.; Scavarda, D.; et al. Permutation entropy-derived parameters to estimate the epileptogenic zone network. *Epilepsia* **2024**, *65*, 389–401. [CrossRef]

entropy

MDPI

Article

The Structure of Bit-String Similarity Networks

David M. Schneider [1,2,†] and Damián H. Zanette [1,2,*,†]

[1] Centro Atómico Bariloche and Instituto Balseiro, Comisión Nacional de Energía Atómica, Universidad Nacional de Cuyo, Av. E. Bustillo 9500, San Carlos de Bariloche 8400, Argentina; davidmarsc@gmail.com
[2] Consejo Nacional de Investigaciones Científicas y Técnicas, Argentina
* Correspondence: damian.zanette@ib.edu.ar
† These authors contributed equally to this work.

Abstract: We study the structural properties of networks formed by random sets of bit strings—namely the ordered arrays of binary variables representing, for instance, genetic information or cultural profiles. Two bit strings are connected by a network link when they are sufficiently similar to each other, i.e., when their Hamming distance is below a certain threshold. Using both analytical and numerical techniques, we determine the degree distribution and the conditions for the existence of a giant component in this kind of network. In addition, we analyze their clustering, assortativity, and mean geodesic distance. We show that these properties combine features specific to random networks with characteristics that derive from the Hamming metrics implicit in the definition of similarity between bit strings.

Keywords: bit-string models; similarity networks; structural properties

1. Introduction

In recent decades, various stylized models of biological evolution—capturing in a simplified manner the elementary genetic mechanisms that drive evolutionary population dynamics, lineage diversification, speciation, and extinction—have been proposed [1–5]. To ease analytical and computational procedures, these models usually represent genomes as binary arrays, or bit strings (BSs), instead of the four-base nucleotide chains that constitute real genetic material. A BS is an ordered array of binary variables, with a given length B. It can be specified as a B-dimensional vector,

$$\mathbf{b} = (b_1, b_2, \dots, b_B), \tag{1}$$

where each component adopts one of two possible values, for instance, $b_i = 0$ or 1 for $i = 1, 2, \dots, B$. This representation for genomes has become a standard tool in the theoretical study of viral RNA evolution [6]. BSs have also been used as the building blocks of models for complex systems of various kinds. For example, Schulze's model of language evolution [7,8] and its variations [9] characterize a language by a collection of dichotomic properties, which naturally calls for a representation in terms of binary variables. Similarly, inspired by Axelrod's model of culture dissemination [10], recent models of culture evolution conceive cultural profiles as sets of binary features represented by BSs [11]. This kind of dichotomic categorization underlies the binary classification of digitized information, a central task in automated data analysis [12].

In many of the evolutionary models referred to above, dynamical rules prescribe that any interaction between two elements is only allowed when the elements are comparable to some degree. For instance, two organisms can become involved in a reproductive event only if they are "compatible", namely if their genomes are sufficiently similar to each other [1,4]. Likewise, cultural interaction is only possible between individuals who already possess a certain degree of cultural similarity [10]. The comparison between two BSs of the same length B, say $\mathbf{b}$ and $\mathbf{b}'$, is usually performed by calculating their Hamming distance,

$$H(\mathbf{b}, \mathbf{b}') = \sum_{i=1}^{B} |b_i - b_i'|, \tag{2}$$

which equals the number of bits where $\mathbf{b}$ and $\mathbf{b}'$ differ. In fact, H is an integer number with $0 \leq H \leq B$. Any interaction between two elements in the above models becomes possible when the Hamming distance of the corresponding BSs is below a prescribed threshold.

In this contribution, we consider a class of networks formed by a set of BSs that are connected depending on their degree of similarity. Specifically, we take N mutually different, randomly chosen BSs of length B and establish links joining each pair when their Hamming distance is lower than or equal to a threshold G, i.e., $H \leq G$, with $1 \leq G \leq B$. Figure 1 shows an example of these BS similarity networks with small values of N, B, and G. In the models described in the preceding paragraphs, where elements represented by BSs interact with each other when their Hamming distance is sufficiently small, this construction yields the corresponding interaction network.

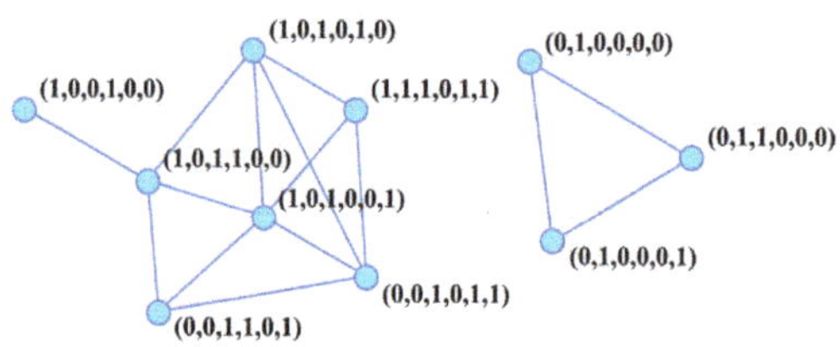

Figure 1. A small bit-string similarity network of size $N = 10$, with BS length $B = 6$, and threshold $G = 2$.

In the following sections, first, we show that random sets of BSs, as those used to build our networks, emerge spontaneously as the result of neutral evolution in populations of constant size. Then, we present analytical and numerical results on the global structural properties of BS similarity networks. We study the degree distribution, the appearance of a giant component, clustering, assortativity, and mean geodesic distance, as functions of N, B, and G. We show that these properties combine the features typically found in both random and non-random networks. Our results are summarized in the last section.

2. Random Bit-String Sets as the Outcome of Neutral Evolution

Before studying the structure of the BS similarity networks defined in the Introduction, it is opportune to show that the sets of BSs on which these networks are built arise spontaneously as the result of simple evolutionary mechanisms based on random mutations and in the absence of selection pressure—i.e., under neutral evolution [13]. Specifically, we consider a population of N BSs of length B that are initially identical, for instance, $\mathbf{b}_n = (0, 0, \ldots, 0)$ for all $n = 1, 2, \ldots, N$. At each time step, a "parent" BS $\mathbf{b}_{\text{parent}}$ is chosen at random, and a "child" BS $\mathbf{b}_{\text{child}}$ is created by mutating $\mathbf{b}_{\text{parent}}$. The mutation amounts to switching a single, randomly chosen bit in $\mathbf{b}_{\text{parent}}$ to its opposite value, either $0 \to 1$ or $1 \to 0$. Then, another randomly selected BS "dies" and is replaced by $\mathbf{b}_{\text{child}}$. In this toy evolutionary model, thus, increasingly dissimilar BSs are created, while the population size remains constant.

A convenient way of characterizing the composition of a given set of BSs—in particular, for comparison with a random set—is to determine the distribution of Hamming distances between all its elements. In a set of random BSs of length B, the expected frequency of a Hamming distance H equals the probability that two random BSs differ by exactly H bits, namely

$$f_H^{\text{random}} = 2^{-B} \binom{B}{H}, \tag{3}$$

for $H = 0, 1, \ldots, B$. In a numerical realization of our evolutionary model, we measure the frequency of each value of the Hamming distance, $f_H(T)$, as the number of time steps T increases, and compare with f_H^{random} by computing the difference

$$D(T) = \sqrt{\sum_{H=0}^{B} \left[f_H(T) - f_H^{\text{random}}\right]^2} \tag{4}$$

between both distributions.

The main panel of Figure 2 shows our measurements of $D(T)$ plotted versus the rescaled time T/NB, for $N = 1024$ and five values of B. For each B, results correspond to an average over 50 realizations of the evolution. The inset shows similar results for $B = 30$ and four values of N. In both plots, a unit in the horizontal scale corresponds to the average waiting time for any of the NB bits in the population to be switched by mutation. We see that, after a short transient, $D(T)$ reaches a regime of exponential decay, decreasing with a rate that is essentially independent of N and B, at least, for large B. Consequently, as may have been expected, the typical number of mutations needed for $D(T)$ to decrease by a given factor in this exponential regime is proportional to both N and B.

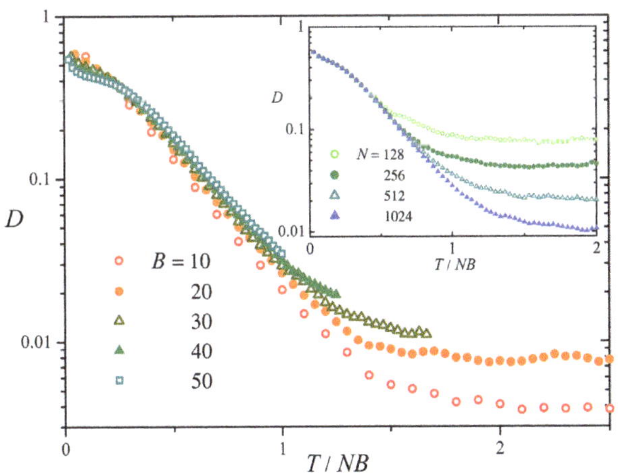

Figure 2. Main panel: Time evolution of the difference D between the distribution of Hamming distances in the evolutionary model described in the text and in a random set of bit strings, Equation (4), for $N = 1024$ and five values of B. In the horizontal axis, the number of steps T is normalized by the product NB. For each B, the results are averaged over 50 realizations of the model. Inset: As in the main panel, for $B = 30$ and four values of N.

For longer times, the difference $D(T)$ ceases to decrease and varies irregularly around a constant value. This asymptotic behavior can be ascribed to the remnant fluctuations of

$f_H(T)$ around the expected value f_H^{random}. As seen in Figure 2, the asymptotic value of D increases with B, as expected from the fact that the calculation of D, Equation (4), involves a sum of $B+1$ terms. On the other hand, it decreases with the population size as N^{-1}. This dependence on N can be understood by taking into account that, by virtue of the law of large numbers, the long-time fluctuations of $f_H(T)$ around f_H^{random} are expected to be of order $Z^{-1/2}$, where Z is the number of data used to evaluate $f_H(T)$. In our case, Z is the number of pairs of BSs in the population, over which we calculate the Hamming distances, so that $Z \propto N^2$. This implies that $D \sim |f_H(T) - f_H^{\text{random}}| \sim N^{-1}$. Thus, up to stochastic fluctuations due to the finite size of the population, the distribution of Hamming distances between BSs converges with time to that of a random set of the same size.

Similar qualitative behavior is observed in a few variants of the above evolutionary model. For instance, if, instead of replacing a randomly chosen BS, $\mathbf{b}_{\text{child}}$ replaces its own parent $\mathbf{b}_{\text{parent}}$, a random set of BSs is obtained at long times just by construction. We have also verified that $D(T)$ decays qualitatively in the same manner as shown in Figure 2, when bit mutations are combined with recombination, where a child is created by concatenating two complementary portions taken from different parents [4].

3. Structural Properties

In this section, we study the global structural properties of BS similarity networks. As shown below, some of them can be evaluated analytically while others require resorting to numerical computation. In qualitative terms, our main conclusion is that networks in this class combine the characteristic features of random graphs, such as the degree distribution and the appearance of a giant component, with properties that cannot be explained as the result of aleatory connections, such as their clustering.

3.1. Degree Distribution, the Giant Component, and Richness

First, we calculate the degree distribution g_k of our networks, namely the probability that a node has exactly k neighbors. Among the $M = 2^B$ possible BSs of length B, there are $\binom{B}{H}$ at a Hamming distance H from any of them (say, $\mathbf{b}$). Therefore, the total number of BSs at a distance less than or equal to G from $\mathbf{b}$, excluding $\mathbf{b}$ itself, is

$$M_G = \sum_{H=1}^{G} \binom{B}{H}. \tag{5}$$

The fraction of BSs at a distance H from $\mathbf{b}$, such that $1 \le H \le G$, is

$$\mu = \frac{M_G}{M-1}. \tag{6}$$

The results presented in the following show that μ is a natural parameter to characterize the connectivity properties of BS similarity networks. Note, from Equations (5) and (6), that μ adopts discrete increasing values as the threshold G varies from 1 to B.

Our networks are built by choosing at random N of the M possible BSs. The probability that k of them are among the M_G at distance H from $\mathbf{b}$, with $1 \le H \le G$, is

$$g_k = \binom{N-1}{k} \mu^k (1-\mu)^{N-1-k}, \tag{7}$$

for $k = 0, 1, \ldots, N-1$. We have numerically verified that, up to a high precision, this is indeed the degree distribution of BS similarity networks of size N and maximal Hamming distance G. Note that it coincides with the degree distribution of an Erdős–Rényi random network with N nodes and $\mu N(N-1)/2$ links [14]. The mean number of links per node is $\langle k \rangle \equiv z = \mu(N-1)$.

Assuming that it is valid to apply the theory of random networks [15] to our case, we can calculate the critical value of μ, which we denote as μ_c, above which a giant component exists in the network. We recall that a giant component is strictly well defined for infinitely large networks, as a connected component that comprises a finite part of the network. In finite, but sufficiently large networks, it is associated with the largest connected component [14]. To evaluate μ_c, we start by computing the generating function of the degree distribution g_k:

$$\Gamma(x) = \sum_{k=0}^{N-1} g_k x^k = (1 - \mu + \mu x)^{N-1}. \tag{8}$$

The critical point μ_c is the minimum value of μ for which the equation

$$u = \frac{\Gamma'(u)}{\Gamma'(1)} = (1 - \mu + \mu u)^{N-2} \tag{9}$$

has a solution $u < 1$ [15], namely

$$\mu_c = \frac{1}{N-2}. \tag{10}$$

For $\mu \geq \mu_c$, the fraction of the network in the giant component is

$$S = 1 - u^{(N-1)/(N-2)}, \tag{11}$$

where u is the solution to Equation (9). For $N \gg 1$, Equations (9) and (11) can be approximated as

$$u = \exp\left[-\frac{\mu}{\mu_c}(1-u)\right], \quad \text{and} \quad S = 1 - u, \tag{12}$$

respectively. For large networks, therefore, it is clear that the relative size of the giant component depends on B, G, and N through the ratio μ/μ_c only. Note that, when N is large, μ/μ_c is a direct measure of the average number of neighbors per site.

$$z = \mu(N-1) = \frac{\mu}{\mu_c}\frac{N-1}{N-2} \approx \frac{\mu}{\mu_c}. \tag{13}$$

The curve in Figure 3 is the solution to Equation (12) for the relative size of the giant component, S, as a function of μ/μ_c. Symbols, in turn, show numerical results for the fraction of the network in the largest component, for two values of N, and three values of B. Error bars represent the standard deviation over sets of 50 realizations for each parameter set. As expected, this dispersion is the largest in the transition zone, $\mu \approx \mu_c$, and becomes negligible elsewhere. Apart from typical finite-size effects for $\mu \lesssim \mu_c$, the agreement with the analytical prediction is excellent, indicating that the theory of random networks correctly describes the appearance and growth of the giant component in BS similarity networks.

An accessory quantity of interest for the evolutionary models quoted in the Introduction is the *richness* R, namely the ratio between the number of connected components in the network and its size N. The richness is a measure of diversity in the population, since each connected component is associated with a different "species" [4,5]. Figure 4 shows our numerical estimation for the richness corresponding to the same realizations as in Figure 3. Note that the collapse of results for different values of B and N onto a single curve, indicating that—at least within this parameter range—R depends on μ/μ_c only, much as the relative size of the giant component S.

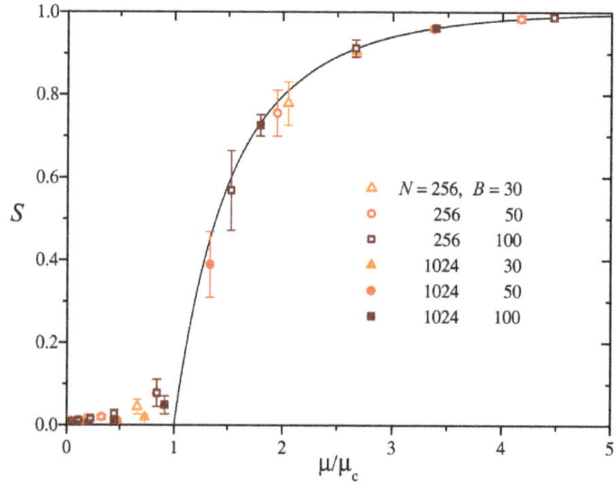

Figure 3. Full curve: The solution to Equation (12) for the fraction of the network inside the giant component, S, as a function of the ratio μ/μ_c. Symbols: Numerical measurement of the relative size of the largest component for various combinations of N and B, averaged over 50 realizations for each parameter set. Error bars represent the standard deviation over realizations.

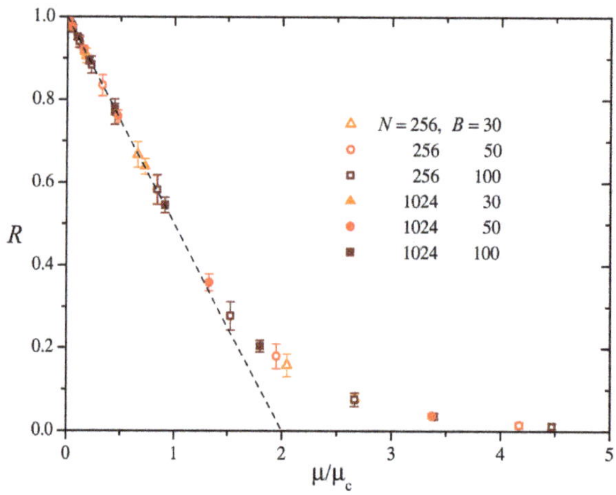

Figure 4. The richness R, given by the number of connected components in the network divided by the network size N, as a function of μ/μ_c, estimated from the same numerical realizations as in Figure 3. Error bars represent the standard deviation over realizations. The dashed straight line is the analytical approximation for small μ/μ_c discussed in the text, Equation (14).

As expected, $R \approx 1$ as $\mu \to 0$. In this limit, in fact, most nodes in the networks are disconnected from each other, and the number of connected components is close to N. For small μ/μ_c, the richness seems to decrease linearly. However, beginning at $\mu/\mu_c \gtrsim 1$, where the giant component first appears and begins to grow, R deviates from the linear behavior to approach N^{-1} for a large μ/μ_c, where the giant component comprises the whole network.

The linear decrease in R for $\mu/\mu_c \lesssim 1$ can be readily understood taking into account that, for sufficiently small μ, the network is formed by isolated nodes and, at most, tiny connected components. From Equation (7), the expected number of isolated nodes is $g_0 N$ which, for a large N, is approximately given by $N \exp(-\mu/\mu_c)$. This is the number of components of size one. Assuming that all the other nodes form connected components of size two, the number of these components would be $N[1-\exp(-\mu/\mu_c)]/2$. In this situation, thus, the richness would be given by the sum of the numbers of components of sizes one and two divided by N:

$$R = \exp(-\mu/\mu_c) + \frac{1-\exp(-\mu/\mu_c)}{2} \approx 1 - \frac{\mu}{2\mu_c}, \tag{14}$$

where the approximation holds for small μ/μ_c. This prediction is plotted in Figure 4 as a dashed line. We see that, actually, it gives a satisfactory explanation of the dependence of R up to $\mu/\mu_c \approx 1$.

3.2. Clustering

The statistical tendency of the neighbors of a given node in a network to be, in turn, mutual neighbors is called clustering or transitivity. This feature, especially ubiquitous in social networks [14,16,17], corresponds to an enhanced probability of finding groups of well-interconnected nodes with respect to the situation where links between nodes are distributed at random. In our BS similarity networks, clustering is expected to differ from that of random networks due to the structure induced by the Hamming metrics involved in their construction. Concretely, suppose that a given BS in the network has two neighbors at a Hamming distance $H=1$. Since the two neighbors are, by construction, not identical, their mutual Hamming distance is necessarily $H=2$. Consequently, if the threshold G is greater than or equal to two, the two neighbors will be mutually connected. Similarly, if the two neighbors are at distances $H=1$ and $H=2$ from the given BS, their mutual distance cannot be larger than $H=3$, and so on. The Hamming metrics thus favor the mutual connection between neighbors of any given BS, with the ensuing growth of clustering.

Several measures have been proposed to quantify clustering [17]. For the BS similarity networks, we here compute the *mean clustering coefficient* C, as originally defined for small-world networks [16]. The mean clustering coefficient is the average over all the nodes in the network of the *local* clustering coefficient, C_i, given by the ratio between the number of pairs of neighbors of node i which are in turn mutual neighbors, and the total number of pairs of neighbors of the same node. In an Erdős–Rényi random network with N nodes and an average of z neighbors per node, the mean clustering coefficient decreases with the network size as $C \sim zN^{-1}$ [17].

To evaluate the mean clustering coefficient of BS similarity networks, we first consider a generic BS $\mathbf{b}_0$ and two other BSs $\mathbf{b}_1$ and $\mathbf{b}_2$ at Hamming distances H_1 and H_2 from $\mathbf{b}_0$, respectively. If the three BSs belong to a similarity network with threshold G, $\mathbf{b}_1$ and $\mathbf{b}_2$ will be neighbors of $\mathbf{b}_0$ when $H_1, H_2 \leq G$. They will contribute to the local clustering of $\mathbf{b}_0$ if they are in turn mutual neighbors or, in other words, if their mutual distance g is lower than or equal to G. Since the number of neighbors at any given distance is the same for all BSs, we can ease the calculations assuming that $\mathbf{b}_0 = (0,0,\ldots,0)$. With this choice, H_1 and H_2 are the number of ones in $\mathbf{b}_1$ and $\mathbf{b}_2$, respectively.

Without generality loss, we take $H_2 \geq H_1$. The distance g between $\mathbf{b}_1$ and $\mathbf{b}_2$ is the number of bits where the ones in $\mathbf{b}_1$ do not coincide with those in $\mathbf{b}_2$. The direct inspection of the possible distributions of ones along both BSs shows that the distance adopts the values $g = H_1 + H_2 - 2h$, with $h = 0,1,\ldots,H_1$, where h is the number of bits where the ones do coincide. For a fixed value of h, the total number of pairs $(\mathbf{b}_1, \mathbf{b}_2)$ whose mutual distance equals $H_1 + H_2 - 2h$ is given by the product of (i) the number of BSs of length B with H_1 ones, times (ii) the number of ways of selecting h bits from H_1 bits with ones, times (iii) the number of ways of selecting $H_2 - h$ bits from $B - H_1$ bits with zeros, namely

$$\pi_h = \binom{B}{H_1}\binom{H_1}{h}\binom{B-H_1}{H_2-h}. \tag{15}$$

Thus, the probability that such a pair does occur in the network equals

$$\Pi_h = \frac{\pi_h}{\sum_{h'=0}^{H_1} \pi_{h'}} = \binom{B}{H_2}^{-1}\binom{H_1}{h}\binom{B-H_1}{H_2-h}. \tag{16}$$

The probability that their mutual distance is lower than or equal to G is then

$$P(H_1, H_2) = \sum_{h=h_0}^{H_1} \Pi_h = \binom{B}{H_2}^{-1} \sum_{h=h_0}^{H_1} \binom{H_1}{h}\binom{B-H_1}{H_2-h}, \tag{17}$$

with $h_0 = \max\{0, [(H_1 + H_2 - G + 1)/2]\}$, where $[\cdot]$ indicates the integer part. It can be shown that the combination of binomial coefficients in Equations (16) and (17) is symmetric with respect to H_1 and H_2,

$$\binom{B}{H_2}^{-1}\binom{H_1}{h}\binom{B-H_1}{H_2-h} = \binom{B}{H_1}^{-1}\binom{H_2}{h}\binom{B-H_2}{H_1-h}. \tag{18}$$

Equation (17) is therefore also valid when $H_2 < H_1$, with the convention that the binomial coefficient $\binom{i}{j}$ vanishes whenever $j > i$.

Taking into account that the total number of BSs at a distance H from any one of them is $\binom{B}{H}$ [cf. Equation (5)], the mean clustering coefficient C, given by the average of local clustering all over the network, can be evaluated as the sum of the contributions coming from all distances H_1 and H_2 of all pairs of mutual neighbors at such distances:

$$\begin{aligned} C &= \sum_{H_1=1}^{G} \sum_{H_2=1}^{G} \binom{B}{H_1}\binom{B}{H_2} P(H_1, H_2) \\ &= \sum_{H_1=1}^{G} \sum_{H_2=1}^{G} \sum_{h=h_0}^{H_1} \frac{B!}{h!(H_1-h)!(H_2-h)!(B-H_1-H_2+h)!}, \end{aligned} \tag{19}$$

with h_0 as in Equation (17). Note that this result is independent of the network size N, since it originates in purely probabilistic considerations. As such, it is expected to hold for large networks.

Symbols in Figure 5 show the numerical results of the mean clustering coefficient C averaged over the 50 realizations of BS similarity networks of size $N = 1024$ and three values of B, as a function of the ratio G/B. For each B, lines join the analytical values of C obtained from Equation (19) varying the threshold G. The agreement is excellent, which validates the above probabilistic arguments. Our numerical results for other network sizes, starting at $N = 128$, also agree very well with the analytical prediction. As expected, clustering increases as G grows and the network connectivity becomes larger. The S-shaped dependence on G shows an inflection point in the zone where the giant component has already appeared. However, even below this zone, mean clustering adopts relatively large values, well above those expected for random networks. The full line in the inset of Figure 5 shows the analytical value of C for $B = 30$ and small values of the threshold G.

For comparison, the dashed line shows the mean clustering coefficient of an Erdős–Rényi random network with the same mean number of neighbors.

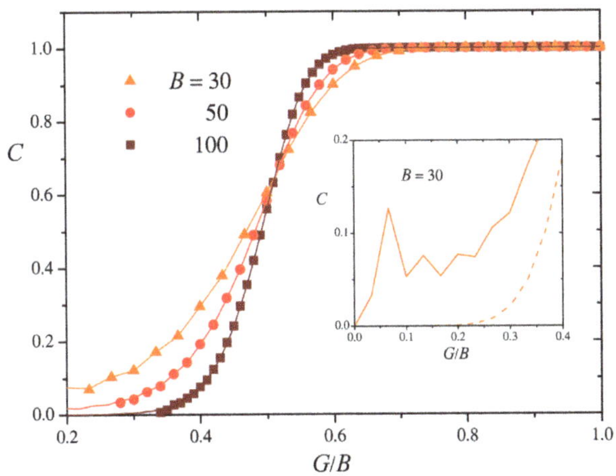

Figure 5. Main panel: Symbols show numerical estimates for the mean clustering coefficient C of BS similarity networks with $N = 1024$ and three values of the bit-string length B, as a function of the ratio G/B, averaged over 50 realizations. The standard deviation of C over realizations is smaller than the size of symbols. Lines join the analytical values for C obtained from Equation (19). Inset: The full line shows C for $B = 30$ and small values of the threshold G. The dashed line shows C for an Erdős–Rényi network with the same mean degree for each value of G/B.

3.3. Degree Assortativity

Given an attribute a assigned to each node in a network, the *assortativity* with respect to a measures the statistical correlation between the values of a in pairs of connected nodes. Assortativity is defined as the Pearson correlation coefficient of a computed over all pairs of neighbors [14,17], and is therefore confined to the interval $[-1,1]$. A generic attribute that allows for the computation of assortativity is the number of neighbors of each node, namely the node degree k. Degree assortativity is large and positive when nodes with either many or few neighbors are preferentially interconnected, while it is large and negative when connections mainly occur between nodes with disparate degrees. On the other hand, assortativity is close to zero in networks with a random distribution of links, such as Erdős–Rényi networks.

We could not find an accurate analytical prediction for the assortativity of BS similarity networks. Our estimations are therefore based on numerical calculations. A convenient way of computing the Pearson correlation coefficient between the degrees of neighbor nodes is [14,18]

$$A = \frac{4\langle k_1 k_2 \rangle - \langle k_1 + k_2 \rangle^2}{2\langle k_1^2 + k_2^2 \rangle - \langle k_1 + k_2 \rangle^2}, \tag{20}$$

where $\langle \cdot \rangle$ indicates the average over all the links in the network, with k_1 and k_2 being the degrees of the two nodes at the ends of each link. In our numerical calculations, we have used this expression to compute the assortativity.

We computed A in a series of 50 realizations of BS similarity networks for various combinations of the sizes N and BS lengths B, as a function of the threshold G. Our first finding is that, as much as the mean clustering coefficient C studied in the preceding section, A becomes independent of N for large network sizes. This is not unexpected, since assortativity is a structural feature defined in terms of statistical properties, as is C. Figure 6 shows our numerical results for $N = 1024$ and four values of B. Symbols and error bars, respectively, represent averages and standard deviations over each series of realizations. We see that, in all cases, the average assortativity reaches small positive values, $A \lesssim 0.17$, with typically large dispersions, which can be much larger than the average itself. As a general trend, A decreases as the BS length B grows. For each value of B, it reaches a maximum at $G/B < 0.5$. Beyond the maximum, it shows a rapid decrease towards zero as the network connectivity grows.

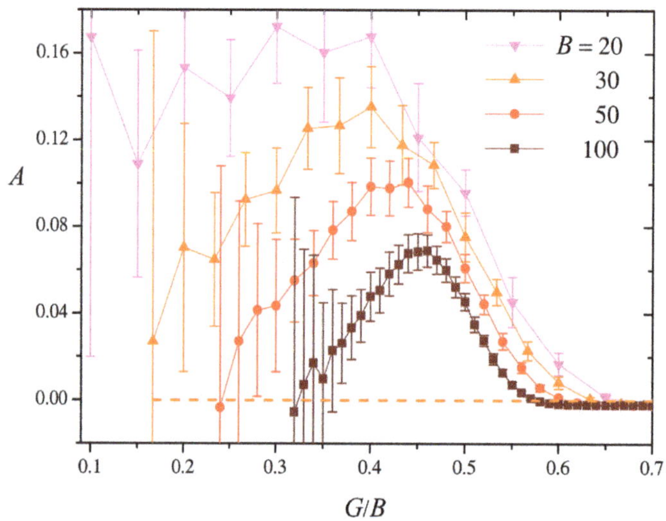

Figure 6. The assortativity A, computed as in Equation (20) as a function of G/B, for a series of 50 realizations with $N = 1024$ and four values of B. Symbols and error bars, respectively, represent the averages and standard deviations over each series of realizations. The dashed line shows numerical results for A in Erdős–Rényi networks of size $N = 1024$ which, for each value of G/B, have the same mean degree as the BS similarity networks with $B = 30$.

The dashed line in Figure 6 shows the assortativity A computed numerically for Erdős–Rényi networks of size $N = 1024$, with the same mean degree as BS similarity networks with $B = 30$. Its value cannot be discerned from zero within the plot scale. We see that, as compared with their random counterparts, the assortativity of BS similarity networks reaches statistically significant levels. However, the overall small values of A suggest that assortativity is not a particularly relevant structural feature of our networks.

3.4. Mean Geodesic Distance

Lastly, we studied the mean geodesic distance in BS similarity networks. The geodesic (or chemical) distance between two nodes in a network is given by the number of links along the shortest path joining the two nodes with each other [14,17]. Naturally, this quantity is well defined when such a path exists, which requires that the two nodes belong to the same connected component. By convention, when the two nodes are in different components, their geodesic distance is taken to be infinite. The *mean geodesic distance* L, therefore, is usually defined as the arithmetic average of geodesic distances over all pairs where the two nodes belong to the same component, i.e., with finite geodesic distance. It is a direct measure of network connectivity, and is related to robustness and percolation properties [19,20].

The main panel of Figure 7 shows our numerical measurements of L for BS similarity networks with $B = 100$ and three values of N as a function of G/B. Symbols stand for averages over 50 realizations and error bars are the corresponding standard deviations. Results for other values of B show the same qualitative behavior. For a small G/B, when the network is poorly connected, we find $L \approx 1$. In this limit, in fact, each node is connected with at most one neighbor, at a unitary geodesic distance. The same value of L is obtained when G/B is sufficiently large, with virtually all nodes connected to each other. Both for small and large G/B, fluctuations of L between realizations are negligible.

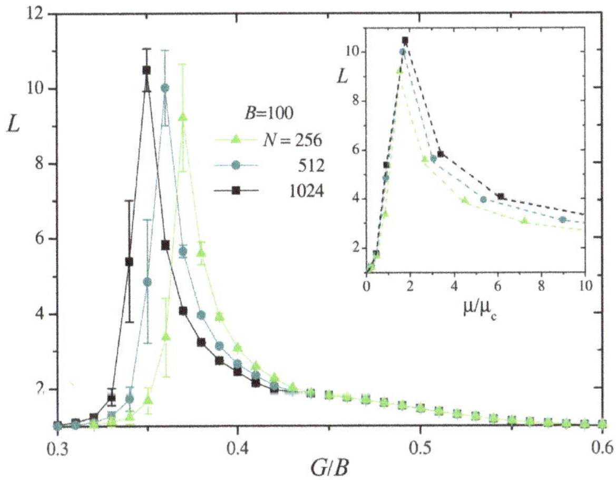

Figure 7. Main panel: Numerical results for the mean geodesic distance L in BS similarity networks with $B = 100$ and three values of N, as a function of G/B. Symbols are averages over 50 realizations and error bars show the corresponding standard deviations. Lines joining symbols are added for clarity. Inset: The same results for L as a function of μ/μ_c in the zone of the maximum. For clarity, error bars have been omitted. Dashed lines join numerical results for L in Erdős–Rényi random networks of the same size and for the same mean degree.

In the intermediate zone, in contrast, L shows a sharp peak, where it reaches significantly larger average values and fluctuations. The height of the peak increases slowly with N. The inset of Figure 7 shows the same numerical results plotted versus the ratio μ/μ_c, in the zone of the maximum of L. We see that the peak is located in the interval $1 \lesssim \mu/\mu_c \lesssim 2$, i.e., just above the appearance of the giant component. In this zone, where several connected components of different sizes are still present and the giant component is just beginning to grow (cf. Figures 3 and 4), the overall network structure is most irregular and geodesic distances can reach large values.

It turns out, however, that this is exactly the same behavior as observed for the mean geodesic distance in random networks. Dashed lines in the inset of Figure 7 join numerical results for L measured in Erdős–Rényi networks of the same size N and the same mean number of neighbors z as the corresponding BS similarity networks, averaged over 50 realizations. Results for both kinds of network are indistinguishable from each other within numerical fluctuations. We conclude that the specific structure of BS similarity networks has no impact on their mean geodesic distance, which only reveals the underlying random choice of the BSs used to build the network.

4. Conclusions

In this paper, we have studied a class of complex networks whose nodes are bit strings—namely, ordered arrays of binary variables—of fixed length B. The set of N bit strings present in the network is chosen at random among the 2^B possibilities. Network links join sufficiently similar bit strings, specifically, those whose Hamming distances are at most equal to a given threshold G. As we have discussed in the Introduction, these bit-string similarity networks coincide with the interaction patterns in a variety of evolutionary models where bit strings represent individual genomes and interactions are reserved to pairs of genetically comparable individuals. We have shown in Section 2 that the random sets of bit strings on which our networks are built arise spontaneously under the action of neutral evolution under various evolutionary mechanisms.

Our main results, reported in Section 3, refer to global structural properties of bit-string similarity networks, which we studied as functions of the parameters N, B, and G. We have obtained exact analytical results for the degree distribution, the appearance of a giant component—which, as in many other network classes, has the character of a critical phenomenon—and the mean clustering coefficient. Meanwhile, the number of connected components, the assortativity, and the mean geodesic distance, have been studied numerically.

In qualitative terms, the most interesting property detected in our bit-string similarity networks is a combination of features which are characteristic of random networks with other properties which disclose non-trivial correlations in their structure. For instance, with a suitable identification of parameters, their degree distribution exactly coincides with that of Erdős–Rényi random networks. Likewise, the appearance of a giant component can be accurately predicted using the generating function formalism for networks with prescribed degree distribution and randomly distributed links [15]. According to numerical results, the mean geodesic distance of bit-string similarity networks also coincides to a large precision with the same quantity measured for their Erdős–Rényi counterparts. On the contrary, clustering and, to a lesser extent, assortativity, reach significantly high levels, in contrast with their negligible values in large random networks [14]. This merging of random and non-random features can be ascribed to the combined impact of the two main ingredients that take part in the construction of bit-string similarity networks. On the one hand, the random choice of the set of bit strings present in each realization of the network determines that the neighbors of a given node will also be a random subset of all the possible bit strings of a given length. On the other, as exemplified at the beginning of Section 3.2, the Hamming metrics implicit in the way nodes are connected establish a statistical interdependence between their neighborhoods. This affects clustering and assortativity which are, precisely, different measures of such correlations. We stress that the joint occurrence of random and non-random attributes is typical of other complex structures, notably, small-world networks, which combine short geodesic distances with large clustering [17]. This feature has already been pointed out in the real-life networks of various origins [16]. It remains to be assessed whether other specific structural properties of bit-string similarity networks are found in the interaction patterns of actual biological and sociocultural systems.

Relaxing the binary nature of the ordered arrays used to build the networks studied here would allow for a more realistic representation of genome chains or, generally, more complex characterizations of individual features. For arrays where each variable can adopt more than two states, or even its own set of states, the Hamming distance can still be defined as the number of variables where two arrays differ from each other. Since the properties of the Hamming metrics are essentially the same as for bit strings, we expect similar structural properties in the resulting similarity networks. We also point out that the Hamming distance naturally provides a way to define a weighted network—a variant not explored in the present contribution—assigning, for example, larger weights to the links between bit strings at shorter distances. Finally, an interesting extension related to evolutionary models would be to study the similarity networks formed by collections of bit strings deriving from evolutionary mechanisms other than neutral evolution, under the action of selection pressure. All these variants may stimulate future work on dynamical processes occurring on such structures, modeling both biological and social phenomena.

Author Contributions: Conceptualization, D.M.S. and D.H.Z.; Investigation, D.M.S. and D.H.Z.; Writing—original draft, D.M.S. and D.H.Z. All authors have read and agreed to the published version of the manuscript.

Funding: This research received no external funding.

Institutional Review Board Statement: Not applicable.

Data Availability Statement: The original contributions presented in this study are included in the article. Further inquiries can be directed to the corresponding author.

Conflicts of Interest: The authors declare no conflicts of interest.

References

1. Higgs, P.G.; Derrida, B. Stochastic models for species formation in evolving populations. *J. Phys. A Math. Gen.* **1991**, *24*, L985. [CrossRef]
2. Penna, T. A bit-string model for biological aging. *J. Stat. Phys.* **1995**, *78*, 1629. [CrossRef]
3. Falkiewicz, D.M.; Makowiec, D. Bit-string model of biological speciation: Revisited. *Phys. A* **2021**, *570*, 125781. [CrossRef]
4. de Aguiar, M.A.M.; Baranger, M.; Baptestini, E.M.; Kaufman, L.; Bar-Yam, Y. Global patterns of speciation and diversity. *Nature* **2009**, *460*, 384. [CrossRef] [PubMed]
5. Schneider, D.; Baptestini, E.M.; de Aguiar, M.A.M. Diploid versus haploid models of neutral speciation. *J. Biol. Phys.* **2016**, *42*, 235. [CrossRef]
6. Elena, S.F.; Solé, R.V.; Sardanyés, J. Simple genomes, complex interactions: Epistasis in RNA virus. *Chaos Interdiscip. J. Nonlinear Sci.* **2010**, *20*, 026106. [CrossRef] [PubMed]
7. Schulze, C.; Stauffer, D. Monte Carlo simulation of the rise and the fall of languages. *Int. J. Mod. Phys. C* **2005**, *16*, 781. [CrossRef]
8. Zanette, D.H. Analytical approach to bit-string models of language evolution. *Int. J. Mod. Phys. C* **2008**, *19*, 569. [CrossRef]
9. de Olivera, P.M.C.; Stauffer, D.; Lima, F.W.S.; Sousa, A.O.; Schulze, C.; Moss de Oliveira, S. Bit-strings and other modifications of Viviane model for language competition. *Phys. A* **2007**, *376*, 609. [CrossRef]
10. Axelrod, R. The dissemination of culture: A model with local convergence and global polarization. *J. Confl. Resolut.* **1997**, *41*, 203. [CrossRef]
11. Pascual, I.; Aguirre, J.; Manrubia, S.; Cuesta, J.A. Epistasis between cultural traits causes paradigm shifts in cultural evolution. *R. Soc. Open Sci.* **2020**, *7*, 191813. [CrossRef]
12. Flach, P. *Machine Learning. The Art and Science of Algorithms that Make Sense of Data*; Cambridge University Press: Cambridge, UK, 2012.
13. Duret, L. Neutral theory: The null hypothesis of molecular evolution. *Nat. Educ.* **2008**, *1*, 218.
14. Newman, M.E.J. *Networks: An Introduction*; Oxford University Press: Cambridge, UK, 2010.
15. Newman, M.E.J.; Strogatz, S.H.; Watts, D.J. Random graphs with arbitrary degree distributions and their applications. *Phys. Rev. E* **2001**, *64*, 026118. [CrossRef] [PubMed]
16. Watts, D.J.; Strogatz, S.H. Colletive dynamics of 'small-world' networks. *Nature* **1998**, *393*, 440. [CrossRef] [PubMed]
17. Newman, M.E.J.; Barabási, A.L.; Watts, D.J. *The Structure and Dynamics of Networks*; Princeton University Press: Princeton, NJ, USA, 2006.
18. Barrenas, F.; Chavali, S.; Holme, P.; Mobini, R.; Benson, M. Network properties of complex human disease genes identified through genome-wide association studies. *PLoS ONE* **2009**, *4*, e8090. [CrossRef]
19. Cohen, R.; Erez, K.; ben-Avraham, D.; Havlin, S. Resilience of the internet to random breakdowns. *Phys. Rev. Lett.* **2000**, *85*, 4626. [CrossRef] [PubMed]
20. Callaway, D.S.; Newman, M.E.J.; Strogatz, S.H.; Watts, D.J. Network robustness and fragility: Percolation on random graphs. *Phys. Rev. Lett.* **2000**, *85*, 5468. [CrossRef] [PubMed]

entropy

MDPI

Article

Exploring Word-Adjacency Networks with Multifractal Time Series Analysis Techniques

Jakub Dec [1], Michał Dolina [1], Stanisław Drożdż [1,2,*], Robert Kluszczyński [2,3], Jarosław Kwapień [2] and Tomasz Stanisz [2]

[1] Faculty of Computer Science and Telecommunications, Cracow University of Technology, 31-155 Kraków, Poland
[2] Complex Systems Theory Department, Institute of Nuclear Physics, Polish Academy of Sciences, 31-342 Kraków, Poland
[3] Faculty of Mathematics and Computer Science, Jagiellonian University, ul. Łojasiewicza 6, 30-348 Kraków, Poland
* Correspondence: stanislaw.drozdz@ifj.edu.pl

Abstract: A novel method of exploring linguistic networks is introduced by mapping word-adjacency networks to time series and applying multifractal analysis techniques. This approach captures the complex structural patterns of language by encoding network properties—such as clustering coefficients and node degrees—into temporal sequences. Using Alice's Adventures in Wonderland by Lewis Carroll as a case study, both traditional word-adjacency networks and extended versions that incorporate punctuation are examined. The results indicate that the time series derived from clustering coefficients, when following the natural reading order, exhibits multifractal characteristics, revealing inherent complexity in textual organization. Statistical validation confirms that observed multifractal properties arise from genuine correlations rather than from spurious effects. Extending this analysis by taking into account punctuation equally with words, however, changes the nature of the global scaling to a more convolved form that is not describable by a uniform multifractal. An analogous analysis based on the node degrees does not show such rich behaviors, however. These findings reveal a new perspective for quantitative linguistics and network science, providing a deeper understanding of the interplay between text structure and complex systems.

Keywords: quantitative linguistics; complex networks; vertex observables; clustring coefficient, time series; multiscale correlations

1. Introduction

The application of the concept of complex networks [1–5] to the study of the organization of written texts is a very natural direction of research [6–11]. Complex networks are mathematical models used to represent interactions within large systems and, when applied to written texts, they provide insights into linguistic patterns, semantics, and the underlying structure of language [7–9,12–15]. In the context of written texts, words, phrases, sentences, or even entire documents can be modeled as nodes of a network, with edges representing relationships between them. Word-adjacency networks, also called word co-occurrence networks or word proximity networks, are a specific type of complex network where nodes represent words in a text, and edges represent their adjacency within a given context. These networks capture the local and global structure of how words are used together in language, revealing linguistic patterns, semantics, and the syntactic properties of texts [6,16–18]. There is strong evidence that, in a linguistic context, such networks have small-world properties [19]. However, there are possible different network architectures that can generate such behavior.

Previous studies have highlighted a potential connection between complex networks and time series [20–23]. Specifically, research provides evidence that time series can be transformed into networks—within the visibility graph scheme [24,25], for instance—and analyzed using techniques developed for complex network analysis [26]. Conversely, complex networks can be mapped onto time series [27], allowing for an alternative analytical approach. This latter transformation can be achieved using random walk frameworks on graphs, where, at each time step, a vertex property is 'emitted' by the walker. As a result, the generated time series may capture the structural organization of the network based on the chosen vertex property [28,29]. Keeping thus in mind the interplay between network theory and time series analysis [30,31], a study of the linguistic properties of word-adjacency networks is carried out here by mapping such networks to time series and analyzing such series using the related methodology.

2. From Word-Adjacency Networks to Time Series

This methodology and approximation are illustrated with the example of a book Alice's Adventures in Wonderland by Lewis Carroll that is widely known and often used in quantitative linguistic studies [32]. The networks to be studied are constructed from such a text sample after linking items of interest that are direct neighbors of each other at least once in a sample. Two cases of such items are considered here. The first one takes the traditional approach of considering all words as nodes in a network. The second one also includes all punctuation marks in addition to all words and is dictated by the fact that punctuation marks appear [33] to obey Zipf's law [34,35] on an equal footing with words. This book contains 2627 different words and includes punctuation 2636 items in total.

The network created in this way for this book, taking into account words only, is shown in Figure 1. It appears to conform quite clearly to a hierarchical organization [1,2]. Top-10 hubs of this network are indicated explicitly. The resulting complete node-degree distribution $P(k)$ follows the power-law trend indicated by the dashed straight line k^{-2} (which corresponds to the Zipf's law) in the bottom-left panel of this figure. Furthermore, as the bottom-right panel of this Figure shows, the correspondence between the distribution of clustering coefficients in the gross structure also follows the theoretical predictions $C(k) \sim k^{-1}$ for hierarchical networks [36,37]. (For a node i with k_i links (edges), these coefficients are defined in the usual way as $C_i = 2n_i/k_i(k_i - 1)$, where n_i is the number of edges between the k_i neighbors of i). Finally, the green line indicates the path across the first 1000 words that corresponds to the actual reading order of this book. As one can see, this line mainly travels around the area of greater concentration of hubs and occasionally moves away to the periphery of the network but keeps returning quickly. Interestingly, even the loop can be seen here (upper left part of the network). In this particular case, it reflects a sequence of the same words separated by commas. However, commas are not present in this network.

In what follows, this network is therefore enriched by incorporating punctuation, and the resulting network is shown in Figure 2 with the same convention as before. Now, there are four punctuation marks among the top-10 hubs, which can significantly affect the dynamic characteristics of the reading trajectory shown, as before, in green. In particular, the loop commented above in Figure 1 has now disappeared. In the topological properties of the network taking into account punctuation marks, here expressed by the distributions of degree $P(k)$ and local clustering coefficient $C(k)$, the changes are not clearly significant.

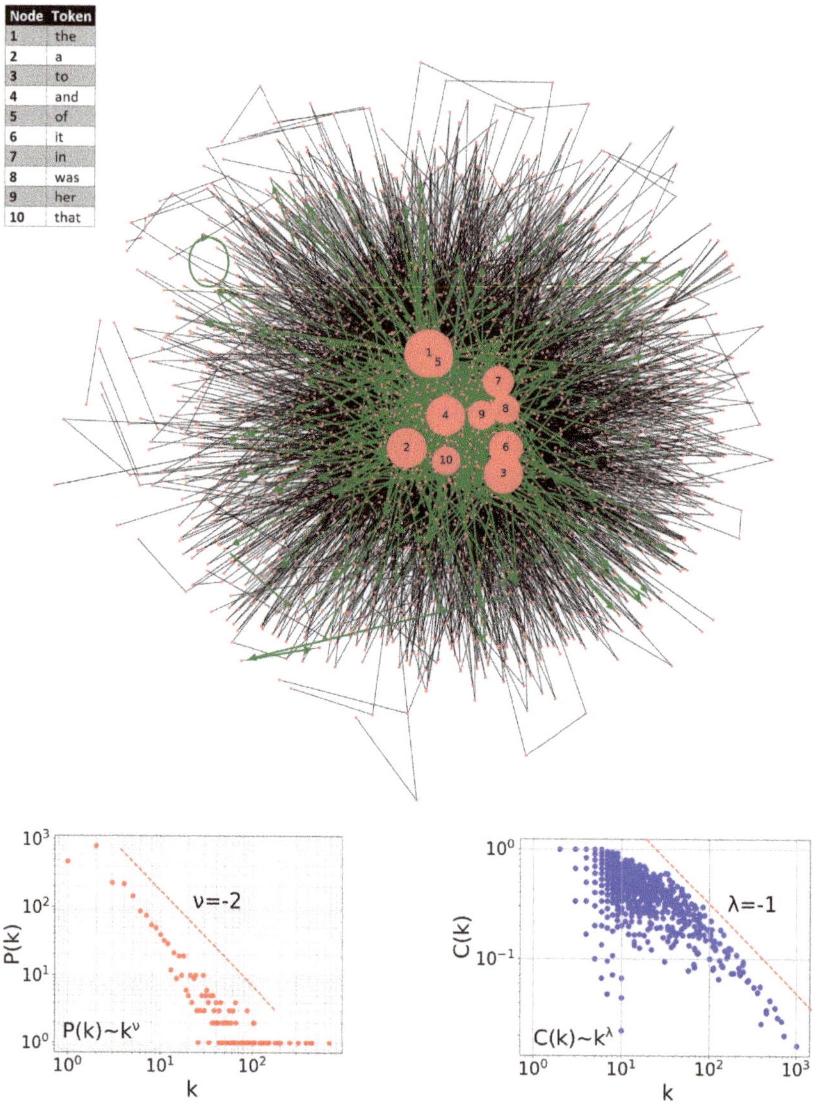

Figure 1. Visualization of the word-adjacency network of a book Alice's Adventures in Wonderland by Lewis Carroll created exclusively from words. Top-10 hubs are pointed out explicitly. The green line indicates the path that corresponds to the actual reading order of this book. In order not to make this image too opaque, only the visit within the first 1000 words is marked here. Bottom-left inset displays the degree k distribution $P(k)$ of this network, and the bottom-right one displays the corresponding distribution of the clustering coefficients $C(k)$. The red dashed lines indicate the corresponding predictions for the model hierarchical networks.

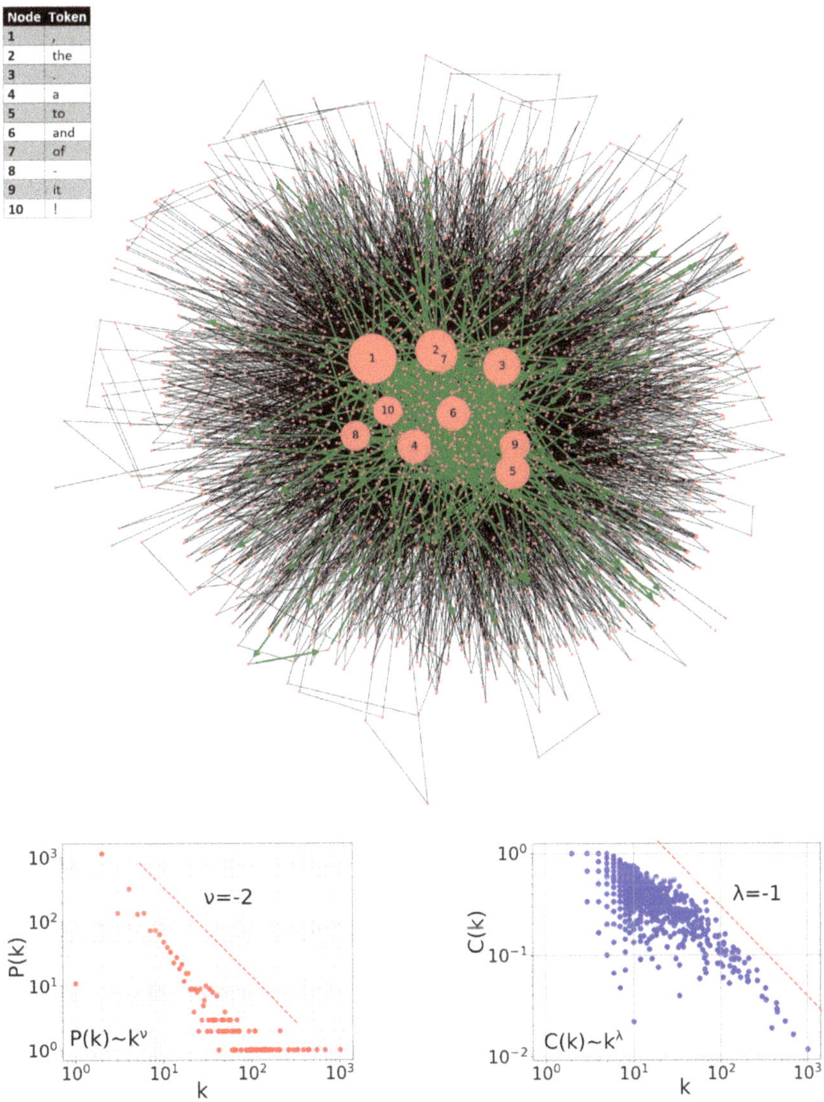

Figure 2. Visualization of the word-adjacency network of a book Alice's Adventures in Wonderland by Lewis Carroll created from both words and punctuation marks. Top-10 hubs are pointed out explicitly. The green line indicates the path that corresponds to the actual reading order of this book. In order not to make this image too opaque, only the visit within the first 1000 items is marked here. Bottom-left inset displays the degree k distribution $P(k)$ of this network, and the bottom-right one displays the corresponding distribution of the clustering coefficients $C(k)$. The red dashed lines indicate the corresponding predictions for the model hierarchical networks.

A more systematic visit to them according to specific rules may provide deeper insight into the organization of such networks also in the sense of how they grow. Such rules are not rigidly defined for all tasks but will be dictated by the specificity of a given issue. In the current linguistic context, the natural order of node visits is in the direction of writing and then reading the text. A natural reference for this way of visiting the word-adjacency network is sampling it by a random walk. Of course, for a more complete, multi-dimensional view, various network characteristics can be used in both cases.

A few sample but representative cases are shown in what follows. Figure 3 shows the time series of node degrees k 'emitted' during visits along the word adjacency network and along the extended network, respectively, so that punctuation marks are also included on par with words. In both cases, a journey is conducted in accordance with the reading order and, independently, determined by the random walk rules. The resulting consecutive 5000 steps in each of the four cases are shown. An analogous composition of cases with the clustering coefficient 'emitted' is shown in Figure 4. The obtained patterns of variability in these series show quite rich organization. A cursory visual inspection shows a slightly more varied organization of this variability when the clustering coefficient is used. In particular, for the case of a network without punctuation and generating a series in accordance with the reading order, it shows clustering of variability resembling the one known, for example, from the dynamics of financial markets, which typically implies self-similarity and fractals [38].

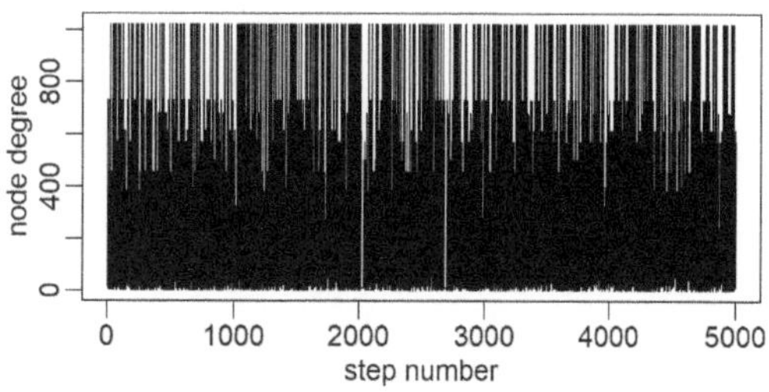

(**a**) Network without punctuation, reading-order-based walk.

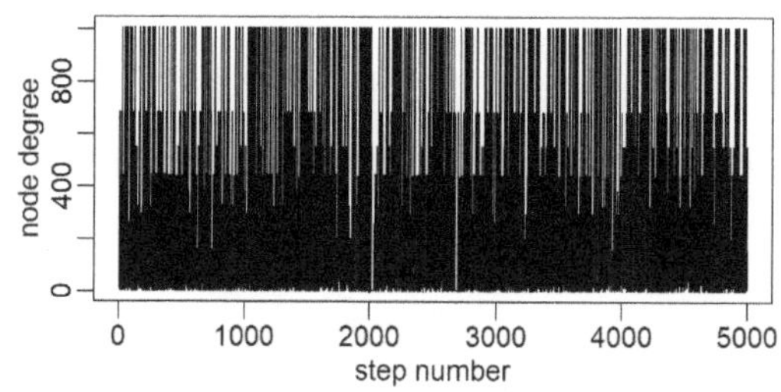

(**b**) Network including punctuation, reading-order-based walk.

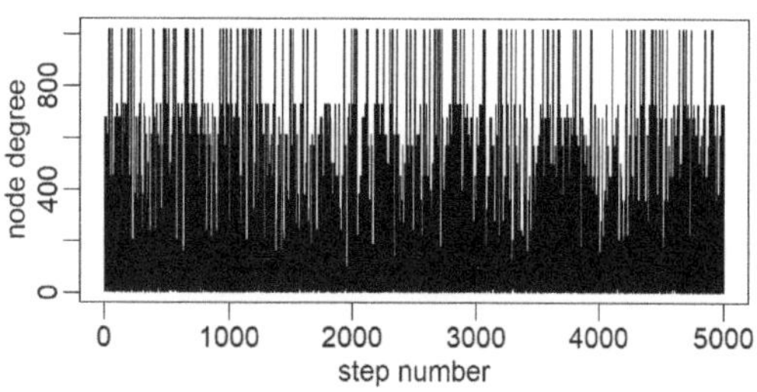

(**c**) Network without punctuation, random walk.

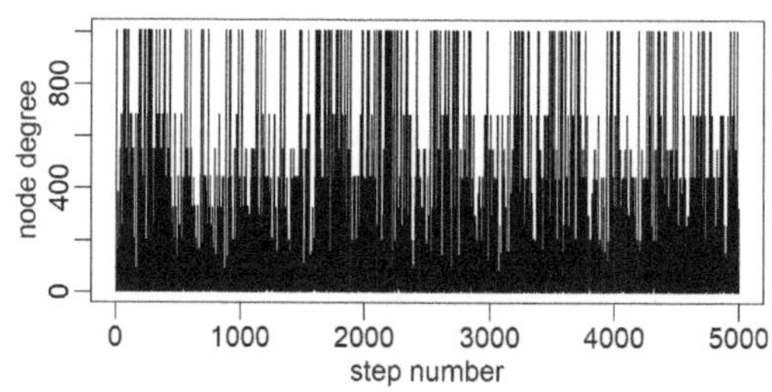

(**d**) Network including punctuation, random walk.

Figure 3. The first 5000 values of time series constructed from the degrees of nodes visited in a network walk. Two variants of the network are considered: without punctuation (**left column**) and with punctuation marks included in the node set (**right column**).

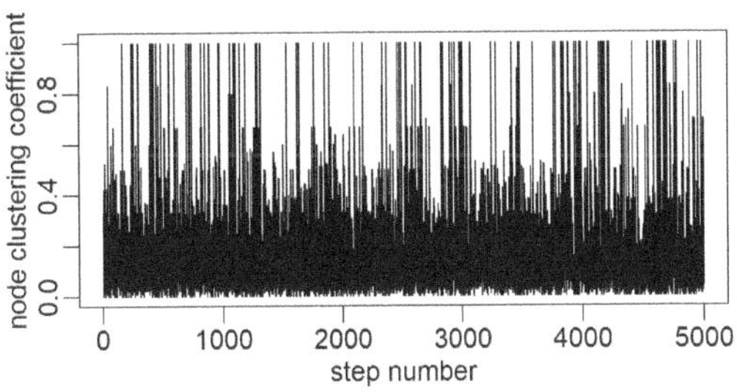

(**a**) Network without punctuation, reading-order-based walk.

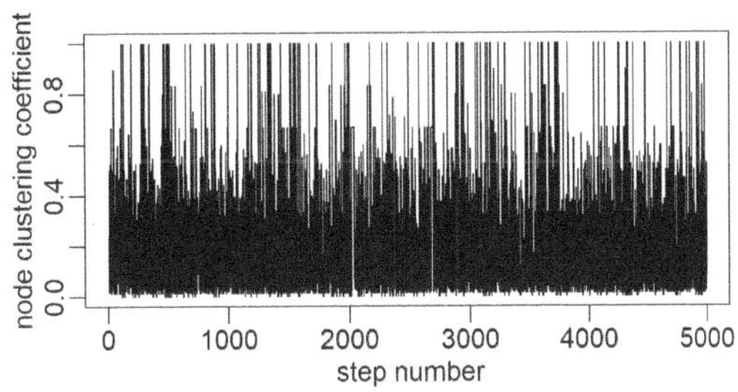

(**b**) Network including punctuation, reading-order-based walk.

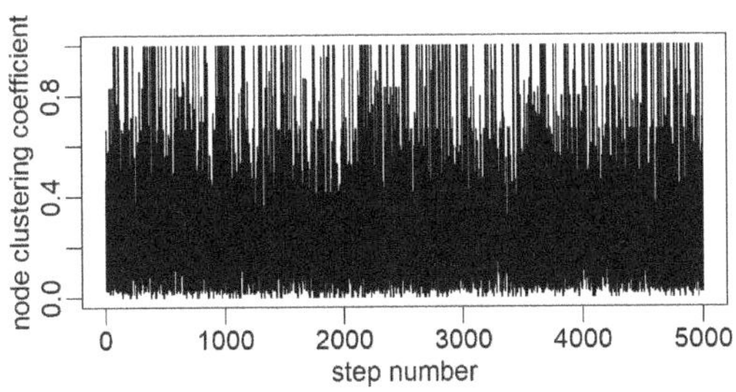

(**c**) Network without punctuation, random walk.

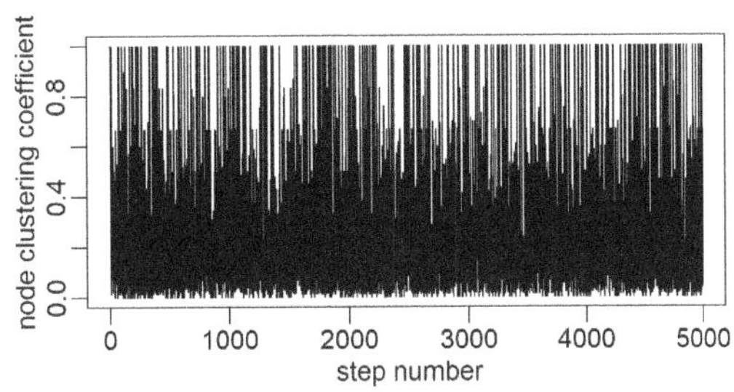

(**d**) Network including punctuation, random walk.

Figure 4. The first 5000 values of time series constructed from the clustering coefficients of nodes visited in a network walk. Two variants of the network are considered: without punctuation (**left column**) and with punctuation marks included in the node set (**right column**).

3. Detrended Fluctuation Analysis

Self-similarity, characterized by the absence of a distinct scale, is a fundamental trait of natural complex systems. Empirically, this property often appears as a distinct temporal structure in measurement outcomes, represented as time series data. In particular, complexity is commonly associated with a hierarchical cascade of data points exhibiting multiscaling, highlighting the need for effective methods to identify such patterns in complex systems research [39].

Extensive studies have documented that multifractal detrended fluctuation analysis (MFDFA) is a highly reliable technique for examining multiscaling patterns [40,41] and in the linguistic context [42–45]. Expanding on the widely utilized detrended fluctuation analysis (DFA) [46], MFDFA offers a multiscale framework for examining hierarchical and multiscaling behaviors. The key steps of the MFDFA algorithm are briefly outlined below.

3.1. Multifractal Formalism

A time series $U = \{u_i\}_{i=1}^{T}$ of T consecutive measurements of an observable u is partitioned into M_s non-overlapping windows of length s, starting from both ends of U. This results in a total of $2M_s$ windows. In order to properly handle potential non-stationarity in the signal, a detrending procedure is conducted within each window on an integrated signal, known as the profile, $X = \{x_i\}_{i=1}^{s}$. Its elements are defined as:

$$x_i = \sum_{j=1}^{i} u_j. \tag{1}$$

The detrending procedure involves fitting a properly matched low-order polynomial $P^{(m)}$ (with m typically between 1 and 3 depending on the degree of signal roughness) to the data X within each window $\nu = 0, \ldots, 2M_s - 1$ and then subtracting it. The variance of such a detrended signal is then determined as:

$$f^2(\nu, s) = \frac{1}{s} \sum_{i=1}^{s} (x_i - P^{(m)}(i))^2. \tag{2}$$

Based on it, a family of fluctuation functions of order r is determined using the average variance computed across all windows:

$$F_r(s) = \left\{ \frac{1}{2M_s} \sum_{\nu=0}^{2M_s-1} \left[f^2(\nu, s) \right]^{r/2} \right\}^{1/r}, \tag{3}$$

where r is a real number.

The fluctuation functions $F_r(s)$ are calculated for different values of the scale s and the index r. Typically, the minimum value of s is chosen to be greater than the longest sequence of constant values in U, while the maximum is set to a factor of a few smaller than T. In contrast to s, there is no universally defined range for r. Since r corresponds to the moments of the signal, extreme values should be avoided in time series with heavy tails to ensure meaningful results. The range of considered values of parameter r is limited by the possibility of the appearance of divergent moments in Equation (3). $-10 \leq r \leq 10$ is a fairly typical range used in the literature and is also appropriate in the present application. If the fluctuation functions follow power-law dependencies on s,

$$F_r(s) \sim s^{h(r)}, \tag{4}$$

within such a range of r and, in addition, $h(r)$ is r-dependent, this indicates that the time series under study is multifractal. Otherwise, when $h(r)$ is constant in r, it is just monofractal. The function $h(r)$ can be considered the generalized Hurst exponent, because $h(r) = H$ for $r = 2$, where H is the standard Hurst exponent [47,48]. Signatures of fractal organization of $F_r(s)$ result in straight lines on double logarithmic plots.

Characteristics of multifractality can thus be quantified directly in terms of $h(r)$-dependence, but another commonly adopted measure is the singularity spectrum $f(\alpha)$, which is derived from $h(r)$ by the Legendre transform:

$$\begin{aligned} \alpha &= h(r) + r h'(r), \\ f(\alpha) &= r[\alpha - h(r)] + 1, \end{aligned} \tag{5}$$

Here, α reflects the data-point singularity, equivalent to the Hölder exponent. Geometrically, $f(\alpha)$ can be interpreted as the fractal dimension of the subset of the data characterized by a specific Hölder exponent α [49]. For a multifractal time series, $f(\alpha)$ typically forms a downward-pointing parabola. The broader the singularity spectrum $f(\alpha)$, the greater the richness of multifractality in the time series, which serves as an indicator of its complexity content. In some cases, the $f(\alpha)$ parabola may appear distorted or asymmetric, suggesting that data points of varying amplitudes exhibit different scaling behaviors [50–55]. For a monofractal time series, the pair $(\alpha, f(\alpha))$ converges to coordinates of a single point.

3.2. Fluctuation Functions and Their Scaling Characteristics

All the most important characteristics of correlation in time series are contained in the fluctuation functions defined by the Equation (3); in their degree of scaling; and, above all, in their mutual relationship for different values of r. Four collections of fluctuation functions of time series constructed from the degrees of nodes visited during walks along networks of Figure 1 (words only) and Figure 2 (words and punctuation marks) are shown in Figure 5. The scaling effects are quite visible here, but this scaling is rather of the monofractal type because it is not diverse enough in r to indicate multifractality. If at all, more dependencies on r can be seen in the case of a random walk in this network, but downward deflections on small scales for negative values of r may result from the appearance of several consecutive similar values in a series and not from the real tendency to multiscaling. It should also be remembered that this type of random walk may induce some correlations because nodes with a higher multiplicity are visited more often.

Finally, an interesting global effect worth noting here is the quite visible presence of two scaling regimes of $F_r(s)$ in two halves of the entire s area considered. It may be significant that in the case of 'text-based walk', an overall slope in the first half of this interval is smaller than in the second half, both without and with punctuation networks. This is indicated by the red dotted lines in Figure 5 generated by fitting $F_2(s)$, which by Equation (4) determines the corresponding Hurst exponents ($H = h(2)$). They are denoted as H_{left} and H_{right} correspondingly. Interestingly, the first ones are less than 0.5, which indicates anti-persistence, while the second ones are greater than 0.5, thus signaling persistence on larger scales. The related crossover observed corresponds to a distance of about 300 words and thus, interestingly and perhaps significantly, to about one page of traditional text. In the second case of the random walk, two regimes also stand out but with the opposite tendencies as far as the numerical values of the Hurst exponents relative to 0.5 are concerned.

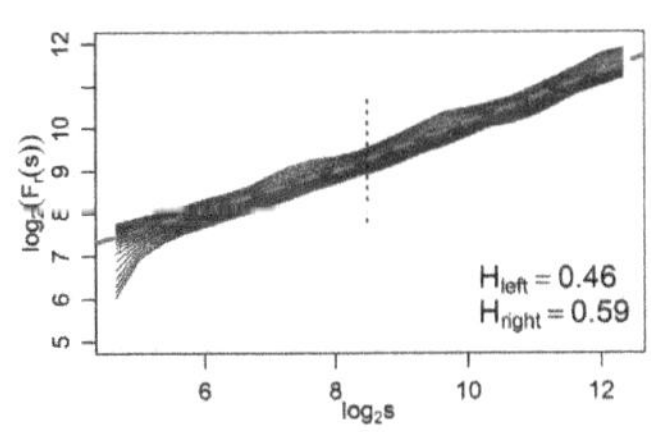

(**a**) No punctuation, text-based walk.

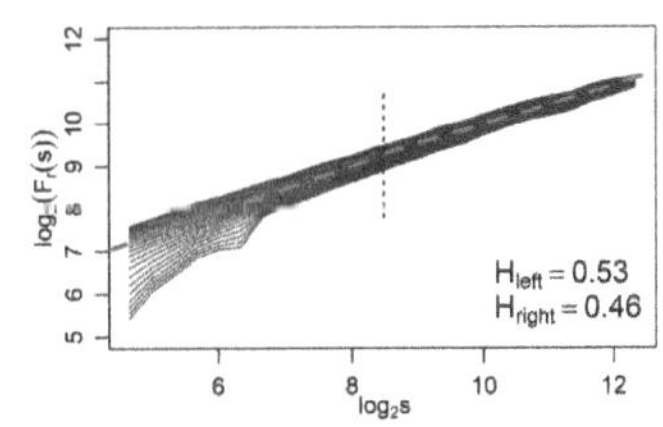

(**b**) No punctuation, random walk.

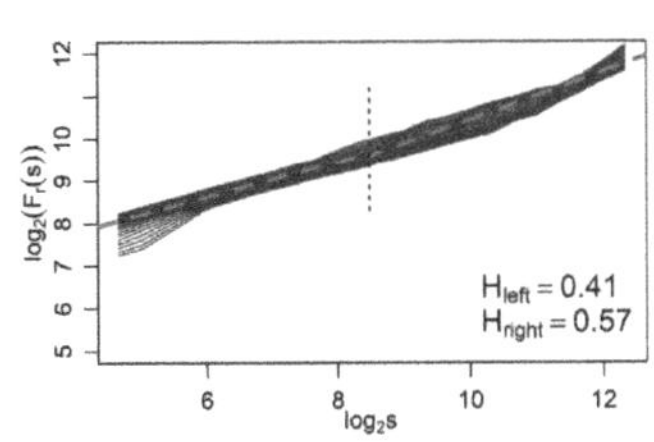

(**c**) With punctuation, text-based walk.

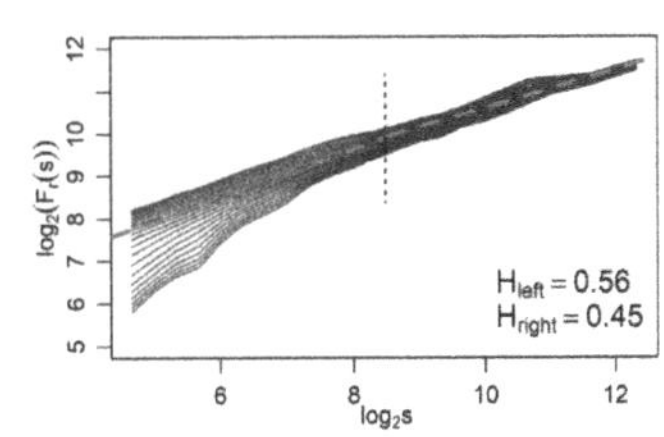

(**d**) With punctuation, random walk.

Figure 5. Fluctuation functions of time series constructed from the degrees of nodes visited in network walks. The linear fits to the points corresponding to $r = 2$, separate for the left and the right halves of the plots, are marked by red dashed lines. The corresponding Hurst exponents are given in the bottom right corner of each plot.

As is shown in Figure 6, the situation changes considerably if, during analogous walks on these two networks, without and with punctuation, the time series are generated by reading the appropriate clustering coefficients instead of the previous node degrees. Now, the scaling does not distinguish between two regimes and develops more in a multifractal direction. For the 'text-based' series on the network without punctuation, the multiscaling is even so rich that it allows us to determine the entire singularity spectrum according to Equation (5), and this spectrum is shown in the inset.

An intriguing related result shown in panel (c) of Figure 6 is that it is definitely not possible to determine an analogous spectrum for the case of a network that also takes into account punctuation. Here, the fluctuation functions $F_r(s)$ for different values of the parameter r move away from each other very significantly as the scale s increases. This is a very unusual picture in this type of analysis and may manifest a different role of punctuation in relation to words in text creation. The behavior of the fluctuation functions in this case may, for instance, originate from a superposition of different scaling components and thus may require an appropriate decomposition procedure. This is a very interesting and significant result worth further investigations. For a random walk counterpart of this series shown in panel (d) of Figure 6, such an effect disappears, and the fluctuation functions look similar to the case (b) of no punctuation.

A question arises as to why multifractality is observed for the clustering-coefficient time series, while it is not observed in the node-degree time series. A possible answer may lie in the fact that, in order to talk about multifractality, long-range correlations must exist [56,57]. In a word-adjacency network, the node degree describes the properties of that specific node and provides little information about the degrees of its neighbors (we neglect the disassortativity effect, which may produce short-range anticorrelations). Therefore, a walker moving on such a network does not generate long-range correlations. On the other hand, the clustering coefficient describes not only a given node but also its surroundings. In this sense, if a walker enters a region with a large (small) number of connections between nodes, it will encounter a sequence of nodes with elevated (lowered) values of the coefficient along its path, which may generate long-range correlations in time and, thus, multifractality.

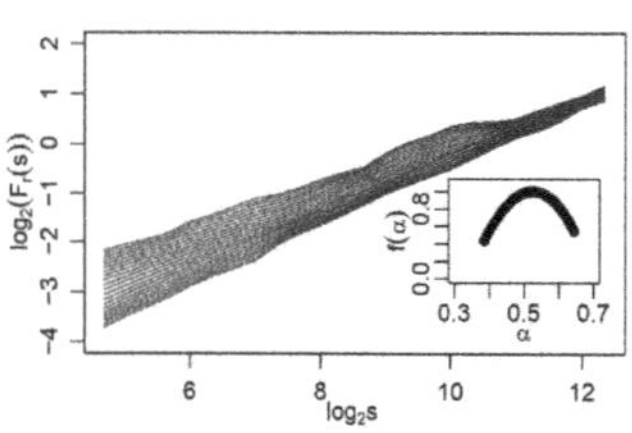

(**a**) No punctuation, text-based walk.

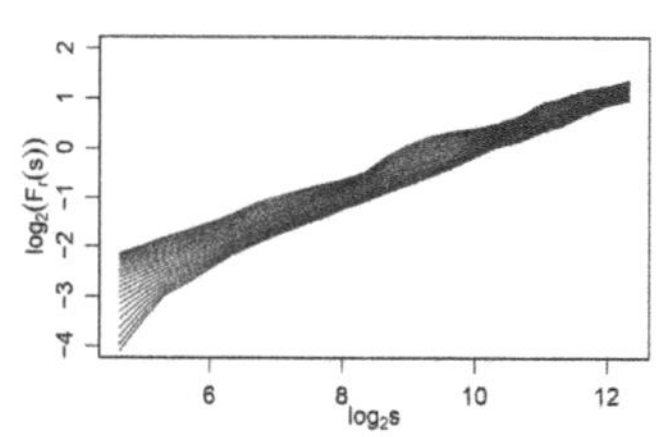

(**b**) No punctuation, random walk.

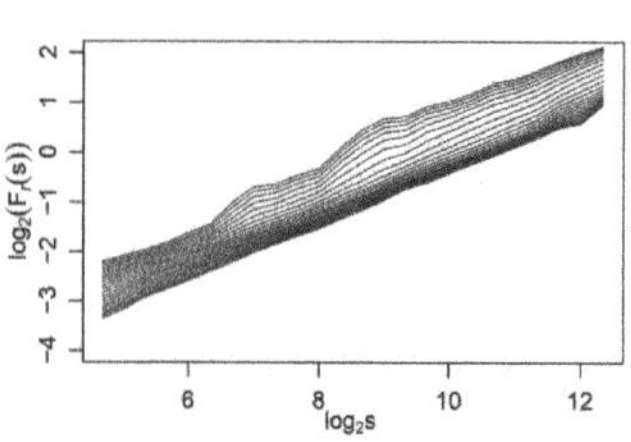

(**c**) With punctuation, text-based walk.

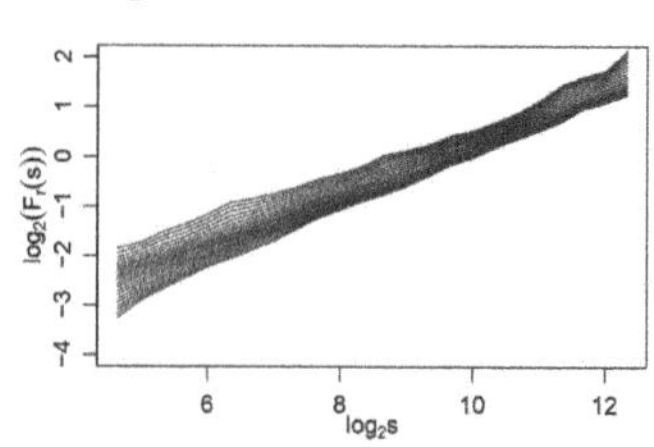

(**d**) With punctuation, random walk.

Figure 6. Fluctuation functions of time series constructed from the clustering coefficients of nodes visited in network walks. In (**a**), the corresponding singularity spectrum, which develops a clear multifractal scaling, is presented in the inset.

3.3. Testing Significance of Multifractality Signals

Obtaining reliable numerical results based on the above MFDFA algorithm is a delicate matter. It is easy to overestimate and misinterpret results, especially for relatively short time series. Standard procedures for creating correlation-free surrogates by shuffling relatively short—even several thousand data points—original series may show the appearance of multifractal scaling, while true multifractality is generated only by correlations [56]. The problem of obtaining a false multifractality signal in the absence of correlations appears particularly persistently when the distributions of data fluctuations in the series are heavy-tailed. Then, very long series are needed to see the true result, indicating the absence of multifractality. In practice, insufficiently long time series, especially those with heavy tails of fluctuations, may give a false signal of multifractality [57]. The newly proposed method [58] of systematically reducing heavy fluctuation tails while maintaining correlation strength provides the most systematic method for verifying the validity of a multifractal signal. The idea is [58] to project the data by using a ranking based probability density transformation in such a way as to maintain their original order in the series but systematically change the distribution of fluctuations towards the Gaussian distribution.

q-Gaussians turn out to be a very practical analytical parametrization of this class of distributions. These distributions may be viewed as a generalization of the Gaussian distribution, in the same way as the Tsallis entropy S_q generalizes the Boltzmann–Gibbs entropy S [59,60]. The q-Gaussian distribution $\mathcal{G}_q$ has two parameters: the shape parameter $q \in (-\infty, 3)$ and the width parameter $\beta > 0$. Its PDF has the following form [59]:

$$p_q(x) = \frac{\sqrt{\beta}}{C_q} e_q(-\beta x^2), \tag{6}$$

where $e_q(x)$ is the *q-exponential* function defined by

$$e_q(x) = \begin{cases} (1 + (1-q)x)^{1/(1-q)} & \text{if } q \neq 1 \text{ and } 1 + (1-q)x > 0 \\ 0 & \text{if } q \neq 1 \text{ and } 1 + (1-q)x \leq 0 \\ e^x & \text{if } q = 1 \end{cases} \tag{7}$$

and $C_q = \int_{-\infty}^{\infty} e_q(x^2)dx$ is a normalization factor. The preservation of clear multifractality characteristics in the limit $q = 1$, i.e., for the Gaussian distribution, is then a convincing signal that we are dealing with true multifractality generated by correlations. It may be recalled here that heavy tails influence the spread of mutlifractality but only when correlations are present. Otherwise, only monofractality may apply or, in extreme situations, bifractality, when the fluctuations are stable in the Lévy region [56].

The case of the promisingly developed multifractality of Figure 6a is now subjected to the above test to check to what extent it is a true multifractality resulting from correlations, i.e., whether it survives the reduction in the fluctuation distribution to the Gaussian distribution. The sequence of generalized Hurst exponent $h(r)$ for successive projections on q-Gaussians of the original series onto correlation-preserving series is shown in Figure 7. Clearly, even at the $q = 1$ limit, thus for the Gaussian distribution, $h(r)$ still depends on r, which means that multifractality applies. In terms of the singularity spectra, the corresponding progression and dependence are illustrated in Figure 8. Applying the same procedure to the shuffled original series, i.e., after destroying the correlations, leads to the disappearance of the manifestations of multifractality.

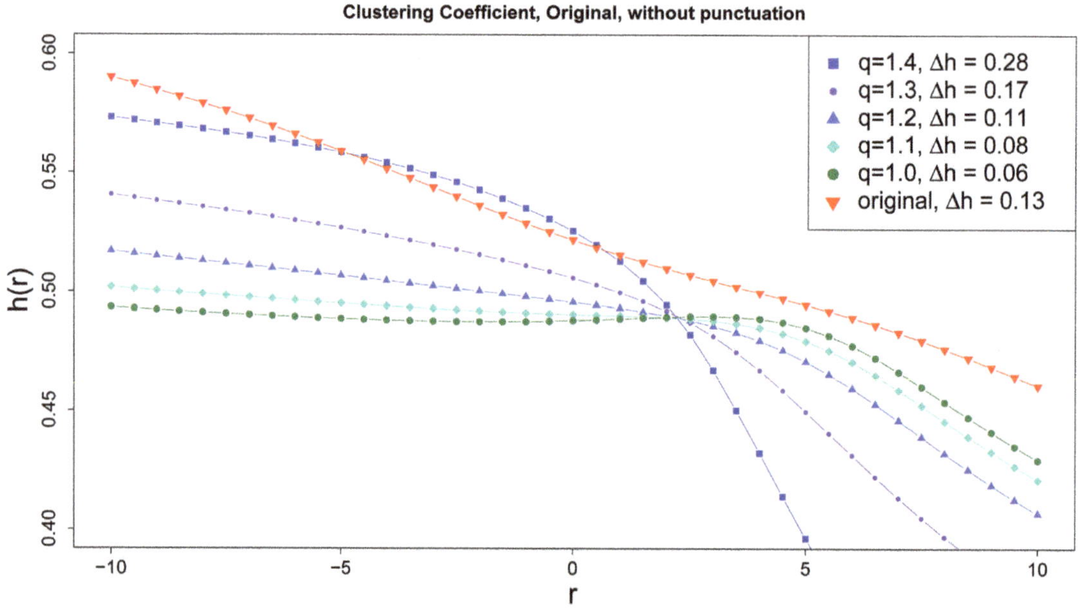

Figure 7. Generalized Hurst exponents for the original case of Figure 6a and for its q-Gaussian projected variants.

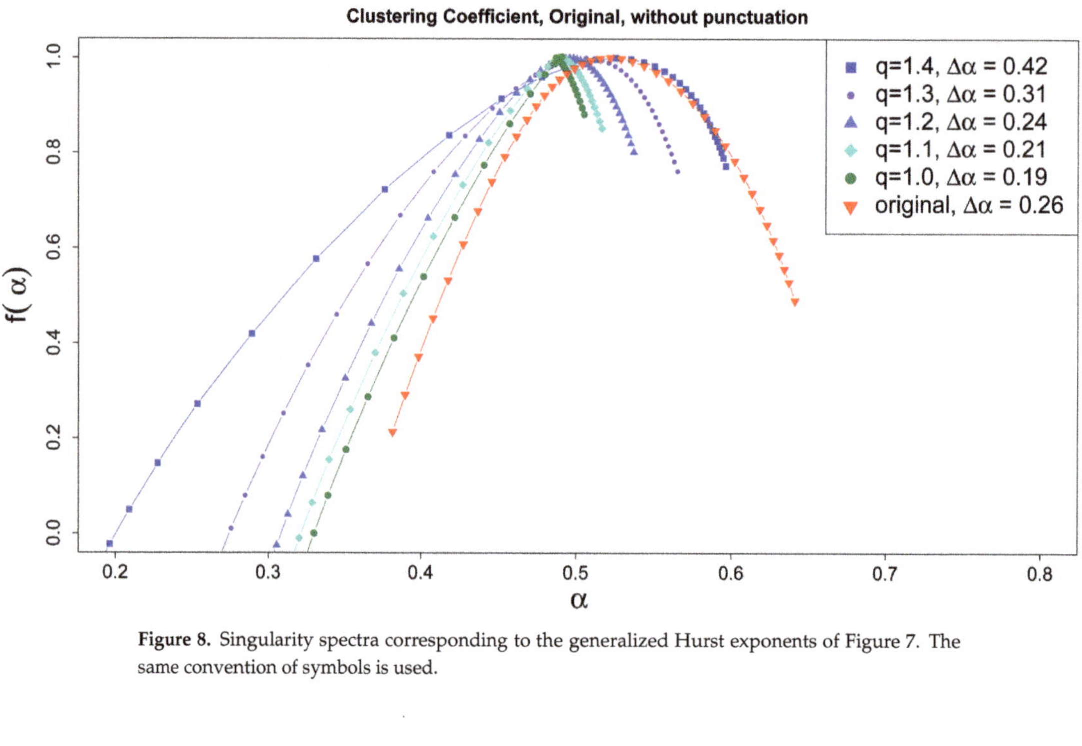

Figure 8. Singularity spectra corresponding to the generalized Hurst exponents of Figure 7. The same convention of symbols is used.

4. Summary

This study proposed and explored the application of multifractal time series analysis to linguistic networks, particularly word-adjacency networks derived from Alice's Adventures in Wonderland by Lewis Carroll. By mapping linguistic structures onto time series and applying multifractal detrended fluctuation analysis (MFDFA), complex scaling behaviors that provide insights into the structural organization of language are uncovered. Notably, the clustering coefficient-based time series exhibited rich multifractal properties when traversed according to the reading order, suggesting an inherent complexity in natural text organization. Furthermore, the analysis confirmed the robustness of multifractal signatures through rigorous significance testing, ruling out the possibility of spurious multifractality arising from heavy-tailed fluctuations.

These findings also demonstrate that punctuation marks play a significant but distinct role in shaping the statistical properties of linguistic networks. Their inclusion alters the scaling characteristics of time series as derived from word-adjacency networks. This is particularly intriguing in light of the fact that punctuation marks obey Zipf's law just like words [33].

The results presented here thus reveal a new perspective for quantitative linguistics and network science by providing a novel approach to studying text complexity through time series transformations. Future research can extend this methodology to diverse textual datasets, languages, and genres to explore whether these patterns are universal features of written communication. Additionally, refining network traversal techniques and integrating deeper semantic layers could further enhance our understanding of linguistic constructs. This work thus underscores the potential of interdisciplinary approaches that merge network theory, time series analysis, and linguistic studies, offering a promising framework for unraveling the intricate organization of language.

Author Contributions: Conceptualization, J.D., M.D., S.D., R.K., J.K., and T.S.; methodology, S.D., J.K., and T.S.; software, J.D., M.D., R.K., and T.S.; validation, J.D., M.D., S.D., R.K., J.K., and T.S.; formal analysis, J.D., M.D., S.D., R.K., J.K., and T.S.; investigation, J.D., M.D., S.D., R.K., J.K., and T.S.; resources, J.D., M.D., and T.S.; data curation, J.D., M.D., and T.S.; writing—original draft preparation, S.D.; writing—review and editing, J.D., M.D., S.D., R.K., J.K., and T.S.; visualization, J.D., M.D., S.D., R.K., and T.S.; and supervision, S.D. and J.K. All authors have read and agreed to the published version of the manuscript.

Funding: This research received no external funding.

Institutional Review Board Statement: Not applicable.

Data Availability Statement: These data are publicly available.

Conflicts of Interest: The authors declare no conflicts of interest.

References

1. Albert, R.; Barabási, A.L. Statistical mechanics of complex networks. *Rev. Mod. Phys.* **2002**, *74*, 47–97. [CrossRef]
2. Dorogovtsev, S.N.; Mendes, J.F. Evolution of networks. *Adv. Phys.* **2002**, *51*, 1079–1187. [CrossRef]
3. Dorogovtsev, S.N.; Goltsev, A.V.; Mendes, J.F. Critical phenomena in complex networks. *Rev. Mod. Phys.* **2008**, *80*, 1275–1335. [CrossRef]
4. Boccaletti, S.; Latora, V.; Moreno, Y.; Chavez, M.; Hwang, D.U. Complex networks: Structure and dynamics. *Phys. Rep.* **2006**, *424*, 175–308. [CrossRef]
5. Ji, P.; Ye, J.; Mu, Y.; Lin, W.; Tian, Y.; Hens, C.; Perc, M.; Tang, Y.; Sun, J.; Kurths, J. Signal propagation in complex networks. *Phys. Rep.* **2023**, *1017*, 1–96. [CrossRef]
6. Amancio, D.R.; Antiqueira, L.; Pardo, T.A.S.; Da, L.; Costa, F.; Oliveira, O.N.; Nunes, M.G.V. Complex networks analysis of manual and machine translations. *Int. J. Mod. Phys. C* **2008**, *19*, 583–598.

7. Choudhury, M.; Mukherjee, A. *The Structure and Dynamics of Linguistic Networks*; Birkhäuser: Basel, Switzerland, 2009; pp. 145–166. [CrossRef]
8. Liu, H.T.; Li, W.W. Language clusters based on linguistic complex networks. *Chin. Sci. Bull.* **2010**, *55*, 3458–3465. [CrossRef]
9. Araújo, T.; Banisch, S. *Multidimensional Analysis of Linguistic Networks*; Springer: Berlin/Heidelberg, Germany, 2015; pp. 107–131. [CrossRef]
10. Ramirez-Arellano, A. Classification of Literary Works: Fractality and Complexity of the Narrative, Essay, and Research Article. *Entropy* **2020**, *22*, 904. [CrossRef]
11. Stanisz, T.; Drożdż, S.; Kwapień, J. Complex systems approach to natural language. *Phys. Rep.* **2024**, *1053*, 1–84. [CrossRef]
12. Antiqueira, L.; Nunes, M.G.; Oliveira, O.N.; da F. Costa, L. Strong correlations between text quality and complex networks features. *Phys. A* **2007**, *373*, 811–820. [CrossRef]
13. Borge-Holthoefer, J.; Arenas, A. Semantic networks: Structure and dynamics. *Entropy* **2010**, *12*, 1264–1302. [CrossRef]
14. de Arruda, H.F.; Silva, F.N.; Marinho, V.Q.; Amancio, D.R.; da Fontoura Costa, L. Representation of texts as complex networks: A mesoscopic approach. *J. Complex Netw.* **2018**, *6*, 125–144. [CrossRef]
15. Oliveira, D.A.; de Barros Pereira, H.B. Modeling texts with networks: Comparing five approaches to sentence representation. *Eur. Phys. J. B* **2024**, *97*, 77. [CrossRef]
16. Dorogovtsev, S.N.; Mendes, J.F. Language as an evolving word web. *Proc. R. Soc. B Biol. Sci.* **2001**, *268*, 2603–2606. [CrossRef]
17. Segarra, S.; Eisen, M.; Ribeiro, A. Authorship attribution through function word adjacency networks. *IEEE Trans. Signal Process.* **2015**, *63*, 5464–5478. [CrossRef]
18. Amancio, D.R. Probing the topological properties of complex networks modeling short written texts. *PLoS ONE* **2015**, *10*, e0118394. [CrossRef]
19. Kulig, A.; Drożdż, S.; Kwapień, J.; Oświęcimka, P. Modeling the average shortest-path length in growth of word-adjacency networks. *Phys. Rev. E* **2015**, *91*, 032810. [CrossRef]
20. Campanharo, A.S.L.O.; Sirer, M.I.; Malmgren, R.D.; Ramos, F.M.; Amaral, L.A.N. Duality between Time Series and Networks. *PLoS ONE* **2011**, *6*, e23378. [CrossRef]
21. Hou, L.; Small, M.; Lao, S. Dynamical systems induced on networks constructed from time series. *Entropy* **2015**, *17*, 6433–6446. [CrossRef]
22. Khor, A.; Small, M. Examining k-nearest neighbour networks: Superfamily phenomena and inversion. *Chaos* **2016**, *26*, 043101. [CrossRef]
23. McCullough, M.; Sakellariou, K.; Stemler, T.; Small, M. Regenerating time series from ordinal networks. *Chaos* **2017**, *27*, 035814. [CrossRef] [PubMed]
24. Lacasa, L.; Luque, B.; Ballesteros, F.; Luque, J.; Nuño, J.C. From time series to complex networks: The visibility graph. *Proc. Natl. Acad. Sci. USA* **2008**, *105*, 4972–4975. [CrossRef] [PubMed]
25. Nuñez, A.M.; Lacasa, L.; Gomez, J.P.; Luque, B. *Visibility Algorithms: A Short Review*; InTech: Houston TX, USA, 2012; pp. 119–152.
26. Donner, R.V.; Small, M.; Donges, J.F.; Marwan, N.; Zou, Y.; Xiang, R.; Kurths, J. Recurrence-based time series analysis by means of complex network methods. *Int. J. Bifurc. Chaos* **2011**, *21*, 1019–1046. [CrossRef]
27. Nicosia, V.; De Domenico, M.; Latora, V. Characteristic exponents of complex networks. *Europhys. Lett.* **2014**, *106*, 58005. [CrossRef]
28. Noh, J.D.; Rieger, H. Random walks on complex networks. *Phys. Rev. Lett.* **2004**, *92*, 118701. [CrossRef]
29. Masuda, N.; Porter, M.A.; Lambiotte, R. Random walks and diffusion on networks. *Phys. Rep.* **2017**, *716–717*, 1–58. [CrossRef]
30. Oświęcimka, P.; Livi, L.; Drożdż, S. Multifractal cross-correlation effects in two-variable time series of complex network vertex observables. *Phys. Rev. E* **2016**, *94*, 042307. [CrossRef]
31. Oświęcimka, P.; Livi, L.; Drożdż, S. Right-side-stretched multifractal spectra indicate small-worldness in networks. *Commun. Nonlinear Sci. Numer. Simul.* **2018**, *57*, 231–245. [CrossRef]
32. Ausloos, M. Generalized Hurst exponent and multifractal function of original and translated texts mapped into frequency and length time series. *Phys. Rev. E* **2012**, *86*, 031108. [CrossRef]
33. Kulig, A.; Kwapień, J.; Stanisz, T.; Drożdż, S. In narrative texts punctuation marks obey the same statistics as words. *Inf. Sci.* **2017**, *375*, 98–113. [CrossRef]
34. Zipf, G.K. *Human Behavior and the Principle of Least Effort: An Introduction to Human Ecology*; Addison-Wesley Press: Cambridge, UK, 1949. [CrossRef]
35. Piantadosi, S.T. Zipf's word frequency law in natural language: A critical review and future directions. *Psychon. Bull. Rev.* **2014**, *21*, 1112–1130. [CrossRef] [PubMed]
36. Ravasz, E.; Barabasi, A.L. Hierarchical organization in complex networks. *Phys. Rev. E* **2003**, *67*, 026112. [CrossRef]

37. Eckmann, J.P.; Moses, E.; Sergi, D. Entropy of dialogues creates coherent structures in e-mail traffic. *Proc. Natl. Acad. Sci. USA* **2004**, *101*, 14333–14337. [CrossRef]
38. Kwapień, J.; Drożdż, S. Physical approach to complex systems. *Phys. Rep.* **2012**, *515*, 115–226. [CrossRef]
39. Jimenez, J. Intermittency and cascades. *J. Fluid Mech.* **2000**, *409*, 99–120.
40. Kantelhardt, J.W.; Zschiegner, S.A.; Koscielny-Bunde, E.; Havlin, S.; Bunde, A.; Stanley, H.E. Multifractal detrended fluctuation analysis of nonstationary time series. *Phys. A* **2002**, *316*, 87–114. [CrossRef]
41. Oświęcimka, P.; Kwapień, J.; Drożdż, S. Wavelet versus detrended fluctuation analysis of multifractal structures. *Phys. Rev. E* **2006**, *74*, 016103. [CrossRef]
42. Ausloos, M. Measuring complexity with multifractals in texts: Translation effects. *Chaos Solitons Fractals* **2012**, *45*, 1349–1357. [CrossRef]
43. Drożdż, S.; Oświęcimka, P.; Kulig, A.; Kwapień, J.; Bazarnik, K.; Grabska-Gradzińska, I.; Rybicki, J.; Stanuszek, M. Quantifying origin and character of long-range correlations in narrative texts. *Inf. Sci.* **2016**, *331*, 32–44. [CrossRef]
44. Dec, J.; Dolina, M.; Drożdż, S.; Kwapień, J.; Stanisz, T. Multifractal Hopscotch in *Hopscotch* by Julio Cortázar. *Entropy* **2024**, *26*, 716. [CrossRef]
45. Bartnicki, K.; Drożdż, S.; Kwapień, J.; Stanisz, T. Punctuation patterns in *Finnegans Wake* by James Joyce are largely translation-invariant. *Entropy* **2025**, *27*, 177. [CrossRef] [PubMed]
46. Peng, C.K.; Buldyrev, S.V.; Havlin, S.; Simons, M.; Stanley, H.E.; Goldberger, A.L. Mosaic organization of DNA nucleotides. *Phys. Rev. E* **1994**, *49*, 1685. [CrossRef]
47. Hurst, H. Long-term storage capacity of reservoirs. *Trans. Am. Soc. Civ. Eng.* **1951**, *116*, 770–799.
48. Heneghan, C.; McDarby, G. Establishing the relation between detrended fluctuation analysis and power spectral density analysis for stochastic processes. *Phys. Rev. E* **2000**, *62*, 6103–6110.
49. Halsey, T.C.; Jensen, M.H.; Kadanoff, L.P.; Procaccia, I.; Shraimant, B.I. Fractal measures and their singularities: The characterization of strange sets. *Phys. Rev. A* **1986**, *33*, 1141–1151.
50. Ohashi, K.; Amaral, L.A.; Natelson, B.H.; Yamamoto, Y. Asymmetrical singularities in real-world signals. *Phys. Rev. E* **2003**, *68*, 065204. [CrossRef]
51. Cao, G.; Cao, J.; Xu, L. Asymmetric multifractal scaling behavior in the Chinese stock market: Based on asymmetric MF-DFA. *Phys. A* **2013**, *392*, 797–807. [CrossRef]
52. Drożdż, S.; Oświęcimka, P. Detecting and interpreting distortions in hierarchical organization of complex time series. *Phys. Rev. E* **2015**, *91*, 030902. [CrossRef]
53. Gómez-Gómez, J.; Carmona-Cabezas, R.; Ariza-Villaverde, A.B.; Gutiérrez de Ravé, E.; Jiménez-Hornero, F.J. Multifractal detrended fluctuation analysis of temperature in Spain (1960–2019). *Phys. A* **2021**, *578*, 126118. [CrossRef]
54. Gomes, L.F.; Gomes, T.F.; Rempel, E.L.; Gama, S. Origin of multifractality in solar wind turbulence: The role of current sheets. *Mon. Not. R. Astron. Soc.* **2023**, *519*, 3623–3634. [CrossRef]
55. Kwapień, J.; Watorek, M.; Bezbradica, M.; Crane, M.; Mai, T.T.; Drożdż, S. Analysis of inter-transaction time fluctuations in the cryptocurrency market. *Chaos* **2022**, *32*, 083142. [CrossRef] [PubMed]
56. Kwapień, J.; Blasiak, P.; Drożdż, S.; Oświęcimka, P. Genuine multifractality in time series is due to temporal correlations. *Phys. Rev. E* **2023**, *107*, 034139. [CrossRef] [PubMed]
57. Drożdż, S.; Kwapień, J.; Oświęcimka, P.; Rak, R. Quantitative features of multifractal subtleties in time series. *EPL* **2009**, *88*, 60003. [CrossRef]
58. Kluszczyński, R.; Drożdż, S.; Kwapień, J.; Stanisz, T.; Wątorek, M. Disentangling sources of multifractality in time series. *Mathematics* **2025**, *13*, 205. [CrossRef]
59. Umarov, S.; Tsallis, C.; Steinberg, S. On a q-Central Limit Theorem Consistent with Nonextensive Statistical Mechanics. *Milan J. Math.* **2008**, *76*, 307–328. [CrossRef]
60. Tsallis, C.; Levy, S.V.F.; Souza, A.M.C.; Maynard, R. Statistical-Mechanical Foundation of the Ubiquity of Lévy Distributions in Nature. *Phys. Rev. Lett.* **1995**, *75*, 3589–3593. [CrossRef]

entropy

MDPI

Review

Applications of Entropy in Data Analysis and Machine Learning: A Review

Salomé A. Sepúlveda-Fontaine and José M. Amigó *

Centro de Investigación Operativa, Universidad Miguel Hernández de Elche, 03202 Elche, Spain; salome.sepulveda.fontaine@gmail.com
* Correspondence: jm.amigo@umh.es

Abstract: Since its origin in the thermodynamics of the 19th century, the concept of entropy has also permeated other fields of physics and mathematics, such as Classical and Quantum Statistical Mechanics, Information Theory, Probability Theory, Ergodic Theory and the Theory of Dynamical Systems. Specifically, we are referring to the classical entropies: the Boltzmann–Gibbs, von Neumann, Shannon, Kolmogorov–Sinai and topological entropies. In addition to their common name, which is historically justified (as we briefly describe in this review), another commonality of the classical entropies is the important role that they have played and are still playing in the theory and applications of their respective fields and beyond. Therefore, it is not surprising that, in the course of time, many other instances of the overarching concept of entropy have been proposed, most of them tailored to specific purposes. Following the current usage, we will refer to all of them, whether classical or new, simply as entropies. In particular, the subject of this review is their applications in data analysis and machine learning. The reason for these particular applications is that entropies are very well suited to characterize probability mass distributions, typically generated by finite-state processes or symbolized signals. Therefore, we will focus on entropies defined as positive functionals on probability mass distributions and provide an axiomatic characterization that goes back to Shannon and Khinchin. Given the plethora of entropies in the literature, we have selected a representative group, including the classical ones. The applications summarized in this review nicely illustrate the power and versatility of entropy in data analysis and machine learning.

Keywords: entropy; entropic measures; data analysis; machine learning; deep learning

1. Introduction

1.1. Aims and Scope

The motivation for this review is two-fold. First, we aim to give researchers in data analysis and machine learning an overview of the applications of the concept of entropy (in a broad sense) in these two fields, both in past and recent times. Needless to say, data analysis and machine learning are hot topics in current research, this being the ultimate reason for choosing them. The second motivation is to familiarize interested researchers with an entropy toolkit. To pave the way thereto, we give the historical and mathematical background of different entropies in use.

Entropy is a concept that appears in different areas of physics and mathematics with different meanings. Thus, entropy is a measure of the following: (i) disorder in Statistical Mechanics, (ii) uncertainty in Information and Probability Theories, (iii) (pseudo)randomness in the Theory of Measure-Preserving Dynamical Systems, and (iv) complexity in Topological Dynamics. This versatility explains why entropy has found extensive applications in various scientific disciplines since its inception in the nineteenth century.

Precisely, this paper aims to provide an up-to-date overview of the applications of entropy in data analysis and machine learning, where entropy stands here not only for the traditional instances, but also for more recent proposals inspired by them. In data analysis, entropy is a powerful tool for the detection of dynamical changes, segmentation, clustering, discrimination, etc. In machine learning, it is used for classification, feature extraction, algorithm optimization, anomaly detection, and more.

The applications of entropy to data analysis date back to Shannon's founding paper of Information Theory [1]. Thus, in Section 7 of [1], Shannon introduces the entropy-based concept of redundancy of a communication source (i.e., a stationary random process) and compares the redundancies of Basic English and James Joyce's prose. With the advent of nonlinear time series analysis in the 1980s [2], entropy gained momentum and became a standard tool for data analysts. The applications of entropy to machine learning may also be traced back to the 1980s at least, when the first energy-based models were formulated [3]. They are also called Boltzmann machines because the probability of finding the model in a certain configuration is given by the Boltzmann distribution (see Section 1.2 for details). Since then, entropy has found many applications in machine learning, especially via cost functions and the maximum entropy principle, as we will see in Section 2.

The ability of entropy to provide insights into data structure and algorithm performance has led to a widespread search for further applications and new proposals tailored to specific needs, both in data analysis and in machine learning. This being the case, the present review will be useful for researchers in the above two fields, interested in the theoretical basics and/or current applications of entropy. Along with established applications, the authors have also taken into account innovative proposals to reflect the intense research activity on entropy that is currently underway.

At this point, the reader may wonder what an entropy is. A search for the word "entropy" on the Internet returns a large number of results, some of them also called entropy metrics, entropy-like measures or entropy-based indices in the literature. So, what is actually entropy?

1.2. Classical Entropies

Historically, the word "entropy" was introduced by the German physicist Clausius in thermodynamics in 1865 to designate the amount of internal energy in a system that cannot be transformed into work. In particular, entropy determines the equilibrium of a thermodynamical system, that is, the state of maximum entropy consistent with the macroscopic constraints. In the second half of the nineteenth century, entropy was given a microscopic interpretation in the foundational works of Boltzmann and Gibbs on Statistical Mechanics. In 1927, von Neumann generalized the Boltzmann–Gibbs concept of entropy to the then-emerging theory of Quantum Mechanics [4]. In 1948, the word entropy appeared in a completely different context: Information Theory. If entropy is a measure of disorder in Statistical Mechanics, in the seminal paper of Shannon [1], the creator of Information Theory, entropy stands for the average uncertainty about the outcome of a random variable (or the information conveyed by knowing it). Albeit in different realms, the coincidence in names is explained because Shannon's formula (see Equation (1) below) is formally the same as Gibbs's for the entropy of a system in thermal equilibrium with a heat bath at constant temperature [5].

This abridged history of entropy continues with Kolmogorov, who crafted Shannon's entropy into a useful invariant in Ergodic Theory [6], and his student Sinai, who adapted Kolmogorov's ideas to the theory of measure-preserving dynamical systems [7]. In turn, Adler, Konheim, and McAndrew [8] generalized the Kolmogorov–Sinai (KS) entropy from measure-preserving dynamics to topological dynamics under the name of topological entropy. According to the Variational Principle, topological entropy is a tight upper bound of the KS entropies of dynamical systems endowed with certain probability measures [9].

To get down to the mathematical formulas, let $\mathcal{P}$ be the set of probability mass distributions $\{p_1, ..., p_W\}$ for all $W \geq 2$. Then, the Shannon entropy of the probability distribution $\mathrm{p}_W = \{p_1, ..., p_W\}$ is defined as

$$S(\mathrm{p}_W) = S(p_1, ..., p_W) = -\sum_{i=1}^{W} p_i \log p_i \tag{1}$$

where the choice of the logarithm base fixes the unit of the entropy, the usual choices being 2 (bit), e (nat) or 10 (dit). If $p_i = 0$, then $0 \log 0 := \lim_{x \to 0+} x \log x = 0$. Mathematically, Equation (1) is the expected value of the *information function* $I(X) = -\log p(X)$, where X is a random variable with probability distribution p_W. Since entropy is the cornerstone of Information Theory, Shannon also justified definition (1) by proving in his seminal paper [1] that it is unique (except for a positive factor) under a few general assumptions. In their modern (equivalent) formulation, these assumptions are called the *Shannon–Khinchin axioms* [10], which we state below.

A positive functional H on $\mathcal{P}$, i.e., a map $H : \mathcal{P} \to \mathbf{R}^+$ ($\mathbf{R}^+$ being the non-negative real numbers), is an entropy if it satisfies the following properties:

SK1 *Continuity*. $H(p_1, ..., p_W)$ depends continuously on all variables for each W.

SK2 *Maximality*. For all W,

$$H(p_1, ..., p_W) \leq H(\tfrac{1}{W}, ..., \tfrac{1}{W}).$$

SK3 *Expansibility*. For all W and $1 \leq i \leq W$,

$$H(0, p_1, ..., p_W) = H(p_1, ..., p_i, 0, p_{i+1}, ..., p_W) = H(p_1, ..., p_i, p_{i+1}, ..., p_W).$$

SK4 *Strong additivity* (or *separability*). For all W and U,

$$\begin{aligned} &H(p_{11}, \ldots, p_{1U}, p_{21}, \ldots p_{2U}, \ldots, p_{W1}, \ldots, p_{WU}) \\ &= H(p_{1*}, p_{2*}, \ldots, p_{W*}) + \sum_{i=1}^{W} p_{i*} H\left(\frac{p_{i1}}{p_{i*}}, \frac{p_{i2}}{p_{i*}}, \ldots, \frac{p_{iU}}{p_{i*}}\right), \end{aligned} \tag{2}$$

where $p_{i*} = \sum_{j=1}^{U} p_{ij}$.

Axiom SK4 can be formulated in a more compact way as

$$H(X, Y) = H(X) + H(Y|X),$$

where X and Y are random variables with probability distributions $\{p_{i*} : 1 \leq i \leq W\}$ and $\{p_{*j} = \sum_{i=1}^{W} p_{ij} : 1 \leq j \leq U\}$, respectively, $H(X, Y) = H(p_{11}, \ldots, p_{1U}, ..., p_{W1}, \ldots, p_{WU})$ and $H(Y|X)$ is the entropy of Y conditional on X, i.e., the expected value of the conditional distributions $p(y|x)$, averaged over the conditioning variable X [11]. In particular, if X and Y are independent (i.e., $p_{ij} = p_{i*}p_{*j}$), then $H(Y|X) = H(Y)$ and

$$H(X, Y) = H(X) + H(Y). \tag{3}$$

If H satisfies Equation (3) for independent random variables X and Y, then it is called *additive*.

It was proved in [1,10] that a positive functional H on $\mathcal{P}$ that fulfills Axioms SK1-SK4 is necessarily of the form

$$H(p_1, ..., p_W) = -k \sum_{i=1}^{W} p_i \log p_i =: S_{BGS}(p_1, ..., p_W) \tag{4}$$

for every $W \geq 2$, where k is a positive constant. For historical reasons, S_{BGS} is usually called the *Boltzmann–Gibbs–Shannon entropy*. In physics, k is the Boltzmann constant

$k_B = 1.3806504(24) \times 10^{-23}$ J/K and log is the natural logarithm. In Information Theory, $k = 1$ and log is the base 2 logarithm when dealing with digital communications. The particular case

$$S_{BGS}(1/W, ..., 1/W) = k \log W, \tag{5}$$

obtained for uniform distributions is sometimes referred to as the *Boltzmann entropy*, although Expression (5) is actually due to Planck [12]. According to Axiom SK2, the Boltzmann entropy is the maximum of S_{BGS}.

The same conclusion about the uniqueness of S_{BGS} can be derived using other equivalent properties [13]. Since we are not interested in physical applications here, we set $k = 1$ and generally refer to S_{BGS} as Shannon's entropy.

In Section 1.1, we mentioned the Boltzmann distribution, not to be confused with the Boltzmann entropy (5). It is defined as $p_i \propto \exp(-E_i)$, $1 \leq i \leq W$, where $E_i \geq 0$ is known as the energy function, so that $p_i > 0$ for all i. In statistical mechanics, $E_i = \beta\varepsilon_i$, where ε_i is the energy of the ith state and $1/\beta = k_B T$, T being the absolute temperature of the system; the normalization factor $Z = \sum_i \exp(-\varepsilon_i / k_B T)$ is called the partition function of the system. While the uniform distribution maximizes the unconstrained Shannon entropy, the Boltzmann distribution is the maximizer of Shannon's entropy under the constraint of a prescribed average energy; that is, $\sum_i p_i \varepsilon_i = \bar{\varepsilon}$ [11,14].

In 1961 Rényi proposed a generalization of Shannon's entropy by using a different, more general definition of expectation value [15,16]: For any real $\alpha > 0$, $\alpha \neq 1$, *Rényi entropy* R_α is defined as

$$R_\alpha(p_1, ..., p_W) = \frac{1}{1-\alpha} \log\left(\sum_{i=1}^{W} p_i^\alpha\right). \tag{6}$$

So, Rényi entropy is actually a family of entropies; in particular, $R_1 := \lim_{\alpha \to 1} R_\alpha = S_{BGS}$. Other limiting cases are $R_0 := \lim_{\alpha \to 0} R_\alpha = \log W$, called *Hartley* or *max-entropy*, which coincides with the Boltzmann entropy (5) except for the value of the constant k, and $R_\infty := \lim_{\alpha \to \infty} R_\alpha = \min_{1 \leq i \leq W}(-\ln p_i)$, called the *min-entropy*. These names are due to the non-increasing monotonicity of Rényi's entropy with respect to the parameter: $R_\alpha \geq R_\beta$ for $\alpha < \beta$.

It is easy to show that Rényi's entropy satisfies Axioms SK1-SK3 but not SK4. Instead of strong additivity, R_α satisfies *additivity*:

$$R_\alpha(\mathrm{p}_U \times \mathrm{q}_W) = R_\alpha(\mathrm{p}_U) + R_\alpha(\mathrm{q}_W);$$

see Equation (3).

A final milestone in this short history of entropy is the introduction of *non-additive entropies* by Havrda and Charvát in Information Theory [17] and Tsallis in Statistical Mechanics [18], which are equivalent and usually called the Tsallis entropy:

$$T_q(p_1, ..., p_W) = \frac{1}{1-q}\left(\sum_{i=1}^{W} p_i^q - 1\right) \tag{7}$$

for any real $q > 0$, $q \neq 1$. Again, Tsallis entropy is a family of entropies that satisfy Axioms SK1-SK3 but not SK4. Instead, T_q is "q-additive", meaning that

$$T_q(\mathrm{p}_U \times \mathrm{q}_W) = T_q(\mathrm{p}_U) + T_q(\mathrm{q}_W) + (1-q)T_q(\mathrm{p}_U)T_q(\mathrm{q}_W).$$

As Rényi's entropy, the Tsallis entropy is a generalization of Shannon's entropy in the sense that $T_1 := \lim_{q \to 1} T_q = S_{BGS}$. Formally, T_q can be obtained from S_{BGS} by replacing the logarithm in Equation (4) with the "q-logarithm" [16].

The appearance of generalizations of the Shannon entropy prompted the weaker concept of *generalized entropy*: a positive functional on probability distributions that satisfies Axioms SK1-SK3. Therefore, the BGS entropy, together with the Rényi and Tsallis entropies, are examples of generalized entropies. Shannon's uniqueness theorem can then be rephrased by saying that the only generalized entropy that is strongly additive is the BGS entropy. Axioms SK1-SK3 are arguably the minimal requirements for a positive functional on probability mass distributions to be called an entropy. Most "entropies" proposed since the formulation of Rényi and Tsallis' entropies are precisely generalized entropies in the axiomatic sense.

To wrap up this short account of the classical and generalized entropies, let us mention that Shannon's, Rényi's and Tsallis' entropies (and other entropies for that matter) have counterparts for continuous-valued random variables and processes (i.e., defined on probability densities). These "differential" versions are formally obtained by replacing probability mass functions by probability densities and summations by integrations in Equations (1), (6) and (7), respectively. Although also useful in applications, differential entropies may lack important properties of their discrete counterparts. For example, the *differential* (Shannon's) *entropy* lacks positivity [11].

1.3. Methodology and Structure of This Review

As said above, the primary objective of this work is to review the applications of entropy in the fields of data analysis and machine learning. In view of the many versions of entropy currently in use, we had to make a selection of them based on their general relevance, and based on interest in their particular applications. Inevitably, this selection may have been influenced by our own experience as the authors, so we apologize for any oversights of relevant applications and references. Apart from the group of classical entropies of Section 1.2, which we will call **G0** for reference purposes, the remaining entropies selected for this review can be classified into the following two groups.

- **G1** Entropies based on (or inspired by) the entropies in Group G0, especially Shannon's entropy (1). This group comprises *bubble entropy, dispersion entropy* and *fluctuation-based dispersion entropy, energy entropy* and *empirical mode decomposition energy entropy, entanglement entropy, Fourier entropy and fractional Fourier entropy, graph entropy, Kaniadakis entropy, permutation entropy, Rao's quadratic entropy, spectral entropy, tone entropy* and *wavelet entropy*.
- **G2** Entropies based on information-theoretical concepts such as the correlation integral, divergences, unconditional or conditional mutual information, or based on other entropies in this group. This group comprises *approximate entropy, cross entropy* and *categorical cross entropy, excess entropy, fuzzy entropy, intrinsic mode entropy, kernel entropy, rank-based entropy, relative entropy* or *Kullback–Leibler divergence, sample entropy* and *transfer entropy*.

In Section 2, each of the 33 selected entropies is assigned a subsection in alphabetical order. For brevity, mathematical definitions are given only when feasible with limited space; otherwise, we give a qualitative account and refer the reader to the original publications or standard bibliography for the formulas. The corresponding applications to data analysis and machine learning are explained with brief but sufficient descriptions and provided with specific references. Practical issues such as the choice of parameters or computational implementations are beyond the scope of this review.

Section 3 contains general comments on the previous sections. Moreover, it presents the most common applications of Section 2 and a citation analysis of the references to identify the most influential papers to date in the field of entropy and its applications. Section 4 begins with concluding remarks for this review and ends with some recent developments and challenges. The aim of our outlook is to attract new researchers to the field. This review ends with a list of references that is long by usual standards, though very short compared to the vast and ever-growing literature on entropy and applications thereof.

There are a number of excellent general reviews on entropy, entropy-like quantities and their applications. Thus, Katok revisits in [19] the fifty years 1958–2007 of entropy in the theory of dynamical systems (i.e., Kolmogorov–Sinai and topological entropies), which we recommend to readers interested not only in the mathematical aspects of those entropies but also in their historical development. The reviews [14,16] are mainly mathematical accounts of the classical and generalized entropies, along with some typical applications. The review "The Entropy Universe" by M. Ribeiro et al. [20] is a fine blend of mathematical descriptions of several entropies and their applications in science and engineering. Unlike the aforementioned reviews, the subject of the present work is the applications of entropy to data analysis and machine learning. Hence, this review differs from [14,16,19] in that the latter focus on the mathematical underpinnings of the considered entropies. And it also differs from [20] in that that review focuses on the definitions of the selected entropies rather than on the applications. Furthermore, both the entropies and applications considered in [20] are more limited in number and scope than those considered in ours; in particular, applications to machine learning are not discussed therein.

2. Applications in Data Analysis and Machine Learning

In this section, the selected entropies are sorted alphabetically. To streamline the exposition, multiscale and weighted versions [21] are included in the same section as the original entropy. The corresponding applications to data analysis and machine learning are tagged with various keywords in alphabetical order so that reverse search (i.e., searching for entropies for a given application) can also be easily performed.

2.1. *Approximate Entropy*

Approximate entropy was proposed by Pincus in 1991 [22] to analyze medical data. Loosely speaking, approximate entropy is a heuristic implementation of the correlation integral with time series data [14]. The approximate entropy depends on a parameter $r > 0$, sometimes called tolerance, which is a cut-off that defines the concept of proximity between points via the Heaviside step function $\Theta(r - \|x_i - x_j\|)$ (where $\Theta(z) = 1$ if $z \geq 0$, and 0 otherwise). It quantifies the change in the relative frequencies of length-k time-delay vectors with increasing k. A modified version of approximate entropy was proposed in 2000 under the name sample entropy (Section 2.25). See [23] for a tutorial.

Applications

- **Alzheimer's disease.** Approximate entropy has been used in the nonlinear analysis of EEGs and MEGs from patients with Alzheimer's disease [24,25].
- **Anesthetic drug effects.** Another field of applications is the quantification of anesthetic drug effects on brain activity as measured by EEGs, including comparative testing of different anesthetics [26].
- **Emotion recognition.** Along with other entropies, approximate entropy has been used for EEG-based human emotion recognition [27].
- **Epileptic seizure detection.** Approximate entropy has also been used as biomarker in algorithms for epileptic EEG analysis, in particular, for epileptic seizure detection [28–30].
- **Physiological time series.** See [31] for an overview of applications of approximate entropy to the analysis of physiological time series.
- **Sleep research.** The applications of approximate entropy include sleep research, in particular, the separation of sleep stages based on EEG data [32].

2.2. *Bubble Entropy*

Bubble entropy is a metric that evaluates changes in the order of data segments in time series when a new element is added. It was proposed by Manis et al. in 2017 [33] as "an entropy almost free of parameters", inspired by permutation entropy (Section 2.20) and rank-based entropy (Section 2.21). Bubble entropy relies on the Bubble Sort algorithm [33], which compares and swaps adjacent items until ordered; see Section 2.7 of [20] for an algorithmic description.

Applications

- **Biomedical applications.** Due to its minimal dependency on parameters, bubble entropy is particularly useful in biomedical applications (e.g., analysis of heart rate variability) to distinguish healthy from pathological conditions [33,34].
- **Fault bearing detection.** Bubble entropy is used to reinforce the accuracy of fault bearing diagnosis through the Gorilla Troops Optimization (GTO) algorithm for classification [35]. A similar application can be found for the so-called Improved Hierarchical Refined Composite Multiscale Multichannel Bubble Entropy [36].
- **Feature extraction.** Bubble entropy is compared with dispersion entropy (Section 2.6) in the extraction of single and double features in [37].

2.3. Categorical Cross Entropy

According to [3,38], using cross entropy (Section 2.4) as an error (cost, loss, objective,...) function speeds up the training of neural networks and yields better results than the mean-squared error function for classification problems. The cross entropy-based error function for multi-class classification is called categorical cross entropy (CCE) [39]. Specifically, if t_{mk}, $1 \leq k \leq K$, are target binary variables for input m, $1 \leq m \leq M$, and y_{mk} are the corresponding predicted output variables, then the categorical cross entropy is given as

$$\text{CCE} = -\sum_{m=1}^{M}\sum_{k=1}^{K} t_{mk} \ln y_{mk}. \tag{8}$$

For clarity, we consider CCE in this section and conventional cross entropy in the next.

Applications

- **Deep Learning.** An improved form of CCE is used in deep neural networks to train them when dealing with noisy labels [40].
- **Multi-class classification.** CCE is used in [39] to train a convolutional neural network tailored to a multi-sensor, multi-channel time series classification of cardiography signals. Likewise, CCE can be used for image classification.
- **Reinforcement learning.** CCE is used as an improvement of value function training (using classification instead of regression) mainly in games [41].
- **Semi-supervised learning.** CCE is used in pseudo-labeling to optimize convolutional neural network parameters [42].

2.4. Cross Entropy

The cross entropy between the probability distributions $\mathrm{p} = (p_1, ..., p_W)$ and $\mathrm{q} = (q_1, ..., q_W)$ is defined as

$$C(\mathrm{p}, \mathrm{q}) = -\sum_{i=1}^{W} p_i \log q_i. \tag{9}$$

It is related to the Shannon entropy $S(\mathrm{p})$ (Equation (1)) and the Kullback–Leibler divergence $D(\mathrm{p}\|\mathrm{q})$ (Section 2.23, Equation (17)) through the following equation:

$$C(\mathrm{p}, \mathrm{q}) = S(\mathrm{p}) + D(\mathrm{p}\|\mathrm{q}). \tag{10}$$

Like $D(\mathrm{p}\|\mathrm{q})$, the cross entropy is used to quantify the difference between two probability distributions p and q. A typical example is determining whether an email is spam or not. In particular, when it comes to selecting parameters θ in a model that predict a distribution q_θ to fit an empirical distribution p, one can minimize $C(\mathrm{p}, \mathrm{q}_\theta)$ instead of $D(\mathrm{p}\|\mathrm{q}_\theta)$ because $S(\mathrm{p})$ does not depend on θ.

Applications

- **Deep learning.** Cross entropy is a standard loss function for training deep neural networks, particularly those involving softmax activation functions. It is very useful for applications such as object detection, language translation and sentiment analysis [43]. In this regard, empirical evidence with limited and noisy data suggests that to measure the top-κ error (a common measure of performance in machine learning performed with deep neural networks trained with the cross entropy loss), the loss function must be smooth, meaning that it should incorporate a smoothing parameter ϵ to handle small probability events [44].
- **Feature selection.** Cross entropy is used to select significant features of binary values from highly imbalanced large datasets via a framework called FMC Selector [45].
- **Image analysis.** Wavelet analysis together with cross entropy is used in image segmentation, object recognition, texture analysis (e.g., fabric defect detection) and pattern classification [46].
- **Learning-to-rank methods.** In [47], the author proposes a learning-to-rank loss function that is based on cross entropy. Learning-to-rank methods form a class of ranking algorithms that are widely applied in information retrieval.
- **Multi-class classification.** Cross entropy is used to enhance the efficiency of solving support vector machines for multi-class classification problems [48].
- **Semi-supervised clustering.** Cross entropy is employed along with the information bottleneck method in semi-supervised clustering. It is robust to noisy labels and automatically determines the optimal number of clusters under mild conditions [49].

2.5. *Differential Entropy*

As said in Section 1.2, differential entropy is the continuous counterpart of Shannon entropy: if X is a continuous random variable with density $\rho(x)$ and support set S, then the differential entropy of X is [11]

$$h(X) = h(\rho) = -\int_S \rho(x) \log \rho(x) dx. \tag{11}$$

Therefore, differential entropy is used in the analysis of continuous random variables and processes, e.g., analog signals.

Applications

- **Anomaly detection.** Differential entropy can measure changes in the probability density function of an analog signal that reveals an anomaly in the source, whether it is a mechanical system or a patient [50].
- **Emotion recognition.** Differential entropy has been used in [27] to extract features in EEG-based human emotion recognition.
- **Feature selection.** A feature selection algorithm based on differential entropy to evaluate feature subsets has been proposed in [38]. This algorithm effectively represents uncertainty in the boundary region of a fuzzy rough model and demonstrates improved performance in selecting optimal feature subsets, thereby enhancing classification accuracy; see [51] for an implementation.
- **Gaussianity of a population.** The differential entropy $h(X)$ reaches its maximum among all random variables with the same variance σ, $h_{\max}(X) = \log(2\pi e\sigma^2)/2$, when X is normal [11]. Therefore, a sample estimate of $h(X)$ that is significantly lower than that upper bound indicates that the underlying population is not Gaussian.
- **Generative models.** Variational Autoencoders and other generative models leverage differential entropy to model the latent space of continuous data distributions. These models can learn better representations of the input data, thus improving performance [52].
- **Mutual information.** Differential entropy is instrumental for computing the mutual information of continuous-valued random variables and processes, e.g., autoregressive

processes. It is used in speech processing (linear prediction), seismic signal processing and biological signal processing [53].

- **Probabilistic models.** Differential entropy is utilized in probabilistic models such as Gaussian Mixture Models to describe the uncertainty and distribution of continuous variables. This approach is applicable to image processing and network inference as well [54].

2.6. Dispersion Entropy

Dispersion entropy was introduced by Rostaghi and Azami in 2016 [55] to solve the shortcoming of permutation entropy (Section 2.20) of only taking into account the ranking of the data amplitudes but not their values. This explains why the definition of dispersion entropy resembles that of permutation entropy.

Indeed, to compute the dispersion entropy of a time series, the series is first transformed into a string of symbols called "dispersion patterns" via a number of linear or nonlinear mapping techniques, then the probabilities of the different dispersion patterns are estimated by their relative frequencies and, finally, the Shannon entropy of the resulting probability distribution is calculated. The dispersion entropy has three parameters: c, the number of letters, running from 1 to c, that build the dispersion patterns; m, the length of the data blocks (or "embedding vectors") used to obtain dispersion patterns of the same length, so that there are c^m possible dispersion patterns of length m; and d, the time delay. See [56] for details.

Applications

- **Feature extraction.** Multiscale fuzzy dispersion entropy is applied in fault diagnosis of rotating machinery to capture the dynamical variability of time series across various scales of complexity [57].
- **Image classification.** A multiscale version of dispersion entropy called $\mathrm{MDispEn}_{2D}$ has been used with biomedical data to measure the impact of key parameters that may greatly influence the entropy values obtained in image classification [58].
- **Signal classification.** Another generalization of dispersion entropy, namely, fractional fuzzy dispersion entropy, has been proposed as a fuzzy membership function for signal classification tasks [59].
- **Signal denoising.** Dispersion entropy is also used in signal denoising via (i) adaptive techniques and group-sparse total variation [60], or (ii) empirical mode decomposition with adaptive noise [61].
- **Time series analysis.** Multiscale graph-based dispersion entropy is a generalization of dispersion entropy used to analyze multivariate time series data in graph and complex network frameworks, e.g., weather and two-phase flow data; it combines temporal dynamics with topological relationships [62].

2.7. Energy Entropy and Empirical Mode Decomposition Energy Entropy

Energy entropy is the Boltzmann–Gibbs–Shannon entropy of a normalized distribution of energy levels or modes. Thus, it is mainly used in data-driven analysis of physical systems, whether in the natural sciences or technology.

In nonlinear time series analysis, one uses empirical mode decomposition (EMD) to decompose a time series into a set of intrinsic mode functions, each representing a simple oscillatory mode inherent to the data. This decomposition is adaptive and data-driven, making it suitable for analyzing complex signals without requiring predefined basis functions [63]. As its name suggests, EMD energy entropy combines energy entropy with EMD to provide information about the energy distribution across different intrinsic mode functions derived from a signal.

Applications

- **Chatter detection.** This application involves detecting vibrations and noise in machining operations that can indicate chattering. In this regard, energy entropy can detect chatter in robotic milling [64].
- **Fault prediction in vibration signal analysis.** EMD energy entropy has been employed to predict early fault of bearings in rotating machinery [65,66].
- **Feature extraction.** Energy entropy is calculated via the empirical decomposition of the signal into intrinsic mode functions and serves as a feature for machine learning models used in chatter detection. The chatter feature extraction method draws on the largest energy entropy [67].
- **Time series forecasting.** EMD energy entropy was used in [68] to predict short-term electricity consumption by taking into account data variability, i.e., the fact that power consumption data are non-stationary, nonlinear and influenced by the season, holidays, and other factors. In [69], this entropy was the tool used to distinguish two kinds of financial markets.

2.8. Entanglement Entropy

Entanglement entropy originated in quantum mechanics as a measure of the degree of quantum entanglement between two subsystems of a quantum system. To be more precise, entanglement entropy is the von Neumann entropy (Section 2.32) of the reduced density matrix for any of the subsystems [70]; see [71] for the estimation of entanglement entropy through supervised learning. In addition to its important role in quantum mechanics and quantum information theory, entanglement entropy is increasingly finding more applications in machine learning as well. This is why we have included entanglement entropy in this review.

Applications

- **Feature extraction.** In quantum machine learning, entanglement entropy is used for feature extraction by representing data in a form that highlights quantum correlations and thus, by leveraging the quantum properties of the data [72].
- **Neural networks in quantum models.** Entanglement entropy is used in quantum models to quantify unknown entanglement by using neural networks to predict entanglement measures of unknown quantum states based on experimentally measurable data: moments or correlation data produced by local measurements [73].

2.9. Excess Entropy

The *mutual information* between random variables X and Y measures the average reduction in uncertainty associated with one of the variables that results from learning the value of the other. It is defined as follows [11]:

$$I(X;Y) = I(Y;X) = S(X) + S(Y) - S(X,Y) \geq 0, \tag{12}$$

where $S(...)$ is Shannon's entropy (1).

Excess entropy (also called dual total correlation) is a non-negative generalization of mutual information to more than two random variables defined as follows [74]:

$$D(X_1, ..., X_N) = S(X_1, ..., X_N) - \sum_{i=1}^{N} S(X_i | X_1, ..., X_{i-1}, X_{i+1}, ..., X_N) \tag{13}$$

where $S(X_i|...)$ is the Shannon conditional entropy of X_i given the other variables. Of course, $D(X_1, X_2) = I(X_1; X_2)$. An alternative definition called *total excess entropy* was introduced by Crutchfield and Packard in [75].

Applications

- **Image segmentation.** Excess entropy is used to measure the structural information of a 2D or 3D image and then determine the optimal threshold in a segmentation algorithm proposed in [76]. The working hypothesis of this thresholding-based segmentation algorithm is that the optimal threshold corresponds to the maximum excess entropy (i.e., to a segmentation with maximum structure).
- **Machine learning.** In [77], the authors present a method called machine learning iterative calculation of entropy for calculating the entropy of physical systems by iteratively dividing the system into smaller subsystems and estimating the mutual information between each pair of halves.
- **Neural estimation in adversarial generative models.** The Mutual Information Neural Estimator is a scalable estimator used in high-dimensional continuous data analysis that optimizes mutual information. The authors apply this estimator to Generative Adversarial Networks (GANs) [78].
- **Time series classification.** In this application, total excess entropy is used for classifying stationary time series into long-term and short-term memory. A stationary sequence with finite block entropy is long-term memory if its excess entropy is infinite [79].

2.10. Fluctuation-Based Dispersion Entropy

Fluctuation-based dispersion entropy (FDispEn) was introduced by Azami and Escudero in 2018 [56] to account for the variability of time series by incorporating information about fluctuations into the definition of dispersion entropy (Section 2.3).

Specifically, FDispEn considers the differences between adjacent symbols (belonging to the alphabet $\{1, 2, ..., c\}$) of the dispersion patterns, termed fluctuation-based dispersion patterns. Therefore, if the dispersion patterns are vectors of length m, the fluctuation-based ones have length $m - 1$ and components that range from $-c + 1$ to $c - 1$. As a result, there are $(2c - 1)^{m-1}$ potential fluctuation-based dispersion patterns. The number of possible patterns for the same parameters is the only difference between the algorithms for the conventional dispersion entropy and FDispEn.

Applications

- **Fault diagnosis.** The so-called refined composite moving average FDispEn is used in machinery fault diagnosis by analysing vibration signals [80].
  Refined composite multiscale FDispEn and supervised manifold mapping are used in fault diagnosis for feature extraction in planetary gearboxes [81].
  Multivariate hierarchical multiscale FDispEn along with multi-cluster feature selection and a Gray Wolf Optimization-based Kernel Extreme Learning Machine helped diagnose faults in rotating machinery. It also captured the high-dimensional fault features hidden in multichannel vibration signals [82].
- **Feature extraction.** FDispEn gives rise to hierarchical refined multi-scale fluctuation-based dispersion entropy, used to extract underwater target features in marine environments and weak target echo signals, thereby improving the detection performance of active sonars [83].
- **Robustness in spectrum sensing.** Reference [84] proposes a machine learning implementation of spectrum sensing using an improved version of FDispEnt as a feature vector. This improved version shows enhanced robustness to noise.
- **Signal classification.** FDispEn helps distinguish various physiological states of biomedical time series and it is commonly used in biomedicine. It is also used to estimate the dynamical variability of the fluctuations of signals applied to neurological diseases [85]. Fluctuation-based reverse dispersion entropy is applied to signal classification combined with k-nearest neighbor [86].
- **Time series analysis.** FDispEn is used to quantify the uncertainty of time series to account for knowledge on parameters sensitivity and studying the effects of linear and

nonlinear mapping on the defined entropy in [56]. FDispEn is defined as a measure for dealing with fluctuations in time series. Then, the performance is compared to complexity measures such as permutation entropy (Section 2.20), sample entropy (Section 2.25) and Lempel–Ziv complexity [11,87].

2.11. Fourier Entropy

The Fourier entropy $h(f)$ of a Boolean function $f : \{-1,1\}^n \to \{-1,1\}$ is the Shannon entropy of its power spectrum $\{\hat{f}(S)^2 : S \subset [n]\}$, where $S \subset [n]$ stands for the 2^n subsets of $\{1,2,...,n\}$ (including the empty set) and $\hat{f}$ is the Fourier transform of f. By Parseval's Theorem, $\sum_{S\subset[n]} \hat{f}(S)^2 = 1$, so the power spectrum of f is a probability distribution.

Applications

- **Decision trees:** The Fourier Entropy–Influence Conjecture, made by Friedgut and Kalai [88], says that the Fourier entropy of any Boolean function f is upper bounded, up to a constant factor, by the total influence (or average sensitivity) of f. This conjecture, applied to decision trees, gives interesting results that boil down to $h(f) = O(\log(L(f)))$, meaning that $|h(f)| < C\log(L(f))$, where $L(f)$ denotes the minimum number of leaves in a decision tree that computes f and $C > 0$ is independent of f and $L(f)$ [89]. Another similar application to decision trees can be found in [90].
- **Learning theory.** The Fourier Entropy–Influence Conjecture is closely related to the problem of learning functions in the membership model. It is said that if a function has low Fourier entropy, that means that its Fourier transform is concentrated on a few characters, i.e., that the function can be approximated by a sparse polynomial, a class that is very important in the context of learning theory [91]. Learning theory provides the mathematical foundation for understanding how algorithms learn from data, guiding the development of machine learning models.

2.12. Fractional Fourier Entropy

Fractional Fourier entropy is calculated in two steps: first, take the fractional Fourier transform of the data [92]; second, compute the Shannon entropy of the probability distribution defined by the normalized frequency spectrum.

Applications

- **Anomaly detection in remote sensing.** Fractional Fourier entropy is used in hyperspectral remote sensing to distinguish signals from background and noise [93,94].
- **Artificial intelligence.** Two-dimensional fractional Fourier entropy helped to diagnose COVID-19 by extracting features from chest CT images [95].
- **Biomedical image classification.** Fractional Fourier entropy has proven helpful in detecting pathological brain conditions. By using it as a new feature in Magnetic Resonance Imaging, the classification of images is improved in time and cost [96].
- **Deep learning.** Fractional Fourier entropy is used in the detection of gingivitis via feed-forward neural networks. It reduces the complexity of image extraction before classification and can obtain better image eigenvalues [97].
- **Emotion recognition.** Fractional Fourier entropy, along with two binary support vector machines, helps improve the accuracy of emotion recognition from physiological signals in electrocardiogram and galvanic skin responses [98].
- **Multilabel classification.** Fractional Fourier entropy has been used in a tea-category identification system, which can automatically determine tea category from images captured by a three charge-coupled device digital camera [99].

2.13. Fuzzy Entropy

Fuzzy entropy quantifies "fuzziness" (uncertainty due to imprecision) in fuzzy set theory [100] and time series analysis. There are different definitions of fuzzy entropy in the literature that draws on the concept of fuzzy sets [101,102]. In 2007, Chen et al. [103] introduced a concept that is akin to approximate and sample entropies (Sections 2.1 and 2.25). In this version, the Heaviside function $\Theta(r - \|x_i - x_j\|)$ used in the latter entropies to define neighborhoods is replaced by the exponential function $\exp(-\delta_{ij}^m/r)$, where $r > 0$ is the tolerance and δ_{ij}^m is a distance between the embedding vectors $(x_i, ..., x_{i+m-1})$ and $(x_j, ..., x_{+m-1})$ of length m. This distance includes also the subtraction of the mean value of the vector components to minimize the effects of non-stationarity.

In 2014, Zheng proposed multiscale fuzzy entropy [104]. Fuzzy entropy was introduced into fuzzy dynamical systems in [105].

Applications

- **Clustering and time series analysis.** Fuzzy entropy is used in problems of robustness against outliers in clustering techniques in [106].
- **Data analysis.** Fuzzy entropy is proposed in [107] to assess the strength of fuzzy rules with respect to a dataset, based on the greatest energy and smallest entropy of a fuzzy relation.
- **Fault detection.** Fuzzy entropy (along with dispersion entropy, Section 2.6) was the best performer in a comparative study of entropy-based methods for detecting motor faults [108]. Multiscale fuzzy entropy is used to measure complexity in time series in rolling bearing fault diagnosis [109].
- **Feature selection and mathematical modeling.** Fuzzy entropy is used in feature selection to evaluate the relevance and contribution of each feature in Picture Fuzzy Sets [110].
- **Image classification.** Fuzzy entropy, in the form of multivariate multiscale fuzzy entropy, is proposed and tested in [111] for the study of texture in color images and their classification.
- **Image segmentation.** Fuzzy entropy is the objective function of a color image segmentation technique based on an improved cuckoo search algorithm [112].

2.14. Graph Entropy

Graph entropy was introduced by Korner in 1971 [113] to quantify the complexity or information content of a graph. It is usually defined as the Shannon entropy (although any other entropy would do) of a probability distribution over the graph's vertex set. In addition to applications in data analysis and machine learning, graph entropy is also applied in combinatorics; see [114] for a survey of graph entropy.

A particular case of graph entropy is the horizontal visibility (HV) graph entropy of a time series, which is the graph entropy of the so-called HV graph of the time series [115]. In particular, this method is useful for distinguishing between different types of dynamical behaviours in nonlinear time series, such as chaotic versus regular dynamics [116].

Applications

- **Dimension reduction and feature selection.** Graph entropy gave rise to Conditional Graph Entropy, which helps in the alternating minimization problem [117].
- **Graph structure.** Graph entropy is used to measure the information content of graphs, as well as to evaluate the complexity of the hierarchical structure of a graph [118].
- **Graph-based time series analysis.** Graph entropy can be used in time series analysis in conjunction with any method that transforms time series into graphs. An example is the HV graph entropy presented above; see [119] and references therein.
- **Node embedding dimension selection.** Graph entropy is applied in Graph Neural Networks through the Minimum Graph Entropy algorithm. It calculates the ideal node embedding dimension of any graph [120].

- **Time series analysis.** HV graph entropy along with sample entropy has been used in [121] to identify abnormalities in the EEGs of alcoholic subjects. HV transfer entropy was proposed in [122] to estimate the direction of the information flow between pairs of coupled time series.

2.15. Havrda–Charvát Entropy

Havrda–Charvát (HC) entropy, also known as the Havrda–Charvát α-entropy, was introduced by Havrda and Charvát [17] in 1967 in Information Theory. It is formally identical to the Tsallis entropy (Section 2.31, introduced by Tsallis in 1988 in Statistical Mechanics [18]. Similar to Rényi entropy (Section 2.24), HC entropy is a family of entropies parameterized by $\alpha > 0$ and generalizes the Shannon entropy in the sense that the HC entropy coincides with the Shannon entropy in the limit $\alpha \to 1$.

Although nowadays the most popular name for this entropy is Tsallis entropy, we present in this section applications published in articles that refer to this entropy as the Havrda–Charvát entropy.

Applications

- **Computer vision.** An HC entropy-based technique for group-wise registration of point sets with unknown correspondence is used in graphics, medical imaging and pattern recognition. By defining the HC entropy for cumulative distribution functions (CDFs), the corresponding CDF-HC divergence quantifies the dissimilarity between CDFs estimated from each point-set in the given population of point sets [123].
- **Financial time series analysis.** Weighted HC entropy outperforms regular HC entropy when used as a complexity measure in financial time series. The weights turn out to be useful for showing amplitude differences between series with the same order mode (i.e., similarities in patterns or specific states) and robust with respect to noise [124].
- **Image segmentation and classification.** HC entropy is applied as a loss function in image segmentation and classification tasks using convolutional neural networks in [125].
- **Loss functions in deep learning.** HC entropy can be used to design loss functions in deep learning models. These loss functions are particularly useful in scenarios with small datasets, common in medical applications [126].

2.16. Intrinsic Mode Entropy

Intrinsic mode entropy (IME) was introduced by Amoud et al. in 2007 [127] to study time series in different scales, in the presence of dominant local trends and low-frequency components. It is obtained by computing the sample entropy (Section 2.25) of the cumulative sums of the intrinsic mode functions extracted by the empirical mode decomposition method (Section 2.7).

Applications

- **Language gesture recognition.** IME is used in [128] to analyze data from a three-dimensional accelerometer and a five-channel surface electromyogram of the user's dominant forearm for automated recognition of Greek sign language gestures.
- **Neural data analysis.** An IME version with improved discriminatory capacity in the analysis of neural data is proposed in [129].
- **Time series analysis.** IME is used in nonlinear time series analysis to efficiently characterize the underlying dynamics [127]. As any multiscale entropy, IME is particularly useful for the analysis of physiological time series [130]. See also [131] for an application to the analysis of postural steadiness.

2.17. Kaniadakis Entropy

Kaniadakis entropy, also known as κ-entropy due to its dependence on a parameter $0 < \kappa < 1$, was introduced by the physicist G. Kaniadakis in 2002 [132] to address the limitations of classical entropy in systems exhibiting relativistic effects. It is defined as

$$S_\kappa(p_1, ..., p_W) = \sum_{i=1}^{W} \frac{p_i^{1-\kappa} - p_i^{1+\kappa}}{2\kappa}. \tag{14}$$

Kaniadakis entropy is a relativistic generalization of the BGS entropy (4) in the sense that the latter is recovered in the $\kappa \to 0$ limit [133]. Like other physical entropies such as Tsallis' and von Neumann's, Kaniadakis entropy has also found interesting applications in applied mathematics.

Applications

- **Image segmentation.** Kaniadakis entropy is used in image thresholding to segment images with long-tailed distribution histograms; the parameter κ is selected via a swarm optimization search algorithm [134].
- **Images threshold selection.** Kaniadakis entropy can be used to construct an objective function for image thresholding. By using the energy curve and the Black Widow optimization algorithm with Gaussian mutation, this approach can be performed on both grayscale and colour images of different modalities and dimensions [135].
- **Seismic imaging.** Application of the Maximum Entropy Principle with S_κ leads to the Kaniadakis distribution, a deformation of the Gaussian distribution that has applications, e.g., in seismic imaging [136].

2.18. Kernel Entropy

Kernel (or kernel-based) entropy is an evolution of the approximate entropy (Section 2.1) that consists of replacing the Heaviside step function in the definition of proximity by other functions (or "kernels") to give more weight to the nearest neighbors. Thus, in the Gaussian kernel entropy, the Gaussian kernel

$$\ker(i, j; r) = \exp\left(-\frac{\|x_i - x_j\|^2}{10r^2}\right) \tag{15}$$

is used. Here x_i and x_j are entries of a time series and r is the parameter of the approximate entropy. Other popular kernels include the spherical, Laplacian and Cauchy functions [137].

Of course, the same refinement can be carried out in the computation of the sample entropy (Section 2.25), an improvement of the approximate entropy. To distinguish between the two resulting kernel entropies, one speaks of kernel-based approximate or sample entropy.

Applications

- **Complexity of time series.** The authors of [138] present experimental evidence that Gaussian kernel entropy outperforms approximate entropy when it comes to analyzing the complexity of time series.
- **Fetal heart rate discrimination.** In [139], the authors compare the performance of several kernel entropies on fetal heart rate discrimination, with the result that the circular and Cauchy kernels outperform other, more popular kernels, such as the Gaussian or the spherical ones.
- **Pathological speech signal analysis.** Reference [137] is a study of several approaches in the field of pathological speech signal analysis. Among the new pathological voice measures, the authors include different kernel-based approximate and sample entropies.

- **Speech signal classification in Parkinson's disease.** Gaussian kernel entropy, along with other nonlinear features, is used in [140] in the task of automatic classification of speech signals from subjects with Parkinson's disease and a control set.

2.19. Kolmogorov–Sinai Entropy

As mentioned in Section 1.2, Kolmogorov–Sinai (KS) entropy is a classical entropy introduced by Kolmogorov in ergodic theory in 1958 [141], and extended by Sinai to the theory of measure-preserving dynamical systems in 1959 [7]. KS entropy is a fundamental invariant in the theory of metric dynamical systems [9].

The calculation of the KS entropy of a dynamical system requires in general three steps: (i) coarse-graining of the state space by a finite partition α, (ii) computation of the Shannon entropy rate of the resulting symbolic dynamics $\mathbf{X}^\alpha$ (a finite-state, stationary random process with so many states as the cardinality of α [14]), called the Shannon entropy with respect to α, $S(\mathbf{X}^\alpha)$, and (iii) taking the supremum of $S(\mathbf{X}^\alpha)$ over all finite partitions α, the latter being equivalent to taking the limit of $S(\mathbf{X}^\alpha)$ for ever finer partitions [9].

Apart from approximating the limits in steps (ii) and (iii), other popular estimators of the KS entropy resort to Pesin's formula [142], which involves the strictly positive Lyapunov exponents of the system; see also [143] for high-dimensional complex systems and [144] for the estimation of the Lyapunov exponents.

Applications

- **Time series analysis.** Perhaps the main practical application of the Kolmogorov–Sinai entropy is the analysis of nonlinear, real-valued time series, where it is used to characterize the underlying dynamical system, in particular, its chaotic behavior. Recent practical examples include short-term heart rate variability [145], physical models of the vocal membranes [143], autonomous driving [146] and EEG-based human emotion recognition [27,147].

2.20. Permutation Entropy

The conventional (or Shannon) permutation entropy of a time series was introduced by Bandt and Pompe in 2002 [148]. It is the Shannon entropy (Section 2.26) of the probability distribution obtained from the ordinal patterns of length $L \geq 2$ in the time series, i.e., the rankings (or permutations) of the series entries in sliding windows of size L. Therefore, permutation entropy depends on the parameter L. Furthermore, permutation entropy is easy to program, relatively robust to noise, and can be computed practically in real time since knowledge of the data range is not needed [149]. See, e.g., [150–153] for references on theoretical and practical aspects of permutation entropy.

If, instead of the Shannon entropy of the ordinal patterns distribution, we use another entropy, e.g., Rényi entropy (Section 2.24) or Tsallis entropy (Section 2.31), then we obtain the corresponding "permutational version": permutation Rényi entropy, permutation Tsallis entropy, and more [154,155]. There are also "weighted versions" that take into account not only the rank order of the entries in a window but also their amplitudes; see, e.g., [156]. In turn, "multiscale versions" (including multiscale permutation Rényi and Tsallis entropy) account for multiple time scales in time series by using different time delays [157–159].

Applications

- **Analysis and classification of EEGs.** One of the first applications of permutation entropy was the analysis of EEGs of subjects with epilepsy because normal and abnormal signals (during epileptic seizures) have different complexities [160]. Furthermore, since permutation entropy can be computed in virtually real time, it has been used to predict seizures in epilepsy patients by tracking dynamical changes in EEGs [151]. Further examples can be found in the [14]. Results can be improved using permutations of Rényi and Tsallis entropy due to their additional, fine-tunable parameter [154,161].

- **Analysis of unstructured data.** Nearest-neighbor permutation entropy is an innovative extension of permutation entropy tailored for unstructured data, irrespective of their spatial or temporal configuration and dimensionality, including, e.g., liquid crystal textures [162].
- **Classification for obstructive sleep apnea.** A combination of permutation entropy-based indices and other entropic metrics was used in [163] to distinguish subjects with obstructive sleep apnea from a control group. The data consisted of heart rate and beat-to-beat blood pressure recordings.
- **Determinism detection.** Time series generated by one-dimensional maps have necessarily forbidden ordinal patterns of all sufficiently large lengths L [150]. Theoretical results under some provisos and numerical results in other cases show that the same happens with higher dimensional maps [164,165]. Therefore, the scaling of permutation entropy with L can distinguish noisy deterministic signals from random signals [150,166].
- **Emotion recognition.** Permutation entropy is used to help in tasks of feature extraction in EEGs [27].
- **Estimation of the Kolmogorov–Sinai (KS) entropy.** Given a piecewise monotone map f of a one-dimensional interval, the permutation entropy rate converges to the KS entropy of f (Section 2.19) when $L \to \infty$ [167]. Therefore, the permutation entropy calculated with a sufficiently large L (and divided by $L - 1$) is a good estimator of the KS entropy of one-dimensional dynamics. This result is remarkable because the computation of the KS entropy requires in general two infinite limits (see Section 2.19), while permutation entropy requires only one.
- **Nonlinear time series analysis.** Permutation entropy has been extensively used in the analysis of continuous-valued time series for its straightforward discretization of the data and ease of calculation. Numerous applications can be found, e.g., in [151–153] and the references therein.
- **Speech signals analysis.** In their seminal paper [148], Bandt and Pompe used precisely permutation entropy to analyze speech signals and showed that it is robust with respect to the window length, sampling frequency and observational noise.
- **The causality–complexity plane.** Permutation entropy together with the so-called statistical complexity builds the causality–complexity plane, which has proven to be a powerful tool for discriminating and classifying time series [168]. By using variants of the permutation entropy and the statistical complexity, the corresponding variants of the causality–complexity plane are obtained, possibly with enhanced discriminatory abilities for the data at hand [153].
- **Wind power prediction.** Permutation entropy has been used along with variational modal decomposition to predict wind power [169].

2.21. Rank-Based Entropy

Rank-based entropy (RbE) was introduced by Citi et al. in 2014 [170] in the framework of multiscale entropy, where traditionally, sample entropy is used. RbE measures the unpredictability of a time series quantifying the "amount of shuffling" that the ranks of the mutual distances between pairs of m-long embedding vectors $(x_i, x_{i+1}, ..., x_{i+m-1})$ and $(x_j, x_{j+1}, ..., x_{j+m-1})$ undergo when considering the next observation, i.e., the corresponding $(m+1)$-long embedding vectors. See, e.g., Section 2.7 of [20] for an algorithmic description.

Applications

- **Anomaly detection.** RbE is applied in mixed data analysis to check the influence of categorical features, using the Jaccard index for anomaly ranking and classification [171].
- **Feature selection.** RbE is used in the Entropy-and-Rank-based-Correlation framework to select features, e.g., in the detection of fruit diseases [172].
- **Mutual information.** RbE is used to rank mutual information in decision trees for monotonic classification [173].

- **Node importance.** RbE is employed in the analysis of graphs to rank nodes taking into account the local and global structure of the information [174].
- **QSAR models.** RbE is employed in Quantitative Structure–Activity Relationship models (QSAR) to analyze their stability via "rank order entropy", suggesting that certain models typically used should be discarded [175].
- **Time series analysis.** RbE is used in terms of correlation entropy to test serial independence in [176]. A multiscale version was used in [170] to study data on heart rate variability.
- **Time series classification.** RbE helps classify order of earliness in time series to generate probability distributions in different stages [177].

2.22. *Rao's Quadratic Entropy*

Rao's quadratic entropy (RQE, not to be confused with Rényi's collision entropy R_2, Section 2.24, sometimes called quadratic entropy, too) was proposed in 1982 [178] as a measure of diversity in biological populations. Given W species, RQE is defined as

$$\mathrm{RQE}(p_1,\ldots,p_W) = \sum_{i,j=1}^{W} \delta_{i,j} p_i p_j \tag{16}$$

where $\delta_{i,j}$ is the difference between the i-th and the j-th species and $\{p_1,\ldots,p_W\}$ is the probability distribution of the W species in the multinomial model.

Applications

- **Environmental monitoring.** RQE helps calculate the environmental heterogeneity index and assist prioritization schemes [179].
- **Genetic diversity metrics.** RQE is used to measure diversity for a whole collection of alleles to accommodate different genetic distance coding schemes and computational tractability in case of large datasets [180].
- **Unsupervised classification in risk management.** RQE is used as a framework in the support vector data description algorithm for risk management, enhancing knowledge in terms of interpretation and optimization, among others [181].

2.23. *Relative Entropy*

Relative entropy, also known as Kullback–Leibler (KL) divergence [11], is an information-theoretical measure that quantifies the difference or "distance" between two probability distributions. Specifically, if $\mathrm{p} = (p_1,\ldots,p_W)$ and $\mathrm{q} = (q_1,\ldots,q_W)$ are two probability distributions, then the relative entropy or KL divergence from p to q, $D(\mathrm{p}\|\mathrm{q})$, is defined as

$$D(\mathrm{p}\|\mathrm{q}) = \sum_{i=1}^{W} p_i \log \frac{p_i}{q_i}. \tag{17}$$

It follows that $D(\mathrm{p}\|\mathrm{q}) \geq 0$ (Gibb's inequality) and $D(\mathrm{p}\|\mathrm{q}) = 0$ if and only if $\mathrm{p} = \mathrm{q}$. Note that $D(\mathrm{p}\|\mathrm{q})$ is not a distance in the strict sense because $D(\mathrm{p}\|\mathrm{q}) \neq D(\mathrm{q}\|\mathrm{p})$ in general, although it can be easily symmetrized by taking any mean (arithmetic, geometric, harmonic,...) of $D(\mathrm{p}\|\mathrm{q})$ and $D(\mathrm{q}\|\mathrm{p})$. Therefore, if p is a true probability distribution approximated with q, then $D(\mathrm{p}\|\mathrm{q})$ is a measure of the approximation error. As another useful example, if p is a bivariate joint distribution and q is the product distribution of the two marginals, then $D(\mathrm{p}\|\mathrm{q})$ is the mutual information between the random variables defined by the marginal distributions; see Equation (12).

See [13] for a generalization of divergence (or relative entropy), where the $\log p_i/q_i$ in Equation (17) is replaced by $f(p_i/q_i)$ where f is a convex function on $(0,\infty)$ with $f(1) = 0$.

Applications

- **Anomaly detection in plane control.** KL divergence has been used for plane control in Software-Defined Networking as a method to detect Denial of Service attacks in [182].
- **Bayesian networks.** The efficient computation of the KL divergence of two probability distributions, each one coming from a different Bayesian network (with possibly different structures),is considered in [183].
- **Feature selection.** The authors of [184] show that the KL divergence is useful in information-theoretic feature selection due to the fact that maximizing conditional likelihood corresponds to minimizing KL divergence between the true and predicted class posterior probabilities.
- **Multiscale errors.** KL divergence is a useful metric in ML for multiscale errors [38]. A recent application to the study and analysis of the behavior of various nonpolar liquids via the Relative Resolution algorithm can be found in [185].
- **Parameter minimization in ML.** Parameters that minimize the KL divergence minimize also the cross entropy and the negative log likelihood. So, the KL divergence is useful in optimization problems where the loss function is a cross entropy [186].

2.24. *Rényi Entropy*

Rényi entropy R_α, where $\alpha > 0$ and $\alpha \neq 1$, was introduced by Alfréd Rényi in 1961 [15] as a generalization of Shannon entropy in the sense that R_1 is set equal to the latter by continuity; see Section 1.2 and the review [16] for detail. The parameter α allows for different emphasis on the probabilities of events, making it a versatile measure in information theory and its applications. Thus, for $\alpha < 1$, the central part of the distribution is flattened, i.e., high-probability events are suppressed, and low-probability events are enhanced. The opposite happens when $\alpha < 1$. As a function of the parameter α, the Rényi entropy is non-increasing. Particular cases include the Hartley entropy or max-entropy $R_0 = \lim_{\alpha\to 0} R_\alpha$, the collision or quadratic entropy R_2, and the min-entropy $R_\infty = \lim_{\alpha\to\infty} R_\alpha$.

Applications

- **Anomaly Detection.** Lower values of α highlight rare events, making this measure useful for identifying anomalies [187,188]. In particular, Rényi entropy is used in network intrusion detection for detecting botnet-like malware based on anomalous patterns [189].
- **Automated identification of EEGs.** Average Renyi entropy, along with other entropic measures, have been used as inputs for SVM algorithms to classify focal or non-focal EEGs of subjects affected by partial epilepsy [190].
- **Clustering.** Rényi entropy can provide robust similarity measures that are less sensitive to outliers [187].
- **Extreme entropy machines.** Rényi's quadratic entropy R_2 is used in the construction of extreme entropy machines to improve classification problems [191].
- **Feature selection and character recognition.** Adjustment of the parameter α can help to emphasize different parts of the underlying probability distribution and, hence, the selection of the most informative features. Rényi entropy is used for feature selection in [187,192]. Max-entropy is used in [193] for convolutional feature extraction and improvement of image perception.
- **Gaussianity of linear random processes.** Differential and conditional Rényi entropy rates were used in [194] to develop a measure of the Gaussianity of a continue-valued linear random process and study heart rate dynamics.
- **Medical time series analysis** Applications of the Rényi entropy in time series analysis reach from epilepsy detection in EEGs (see, e.g., [28]) and artifact rejection in multichannel scalp EEGs (see [195] and references therein) to early diagnosis of Alzheimer's disease in MEG data (see, e.g., [196]).

2.25. Sample Entropy

Sample Entropy was introduced by Richman and Moorman in 2000 [31] as an improvement over approximate entropy (Section 2.1); namely, its calculation is easier and independent of the time series length. Sample entropy is the negative natural logarithm of the conditional probability that close sequences of m points remain close when one more point is added, within a tolerance $r > 0$.

Applications

- **Automated identification.** Average sample entropy and other entropy measures are used as input for an SVM algorithm to classify focal and non-focal EEG signals of subjects with epilepsy [190].
- **Fault diagnosis.** Sample entropy has been used for multi-fault diagnosis in lithium batteries [197].
- **Image classification.** Sample entropy, in the form of multivariate multiscale sample entropy, is used for classifying RGB colour images to compare textures, based on a threshold to measure similarity [111].
- **Image texture analysis.** Two-dimensional sample entropy has shown to be a useful texture feature quantifier for the analysis of biomedical images [198].
- **Mutual information.** Modified sample entropy has been used in skin blood flow signals to analyze mutual information and, hence, study the association of microvascular dysfunction in different age groups [199].
- **Neonatal heart rate variability.** Sample entropy was used in [200] to analyze neonatal heart rate variability. The authors also address practical issues such as the selection of optimal tolerance and embedding dimension, and the impact of missing data. They found that entropy falls before clinical signs of neonatal sepsis and that missing points are well tolerated.
- **Neurodegenerative disease classification.** Sample entropy is used to classify neurodegenerative diseases. Gait signals, support vector machines and nearest neighbours are employed to process the features extracted using sample entropy [201].
- **Short signal analysis.** The "coefficient of sample entropy" (COSEn) was introduced in [202] for the analysis of short-length physiological time series.
- **Time series analysis.** Sample entropy, often in the form of multiscale sample entropy, is a popular tool in time series analysis, in particular with biomedical data [203]. For example, it is used for the fast diagnosis and monitoring of Parkinson's disease [204] and human emotion recognition [27] using EEGs. A modified version of multiscale sample entropy has recently been used for diagnosing epilepsy [205]. See [31] for an overview of applications of sample entropy to the analysis of physiological time series.
- **Weather forecasting.** Sample entropy is applied in weather forecasting by using transductive feature selection methods based on clustering-based sample entropy [206].

2.26. Shannon Entropy

Shannon entropy is the prototypical entropy for characterizing known or empirical probability distributions, predicting unknown or unobservable distributions via the maximum entropy principle (possibly under constraints) [207] and optimizing models by minimizing entropy-based cost functions (see e.g., Sections 2.4 and 2.23). As explained in Section 1.2, the concept was introduced by Claude Shannon in his foundational 1948 paper "A Mathematical Theory of Communication" [1] as the cornerstone of digital and analog Information Theory. Indeed, Shannon entropy informs the core theorems of Information Theory [11]. In the case of discrete probability distributions, the Shannon entropy measures the uncertainty associated with the outcome of a random variable with the given distribution or, alternatively, the expected information conveyed by that outcome, with the maximum uncertainty (or minimum information) being achieved by uniform distributions. It also quantifies the rate of information growth produced by a data source modeled as a stationary random process. In the case of continuous probability distributions, the Shannon entropy is called differential entropy and its applications to data analysis and machine learning are the subject of Section 2.5.

In this section, we only consider discrete probability distributions, i.e., finite state random variables and processes (possibly after a discretization or symbolization of the data). A typical example are real-valued time series, where the Shannon entropy is used to measure their complexity and, hence, distinguish between different dynamics. The Shannon entropy of certain probability mass distributions (e.g., ordinal patterns of a time series, power spectrum of a signal, eigenvalues of a matrix) may have particular names (permutation, spectral, von Neumann entropies); in this case, the applications of such entropies are presented in the corresponding sections.

Applications

- **Accurate prediction.** Shannon entropy is employed in machine learning models to improve the accuracy of predictions of molecular properties in the screening and development of drug molecules and other functional materials [208].
- **Anomaly detection.** Shannon entropy is employed in sensors (Internet of Things) to identify anomalies using the CorrAUC algorithm [209].
- **Artificial intelligence.** Shannon entropy contributes to the creation of the Kolmogorov Learning Cycle, which acts as a framework for optimizing the "Entropy Economy", helped by the intersection of Algorithmic Information Theory (AIT) and machine learning (ML). This framework enhances the performance of the Kolmogorov Structure Function, leading to the development of "Additive AI". By integrating principles from both AIT and ML, this approach aims to improve algorithmic efficiency and effectiveness, driving innovation in AI by balancing information theory with practical machine learning applications [210].
- **Automated identification of EEG signals.** Average Shannon entropy and other entropy measures are used as inputs of an SVM algorithm to classify focal or non-focal EEG signals of subjects with epilepsy [190].
- **Classification and feature detection.** The restricted Boltzmann machine is a probabilistic graphic network [3,211,212] that is widely used for classification and feature detection [213]. The name is due to the fact that its connectivity is constrained [3] and the probability of finding the network in a certain configuration is given by the Boltzmann distribution (Section 1.2). If considered as cognitive processes (i.e., with efficient learning and information retrieval), restricted Boltzmann machines are equivalent to Hopfield networks [3], meaning that each one can be mapped onto the other [213,214]. Both of them are among the most popular examples of neural networks.
- **Classification in the small data regime.** eSPA+ is an evolution of the entropy-optimal scalable probabilistic approximation algorithm (eSPA) [215], which has better stability and a lower iteration cost with respect to scaling than its precursor. According to [216], eSPA+ outperforms other popular approaches like support vector machines, random forest, gradient boosting machine and long short-term memory (LSTM) networks in classification problems in the small data regime, i.e., when handling datasets where the number of feature dimension is considerably higher than the number of observations.
- **Fault bearing diagnosis.** Multi-scale stationary wavelet packet analysis and the Fourier amplitude spectrum are combined to obtain a new discriminative Shannon entropy feature that is called stationary wavelet packet Fourier entropy in [217]. Features extracted by this method are then used to diagnose bearing failure.
- **Feature selection.** Shannon's mutual information entropy is used to design an entire information-theoretic framework that improves the selection of features in [184]. Shannon entropy is employed in biological science to improve the classification of data regarding the clustering of genes using microarray data [218].
- **Hard clustering.** Shannon entropy is used as a criterion to measure the confidence in unsupervised clustering tasks [219].

- **Maximum entropy and feature extraction.** By sampling from the maximum entropy distribution over possible sequences in multilayer artificial neural networks, the authors of [220] present a method for interpreting neural networks and extracting the features that the network has learned from the input data. The authors also apply their approach to biological sequence analysis.
- **Maximum entropy and scalability in AI/ML models.** In [221], the authors propose a new Maximum Entropy Method (MEMe) that improves upon the scalability of existing machine learning algorithms by efficiently approximating computational bottlenecks using maximum entropy and fast moment estimation techniques.
- **Mislabeled data in supervised classification.** Entropic outlier sparsification (EOS) is used in robust learning when data present anomalies and outliers [222]. EOS leverages the analytic solution of the (weighted) expected loss minimization problem subject to Shannon entropy regularization. EOS is tested with biomedical datasets in [222] with favorable results as compared to other methods. EOS can be also applied to feature selection and novelty detection problems.
- **Natural language processing.** Shannon entropy quantifies the predictability (or redundancy) of a text. Therefore, it is instrumental in language modelling, text compression and information retrieval, among others [1,11]. For example, it is used in [223] for keyword extraction, i.e., to rank the relevance of words.
- **Policy learning.** Shannon entropy acts as a regularization inside of an iterative policy optimization method for certain quadratic linear control scenarios [224].
- **Regression learning.** The principle of entropy maximization is implemented in the sparse probabilistic approximation for regression task analysis (SPARTAn) algorithm, which is a computationally cheap and robust algorithm for regression learning [225]. SPARTAn has been applied to predict the El Niño Southern Oscillation in [225], providing more predictive, sparse and physically explainable data descriptions than other alternative methods.
- **Signal analysis.** Shannon entropy is used as the cost functional of compression algorithms in sound and image processing [226].
- **Statistical inference.** According to the Maximum Entropy Principle of Jaynes [207], "in making inferences on the basis of partial information we must use the probability distribution which has maximum entropy subject to whatever is known". This principle has been traditionally applied with the Shannon entropy and several moment constraints of a probability distribution to infer the actual distribution [11,14].

2.27. Spectral Entropy

Spectral entropy, proposed by Kapur and Kesavan in 1992 [227], is an entropy based on the Shannon entropy. Here, the probability distribution is the (continuous or discrete) power spectrum in a representative frequency band, obtained from a signal or time series via Fourier transform, and conveniently normalized. Hence, the spectral entropy characterizes a signal by the distribution of power among its frequency components.

Applications

- **Audio analysis.** Spectral entropy has been applied for robust audio content classification in noisy signals [228]. Specifically, spectral entropy is used to segment input signals into noisy audio and noise. Also, spectral entropy (in the form of Multiband Spectral Entropy Signature) has been shown to outperform other approaches in the task of sound recognition [229].
- **Damage event detection.** Spectral entropy detects damage in vibration recordings from a wind turbine gearbox [230].
- **Data time compression.** Spectral entropy has been successfully applied to identify important segments in speech, enabling time-compression of speech for skimming [231].
- **Deep learning synchronization.** Spectral entropy evaluates synchronization in neuronal networks, providing analysis of possibly noisy recordings collected with microelectrode arrays [232].

- **Feature extraction.** Spectral entropy is used to extract features from EEG signals in [27] (emotion recognition) and [233] (assessment of the depth of anaesthesia).
- **Hyperspectral anomaly detection.** Hyperspectral Conditional Entropy features in the Entropy Rate Superpixel Algorithm, which is used in hyperspectral–spatial data to recognize unusual patterns [234].
- **Signal detection.** Spectral entropy has been used to detect cetacean vocalization in marine audio data [235]. The time–frequency decomposition was carried out with short-time Fourier transform and continuous wavelet transform.

2.28. Tone Entropy

Given a time series $(x_i)_{1\leq i\leq N}$, its tone entropy is the Shannon entropy of the probability distribution derived from the percentage indices $\mathrm{PI}(i) = 100(x_i - x_{i+1})/x_i, 1 \leq i \leq N-1$. Tone entropy was proposed by Oida et al. in 1997 [236] to study heart period fluctuations in electrocardiograms. Therefore, its applications are mainly in cardiology.

Applications

- **Biomedical analysis.** Tone entropy has been employed to study the autonomic nervous system in age groups at high-risk of cardiovascular diseases [237]. In [238], tone entropy was used to study the influence of gestational ages on the development of the fetal autonomic nervous system by analyzing fetal heart rate variability.
- **Time series.** Tone entropy has been used in time series analysis to differentiate between physiologic and synthetic interbeat time series [239].

2.29. Topological and Topology-Based Entropies

Topological entropy was introduced in [8] to measure the complexity of continuous dynamics on topological spaces. On metric spaces, topological entropy measures the exponential growth rate of the number of distinguishable orbits with finite precision. See, e.g., [240,241] for exact formulas and fast algorithms to compute the topological entropy of piecewise monotone maps and multimodal maps.

In time series analysis and digital communication technology, one is mostly interested in spaces with a finite number of states. Such time series can be the result symbolizing a continuous-valued time series. In this case, the states are usually called letters (or symbols), the state space is called ab alphabet and the blocks of letters are called words (which correspond to the "admissible" or "allowed" strings of letters). If $A(n)$ is the number of words of length n, then the topological entropy of the time series is defined as

$$h_{top} = \lim_{n\to\infty} \log(A(n)/n), \tag{18}$$

where the base of the logarithm is usually 2 or e.

Along with the above "conventional" topological entropies in dynamical systems and time series analysis, there are a number of ad hoc entropies in time series analysis, sometimes also called topological entropies. This name is due to the fact that these entropies draw on topological properties extracted from data, for example, via graphs or persistent homology. For clarity, here we refer to them as topology-based entropies. Examples include graph entropy and horizontal visibility graph entropy (Section 2.14). See [242] for an account of topological methods in data analysis.

Applications

- **Cardiac dynamics classification.** Given a (finite) time series, the *out-link entropy* is derived from the adjacency matrix of its ordinal network. This entropy has been used to classify cardiac dynamics in [243].
- **Convolutional neural networks.** The authors of [244] propose a method for quantitatively clarifying the status of single-unit convolutional neural networks using algebraic topological tools. Unit status is indicated via the calculation of a topology-based entropy, called *feature entropy*.

- **Damage detection in civil engineering.** Persistent entropy can also address the damage detection problem in civil engineering structures. In particular, it was used to solve the supervised classification damage detection problem [245].
- **Detection of determinism in time series.** Permutation topological entropy (i.e., the topological entropy of the distribution of ordinal patterns of length $L \geq 2$ obtained from a time series) can be used to detect determinism in continuous-valued time series. Actually, it suffices to check the growth of ordinal patterns with increasing L values, since this growth is exponential for deterministic signals (h_{top} converges to a finite number) and factorial for random ones (h_{top} diverges) [150,164].
- **Financial time series analysis.** Topological entropy has been applied to horizontal visibility graphs for financial time series in [246] to help quantify changes in complex stock market data.
- **Similarity of piecewise linear functions.** Piecewise linear functions are a useful mathematical tool in different areas of applied mathematics, including signal processing and machine learning methods. In this regard, persistent entropy (a topological entropy based on persistent homology) can be used to measure their similarity [247].

2.30. *Transfer Entropy*

A relevant question in time series analysis of coupled random or deterministic processes is the causality relation, i.e., which process is driving and which is responding. Transfer entropy, introduced by Schreiber in 2000 [248], measures the information exchanged between two processes in both directions separately. It can be considered as an information-theoretical (or nonlinear) implementation of Granger causality [249].

Given two stationary random processes $\mathbf{X} = (X_t)_{t\geq 0}$ and $\mathbf{Y} = (Y_t)_{t\geq 0}$, the transfer entropy from $\mathbf{Y}$ to $\mathbf{X}$, $T_{\mathbf{Y}\to\mathbf{X}}$, is the reduction of uncertainty in future values of $\mathbf{X}$, given past values of $\mathbf{X}$, due to the additional knowledge of past values of $\mathbf{Y}$. For simplicity, we consider here the simplest (lowest dimensional) case:

$$T_{\mathbf{Y}\to\mathbf{X}} = S(X_{t+1}|X_t) - S(X_{t+1}|X_t, Y_t) \tag{19}$$

where $S(X_{t+1}|...)$ is the conditional Shannon entropy of the variable X_{t+1} on the other variable(s) [11]. If the process $\mathbf{Y}$ is not causal with respect to $\mathbf{X}$ (i.e., they are independent), then $S(X_{t+1}|X_t, Y_t) = S(X_{t+1}|X_t)$ and $T_{\mathbf{Y}\to\mathbf{X}} = 0$; otherwise, $S(X_{t+1}|X_t, Y_t) < S(X_{t+1}|X_t)$ and $T_{\mathbf{Y}\to\mathbf{X}} > 0$. Observe that $T_{\mathbf{Y}\to\mathbf{X}}$ is not an entropy proper but a conditional mutual information, namely, $T_{\mathbf{Y}\to\mathbf{X}} = I(X_{t+1}; Y_t \mid X_t)$ [11,14].

Applications

- **Accelerated training in convolutional neural networks (CNNs).** The authors of [250] propose a training mechanism for CNN architectures that integrates transfer entropy feedback connections. In this way, the training process is accelerated as fewer epochs are needed. Furthermore, it generates stability; hence, it can be considered a smoothing factor.
- **Improving accuracy in graph convolutional neural networks (GCN).** The accuracy of a GCN can be improved by using node relational characteristics (such as heterophily), degree information and feature-based transfer entropy calculations. However, depending on the number of graph nodes, the computation of the transfer entropy can significantly increase the computational load [251].
- **Improving neural network performance.** A small, few-layer artificial neural network that employs feedback can reach top-level performance on standard benchmark tasks, otherwise only obtained by large feed-forward structures. To show this, the authors of [252] use feed-forward transfer entropy between neurons to structure feedback connectivity.

- **Multivariate time series forecasting.** Transfer entropy is used to establish causal relationships in multivariate time series converted into graph neural networks, each node corresponding to a variable and edges representing the casual relationships between the variables. Such neural networks are then used for prediction [253].
- **Time series analysis.** The main application of transfer entropy since its formulation has been the analysis of multivariate time series (whether biomedical, physical, economical, financial, ...) for revealing causal relationships via information directionality. See [254] and the references therein for the conceptual underpinnings and practical applications.

2.31. Tsallis Entropy

Tsallis entropy T_q, defined in Section 1.2, Equation (7), was introduced by Tsallis in 1988 in Statistical Mechanics [18]. The parameter q takes the real values $q > 0$, $q \neq 1$; T_q converges to the Shannon entropy when $q \to 1$ [16]. Tsallis entropy is identical in form to the Havrda–Charvát entropy (Section 2.15).

Tsallis entropy is particularly useful for describing systems with non-extensive properties, such as long-range interactions, non-Markovian processes and fractal structures. In machine learning, Tsallis entropy is used to improve algorithms in areas such as clustering, image segmentation and anomaly detection by changing and fine tuning the parameter q. See [255] for the general properties of Tsallis entropy.

Applications

- **Anomaly detection.** Tsallis entropy is used in network intrusion detection by detecting botnet-like malware based on anomalous patterns in the network [189].
- **Clustering.** A Tsallis entropy based categorical data clustering algorithm is proposed in [256]. It is shown that when the attributes have a power law behavior, the proposed algorithm outperforms existing Shannon entropy-based clustering algorithms.
- **Feature selection.** Tsallis entropy-based feature selection is used in [257] to identify significant features, which boosts the classification performance in machine learnin g. The authors propose an algorithm to optimize both the classifier (a Support Vector Machine) and Tsallis entropy parameters, thus improving the classification accuracy.
- **Image segmentation.** Tsallis entropy can be used for segmenting images by maximizing the entropy within different regions of images [258].
- **Pre-seismic signals.** Tsallis entropy has been used in [259] to analyze pre-seismic electromagnetic signals.

2.32. Von Neumann Entropy

Von Neumann entropy [4] is the equivalent of Shannon entropy in Quantum Statistical Mechanics and Quantum Information Theory. It is defined via the density matrix of an ensemble of quantum states, which (i) is Hermitian, (ii) has unit trace and (iii) is positive semidefinite. Therefore, the eigenvalues of the density matrix build a probability distribution and, precisely, the von Neumann entropy of the ensemble is the Shannon entropy of the probability distribution defined by the eigenvalues of the corresponding density matrix.

The same approach can be used with any matrix with the properties (i)–(iii), for example, the Pearson correlation matrix of a Markov chain (divided by the dimension of the matrix) or the Laplacian matrix of a graph. This explains the use of von Neumann entropy in classical data analysis as well.

Applications

- **Feature selection and dimensionality reduction.** Von Neumann entropy is employed in the case of kernelized relevance vector machines to assess dimensionality reduction for better model performance [260].
- **Graph-based learning.** In [261], the authors propose a method to identify vital nodes in hypergraphs that draws on von Neumann entropy. More precisely, this method is based on the high-order line graph structure of hypergraphs and measures changes in network complexity using von Neumann entropy.
- **Graph similarity and anomaly detection.** Von Neumann graph entropy (VNGE) is used to measure the information divergence and distance between graphs in a sequence. This is used for various learning tasks involving network-based data. The Fast Incremental von Neumann Graph Entropy algorithm reduces the computation time of the VNGE, making it feasible for real-time applications and large datasets [262].
- **Network analysis.** Von Neumann entropy is used in [263] to build visualization histograms from the edges of networks and then component analysis is performed on a sample for different networks.
- **Pattern recognition in neurological time series.** Von Neumann entropy was used (together with other entropies) in [264] for automated pattern recognition in neurological conditions, a crucial task in patient monitoring and medical diagnosis.

2.33. *Wavelet Entropy*

Wavelet entropy was introduced by Rosso et al. in 2001 [265]. It combines the wavelet transform with the concept of entropy to analyze the complexity and information content of multi-frequency signals. More precisely, the wavelet transform decomposes a signal into components at various scales, capturing both time and frequency information, while wavelet entropy quantifies the degree of disorder or unpredictability in those components, thus providing insights into the signal structure and complexity. Wavelet entropy is defined by the Shannon formula, but here the probability distribution corresponds to different resolution levels.

Applications

- **Emotion recognition.** Wavelet entropy can detect little variations in signals and was used in [27] to develop an automatic EEG classifier.
- **Fault detection.** Wavelet entropy was applied to monitor the condition of machinery and detect faults by analyzing vibration signals in [266].
- **Feature extraction.** Wavelet entropy was used to extract features from biomedical signals such as EEGs and ECGs to identify different physiological states or detect abnormalities in [265].

3. Discussion

In Section 1 of this review, we provided a brief historical account of the concept of entropy. Due to its seemingly unconnected appearances in thermodynamics, statistical mechanics, information theory, dynamical systems, etc., this was necessary to clarify its role in data analysis and machine learning, the subject of the present review. To this end, we gave an axiomatic characterization of entropy and generalized entropies, and acknowledged the inclusion in our review of several entropy-based probability functionals that also go by the name of entropy in the literature and are useful for the mentioned applications. In the case of univariate arguments, entropy quantifies the uncertainty, complexity and information content of the data. In the case of bivariate and multivariate arguments, entropy quantifies similarity, distance and information flow.

In Section 2, we collected a representative sample of 33 entropies (including the classical entropies visited in Section 1.2) to show the versatility and potential of the general concept of entropy as applied to practical issues. Indeed, applications such as biomedical signal analysis, fault diagnosis, feature extraction, anomaly detection, cost optimization and more highlight the diversity of the applications of entropy in data analysis and machine learning.

Section 2 also showed that, more than 150 years after its formulation, entropy and its applications remain the subject of intense research. In fact, new concepts of entropy, evolutions and generalizations are constantly being proposed in the literature to address new challenges. As a result, entropy is being applied to current topics in applied mathematics, in particular, data analysis and machine learning.

Next we turn our attention to a more quantitative description of the entropies used in this review, their applications and the associated references.

First of all, the groups G0, G1 and G2 defined in Section 1.3 (G0 comprising the classical entropies of Section 1.2) provide a coarse classification of the different entropies that is sufficient for the following analysis of their relationship with different applications and the number of citations.

Furthermore, Figure 1 (a WordCloud for ease of visualization) shows that data analysis is the most encountered application in Section 2, the reason being that this concept brings together all types of data analysis. The rest are applications in machine learning, where the most common ones are feature selection/extraction, series classification and anomaly detection.

The relationship between applications and entropy groups is visualized in Figure 2 by a Sankey diagram, where the applications chosen (leftmost column) are the top 10 applications across entropies according to Figure 1. The diagram shows a satisfactory balance between the most popular applications of entropy and the three entropy groups, with group G0 ranking slightly higher than groups G1 and G2.

As for the number of citations and publications, Table 1 ranks the 10 most cited references, according to a DOI citation analysis to date. Along with Shannon's and Tsallis' seminal papers (G0), there are four references to entropies from group G1 and another four from group G2, which again shows a tie between the two groups. Among these most influential papers, we see two G2 papers published in 2000 (authored by Richmann and Moorland, and Schreiber) and two G1 papers published shortly afterwards (by Bandt and Pompe, and Costa et al.), all of them related to signal analysis and, more specifically, two of them to the analysis of biological signals.

Figure 1. Top 10 applications across entropies.

Table 1. Most referenced publications by DOI. Total citations are based on Semantic Scholar, CrossRef and OpenCitations.

Reference	Year	Authors	Total Citations	Taxonomy
[1]	1948	Shannon	43,150	G0
[31]	2000	Richman, Moorman	16,606	G2
[18]	1988	Tsallis	14,855	G0
[22]	1991	Pincus	13,763	G2
[148]	2002	Bandt, Pompe	10,260	G1
[248]	2000	Schreiber	9743	G2
[226]	1992	Coifman, Wickerhauser	8003	G1
[21]	2005	Costa, Goldberger, Peng	5624	G1
[101]	1972	De Luca, Termini	5217	G2
[178]	1982	Rao	4027	G1

Finally, Figure 3 is a bar chart of the number of publications per year from 2000 on. In fact, it was around the year 2000 when the number of papers on entropy began to increase, after an initial period where that number was small and practically constant. This shows the growing interest that the concept of entropy and its applications have generated in the last two decades or so among researchers in general and data analysts in particular.

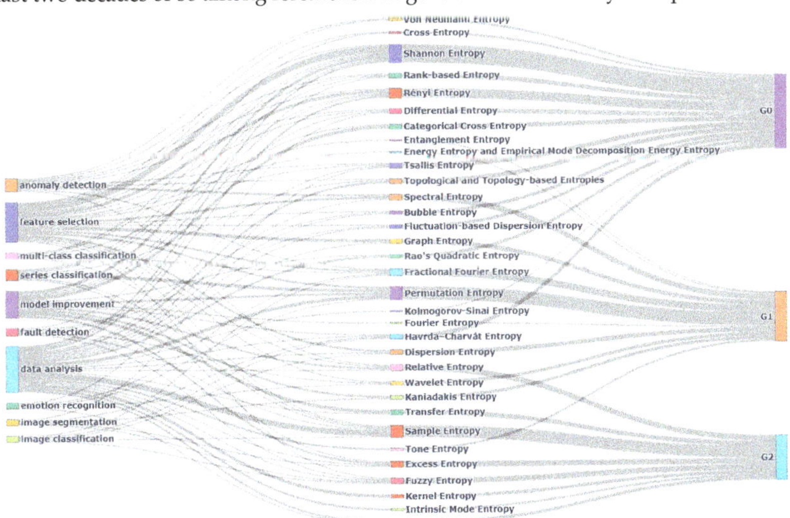

Figure 2. Relationship between applications and entropies (categorized as groups G0/G1/G2).

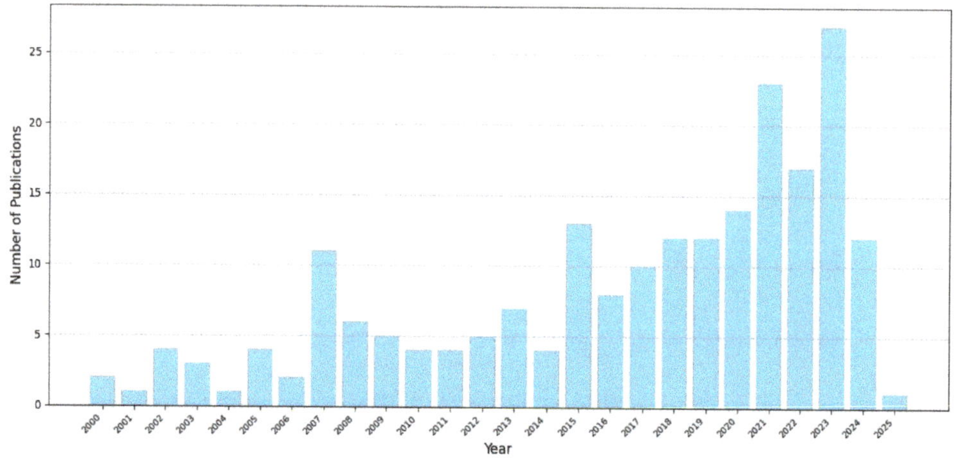

Figure 3. Number of publications per year on entropy and its applications (2000 onwards).

4. Conclusions and Outlook

To conclude this review, we begin with a few remarks on the materials presented or omitted in the previous chapters.

- The choice of a particular entropy-based method depends in general on the relevant application. Some methods may be more popular than others because they are designed for the purpose or dataset at hand, or simply because they have some computational advantage. In this regard, Section 2 presented possible candidates for different applications but no performance comparison between them was discussed. In fact, such a comparison would require a case-by-case test as, for example, in Reference [108], where the authors study motor fault detection with the approximate, dispersion, energy, fuzzy, permutation, sample and Shannon entropies (see Section 2.13 for the best performer). Along with the selection of the "right" entropy, a common concern among practitioners is the choice of parameters and hyperparameters. A combination of methods and parameter settings may be also a good approach in practice [163,264].
- We have not included applications of entropy to cryptography in this review because they belong to the general field of Data Science (through data security) rather than to Data Analysis. Entropy (mainly Shannon's and Rényi's entropies) has been applied to measure the "randomness" of encrypted messages in so-called chaotic cryptography, which applies ideas from Chaos Theory and chaos synchronization to the masking of analog signals [267]. To deal with digital signals, new tools such as discrete entropy [268] and discrete Lyapunov exponents [269] have also been developed for application to chaotic cryptography, inspired by their conventional counterparts. For a general review of the applications of entropy and related information-theoretical concepts in cryptography, see the review [270].
- We have not delved into the numerical methods used to compute entropies from data. As functionals of probability distributions, most methods for computing entropy are based on the estimation of data probabilities. In the case of discrete-valued data, the probabilities are usually estimated via relative frequencies (the maximum likelihood estimator) and possibly extrapolation methods in case of undersampling [150]. In the case of continuous-valued data, the probability densities are usually estimated via kernel density estimation (also called Parzen–Rosenblatt windowing) [271]. Furthermore, there are particular methods that do not rely on probability estimation, e.g., Lempel–Ziv complexity, which resorts to pattern matching and approximates the Shannon entropy rate of stationary binary sequences [87]. Also, in some particular cases, the entropy can be estimated via spectral information. For example, the

quadratic Rényi entropy R_2 can be estimated via spectral data transformations and kernel matrices [272]. See [273] for a general review on the estimation of entropy.

Thus far in this review, we have looked back and reviewed the applications of entropy in data analysis and machine learning to date. As we look forward, unsurprisingly, we see many challenging topics waiting to be explored and further developed. To conclude this review, we comment on some promising future research directions.

- **Algebraic Representations of Entropy**. The study and application of information-theoretical tools and, in particular, entropies can benefit from an algebraic approach in different ways. A first example is permutation entropy (Section 2.20), which is the Shannon entropy of the symbolic representation of a continuous-valued stochastic process obtained using sliding windows and ordinal patterns. Since ordinal patterns can be interpreted as permutations, such representations can harness the algebraic structure of the symmetric group. This is what happens, e.g., with the concept of transcript; see [274] for definitions and applications. A second example is the correspondence between Shannon entropy (as well as other multivariate entropic quantities for that matter) and set-theoretic expressions [275]. This correspondence provides a handle to apply algebraic concepts and tools to information-theoretical quantities such as several multivariate generalizations of mutual information; see, e.g., [276,277].
- **Data Dimensional Reduction**. In general data analysis it is crucial to transform potentially high-dimensional data into a lower dimensional representation. Along with linear methods, such as principal component analysis (PCA) and metric multidimensional scaling, there are also nonlinear spectral methods such as kernel PCA [278]. Interestingly, some of these methods involve entropies. Thus, in kernel entropy component analysis (ECA), the data transformation is related to the Rényi entropy of the input space dataset [272]. This and similar approaches could be generalized with the help of other entropies.
- **Partial Information Decomposition** (PID) decomposes the mutual information $I(X_1, X_2, ..., X_n; T)$ (Equation (12) with a vector variable in the first argument) between a series of random source variables $X_1, ..., X_n$ and a target variable T into the amount of influence that each X_i has on T. In the simplest case of two sources X_1 and X_2, this influence splits into four components, although there is much controversy as to how these four components should be quantified [279]. Other issues such as the applications of PID [280] and the generalization of PID to continuous-valued variables offer new topics worth researching; see, e.g., [281].
- **Quantum generalized entropies.** In our list of entropies, we included two that arise from quantum mechanics, namely, von Neumann entropy (Section 2.32) and entanglement entropy (Section 2.8). With the advent of quantum information theory and quantum neural networks, the possible applications of conventional entropic quantities and, above all, quantum generalized entropies is a wide open field for further research. See [282] for the definition, properties and applications of quantum generalized entropies.
- Finally, we should point out that many of the applications of entropy mentioned in Section 2 can be refined and diversified, for example, in the formulation and/or regularization of cost functions, decision trees or the generalization of concepts. Another interesting research field is the improvement of existing algorithms for computing entropies and entropic quantities (especially of multivariate quantities and time-continuous signals) and the introduction of new numerical techniques, possibly adapted to new progress in deep learning, reservoir computing and more.

Author Contributions: Conceptualization, S.A.S.-F.; methodology, S.A.S.-F. and J.M.A.; writing—original draft preparation, S.A.S.-F.; writing—review and editing, S.A.S.-F. and J.M.A.; funding acquisition, J.M.A. All authors have read and agreed to the published version of the manuscript.

Funding: This work was financially supported by Generalitat Valenciana, Spain, grant PROMETEO/2021/063.

Data Availability Statement: No new data were created or analyzed in this study.

Acknowledgments: We sincerely thank the reviewers for their constructive criticism, which helped us improve the presentation and scope of our work. We also thank Joaquín Sánchez-Soriano and Ulrich Parlitz for helpful comments on different drafts of this review.

Conflicts of Interest: The authors declare no conflicts of interests.

References

1. Shannon, C.E. A mathematical theory of communication. *Bell Syst. Tech. J.* **1948**, *27*, 379–423. [CrossRef]
2. Kantz, H.; Schreiber, T. *Nonlinear Time Series Analysis*; Cambridge University Press: Cambridge, UK, 1999. [CrossRef]
3. Goodfellow, I.; Bengio, Y.; Courville, A. *Deep Learning*; The MIT Press: Cambridge, MA, USA, 2016. [CrossRef]
4. von Neumann, J. *Mathematische Grundlagen der Quantenmechanik*; Springer: Berlin/Heidelberg, Germany; New York, NY, USA, 1971; p. 150. [CrossRef]
5. Gibbs, J.W. *Elementary Principles in Statistical Mechanics: Developed with Especial Reference to the Rational Foundation of Thermodynamics*; Charles Scribner's Sons: Farmington Hills, MI, USA, 1902.
6. Kolmogorov, A.N. Entropy per unit time as a metric invariant of automorphisms. *Dokl. Russ. Acad. Sci.* **1959**, *124*, 754–755.
7. Sinai, Y.G. On the Notion of Entropy of a Dynamical System. *Dokl. Russ. Acad. Sci.* **1959**, *124*, 768–771.
8. Adler, R.L.; Konheim, A.G.; McAndrew, M.H. Topological entropy. *Trans. Am. Math. Soc.* **1965**, *114*, 309–319. [CrossRef]
9. Walters, P. *An Introduction to Ergodic Theory*; Springer: New York, NY, USA, 2000.
10. Khinchin, A.I. *Mathematical Foundations of Information Theory*; Dover: New York, NY, USA, 1957.
11. Cover, T.M.; Thomas, J.A. *Elements of Information Theory*, 2nd ed.; John Wiley & Sons, Inc.: Hoboken, NJ, USA, 2005. [CrossRef]
12. Planck, M. Zur Theorie des Gesetzes der Energieverteilung im Normalspectrum. *Verhandlungen Dtsch. Phys. Ges.* **1900**, *2*, 237–245.
13. Csiszár, I. Axiomatic Characterization of Information Measures. *Entropy* **2008**, *10*, 261–273. [CrossRef]
14. Amigó, J.M.; Keller, K.; Unakafova, V. On entropy, entropy-like quantities, and applications. In *Frontiers in Entropy Across the Disciplines*; Freeden, W., Nashed, M., Eds.; World Scientific: Singapore, 2022; Chapter 8. [CrossRef]
15. Rényi, A. On measures of entropy and information. In Proceedings of the 4th Berkeley Symposium on Mathematical Statistics and Probability, Davis, CA, USA, 20 June–30 July 1960; Volume 1, pp. 547–561.
16. Amigó, J.M.; Balogh, S.G.; Hernández, S. A Brief Review of Generalized Entropies. *Entropy* **2018**, *20*, 813. [CrossRef] [PubMed]
17. Havrda, J.; Charvát, F. Quantification method of classification processes. *Kybernetika* **1967**, *3*, 30–35.
18. Tsallis, C. Possible generalization of Boltzmann-Gibbs statistics. *J. Stat. Phys.* **1988**, *52*, 479–487. [CrossRef]
19. Katok, A. Fifty years of entropy in dynamics: 1958–2007. *J. Mod. Dyn.* **2007**, *1*, 545–596. [CrossRef]
20. Ribeiro, M.; Henriques, T.; Castro, L.; Souto, A.; Antunes, L.; Costa-Santos, C.; Teixeira, A. The Entropy Universe. *Entropy* **2021**, *23*, 222. [CrossRef] [PubMed]
21. Costa, M.; Goldberger, A.L.; Peng, C.K. Multiscale entropy analysis of biological signals. *Phys. Rev. E* **2005**, *71*, 021906. [CrossRef] [PubMed]
22. Pincus, J.M. Approximate entropy as a measure of system complexity. *Proc. Natl. Acad. Sci. USA* **1991**, *88*, 2297–2301. [CrossRef]
23. Delgado-Bonal, A.; Marshak, A. Approximate Entropy and Sample Entropy: A Comprehensive Tutorial. *Entropy* **2019**, *21*, 541. [CrossRef] [PubMed]
24. Hornero, R.; Abásolo, D.; Escudero, J.; Gómez, C. Nonlinear analysis of electroencephalogram and magnetoencephalogram recordings in patients with Alzheimer's disease. *Philos. Trans. R. Soc. Lond. Ser. A* **2009**, *367*, 317–336. [CrossRef] [PubMed]
25. Morabito, F.C.; Labate, D.; La Foresta, F.; Bramanti, A.; Morabito, G.; Palamara, I. Multivariate Multi-Scale Permutation Entropy for Complexity Analysis of Alzheimer's Disease EEG. *Entropy* **2012**, *14*, 1186–1202. [CrossRef]
26. Liang, Z.; Wang, Y.; Sun, X.; Li, D.; Voss, L.J.; Sleigh, J.W.; Hagihira, S.; Li, X. EEG Entropy Measures in Anesthesia. *Front. Comput. Neurosci.* **2015**, *9*, 16. [CrossRef] [PubMed]
27. Patel, P.R.; Annavarapu, R.N. EEG-based human emotion recognition using entropy as a feature extraction measure. *Brain Inform.* **2021**, *8*, 20. [CrossRef] [PubMed]
28. Kannathal, N.; Choo, M.L.; Acharya, U.R.; Sadasivan, P.K. Entropies for Detection of Epilepsy in EEG. *Comput. Methods Programs Biomed.* **2005**, *80*, 187–194. [CrossRef] [PubMed]

29. Srinivasan, V.; Eswaran, C.; Sriraam, N. Approximate Entropy-Based Epileptic EEG Detection Using Artificial Neural Networks. *IEEE Trans. Inf. Technol. Biomed.* **2007**, *11*, 288–295. [CrossRef] [PubMed]
30. Jouny, C.C.; Bergey, G.K. Characterization of Early Partial Seizure Onset: Frequency, Complexity and Entropy. *Clin. Neurophysiol.* **2012**, *123*, 658–669. [CrossRef] [PubMed]
31. Richman, J.; Moorman, J.R. Physiological Time-Series Analysis Using Approximate Entropy and Sample Entropy. *Am. J. Physiol.-Heart Circ. Physiol.* **2000**, *278*, H2039–H2049. [CrossRef]
32. Acharya, U.R.; Faust, O.; Kannathal, N.; Chua, T.; Laxminarayan, S. Non-linear Analysis of EEG Signals at Various Sleep Stages. *Comput. Methods Programs Biomed.* **2005**, *80*, 37–45. [CrossRef] [PubMed]
33. Manis, G.; Aktaruzzaman, M.; Sassi, R. Bubble Entropy: An Entropy Almost Free of Parameters. *IEEE Trans. Biomed. Eng.* **2017**, *64*, 2711–2718. [CrossRef] [PubMed]
34. Manis, G.; Bodini, M.; Rivolta, M.W.; Sassi, R. A Two-Steps-Ahead Estimator for Bubble Entropy. *Entropy* **2021**, *23*, 761. [CrossRef] [PubMed]
35. Gong, J.; Yang, X.; Wang, H.; Shen, J.; Liu, W.; Zhou, F. Coordinated method fusing improved bubble entropy and artificial Gorilla Troops Optimizer optimized KELM for rolling bearing fault diagnosis. *Appl. Acoust.* **2022**, *195*, 108844. [CrossRef]
36. Gong, J.; Yang, X.; Qian, K.; Chen, Z.; Han, T. Application of improved bubble entropy and machine learning in the adaptive diagnosis of rotating machinery faults. *Alex. Eng. J.* **2023**, *80*, 22–40. [CrossRef]
37. Jiang, X.; Yi, Y.; Wu, J. Analysis of the synergistic complementarity between bubble entropy and dispersion entropy in the application of feature extraction. *Front. Phys.* **2023**, *11*, 1163767. [CrossRef]
38. Bishop, C.M. *Pattern Recognition and Machine Learning*; Springer Science + Business Media: Berlin/Heidelberg, Germany, 2006.
39. Spindelböck, T.; Ranftl, S.; von der Linden, W. Cross-Entropy Learning for Aortic Pathology Classification of Artificial Multi-Sensor Impedance Cardiography Signals. *Entropy* **2021**, *23*, 1661. [CrossRef] [PubMed]
40. Li, P.; He, X.; Song, D.; Ding, Z.; Qiao, M.; Cheng, X.; Li, R. Improved Categorical Cross-Entropy Loss for Training Deep Neural Networks with Noisy Labels. In *Pattern Recognition and Computer Vision, Proceedings of the 4th Chinese Conference, PRCV 2021, Beijing, China, 29 October–1 November 2021*; Ma, H., Wang, L., Zhang, C., Wu, F., Tan, T., Wang, Y., Lai, J., Zhao, Y., Eds.; Springer International Publishing: Berlin/Heidelberg, Germany, 2021; pp. 78–89. [CrossRef]
41. Farebrother, J.; Orbay, J.; Vuong, Q.; Taïga, A.A.; Chebotar, Y.; Xiao, T.; Irpan, A.; Levine, S.; Castro, P.S.; Faust, A.; et al. Stop Regressing: Training Value Functions via Classification for Scalable Deep RL. *arXiv* **2024**, arXiv:2403.03950. [CrossRef]
42. Arazo, E.; Ortego, D.; Albert, P.; O'Connor, N.E.; McGuinness, K. Pseudo-Labeling and Confirmation Bias in Deep Semi-Supervised Learning. *arXiv* **2020**, arXiv:1908.02983. [CrossRef]
43. Mao, A.; Mohri, M.; Zhong, Y. Cross-entropy loss functions: Theoretical analysis and applications. *arXiv* **2023**, arXiv:2304.07288. [CrossRef]
44. Berrada, L.; Zisserman, A.; Kumar, M.P. Smooth Loss Functions for Deep Top-k Classification. *arXiv* **2018**, arXiv:1802.07595. [CrossRef]
45. Wang, Z.; Zhu, Q. A Cross-Entropy Based Feature Selection Method for Binary Valued Data Classification. In *Intelligent Systems Design and Applications, Proceedings of the 21st International Conference on Intelligent Systems Design and Applications (ISDA 2021), Online, 13–15 December 2021*; Abraham, A., Gandhi, N., Hanne, T., Hong, T., Nogueira Rios, T., Ding, W., Eds.; Springer International Publishing: Berlin/Heidelberg, Germany, 2022; pp. 1406–1416. [CrossRef]
46. Kim, S.C.; Kang, T.J. Texture classification and segmentation using wavelet packet frame and Gaussian mixture model. *Pattern Recognit.* **2007**, *40*, 1207–1221. [CrossRef]
47. Bruch, S. An Alternative Cross Entropy Loss for Learning-to-Rank. *arXiv* **2021**, arXiv:1911.09798. [CrossRef]
48. Santosa, B. Multiclass Classification with Cross Entropy-Support Vector Machines. *Procedia Comput. Sci.* **2015**, *72*, 345–352. [CrossRef]
49. Śmieja, M.; Geiger, B.C. Semi-supervised cross-entropy clustering with information bottleneck constraint. *Inf. Sci.* **2017**, *421*, 254–271. [CrossRef]
50. Orchard, M.E.; Olivares, B.; Cerda, M.; Silva, J.F. Anomaly Detection based on Information-Theoretic Measures and Particle Filtering Algorithms. In Proceedings of the Annual Conference of the Prognostics and Health Management (PHM) Society, Minneapolis, MN, USA, 23–27 September 2012; Volume 4. [CrossRef]
51. Qu, Y.; Li, R.; Deng, A.; Shang, C.; Shen, Q. Non-unique Decision Differential Entropy-Based Feature Selection. *Neurocomputing* **2020**, *393*, 187–193. [CrossRef]
52. Grassucci, E.; Comminiello, D.; Uncini, A. An Information-Theoretic Perspective on Proper Quaternion Variational Autoencoders. *Entropy* **2021**, *23*, 856. [CrossRef] [PubMed]
53. Gibson, J. Entropy Power, Autoregressive Models, and Mutual Information. *Entropy* **2018**, *20*, 750. [CrossRef]
54. Robin, S.; Scrucca, L. Mixture-based estimation of entropy. *Comput. Stat. Data Anal.* **2023**, *177*, 107582. [CrossRef]
55. Rostaghi, M.; Azami, H. Dispersion Entropy: A Measure for Time-Series Analysis. *IEEE Signal Process. Lett.* **2016**, *23*, 610–614. [CrossRef]

56. Azami, H.; Escudero, J. Amplitude- and Fluctuation-Based Dispersion Entropy. *Entropy* **2018**, *20*, 210. [CrossRef] [PubMed]
57. Rostaghi, M.; Khatibi, M.M.; Ashory, M.R.; Azami, H. Refined Composite Multiscale Fuzzy Dispersion Entropy and Its Applications to Bearing Fault Diagnosis. *Entropy* **2023**, *25*, 1494. [CrossRef] [PubMed]
58. Furlong, R.; Hilal, M.; O'Brien, V.; Humeau-Heurtier, A. Parameter Analysis of Multiscale Two-Dimensional Fuzzy and Dispersion Entropy Measures Using Machine Learning Classification. *Entropy* **2021**, *23*, 1303. [CrossRef] [PubMed]
59. Hu, B.; Wang, Y.; Mu, J. A new fractional fuzzy dispersion entropy and its application in muscle fatigue detection. *Math. Biosci. Eng.* **2024**, *21*, 144–169. [CrossRef] [PubMed]
60. Dhandapani, R.; Mitiche, I.; McMeekin, S.; Mallela, V.S.; Morison, G. Enhanced Partial Discharge Signal Denoising Using Dispersion Entropy Optimized Variational Mode Decomposition. *Entropy* **2021**, *23*, 1567. [CrossRef] [PubMed]
61. Li, G.; Yang, Z.; Yang, H. A Denoising Method of Ship Radiated Noise Signal Based on Modified CEEMDAN, Dispersion Entropy, and Interval Thresholding. *Electronics* **2019**, *8*, 597. [CrossRef]
62. Fabila-Carrasco, J.S.; Tan, C.; Escudero, J. Graph-Based Multivariate Multiscale Dispersion Entropy: Efficient Implementation and Applications to Real-World Network Data. *arXiv* **2024**, arXiv:2405.00518. [CrossRef]
63. Ge, H.; Chen, G.; Yu, H.; Chen, H.; An, F. Theoretical Analysis of Empirical Mode Decomposition. *Symmetry* **2018**, *10*, 623. [CrossRef]
64. Liu, C.; Zhu, L.; Ni, C. Chatter detection in milling process based on VMD and energy entropy. *Mech. Syst. Signal Process.* **2018**, *105*, 169–182. [CrossRef]
65. Gao, Z.; Liu, Y.; Wang, Q.; Wang, J.; Luo, Y. Ensemble empirical mode decomposition energy moment entropy and enhanced long short-term memory for early fault prediction of bearing. *Measurement* **2022**, *188*, 110417. [CrossRef]
66. Yu, Y.; Dejie, Y.; Junsheng, C. A roller bearing fault diagnosis method based on EMD energy entropy and ANN. *J. Sound Vib.* **2006**, *294*, 269–277. [CrossRef]
67. Yang, Z.; Luo, S.; Zhong, P.; Chen, R.; Pan, C.; Li, K. An EMD and IMF Energy Entropy-Based Optimized Feature Extraction and Classification Scheme for Single Trial EEG Signal. *J. Mech. Med. Biol.* **2023**, *23*, 2340063. [CrossRef]
68. Zhu, G.; Peng, S.; Lao, Y.; Su, Q.; Sun, Q. Short-Term Electricity Consumption Forecasting Based on the EMD-Fbprophet-LSTM Method. *Math. Probl. Eng.* **2021**, *2021*, 6613604. [CrossRef]
69. Gao, J.; Shang, P. Analysis of complex time series based on EMD energy entropy plane. *Nonlinear Dyn.* **2019**, *96*, 465–482. [CrossRef]
70. Headrick, M. Lectures on entanglement entropy in field theory and holography. *arXiv* **2019**, arXiv:1907.08126. [CrossRef]
71. Rieger, M.; Reh, M.; Gärtner, M. Sample-efficient estimation of entanglement entropy through supervised learning. *Phys. Rev. A* **2024**, *109*, 012403. [CrossRef]
72. Liu, Y.; Li, W.J.; Zhang, X.; Lewenstein, M.; Su, G.; Ran, S.J. Entanglement-Based Feature Extraction by Tensor Network Machine Learning. *Front. Appl. Math. Stat.* **2021**, *7*, 716044. [CrossRef]
73. Lin, X.; Chen, Z.; Wei, Z. Quantifying Unknown Quantum Entanglement via a Hybrid Quantum-Classical Machine Learning Framework. *Phys. Rev. A* **2023**, *107*, 062409. [CrossRef]
74. Abdallah, S.A.; Plumbley, M.D. A measure of statistical complexity based on predictive information with application to finite spin systems. *Phys. Lett. A* **2012**, *376*, 275–281. [CrossRef]
75. Crutchfield, J.P.; Packard, N.H. Symbolic dynamics of noisy chaos. *Phys. D* **1983**, *7*, 201–223. [CrossRef]
76. Bardera, A.; Boada, I.; Feixas, M.; Sbert, M. Image Segmentation Using Excess Entropy. *J. Signal Process. Syst.* **2009**, *54*, 205–214. [CrossRef]
77. Nir, A.; Sela, E.; Beck, R.; Bar-Sinai, Y. Machine-Learning Iterative Calculation of Entropy for Physical Systems. *Proc. Natl. Acad. Sci. USA* **2020**, *117*, 30234–30240. [CrossRef] [PubMed]
78. Belghazi, M.I.; Baratin, A.; Rajeswar, S.; Ozair, S.; Bengio, Y.; Courville, A.; Hjelm, R.D. MINE: Mutual Information Neural Estimation. *arXiv* **2021**, arXiv:1801.04062. [CrossRef]
79. Xiang, X.; Zhou, J. An Excess Entropy Approach to Classify Long-Term and Short-Term Memory Stationary Time Series. *Mathematics* **2023**, *11*, 2448. [CrossRef]
80. Chen, Y.; Chen, J.; Qiang, Y.; Yuan, Z.; Yang, J. Refined composite moving average fluctuation dispersion entropy and its application on rolling bearing fault diagnosis. *Rev. Sci. Instrum.* **2023**, *94*, 105110. [CrossRef] [PubMed]
81. Su, H.; Wang, Z.; Cai, Y.; Ding, J.; Wang, X.; Yao, L. Refined Composite Multiscale Fluctuation Dispersion Entropy and Supervised Manifold Mapping for Planetary Gearbox Fault Diagnosis. *Machines* **2023**, *11*, 47. [CrossRef]
82. Zhou, F.; Han, J.; Yang, X. Multivariate hierarchical multiscale fluctuation dispersion entropy: Applications to fault diagnosis of rotating machinery. *Appl. Acoust.* **2021**, *182*, 108271. [CrossRef]
83. Li, Z.; Lan, T.; Li, Z.; Gao, P. Exploring Relationships between Boltzmann Entropy of Images and Building Classification Accuracy in Land Cover Mapping. *Entropy* **2023**, *25*, 1182. [CrossRef] [PubMed]

84. Baldini, G.; Chareau, J.M.; Bonavitacola, F. Spectrum Sensing Implemented with Improved Fluctuation-Based Dispersion Entropy and Machine Learning. *Entropy* **2021**, *23*, 1611. [CrossRef]
85. Azami, H.; Arnold, S.E.; Sanei, S.; Chang, Z.; Sapiro, G.; Escudero, J.; Gupta, A.S. Multiscale Fluctuation-based Dispersion Entropy and its Applications to Neurological Diseases. *arXiv* **2019**, arXiv:1902.10825. [CrossRef]
86. Jiao, S.; Geng, B.; Li, Y.; Zhang, Q.; Wang, Q. Fluctuation-based reverse dispersion entropy and its applications to signal classification. *Appl. Acoust.* **2021**, *175*, 107857. [CrossRef]
87. Amigó, J.M.; Szczepanski, J.; Wajnryb, E.; Sanchez-Vives, M.V. Estimating the entropy of spike trains via Lempel-Ziv complexity. *Neural Comput.* **2004**, *16*, 717–736. [CrossRef] [PubMed]
88. Friedgut, E.; Kalai, G. Every Monotone Graph Property Has A Sharp Threshold. *Proc. Am. Math. Soc.* **1999**, *124*, 2993–3002. [CrossRef]
89. Chakraborty, S.; Kulkarni, R.; Lokam, S.; Saurabh, N. Upper bounds on Fourier entropy. *Theor. Comput. Sci.* **2016**, *654*, 92–112. [CrossRef]
90. O'Donnell, R.; Wright, J.; Zhou, Y. The Fourier Entropy–Influence Conjecture for Certain Classes of Boolean Functions. In *Automata, Languages and Programming*; Aceto, L., Henzinger, M., Sgall, J., Eds.; Lecture Notes in Computer Science; Springer: Berlin/Heidelberg, Germany, 2011; Volume 6755. [CrossRef]
91. Kelman, E.; Kindler, G.; Lifshitz, N.; Minzer, D.; Safra, M. Towards a Proof of the Fourier–Entropy Conjecture? In Proceedings of the 2020 IEEE 61st Annual Symposium on Foundations of Computer Science (FOCS), Durham, NC, USA, 16–19 November 2020; pp. 247–258. [CrossRef]
92. Almeida, L.B. The fractional Fourier transform and time-frequency representations. *IEEE Trans. Signal Process.* **1994**, *42*, 3084–3091. [CrossRef]
93. Tao, R.; Zhao, X.; Li, W.; Li, H.C.; Du, Q. Hyperspectral Anomaly Detection by Fractional Fourier Entropy. *IEEE J. Sel. Top. Appl. Earth Obs. Remote. Sens.* **2019**, *12*, 4920–4929. [CrossRef]
94. Zhang, L.; Ma, J.; Cheng, B.; Lin, F. Fractional Fourier Transform-Based Tensor RX for Hyperspectral Anomaly Detection. *Remote Sens.* **2022**, *14*, 797. [CrossRef]
95. Wang, S.H.; Zhang, X.; Zhang, Y.D. DSSAE: Deep Stacked Sparse Autoencoder Analytical Model for COVID-19 Diagnosis by Fractional Fourier Entropy. *ACM Trans. Manag. Inf. Syst. (TMIS)* **2021**, *13*, 1–20. [CrossRef]
96. Wang, S.; Zhang, Y.; Yang, X.; Sun, P.; Dong, Z.; Liu, A.; Yuan, T.F. Pathological Brain Detection by a Novel Image Feature—Fractional Fourier Entropy. *Entropy* **2015**, *17*, 8278–8296. [CrossRef]
97. Yan, Y. Gingivitis detection by fractional Fourier entropy with optimization of hidden neurons. *Int. J. Cogn. Comput. Eng.* **2020**, *1*, 36–44. [CrossRef]
98. Panahi, F.; Rashidi, S.; Sheikhani, A. Application of fractional Fourier transform in feature extraction from Electrocardiogram and Galvanic Skin RESPONSE for emotion recognition. *Biomed. Signal Process. Control.* **2021**, *69*, 102863. [CrossRef]
99. Zhang, Y.; Yang, X.; Cattani, C.; Rao, R.V.; Wang, S.; Phillips, P. Tea Category Identification Using a Novel Fractional Fourier Entropy and Jaya Algorithm. *Entropy* **2016**, *18*, 77. [CrossRef]
100. Zadeh, L.A. Fuzzy sets. *Inf. Control* **1965**, *8*, 338–353. [CrossRef]
101. De Luca, A.; Termini, S. A definition of a nonprobabilistic entropy in the setting of fuzzy sets theory. *Inf. Control* **1972**, *20*, 301–312. [CrossRef]
102. Ishikawa, A.; Mieno, H. The fuzzy entropy concept and its application. *Fuzzy Sets Syst.* **1979**, *2*, 113–123. [CrossRef]
103. Chen, W.; Wang, Z.; Xie, H.; Yu, W. Characterization of surface EMG signal based on fuzzy entropy. *IEEE Trans. Neural Syst. Rehabil. Eng.* **2007**, *15*, 266–272. [CrossRef] [PubMed]
104. Zheng, J.; Cheng, J.; Yang, Y.; Luo, S. A rolling bearing fault diagnosis method based on multi-scale fuzzy entropy and variable predictive model-based class discrimination. *Mech. Mach. Theory* **2014**, *78*, 187–200. [CrossRef]
105. Markechová, D.; Riečan, B. Entropy of Fuzzy Partitions and Entropy of Fuzzy Dynamical Systems. *Entropy* **2016**, *18*, 19. [CrossRef]
106. D'Urso, P.; De Giovanni, L.; Vitale, V. Robust DTW-based entropy fuzzy clustering of time series. *Ann. Oper. Res.* **2023**. [CrossRef]
107. Di Martino, F.; Sessa, S. Energy and Entropy Measures of Fuzzy Relations for Data Analysis. *Entropy* **2018**, *20*, 424. [CrossRef] [PubMed]
108. Aguayo-Tapia, S.; Avalos-Almazan, G.; Rangel-Magdaleno, J.J. Entropy-Based Methods for Motor Fault Detection: A Review. *Entropy* **2024**, *26*, 299. [CrossRef]
109. Jinde, Z.; Chen, M.J.; Junsheng, C.; Yang, Y. Multiscale fuzzy entropy and its application in rolling bearing fault diagnosis. *Zhendong Gongcheng Xuebao/J. Vib. Eng.* **2014**, *27*, 145–151.
110. Kumar, R.; Bisht, D.C.S. Picture fuzzy entropy: A novel measure for managing uncertainty in multi-criteria decision-making. *Decis. Anal. J.* **2023**, *9*, 100351. [CrossRef]

111. Lhermitte, E.; Hilal, M.; Furlong, R.; O'Brien, V.; Humeau-Heurtier, A. Deep Learning and Entropy-Based Texture Features for Color Image Classification. *Entropy* **2022**, *24*, 1577. [CrossRef] [PubMed]
112. Tan, Z.; Li, K.; Wang, Y. An improved cuckoo search algorithm for multilevel color image thresholding based on modified fuzzy entropy. *J. Ambient. Intell. Humaniz. Comput.* **2021**. [CrossRef]
113. Korner, J. Coding of an information source having ambiguous alphabet and the entropy of graphs. In Proceedings of the Transactions of the 6th Prague conference on Information Theory, Prague, Czech Republic, 19–25 September 1971; pp. 411–425.
114. Simonyi, G. Graph Entropy: A Survey. *Comb. Optim.* **1995**, *20*, 399–441. [CrossRef]
115. Luque, B.; Lacasa, L.; Ballesteros, F.; Luque, J. Horizontal visibility graphs: Exact results for random time series. *Phys. Rev. E* **2009**, *80*, 046103. [CrossRef] [PubMed]
116. Lacasa, L.; Just, W. Visibility graphs and symbolic dynamics. *Phys. D Nonlinear Phenom.* **2018**, *374–375*, 35–44. [CrossRef]
117. Harangi, V.; Niu, X.; Bai, B. Conditional graph entropy as an alternating minimization problem. *arXiv* **2023**, arXiv:2209.00283. [CrossRef]
118. Wu, J.; Chen, X.; Xu, K.; Li, S. Structural Entropy Guided Graph Hierarchical Pooling. *arXiv* **2022**, arXiv:2206.13510. [CrossRef]
119. Juhnke-Kubitzke, M.; Köhne, D.; Schmidt, J. Counting Horizontal Visibility Graphs. *arXiv* **2021**, arXiv:2111.02723. [CrossRef]
120. Luo, G.; Li, J.; Su, J.; Peng, H.; Yang, C.; Sun, L.; Yu, P.S.; He, L. Graph Entropy Guided Node Embedding Dimension Selection for Graph Neural Networks. *arXiv* **2021**, arXiv:2105.03178. [CrossRef]
121. Zhu, G.; Li, Y.; Wen, P. Analysis of alcoholic EEG signals based on horizontal visibility graph entropy. *Brain Inform.* **2014**, *1*, 19–25. [CrossRef] [PubMed]
122. Yu, M.; Hillebrand, A.; Gouw, A.; Stam, C.J. Horizontal visibility graph transfer entropy (HVG-TE): A novel metric to characterize directed connectivity in large-scale brain networks. *NeuroImage* **2017**, *156*, 249–264. [CrossRef] [PubMed]
123. Chen, T.; Vemuri, B.; Rangarajan, A.; Eisenschenk, S.J. Group-Wise Point-Set Registration Using a Novel CDF-Based Havrda-Charvát Divergence. *Int. J. Comput. Vis.* **2010**, *86*, 111–124. [CrossRef]
124. Shi, Y.; Wu, Y.; Shang, P. Research on weighted Havrda—Charvat's entropy in financial time series. *Phys. A* **2021**, *572*, 125914. [CrossRef]
125. Brochet, T.; Lapuyade-Lahorgue, J.; Bougleux, S.; Salaun, M.; Ruan, S. Deep learning using Havrda-Charvat entropy for classification of pulmonary endomicroscopy. *arXiv* **2021**, arXiv:2104.05450. [CrossRef]
126. Brochet, T.; Lapuyade-Lahorgue, J.; Huat, A.; Thureau, S.; Pasquier, D.; Gardin, I.; Modzelewski, R.; Gibon, D.; Thariat, J.; Grégoire, V.; et al. A Quantitative Comparison between Shannon and Tsallis–Havrda–Charvat Entropies Applied to Cancer Outcome Prediction. *Entropy* **2022**, *24*, 436. [CrossRef] [PubMed]
127. Amoud, H.; Snoussi, H.; Hewson, D.; Doussot, M.; Duchene, J. Intrinsic Mode Entropy for Nonlinear Discriminant Analysis. *IEEE Signal Process. Lett.* **2007**, *14*, 297–300. [CrossRef]
128. Kosmidou, V.E.; Hadjileontiadis, L.J. Intrinsic mode entropy: An enhanced classification means for automated Greek Sign Language gesture recognition. In Proceedings of the 2008 30th Annual International Conference of the IEEE Engineering in Medicine and Biology Society, Vancouver, BC, Canada, 20–25 August 2008; pp. 5057–5060. [CrossRef]
129. Hu, M.; Liang, H. Intrinsic mode entropy based on multivariate empirical mode decomposition and its application to neural data analysis. *Cogn. Neurodyn.* **2011**, *5*, 277–284. [CrossRef] [PubMed]
130. Hu, M.; Liang, H., Multiscale Entropy: Recent Advances. In *Complexity and Nonlinearity in Cardiovascular Signals*; Barbieri, R., Scilingo, E., Valenza, G., Eds.; Springer International Publishing: Berlin/Heidelberg, Germany, 2017; pp. 115–138. [CrossRef]
131. Amoud, H.; Snoussi, H.; Hewson, D.; Duchêne, J. Intrinsic Mode Entropy for postural steadiness analysis. In Proceedings of the 4th European Conference of the International Federation for Medical and Biological Engineering, Antwerp, Belgium, 23–27 November 2008; IFMBE Proceedings; Springer: Berlin/Heidelberg, Germany, 2009; Volume 22, pp. 212–215. [CrossRef]
132. Kaniadakis, G. Statistical mechanics in the context of special relativity. *Phys. Rev. E* **2002**, *66*, 056125. [CrossRef] [PubMed]
133. Kaniadakis, G. Relativistic Roots of κ-Entropy. *Entropy* **2024**, *26*, 406. [CrossRef] [PubMed]
134. Lei, B.; Fan, J.I. Adaptive Kaniadakis entropy thresholding segmentation algorithm based on particle swarm optimization. *Soft Comput.* **2020**, *24*, 7305–7318. [CrossRef]
135. Jena, B.; Naik, M.K.; Panda, R. A novel Kaniadakis entropy-based multilevel thresholding using energy curve and Black Widow optimization algorithm with Gaussian mutation. In Proceedings of the 2023 International Conference in Advances in Power, Signal, and Information Technology (APSIT), Bhubaneswar, India, 9–11 June 2023; pp. 86–91. [CrossRef]
136. da Silva, S.L.E.F.; de Araújo, J.M.; de la Barra, E.; Corso, G. A Graph-Space Optimal Transport Approach Based on Kaniadakis κ-Gaussian Distribution for Inverse Problems Related to Wave Propagation. *Entropy* **2023**, *25*, 990. [CrossRef] [PubMed]
137. Mekyska, J.; Janousova, E.; Gomez-Vilda, P.; Smekal, Z.; Rektorova, I.; Eliasova, I.; Kostalova, M.; Mrackova, M.; Alonso-Hernandez, J.B.; Faundez-Zanuy, M.; et al. Robust and complex approach of pathological speech signal analysis. *Neurocomputing* **2015**, *167*, 94–111. [CrossRef]

138. Xu, L.S.; Wang, K.Q.; Wang, L. Gaussian kernel approximate entropy algorithm for analyzing irregularity of time-series. In Proceedings of the 2005 International Conference on Machine Learning and Cybernetics, Guangzhou, China, 18–21 August 2005; Volume 9, pp. 5605–5608. [CrossRef]
139. Zaylaa, A.; Saleh, S.; Karameh, F.; Nahas, Z.; Bouakaz, A. Cascade of nonlinear entropy and statistics to discriminate fetal heart rates. In Proceedings of the 2016 3rd International Conference on Advances in Computational Tools for Engineering Applications (ACTEA), Zouk Mosbeh, Lebanon, 13–15 July 2016; pp. 152–157. [CrossRef]
140. Orozco-Arroyave, J.R.; Arias-Londono, J.D.; Vargas-Bonilla, J.F.; Nöth, E. Analysis of speech from people with Parkinson's disease through nonlinear dynamics. In *Advances in Nonlinear Speech Processing, Proceedings of the 6th International Conference, NOLISP 2013, Mons, Belgium, 19–21 June 2013*; Drugman, T., Dutoit, T., Eds.; Lecture Notes in Computer Science; Springer: Berlin/Heidelberg, Germany, 2013; Volume 7911. [CrossRef]
141. Kolmogorov, A.N. A New Metric Invariant of Transitive Dynamical Systems and Automorphisms of Lebesgue Spaces. *Proc. Steklov Inst. Math.* **1986**, *169*, 97–102.
142. Pesin, Y.B. Characteristic Lyapunov exponents and smooth ergodic theory. *Russ. Math. Surv.* **1977**, *32*, 55. [CrossRef]
143. Shiozawa, K.; Tokuda, I. Estimating Kolmogorov-Sinai entropy from time series of high-dimensional complex systems. *Phys. Lett. A* **2024**, *510*, 129531. [CrossRef]
144. Parlitz, U. Estimating Lyapunov Exponents from Time Series. In *Chaos Detection and Predictability*; Skokos, C., Gottwald, G., Laskar, J., Eds.; Springer: Berlin/Heidelberg, Germany, 2016; Chapter 1. [CrossRef]
145. Karmakar, C.; Udhayakumar, R.; Palaniswami, M. Entropy Profiling: A Reduced-Parametric Measure of Kolmogorov-Sinai Entropy from Short-Term HRV Signal. *Entropy* **2020**, *22*, 1396. [CrossRef] [PubMed]
146. Kiss, G.; Bakucz, P. Using Kolmogorov Entropy to Verify the Description Completeness of Traffic Dynamics of Highly Autonomous Driving. *Appl. Sci.* **2024**, *14*, 2261. [CrossRef]
147. Aftanas, L.I.; Lotova, N.V.; Koshkarov, V.I.; Pokrovskaja, V.L.; Popov, S.A.; Makhnev, V.P. Non-linear analysis of emotion EEG: Calculation of Kolmogorov entropy and the principal Lyapunov exponent. *Neurosci. Lett.* **1997**, *226*, 13–16. [CrossRef] [PubMed]
148. Bandt, C.; Pompe, B. Permutation Entropy: A Natural Complexity Measure for Time Series. *Phys. Rev. Lett.* **2002**, *88*, 174102. [CrossRef] [PubMed]
149. Riedl, M.; Müller, A.; Wessel, N. Practical considerations of permutation entropy. *Eur. Phys. J. Spec. Top.* **2013**, *222*, 249–262. [CrossRef]
150. Amigó, J.M. *Permutation Complexity in Dynamical Systems*; Springer: Berlin/Heidelberg, Germany, 2010. [CrossRef]
151. Zanin, M.; Zunino, L.; Rosso, O.A.; Papo, D. Permutation Entropy and Its Main Biomedical and Econophysics Applications: A Review. *Entropy* **2012**, *14*, 1553–1577. [CrossRef]
152. Amigó, J.M.; Keller, K.; Kurths, J. Recent Progress in Symbolic Dynamics and Permutation Complexity—Ten Years of Permutation Entropy. *Eur. Phys. J. Spec. Top.* **2013**, *222*, 241–247. [CrossRef]
153. Amigó, J.M.; Rosso, O.A. Ordinal methods: Concepts, applications, new developments, and challenges—In memory of Karsten Keller (1961–2022). *Chaos* **2023**, *33*, 080401. [CrossRef]
154. Mammone, N.; Duun-Henriksen, J.; Kjaer, T.W.; Morabito, F.C. Differentiating Interictal and Ictal States in Childhood Absence Epilepsy through Permutation Rényi Entropy. *Entropy* **2015**, *17*, 4627–4643. [CrossRef]
155. Zunino, L.; Pérez, D.G.; Kowalski, A.; Martin, M.T.; Garavaglia, M.; Plastino, A.; Rosso, O.A. Fractional Brownian motion, fractional Gaussian noise, and Tsallis permutation entropy. *Phys. A* **2008**, *387*, 6057–6068. [CrossRef]
156. Stosic, D.; Stosic, D.; Stosic, T.; Stosic, B. Generalized weighted permutation entropy. *Chaos* **2022**, *32*, 103105. [CrossRef] [PubMed]
157. Yin, Y.; Sun, K.; He, S. Multiscale permutation Rényi entropy and its application for EEG signals. *PLoS ONE* **2018**, *13*, e202558. [CrossRef]
158. Li, C.; Shang, P. Multiscale Tsallis permutation entropy analysis for complex physiological time series. *Phys. A* **2019**, *523*, 10–20. [CrossRef]
159. Azami, H.; Escudero, J. Improved multiscale permutation entropy for biomedical signal analysis: Interpretation and application to electroencephalogram recordings. *Biomed. Signal Process. Control* **2016**, *23*, 28–41. [CrossRef]
160. Keller, K.; Lauffer, H. Symbolic Analysis of High-Dimensional Time Series. *Int. J. Bifurc. Chaos* **2003**, *13*, 2657–2668. [CrossRef]
161. Keller, K.; Mangold, T.; Stolz, L.; Werner, J. Permutation Entropy: New Ideas and Challenges. *Entropy* **2017**, *19*, 134. [CrossRef]
162. Voltarelli, L.G.J.M.; Pessa, A.A.B.; Zunino, L.; Zola, R.S.; Lenzi, E.K.; Perc, M.; Ribeiro, H.V. Characterizing unstructured data with the nearest neighbor permutation entropy. *Chaos* **2024**, *34*, 053130. [CrossRef] [PubMed]
163. Pilarczyk, P.; Graff, G.; Amigó, J.M.; Tessmer, K.; Narkiewicz, K.; Graff, B. Differentiating patients with obstructive sleep apnea from healthy controls based on heart rate-blood pressure coupling quantified by entropy-based indices. *Chaos* **2023**, *33*, 103140. [CrossRef] [PubMed]
164. Amigó, J.M.; Kennel, M.B. Forbidden ordinal patterns in higher dimensional dynamics. *Phys. D* **2008**, *237*, 2893–2899. [CrossRef]

165. Amigó, J.M.; Keller, K. Permutation entropy: One concept, two approaches. *Eur. Phys. J. Spec. Top.* **2013**, *222*, 263–273. [CrossRef]
166. Carpi, L.C.; Saco, P.M.; Rosso, O.A. Missing ordinal patterns in correlated noises. *Phys. A* **2010**, *389*, 2020–2029. [CrossRef]
167. Bandt, C.; Keller, G.; Pompe, B. Entropy of interval maps via permutations. *Nonlinearity* **2002**, *15*, 1595. [CrossRef]
168. Rosso, O.A.; Larrondo, H.A.; Martin, M.T.; Plastino, A.; Fuentes, M.A. Distinguishing Noise from Chaos. *Phys. Rev. Lett.* **2007**, *99*, 154102. [CrossRef] [PubMed]
169. Qu, Z.; Hou, X.; Hu, W.; Yang, R.; Ju, C. Wind power forecasting based on improved variational mode decomposition and permutation entropy. *Clean Energy* **2023**, *7*, 1032–1045. [CrossRef]
170. Citi, L.; Guffanti, G.; Mainardi, L. Rank-based Multi-Scale Entropy analysis of heart rate variability. In Proceedings of the Computing in Cardiology, Cambridge, MA, USA, 7–10 September 2014; pp. 597–600.
171. Garchery, M.; Granitzer, M. On the influence of categorical features in ranking anomalies using mixed data. *Procedia Comput. Sci.* **2018**, *126*, 77–86. [CrossRef]
172. Khan, M.A.; Akram, T.; Sharif, M.; Alhaisoni, M.; Saba, T.; Nawaz, N. A probabilistic segmentation and entropy-rank correlation-based feature selection approach for the recognition of fruit diseases. *EURASIP J. Image Video Process.* **2021**, *2021*, 14. [CrossRef]
173. Hu, Q.; Che, X.; Zhang, L.; Zhang, D.; Guo, M.; Yu, D. Rank Entropy-Based Decision Trees for Monotonic Classification. *IEEE Trans. Knowl. Data Eng.* **2012**, *24*, 2052–2064. [CrossRef]
174. Liu, S.; Gao, H. The Structure Entropy-Based Node Importance Ranking Method for Graph Data. *Entropy* **2023**, *25*, 941. [CrossRef] [PubMed]
175. McLellan, M.R.; Ryan, M.D.; Breneman, C.M. Rank order entropy: Why one metric is not enough. *J. Chem. Inf. Model.* **2011**, *51*, 2302–2319. [CrossRef] [PubMed]
176. Diks, C.; Panchenko, V. Rank-based Entropy Tests for Serial Independence. *Stud. Nonlinear Dyn. Econom.* **2008**, *12*. [CrossRef]
177. Sun, C.; Li, H.; Song, M.; Hong, S. A Ranking-Based Cross-Entropy Loss for Early Classification of Time Series. *IEEE Trans. Neural Netw. Learn. Syst.* **2024**, *35*, 11194–11203. [CrossRef] [PubMed]
178. Rao, C.R. Diversity and Dissimilarity Coefficients: A Unified Approach. *Theor. Popul. Biol.* **1982**, *21*, 24–43. [CrossRef]
179. Doxa, A.; Prastacos, P. Using Rao's quadratic entropy to define environmental heterogeneity priority areas in the European Mediterranean biome. *Biol. Conserv.* **2020**, *241*, 108366. [CrossRef]
180. Smouse, P.E.; Banks, S.C.; Peakall, R. Converting quadratic entropy to diversity: Both animals and alleles are diverse, but some are more diverse than others. *PLoS ONE* **2017**, *12*, e185499. [CrossRef] [PubMed]
181. Dionne, G.; Koumou, G. Machine Learning and Risk Management: SVDD Meets RQE. Technical Report 18-6. 2018. Available online: https://chairegestiondesrisques.hec.ca/en/wp-content/uploads/sites/2/2023/04/18-06.pdf (accessed on 30 September 2024).
182. Niknami, N.; Wu, J. Entropy-KL-ML: Enhancing the Entropy-KL-Based Anomaly Detection on Software-Defined Networks. *IEEE Trans. Netw. Sci. Eng.* **2022**, *9*, 4458–4467. [CrossRef]
183. Moral, S.; Cano, A.; Gómez-Olmedo, M. Computation of Kullback–Leibler Divergence in Bayesian Networks. *Entropy* **2021**, *23*, 1122. [CrossRef] [PubMed]
184. Brown, G.; Pocock, A.; Zhao, M.J.; Luján, M. Conditional likelihood maximisation: A unifying framework for information theoretic feature selection. *J. Mach. Learn. Res.* **2012**, *13*, 27–66.
185. Chaimovich, M.; Chaimovich, A. Relative Resolution: An Analysis with the Kullback–Leibler Entropy. *J. Chem. Theory Comput.* **2024**, *20*, 2074–2087. [CrossRef] [PubMed]
186. Draelos, R. Connections: Log-Likelihood, Cross-Entropy, KL-Divergence, Logistic Regression, and Neural Networks. GlassBox Medicine 2019. Available online: https://glassboxmedicine.com/2019/12/07/connections-log-likelihood-cross-entropy-kl-divergence-logistic-regression-and-neural-networks/ (accessed on 30 September 2024).
187. De La Pava Panche, I.; Alvarez-Meza, A.M.; Orozco-Gutierrez, A. A Data-Driven Measure of Effective Connectivity Based on Renyi's α-Entropy. *Front. Neurosci.* **2019**, *13*, 1277. [CrossRef]
188. Rioul, O. The Interplay between Error, Total Variation, Alpha-Entropy and Guessing: Fano and Pinsker Direct and Reverse Inequalities. *Entropy* **2023**, *25*, 978. [CrossRef] [PubMed]
189. Berezinski, P.; Jasiul, B.; Szpyrka, M. An Entropy-Based Network Anomaly Detection Method. *Entropy* **2015**, *17*, 2367–2408. [CrossRef]
190. Sharma, R.; Pachori, R.B.; Acharya, U.R. Application of Entropy Measures on Intrinsic Mode Functions for the Automated Identification of Focal Electroencephalogram Signals. *Entropy* **2015**, *17*, 669–691. [CrossRef]
191. Czarnecki, W.M.; Tabor, J. Extreme entropy machines: Robust information theoretic classification. *Pattern Anal. Appl.* **2017**, *20*, 383–400. [CrossRef]
192. Sluga, D.; Lotrič, U. Quadratic Mutual Information Feature Selection. *Entropy* **2017**, *19*, 157. [CrossRef]

193. Gowdra, N.; Sinha, R.; MacDonell, S. Examining convolutional feature extraction using Maximum Entropy (ME) and Signal-to-Noise Ratio (SNR) for image classification. In Proceedings of the IECON 2020 The 46th Annual Conference of the IEEE Industrial Electronics Society, Singapore, 18–21 October 2020; pp. 471–476. [CrossRef]
194. Lake, D.E. Renyi Entropy Measures of Heart Rate Gaussianity. *IEEE Trans. Biomed. Eng.* **2005**, *53*, 21–27. [CrossRef] [PubMed]
195. Mammone, N.; La Foresta, F.; Morabito, F.C. Automatic Artifact Rejection from Multichannel Scalp EEG by Wavelet ICA. *IEEE Sens. J.* **2012**, *12*, 533–542. [CrossRef]
196. Poza, J.; Escudero, J.; Hornero, R.; Fernandez, A.; Sanchez, C.I. Regional Analysis of Spontaneous MEG Rhythms in Patients with Alzheimer's Disease Using Spectral Entropies. *Ann. Biomed. Eng.* **2008**, *36*, 141–152. [CrossRef] [PubMed]
197. Shang, Y.; Lu, G.; Kang, Y.; Zhou, Z.; Duan, B.; Zhang, C. A multi-fault diagnosis method based on modified Sample Entropy for lithium-ion battery strings. *J. Power Sources* **2020**, *446*, 227275. [CrossRef]
198. Silva, L.E.V.; Senra Filho, A.C.S.; Fazan, V.P.S.; Felipe, J.C.; Murta Junior, L.O. Two-dimensional sample entropy: Assessing image texture through irregularity. *Biomed. Phys. Eng. Express* **2016**, *2*, 045002. [CrossRef]
199. Liao, F.; Jan, Y.K. Using Modified Sample Entropy to Characterize Aging-Associated Microvascular Dysfunction. *Front. Physiol.* **2016**, *7*, 126. [CrossRef] [PubMed]
200. Lake, D.E.; Richman, J.S.; Griffin, M.P.; Moorman, J.R. Sample entropy analysis of neonatal heart rate variability. *Am. J. Physiol.-Regul. Integr. Comp. Physiol.* **2002**, *283*, 301–312. [CrossRef] [PubMed]
201. Nam, N.Q.D.; Liu, A.B.; Lin, C.W. Development of a Neurodegenerative Disease Gait Classification Algorithm Using Multiscale Sample Entropy and Machine Learning Classifiers. *Entropy* **2020**, *22*, 1340. [CrossRef] [PubMed]
202. Lake, D.E.; Moorman, J.R. Accurate estimation of entropy in very short physiological time series: The problem of atrial fibrillation detection in implanted ventricular devices. *Am. J. Physiol.-Heart Circ. Physiol.* **2011**, *300*, H319–H325. [CrossRef] [PubMed]
203. Humeau-Heurtier, A. Evaluation of Systems' Irregularity and Complexity: Sample Entropy, Its Derivatives, and Their Applications across Scales and Disciplines. *Entropy* **2018**, *20*, 794. [CrossRef] [PubMed]
204. Belyaev, M.; Murugappan, M.; Velichko, A.; Korzun, D. Entropy-Based Machine Learning Model for Fast Diagnosis and Monitoring of Parkinson's Disease. *Sensors* **2023**, *23*, 8609. [CrossRef] [PubMed]
205. Lin, G.; Lin, A. Modified multiscale sample entropy and cross-sample entropy based on horizontal visibility graph. *Chaos Solitons Fractals* **2022**, *165*, 112802. [CrossRef]
206. Karevan, Z.; Suykens, J.A.K. Transductive Feature Selection Using Clustering-Based Sample Entropy for Temperature Prediction in Weather Forecasting. *Entropy* **2018**, *20*, 264. [CrossRef] [PubMed]
207. Jaynes, E.T. Information Theory and Statistical Mechanics. *Phys. Rev.* **1957**, *106*, 620–630. [CrossRef]
208. Guha, R.; Velegol, D. Harnessing Shannon entropy-based descriptors in machine learning models to enhance the prediction accuracy of molecular properties. *J. Cheminform.* **2023**, *15*, 54. [CrossRef]
209. DeMedeiros, K.; Hendawi, A.; Alvarez, M. A Survey of AI-Based Anomaly Detection in IoT and Sensor Networks. *Sensors* **2023**, *23*, 1352. [CrossRef]
210. Evans, S.C.; Shah, T.; Huang, H.; Ekanayake, S.P. The Entropy Economy and the Kolmogorov Learning Cycle: Leveraging the intersection of Machine Learning and Algorithmic Information Theory to jointly optimize energy and learning. *Phys. D* **2024**, *461*, 134051. [CrossRef]
211. Hinton, G.E.; Sejnowski, T.J. Learning and Relearning in Boltzmann Machines. In *Parallel Distributed Processing: Explorations in the Microstructure of Cognition, Volume 1: Foundations*; Rumelhart, D.E., McClelland, J.L., Eds.; Springer: Berlin/Heidelberg, Germany, 2016; p. 282–317.
212. Oh, S.; Baggag, A.; Nha, H. Entropy, Free Energy, and Work of Restricted Boltzmann Machines. *Entropy* **2020**, *22*, 538. [CrossRef]
213. Marullo, C.; Agliari, E. Boltzmann Machines as Generalized Hopfield Networks: A Review of Recent Results and Outlooks. *Entropy* **2021**, *23*, 34. [CrossRef] [PubMed]
214. Smart, M.; Zilman, A. On the mapping between Hopfield networks and Restricted Boltzmann Machines. *arXiv* **2021**, arXiv:2101.11744. [CrossRef]
215. Horenko, I. On a Scalable Entropic Breaching of the Overfitting Barrier for Small Data Problems in Machine Learning. *Neural Comput.* **2020**, *32*, 1563–1579. [CrossRef]
216. Vecchi, E.; Pospíšil, L.; Albrecht, S.; O'Kane, T.J.; Horenko, I. eSPA+: Scalable Entropy-Optimal Machine Learning Classification for Small Data Problems. *Neural Comput.* **2022**, *34*, 1220–1255. [CrossRef] [PubMed]
217. Rodriguez, N.; Barba, L.; Alvarez, P.; Cabrera-Guerrero, G. Stationary Wavelet-Fourier Entropy and Kernel Extreme Learning for Bearing Multi-Fault Diagnosis. *Entropy* **2019**, *21*, 540. [CrossRef] [PubMed]
218. Soltanian, A.R.; Rabiei, N.; Bahreini, F. Feature Selection in Microarray Data Using Entropy Information. In *Computational Biology*; Husi, H., Ed.; Exon Publications: Brisbane City, QLD, Australia, 2019; Chapter 10. [CrossRef]
219. Hoayek, A.; Rullière, D. Assessing Clustering Methods Using Shannon's Entropy. *Inf. Sci.* **2025**, *689*, 121510. [CrossRef]
220. Finnegan, A.; Song, J.S. Maximum entropy methods for extracting the learned features of deep neural networks. *PLoS Comput. Biol.* **2017**, *13*, e1005836. [CrossRef]
221. Granziol, D.; Ru, B.; Zohren, S.; Dong, X.; Osborne, M.; Roberts, S. MEMe: An Accurate Maximum Entropy Method for Efficient Approximations in Large-Scale Machine Learning. *Entropy* **2019**, *21*, 551. [CrossRef] [PubMed]
222. Horenko, I. Cheap robust learning of data anomalies with analytically solvable entropic outlier sparsification. *Proc. Natl. Acad. Sci. USA* **2022**, *119*, e2119659119. [CrossRef] [PubMed]

223. Yang, Z.; Lei, J.; Fan, K.; Lai, Y. Keyword extraction by entropy difference between the intrinsic and extrinsic mode. *Phys. A* **2013**, *392*, 4523–4531. [CrossRef]
224. Guo, X.; Li, X.; Xu, R. Fast Policy Learning for Linear Quadratic Control with Entropy Regularization. *arXiv* **2023**, arXiv:2311.14168. [CrossRef]
225. Horenko, I.; Vecchi, E.; Kardoš, J.; Wächter, A.; Schenk, O.; O'Kane, T.; Gagliardine, P.; Gerber, S. On cheap entropy-sparsified regression learning. *Proc. Natl. Acad. Sci. USA* **2023**, *120*, e2214972120. [CrossRef] [PubMed]
226. Coifman, R.; Wickerhauser, M. Entropy-Based Algorithms for Best Basis Selection. *IEEE Trans. Inf. Theory* **1992**, *38*, 713–718. [CrossRef]
227. Kapur, J.N.; Kesavan, H.K. Entropy Optimization Principles and Their Applications. In *Entropy and Energy Dissipation in Water Resources*; Springer: Berlin/Heidelberg, Germany, 1992; pp. 3–20. [CrossRef]
228. Wang, K.C. Robust Audio Content Classification Using Hybrid-Based SMD and Entropy-Based VAD. *Entropy* **2020**, *22*, 183. [CrossRef] [PubMed]
229. Manzo-Martínez, A.; Gaxiola, F.; Ramírez-Alonso, G.; Martínez-Reyes, F. A Comparative Study in Machine Learning and Audio Features for Kitchen Sounds Recognition. *Comput. Sist.* **2022**, *26*, 4244. [CrossRef]
230. Civera, M.; Surace, C. An Application of Instantaneous Spectral Entropy for the Condition Monitoring of Wind Turbines. *Appl. Sci.* **2022**, *12*, 1059. [CrossRef]
231. Ajmal, M.; Kushki, A.; Plataniotis, K.N. Time-Compression of Speech in Information Talks Using Spectral Entropy. In Proceedings of the Eighth International Workshop on Image Analysis for Multimedia Interactive Services (WIAMIS '07), Santorini, Greece, 6–8 June 2007; p. 80. [CrossRef]
232. Kapucu, F.E.; Välkki, I.; Mikkonen, J.E.; Leone, C.; Lenk, K.; Tanskanen, J.M.A.; Hyttinen, J.A.K. Spectral Entropy Based Neuronal Network Synchronization Analysis Based on Microelectrode Array Measurements. *Front. Comput. Neurosci.* **2016**, *10*, 112. [CrossRef] [PubMed]
233. Ra, J.S.; Li, T.; Li, Y. A novel spectral entropy-based index for assessing the depth of anaesthesia. *Brain Inform.* **2021**, *8*, 1–12. [CrossRef] [PubMed]
234. Liu, S.; Li, Z.; Wang, G.; Qiu, X.; Liu, T.; Cao, J.; Zhang, D. Spectral–Spatial Feature Fusion for Hyperspectral Anomaly Detection. *Sensors* **2024**, *24*, 1652. [CrossRef] [PubMed]
235. Rademan, M.W.; Versfeld, D.J.J.; du Preez, J.A. Soft-Output Signal Detection for Cetacean Vocalizations Using Spectral Entropy, K-Means Clustering and the Continuous Wavelet Transform. *Ecol. Inform.* **2023**, *74*, 101990. [CrossRef]
236. Oida, E.; Moritani, T.; Yamori, Y. Tone-Entropy Analysis on Cardiac Recovery After Dynamic Exercise. *J. Appl. Physiol.* **1997**, *82*, 1794–1801. [CrossRef] [PubMed]
237. Khandoker, A.H.; Al Zaabi, Y.; Jelinek, H.F. What Can Tone and Entropy Tell Us About Risk of Cardiovascular Diseases? In Proceedings of the 2019 Computing in Cardiology (CinC), Singapore, 8–11 September 2019; pp. 1–4. [CrossRef]
238. Khandoker, A.; Karmakar, C.; Kimura, Y.; Endo, M.; Oshio, S.; Palaniswami, M. Tone Entropy Analysis of Foetal Heart Rate Variability. *Entropy* **2015**, *17*, 1042–1053. [CrossRef]
239. Karmakar, C.K.; Khandoker, A.H.; Palaniswami, M. Multi-scale Tone Entropy in differentiating physiologic and synthetic RR time series. In Proceedings of the 2013 35th Annual International Conference of the IEEE Engineering in Medicine and Biology Society (EMBC), Osaka, Japan, 3–7 July 2013; pp. 6135–6138. [CrossRef]
240. Amigó, J.M.; Giménez, Á. A Simplified Algorithm for the Topological Entropy of Multimodal Maps. *Entropy* **2014**, *16*, 627–644. [CrossRef]
241. Amigó, J.M.; Giménez, A. Formulas for the topological entropy of multimodal maps based on min-max symbols. *Discret. Contin. Dyn. Syst. B* **2015**, *20*, 3415–3434. [CrossRef]
242. Lum, P.; Singh, G.; Lehman, A.; Ishkanov, T.; Vejdemo-Johansson, M.; Alagappan, M.; Carlsson, J.; Carlsson, G. Extracting insights from the shape of complex data using topology. *Sci. Rep.* **2013**, *3*, 1236. [CrossRef] [PubMed]
243. McCullough, M.; Small, M.; Iu, H.H.C.; Stemler, T. Multiscale Ordinal Network Analysis of Human Cardiac Dynamics. *Philos. Trans. R. Soc. A* **2017**, *375*, 20160292. [CrossRef] [PubMed]
244. Zhao, Y.; Zhang, H. Quantitative Performance Assessment of CNN Units via Topological Entropy Calculation. *arXiv* **2022**, arXiv:2103.09716. [CrossRef]
245. Jiménez-Alonso, J.F.; López-Martínez, J.; Blanco-Claraco, J.L.; González-Díaz, R.; Sáez, A. A topological entropy-based approach for damage detection of civil engineering structures. In Proceedings of the 5th International Conference on Mechanical Models in Structural Engineering (CMMoST 2019), Alicante, Spain, 23–25 October 2019; pp. 55–62.
246. Rong, L.; Shang, P. Topological entropy and geometric entropy and their application to the horizontal visibility graph for financial time series. *Nonlinear Dyn.* **2018**, *92*, 41–58. [CrossRef]
247. Rucco, M.; Gonzalez-Diaz, R.; Jimenez, M.; Atienza, N.; Cristalli, C.; Concettoni, E.; Ferrante, A.; Merelli, E. A new topological entropy-based approach for measuring similarities among piecewise linear functions. *Signal Process.* **2017**, *134*, 130–138. [CrossRef]
248. Schreiber, T. Measuring Information Transfer. *Phys. Rev. Lett.* **2000**, *85*, 461–464. [CrossRef]
249. Granger, C.W.J. Investigating Causal Relations by Econometric Models and Cross-Spectral Methods. *Econometrica* **1969**, *37*, 424–438. [CrossRef]

250. Moldovan, A.; Caţaron, A.; Răzvan, A. Learning in Convolutional Neural Networks Accelerated by Transfer Entropy. *Entropy* **2021**, *23*, 1218. [CrossRef] [PubMed]
251. Moldovan, A.; Caţaron, A.; Andonie, R. Transfer Entropy in Graph Convolutional Neural Networks. *arXiv* **2024**, arXiv:2406.06632. [CrossRef]
252. Herzog, S.; Tetzlaff, C.; Wörgötter, F. Transfer entropy-based feedback improves performance in artificial neural networks. *arXiv* **2017**, arXiv:1706.04265. [CrossRef]
253. Duan, Z.; Xu, H.; Huang, Y.; Feng, J.; Wang, Y. Multivariate Time Series Forecasting with Transfer Entropy Graph. *Tsinghua Sci. Technol.* **2023**, *28*, 141–149. [CrossRef]
254. Amblard, P.O.; Michel, O.J.J. The relation between Granger causality and directed information theory: A Review. *Entropy* **2013**, *15*, 113–143. [CrossRef]
255. Alomani, G.; Kayid, M. Further Properties of Tsallis Entropy and Its Application. *Entropy* **2023**, *25*, 199. [CrossRef] [PubMed]
256. Sharma, S.; Bassi, I. Efficacy of Tsallis Entropy in Clustering Categorical Data. In Proceedings of the 2019 IEEE Bombay Section Signature Conference (IBSSC), Mumbai, India, 26–28 July 2019; pp. 1–5. [CrossRef]
257. Wu, D.; Jia, H.; Abualigah, L.; Xing, Z.; Zheng, R.; Wang, H.; Altalhi, M. Enhance Teaching-Learning-Based Optimization for Tsallis-Entropy-Based Feature Selection Classification Approach. *Processes* **2022**, *10*, 360. [CrossRef]
258. Naidu, M.S.R.; Kumar, P.R. Tsallis Entropy Based Image Thresholding for Image Segmentation. In *Computational Intelligence in Data Mining*; Springer: Singapore, 2017; Volume 556, pp. 371–379. [CrossRef]
259. Kalimeri, M.; Papadimitriou, C.; Balasis, G.; Eftaxias, K. Dynamical complexity detection in pre-seismic emissions using nonadditive Tsallis entropy. *Phys. A* **2008**, *387*, 1161–1172. [CrossRef]
260. Belanche-Muñoz, L.A.; Wiejacha, M. Analysis of Kernel Matrices via the von Neumann Entropy and Its Relation to RVM Performances. *Entropy* **2023**, *25*, 154. [CrossRef] [PubMed]
261. Hu, F.; Tian, K.; Zhang, Z.K. Identifying Vital Nodes in Hypergraphs Based on Von Neumann Entropy. *Entropy* **2023**, *25*, 1263. [CrossRef] [PubMed]
262. Chen, P.Y.; Wu, L.; Liu, S.; Rajapakse, I. Fast Incremental von Neumann Graph Entropy Computation: Theory, Algorithm, and Applications. *arXiv* **2019**, arXiv:1805.11769. [CrossRef]
263. Ye, C.; Wilson, R.C.; Hancock, E.R. Network analysis using entropy component analysis. *J. Complex Netw.* **2017**, *6*, 831. [CrossRef]
264. Huang, Y.; Zhao, Y.; Capstick, A.; Palermo, F.; Haddadi, H.; Barnaghi, P. Analyzing entropy features in time-series data for pattern recognition in neurological conditions. *Artif. Intell. Med.* **2024**, *150*, 102821. [CrossRef] [PubMed]
265. Rosso, O.A.; Blanco, S.; Yordanova, J.; Kolev, V.; Figliola, A.; Schürmann, M.; Başar, E. Wavelet entropy: A new tool for analysis of short duration brain electrical signals. *J. Neurosci. Methods* **2001**, *105*, 65–75. [CrossRef] [PubMed]
266. Hu, P.; Zhao, C.; Huang, J.; Song, T. Intelligent and Small Samples Gear Fault Detection Based on Wavelet Analysis and Improved CNN. *Processes* **2023**, *11*, 2969. [CrossRef]
267. Cuomo, K.M.; Oppenheim, A.V.; Strogatz, S.H. Synchronization of Lorenz-Based Chaotic Circuits with Applications to Communications. *IEEE Trans. Circuits Syst. II* **1993**, *40*, 626–633. [CrossRef]
268. Amigó, J.M.; Kocarev, L.; Tomovski, I. Discrete entropy. *Phys. D* **2007**, *228*, 77–85. [CrossRef]
269. Amigó, J.M.; Kocarev, L.; Szczepanski, J. Discrete Lyapunov exponent and resistance to differential cryptanalysis. *IEEE Trans. Circuits Syst. II* **2007**, *54*, 882–886. [CrossRef]
270. Zolfaghari, B.; Bibak, K.; Koshiba, T. The Odyssey of Entropy: Cryptography. *Entropy* **2022**, *24*, 266. [CrossRef] [PubMed]
271. Hastie, T.; Tibshirani, R.; Friedman, J. *The Elements of Statistical Learning*, 2nd ed.; Springer: New York, NY, USA, 2009. [CrossRef]
272. Jenssen, R. Kernel entropy component analysis. *IEEE Trans. Pattern Anal. Mach. Intell.* **2010**, *32*, 847–860. [CrossRef] [PubMed]
273. Paninski, L. Estimation of Entropy and Mutual Information. *Neural Comput.* **2003**, *15*, 1191–1253. [CrossRef]
274. Amigó, J.M.; Monetti, R.; Graff, B.; Graff, G. Computing algebraic transfer entropy and coupling directions via transcripts. *Chaos* **2019**, *26*, 113115. [CrossRef]
275. Yeung, R.W. A new outlook on Shannon's information measures. *IEEE Trans. Inf. Theory* **1991**, *37*, 466–474. [CrossRef]
276. Down, K.J.A.; Mediano, P.A.M. Algebraic Representations of Entropy and Fixed-Parity Information Quantities. *arXiv* **2024**, arXiv:2409.04845. [CrossRef]
277. Baudot, P.; Bennequin, D. The homological nature of entropy. *Entropy* **2015**, *17*, 3253–3318. [CrossRef]
278. Schölkopf, B.; Smola, A.; Müller, K.R. Nonlinear component analysis as a kernel eigenvalue problem. *Neural Comput.* **1998**, *10*, 1299–1319. [CrossRef]
279. James, R.G.; Crutchfield, P. Multivariate Dependence beyond Shannon Information. *Entropy* **2017**, *19*, 531. [CrossRef]
280. James, R.G.; Emenheiser, J.; Crutchfield, P. Unique information and secret key agreement. *Entropy* **2019**, *21*, 12. [CrossRef] [PubMed]
281. Ehrlich, D.A.; Schick-Poland, K.; Makkeh, A.; Lanfermann, F.; Wollstadt, P.; Wibral, M. Partial information decomposition for continuous variables based on shared exclusions: Analytical formulation and estimation. *Phys. Rev. E* **2024**, *110*, 014115. [CrossRef] [PubMed]
282. Bosyk, G.M.; Zozor, S.; Holik, F.; Portesi, M.; Lamberti, P.W. A family of generalized quantum entropies: Definition and properties. *Quantum Inf. Process.* **2016**, *15*, 3393–3420. [CrossRef]

MDPI AG
Grosspeteranlage 5
4052 Basel
Switzerland
Tel.: +41 61 683 77 34

Entropy Editorial Office
E-mail: entropy@mdpi.com
www.mdpi.com/journal/entropy

www.ingramcontent.com/pod-product-compliance
Lightning Source LLC
Chambersburg PA
CBHW061603220926
51850CB00005B/412
* 9 7 8 3 7 2 5 8 4 7 3 5 8 *